T0180345

Lecture Notes in Computer Science   13019

More information about this subseries at http://www.springer.com/series/7412

Huimin Ma · Liang Wang · Changshui Zhang ·
Fei Wu · Tieniu Tan · Yaonan Wang ·
Jianhuang Lai · Yao Zhao (Eds.)

Pattern Recognition and Computer Vision

4th Chinese Conference, PRCV 2021
Beijing, China, October 29 – November 1, 2021
Proceedings, Part I

 Springer

Editors
Huimin Ma (iD)
University of Science and Technology Beijing
Beijing, China

Changshui Zhang
Tsinghua University
Beijing, China

Tieniu Tan
Chinese Academy of Sciences
Beijing, China

Jianhuang Lai
Sun Yat-Sen University
Guangzhou, Guangdong, China

Liang Wang
Chinese Academy of Sciences
Beijing, China

Fei Wu (iD)
Zhejiang University
Hangzhou, China

Yaonan Wang
Hunan University
Changsha, China

Yao Zhao (iD)
Beijing Jiaotong University
Beijing, China

ISSN 0302-9743   ISSN 1611-3349 (electronic)
Lecture Notes in Computer Science
ISBN 978-3-030-88003-3   ISBN 978-3-030-88004-0 (eBook)
https://doi.org/10.1007/978-3-030-88004-0

LNCS Sublibrary: SL6 – Image Processing, Computer Vision, Pattern Recognition, and Graphics

This Springer imprint is published by the registered company Springer Nature Switzerland AG
The registered company address is: Gewerbestrasse 11, 6330 Cham, Switzerland

Preface

Welcome to the proceedings of the 4th Chinese Conference on Pattern Recognition and Computer Vision (PRCV 2021) held in Beijing, China!

PRCV was established to further boost the impact of the Chinese community in pattern recognition and computer vision, which are two core areas of artificial intelligence, and further improve the quality of academic communication. Accordingly, PRCV is co-sponsored by four major academic societies of China: the China Society of Image and Graphics (CSIG), the Chinese Association for Artificial Intelligence (CAAI), the China Computer Federation (CCF), and the Chinese Association of Automation (CAA).

PRCV aims at providing an interactive communication platform for researchers from academia and from industry. It promotes not only academic exchange but also communication between academia and industry. In order to keep track of the frontier of academic trends and share the latest research achievements, innovative ideas, and scientific methods, international and local leading experts and professors are invited to deliver keynote speeches, introducing the latest advances in theories and methods in the fields of pattern recognition and computer vision.

PRCV 2021 was hosted by University of Science and Technology Beijing, Beijing Jiaotong University, and the Beijing University of Posts and Telecommunications. We received 513 full submissions. Each submission was reviewed by at least three reviewers selected from the Program Committee and other qualified researchers. Based on the reviewers' reports, 201 papers were finally accepted for presentation at the conference, including 30 oral and 171 posters. The acceptance rate was 39.2%. PRCV took place during October 29 to November 1, 2021, and the proceedings are published in this volume in Springer's Lecture Notes in Computer Science (LNCS) series.

We are grateful to the keynote speakers, Larry Davis from the University of Maryland, USA, Yoichi Sato from the University of Tokyo, Japan, Michael Black from the Max Planck Institute for Intelligent Systems, Germany, Songchun Zhu from Peking University and Tsinghua University, China, and Bo Xu from the Institute of Automation, Chinese Academy of Sciences, China.

We give sincere thanks to the authors of all submitted papers, the Program Committee members and the reviewers, and the Organizing Committee. Without their contributions, this conference would not have been possible. Special thanks also go to all of the sponsors and the organizers of the special forums; their support helped to make the conference a success. We are also grateful to Springer for publishing the proceedings.

October 2021

Tieniu Tan
Yaonan Wang
Jianhuang Lai
Yao Zhao
Huimin Ma
Liang Wang
Changshui Zhang
Fei Wu

Organization

Steering Committee Chair

Tieniu Tan   Institute of Automation, Chinese Academy of Sciences, China

Steering Committee

Xilin Chen	Institute of Computing Technology, Chinese Academy of Sciences, China
Chenglin Liu	Institute of Automation, Chinese Academy of Sciences, China
Yong Rui	Lenovo, China
Hongbing Zha	Peking University, China
Nanning Zheng	Xi'an Jiaotong University, China
Jie Zhou	Tsinghua University, China

Steering Committee Secretariat

Liang Wang   Institute of Automation, Chinese Academy of Sciences, China

General Chairs

Tieniu Tan	Institute of Automation, Chinese Academy of Sciences, China
Yaonan Wang	Hunan University, China
Jianhuang Lai	Sun Yat-sen University, China
Yao Zhao	Beijing Jiaotong University, China

Program Chairs

Huimin Ma	University of Science and Technology Beijing, China
Liang Wang	Institute of Automation, Chinese Academy of Sciences, China
Changshui Zhang	Tsinghua University, China
Fei Wu	Zhejiang University, China

Organizing Committee Chairs

Xucheng Yin	University of Science and Technology Beijing, China
Zhanyu Ma	Beijing University of Posts and Telecommunications, China
Zhenfeng Zhu	Beijing Jiaotong University, China
Ruiping Wang	Institute of Computing Technology, Chinese Academy of Sciences, China

Sponsorship Chairs

Nenghai Yu   University of Science and Technology of China, China
Xiang Bai   Huazhong University of Science and Technology, China
Yue Liu   Beijing Institute of Technology, China
Jinfeng Yang   Shenzhen Polytechnic, China

Publicity Chairs

Xiangwei Kong   Zhejiang University, China
Tao Mei   JD.com, China
Jiaying Liu   Peking University, China
Dan Zeng   Shanghai University, China

International Liaison Chairs

Jingyi Yu   ShanghaiTech University, China
Xuelong Li   Northwestern Polytechnical University, China
Bangzhi Ruan   Hong Kong Baptist University, China

Tutorial Chairs

Weishi Zheng   Sun Yat-sen University, China
Mingming Cheng   Nankai University, China
Shikui Wei   Beijing Jiaotong University, China

Symposium Chairs

Hua Huang   Beijing Normal University, China
Yuxin Peng   Peking University, China
Nannan Wang   Xidian University, China

Doctoral Forum Chairs

Xi Peng   Sichuan University, China
Hang Su   Tsinghua University, China
Huihui Bai   Beijing Jiaotong University, China

Competition Chairs

Nong Sang   Huazhong University of Science and Technology, China
Wangmeng Zuo   Harbin Institute of Technology, China
Xiaohua Xie   Sun Yat-sen University, China

Special Issue Chairs

Jiwen Lu	Tsinghua University, China
Shiming Xiang	Institute of Automation, Chinese Academy of Sciences, China
Jianxin Wu	Nanjing University, China

Publication Chairs

Zhouchen Lin	Peking University, China
Chunyu Lin	Beijing Jiaotong University, China
Huawei Tian	People's Public Security University of China, China

Registration Chairs

Junjun Yin	University of Science and Technology Beijing, China
Yue Ming	Beijing University of Posts and Telecommunications, China
Jimin Xiao	Xi'an Jiaotong-Liverpool University, China

Demo Chairs

Xiaokang Yang	Shanghai Jiaotong University, China
Xiaobin Zhu	University of Science and Technology Beijing, China
Chunjie Zhang	Beijing Jiaotong University, China

Website Chairs

Chao Zhu	University of Science and Technology Beijing, China
Zhaofeng He	Beijing University of Posts and Telecommunications, China
Runmin Cong	Beijing Jiaotong University, China

Finance Chairs

Weiping Wang	University of Science and Technology Beijing, China
Lifang Wu	Beijing University of Technology, China
Meiqin Liu	Beijing Jiaotong University, China

Program Committee

Jing Dong	Chinese Academy of Sciences, China
Ran He	Institute of Automation, Chinese Academy of Sciences, China
Xi Li	Zhejiang University, China
Si Liu	Beihang University, China
Xi Peng	Sichuan University, China
Yu Qiao	Chinese Academy of Sciences, China
Jian Sun	Xi'an Jiaotong University, China
Rongrong Ji	Xiamen University, China
Xiang Bai	Huazhong University of Science and Technology, China
Jian Cheng	Institute of Automation, Chinese Academy of Sciences, China
Mingming Cheng	Nankai University, China
Junyu Dong	Ocean University of China, China
Weisheng Dong	Xidian University, China
Yuming Fang	Jiangxi University of Finance and Economics, China
Jianjiang Feng	Tsinghua University, China
Shenghua Gao	ShanghaiTech University, China
Maoguo Gong	Xidian University, China
Yahong Han	Tianjin University, China
Huiguang He	Institute of Automation, Chinese Academy of Sciences, China
Shuqiang Jiang	Institute of Computing Technology, China Academy of Science, China
Lianwen Jin	South China University of Technology, China
Xiaoyuan Jing	Wuhan University, China
Haojie Li	Dalian University of Technology, China
Jianguo Li	Ant Group, China
Peihua Li	Dalian University of Technology, China
Liang Lin	Sun Yat-sen University, China
Zhouchen Lin	Peking University, China
Jiwen Lu	Tsinghua University, China
Siwei Ma	Peking University, China
Deyu Meng	Xi'an Jiaotong University, China
Qiguang Miao	Xidian University, China
Liqiang Nie	Shandong University, China
Wanli Ouyang	The University of Sydney, Australia
Jinshan Pan	Nanjing University of Science and Technology, China
Nong Sang	Huazhong University of Science and Technology, China
Shiguang Shan	Institute of Computing Technology, Chinese Academy of Sciences, China
Hongbin Shen	Shanghai Jiao Tong University, China
Linlin Shen	Shenzhen University, China
Mingli Song	Zhejiang University, China
Hanli Wang	Tongji University, China
Hanzi Wang	Xiamen University, China
Jingdong Wang	Microsoft, China

Nannan Wang	Xidian University, China
Jianxin Wu	Nanjing University, China
Jinjian Wu	Xidian University, China
Yihong Wu	Institute of Automation, Chinese Academy of Sciences, China
Guisong Xia	Wuhan University, China
Yong Xia	Northwestern Polytechnical University, China
Shiming Xiang	Chinese Academy of Sciences, China
Xiaohua Xie	Sun Yat-sen University, China
Jufeng Yang	Nankai University, China
Wankou Yang	Southeast University, China
Yang Yang	University of Electronic Science and Technology of China, China
Yilong Yin	Shandong University, China
Xiaotong Yuan	Nanjing University of Information Science and Technology, China
Zhengjun Zha	University of Science and Technology of China, China
Daoqiang Zhang	Nanjing University of Aeronautics and Astronautics, China
Zhaoxiang Zhang	Institute of Automation, Chinese Academy of Sciences, China
Weishi Zheng	Sun Yat-sen University, China
Wangmeng Zuo	Harbin Institute of Technology, China

Reviewers

Bai Xiang	Feng Jiachang	He Hongliang
Bai Xiao	Feng Jiawei	Hong Jincheng
Cai Shen	Fu Bin	Hu Shishuai
Cai Yinghao	Fu Ying	Hu Jie
Chen Zailiang	Gao Hongxia	Hu Yang
Chen Weixiang	Gao Shang-Hua	Hu Fuyuan
Chen Jinyu	Gao Changxin	Hu Ruyun
Chen Yifan	Gao Guangwei	Hu Yangwen
Cheng Gong	Gao Yi	Huang Lei
Chu Jun	Ge Shiming	Huang Sheng
Cui Chaoran	Ge Yongxin	Huang Dong
Cui Hengfei	Geng Xin	Huang Huaibo
Cui Zhe	Gong Chen	Huang Jiangtao
Deng Hongxia	Gong Xun	Huang Xiaoming
Deng Cheng	Gu Guanghua	Ji Fanfan
Ding Zihan	Gu Yu-Chao	Ji Jiayi
Dong Qiulei	Guo Chunle	Ji Zhong
Dong Yu	Guo Jianwei	Jia Chuanmin
Dong Xue	Guo Zhenhua	Jia Wei
Duan Lijuan	Han Qi	Jia Xibin
Fan Bin	Han Linghao	Jiang Bo
Fan Yongxian	He Hong	Jiang Peng-Tao
Fan Bohao	He Mingjie	Kan Meina
Fang Yuchun	He Zhaofeng	Kang Wenxiong

Lei Na
Lei Zhen
Leng Lu
Li Chenglong
Li Chunlei
Li Hongjun
Li Shuyan
Li Xia
Li Zhiyong
Li Guanbin
Li Peng
Li Ruirui
Li Zechao
Li Zhen
Li Ce
Li Changzhou
Li Jia
Li Jian
Li Shiying
Li Wanhua
Li Yongjie
Li Yunfan
Liang Jian
Liang Yanjie
Liao Zehui
Lin Zihang
Lin Chunyu
Lin Guangfeng
Liu Heng
Liu Li
Liu Wu
Liu Yiguang
Liu Zhiang
Liu Chongyu
Liu Li
Liu Qingshan
Liu Yun
Liu Cheng-Lin
Liu Min
Liu Risheng
Liu Tiange
Liu Weifeng
Liu Xiaolong
Liu Yang
Liu Zhi

Liu Zhou
Lu Shaoping
Lu Haopeng
Luo Bin
Luo Gen
Ma Chao
Ma Wenchao
Ma Cheng
Ma Wei
Mei Jie
Miao Yongwei
Nie Liqiang
Nie Xiushan
Niu Xuesong
Niu Yuzhen
Ouyang Jianquan
Pan Chunyan
Pan Zhiyu
Pan Jinshan
Peng Yixing
Peng Jun
Qian Wenhua
Qin Binjie
Qu Yanyun
Rao Yongming
Ren Wenqi
Rui Song
Shen Chao
Shen Haifeng
Shen Shuhan
Shen Tiancheng
Sheng Lijun
Shi Caijuan
Shi Wu
Shi Zhiping
Shi Hailin
Shi Lukui
Song Chunfeng
Su Hang
Sun Xiaoshuai
Sun Jinqiu
Sun Zhanli
Sun Jun
Sun Xian
Sun Zhenan

Tan Chaolei
Tan Xiaoyang
Tang Jin
Tu Zhengzheng
Wang Fudong
Wang Hao
Wang Limin
Wang Qinfen
Wang Xingce
Wang Xinnian
Wang Zitian
Wang Hongxing
Wang Jiapeng
Wang Luting
Wang Shanshan
Wang Shengke
Wang Yude
Wang Zilei
Wang Dong
Wang Hanzi
Wang Jinjia
Wang Long
Wang Qiufeng
Wang Shuqiang
Wang Xingzheng
Wei Xiu-Shen
Wei Wei
Wen Jie
Wu Yadong
Wu Hong
Wu Shixiang
Wu Xia
Wu Yongxian
Wu Yuwei
Wu Xinxiao
Wu Yihong
Xia Daoxun
Xiang Shiming
Xiao Jinsheng
Xiao Liang
Xiao Jun
Xie Xingyu
Xu Gang
Xu Shugong
Xu Xun

Xu Zhenghua
Xu Lixiang
Xu Xin-Shun
Xu Mingye
Xu Yong
Xue Nan
Yan Bo
Yan Dongming
Yan Junchi
Yang Dong
Yang Guan
Yang Peipei
Yang Wenming
Yang Yibo
Yang Lu
Yang Jinfu
Yang Wen
Yao Tao
Ye Mao
Yin Ming
Yin Fei

You Gexin
Yu Ye
Yu Qian
Yu Zhe
Zeng Lingan
Zeng Hui
Zhai Yongjie
Zhang Aiwu
Zhang Chi
Zhang Jie
Zhang Shu
Zhang Wenqiang
Zhang Yunfeng
Zhang Zhao
Zhang Hui
Zhang Lei
Zhang Xuyao
Zhang Yongfei
Zhang Dingwen
Zhang Honggang
Zhang Lin

Zhang Mingjin
Zhang Shanshan
Zhang Xiao-Yu
Zhang Yanming
Zhang Yuefeng
Zhao Cairong
Zhao Yang
Zhao Yuqian
Zhen Peng
Zheng Wenming
Zheng Feng
Zhong Dexing
Zhong Guoqiang
Zhou Xiaolong
Zhou Xue
Zhou Quan
Zhou Xiaowei
Zhu Chaoyang
Zhu Xiangping
Zou Yuexian
Zuo Wangmeng

Contents – Part I

Object Detection, Tracking and Recognition

High-Performance Discriminative Tracking with Target-Aware Feature Embeddings

Bin Yu[1,2(✉)], Ming Tang[2], Linyu Zheng[1,2], Guibo Zhu[1,2], Jinqiao Wang[1,2,3], and Hanqing Lu[1,2]

[1] School of Artificial Intelligence, University of Chinese Academy of Sciences, Beijing 100049, China
[2] National Laboratory of Pattern Recognition, Institute of Automation, Chinese Academy of Sciences, No.95, Zhongguancun East Road, Beijing 100190, China
{bin.yu,tangm,linyu.zheng,gbzhu,jqwang,luhq}@nlpr.ia.ac.cn
[3] ObjectEye Inc., Beijing, China

Abstract. Discriminative model-based trackers have made remarkable progress recently. However, due to the extreme imbalance of foreground and background samples, the learned model is hard to fit the training samples well in the online tracking. In this paper, to alleviate the negative influence caused by the imbalance issue, we propose a novel construction scheme of target-aware features for online discriminative tracking. Specifically, we design a sub-network to generate target-aware feature embeddings of foregrounds and backgrounds by projecting the learned feature embeddings into the target-aware feature space. Then, a model solver, which is integrated into our networks, is applied to learn the discriminative model. Based on such feature construction, the learned model is able to fit training samples well in the online tracking. Experimental results on four benchmarks, OTB-2015, VOT-2018, NfS, and GOT-10k, show that the proposed target-aware feature construction is effective for visual tracking, leading to the high-performance of our tracker.

Keywords: Visual tracking · Data imbalance · Discriminative model

1 Introduction

Visual tracking is one of the fundamental problems in computer vision, spanning a wide range of applications including video understanding, surveillance, and robotics. Despite significant progress in recent years [2,5,17,30,35], visual

---

B. Yu—The first author is a student.

Electronic supplementary material The online version of this chapter (https://doi.org/10.1007/978-3-030-88004-0_1) contains supplementary material, which is available to authorized users.

H. Ma et al. (Eds.): PRCV 2021, LNCS 13019, pp. 3–15, 2021.
https://doi.org/10.1007/978-3-030-88004-0_1

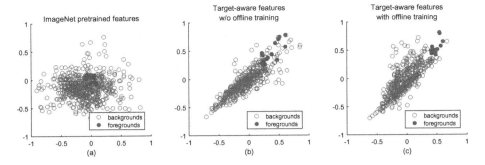

Fig. 1. Distributions of the foreground and background samples with three different kinds of features. It can be seen that from (a) to (c), it is from hard to easy for a discriminative model to fit training samples well in online tracking.

tracking is still a challenge due to several severe interferences (*i.e.*, large appearance changes, occlusions, and background clutters) and very limited training samples.

Different from Siamese based trackers [17,30,32], online discriminative trackers [2,6,23,27–29,33–35] can effectively exploit the background information online, and thus have achieved the state-of-the-art results on multiple challenging benchmarks [11,16,31]. However, in the online tracking, the extreme imbalance of foreground and background samples causes the learned discriminative model to pay more attention to backgrounds while less to foregrounds, and further to be difficult to fit training samples well (see Fig. 1(a)). These problems negatively affect the discriminative power of the trackers. In fact, there are already methods before attempting to solve the problems by mitigating the emphasis on background information in the learned models, such as sample re-weighting using Gaussian-like maps [7], spatial reliability maps [22] or binary maps [14]. However, they inevitably increase the complexity of model learning, and thus affect the tracking efficiency. Besides, they are hard to take advantage of the end-to-end CNNs training like [2,5,29,35] to improve their accuracy and robustness further.

Different from the previous methods, in this paper, we attempt to alleviate the ahead-mentioned problems from the perspective of feature construction. We consider the feature embeddings used in most previous trackers are target-unaware ones where the construction of feature is independent of whether a sample is from the foreground or not. With such features, the distributions of foreground and background samples which are all extracted over the image region overlap severely (as shown in Fig. 1(a)), making it hard for the discriminative model to fit foreground and background samples well. Note that although there exists methods [4,19] that aim at constructing target-aware features for visual tracking, both are designed for Siamese-based tracking schemes and may not be suitable for online discriminative ones.

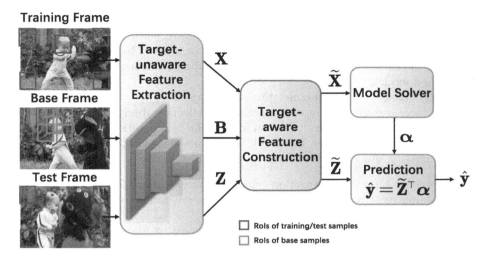

Fig. 2. An illustration of the proposed discriminative tracking scheme with target-aware feature construction.

Based on the above observation, we propose a simple yet powerful scheme of target-aware feature construction for online discriminative tracking. First, we introduce a set of *base samples*, $\{\mathbf{b}_i\}$, which are extracted around the target over a smaller region than that where training samples are extracted (see Fig. 2), to obtain target-specific information directly. Then, target-aware feature embeddings are constructed by mapping the target-unaware ones of training samples to the target-aware space with the dot products $\langle \cdot, \mathbf{b}_i \rangle$. In this way, the generated features are composed of similarities to base samples, causing the overlap between the distributions of foreground samples and those of background samples decreases in the feature space (see Fig. 1(b)). Based on such feature construction, the learned discriminative model is able to distinguish between foreground and background samples better.

However, off-the-shelf CNNs features of training samples and base ones may not be suitable enough for our target-aware feature construction since the extractions of those features are independent of our scheme. To this end, we design a sub-network to construct the target-aware features, enabling an end-to-end manner to learn target-unaware feature embeddings. Concretely, as shown in Fig. 2, first we extract the target-unaware features of training samples and base ones from the input frames, respectively, with separate branches in the proposed networks. Then, the target-aware feature embeddings are obtained through a target-aware feature construction sub-network, and a differentiable model solver [1,35], which is integrated into our networks, is applied to learn the discriminative model. Given a test frame, the target-aware feature embeddings of test samples are obtained in the same way and the labels of these samples are predicted with the learned discriminative model. In this way, our networks can be trained in an end-to-end manner, enabling learning the feature embeddings that are tightly coupled with our novel target-aware feature construction and the discriminative tracking scheme. With such target-aware feature embeddings, the overlap between the distributions of foregrounds and that of backgrounds further decreases in the feature space (see Fig. 1(c)), leading to strong discrimination of the learned model.

Finally, based on the learned target-unaware feature extraction network and the target-aware feature construction, an effective discriminative tracker, TADFT, is developed and evaluated on four public benchmarks, OTB-2015 [31], VOT-2018 [16], NfS [13], and GOT-10k [11], achieving state-of-the-art results.

In summary, our main contributions are in three folds.

1. We present a novel construction scheme of target-aware features for online discriminative tracking to alleviate the negative influence of imbalance of foreground and background samples.
2. We design a sub-network to construct the target-aware features of foregrounds and backgrounds, benefiting an end-to-end way to learn the target-unaware feature embeddings.
3. We develop a discriminative tracker TAFDT based on our target-aware feature construction and extensively evaluate TAFDT on four public benchmarks.

2 Discriminative Tracking with Target-Aware Feature Embeddings

2.1 Target-Unaware Feature Extraction

To achieve strong discrimination of target-unaware features, each image is processed by a target-unaware feature extraction network consisting of a backbone, *i.e.*, ResNet [9], and a head network [35]. Concretely, for each input training/test frame, N RoIs are first generated by uniform sampling across the whole image region. Then we extract Block-3 and Block-4 feature maps of ResNet and pass them through two convolutional layers to obtain two feature maps. Two 512-dimensional feature vectors of each RoI are finally obtained by using PrPool layers [12] and fully-connected layers. Thus, the feature dimension D of each sample is 1024 (concatenated by the two feature vectors). As illustrated in Fig. 2, we obtain $\mathbf{X} = [\mathbf{x}_1, \mathbf{x}_2, ..., \mathbf{x}_N] \in \mathbb{R}^{D \times N}$ and $\mathbf{Z} = [\mathbf{z}_1, \mathbf{z}_2, ..., \mathbf{z}_N] \in \mathbb{R}^{D \times N}$ consisting of N D-dimensional training samples and test ones, repectively.

2.2 Target-Aware Feature Construction

In order to obtain target-specific information, we extract base samples around the target object over a smaller region than that where training samples are extracted (see Fig. 5). Given an additional input frame, called base frame, we obtain $\mathbf{B} = [\mathbf{b}_1, \mathbf{b}_2, ..., \mathbf{b}_M] \in \mathbb{R}^{D \times M}$ through the target-unaware feature extraction network, where $\mathbf{b}_i$ is a D-dimensional base sample.

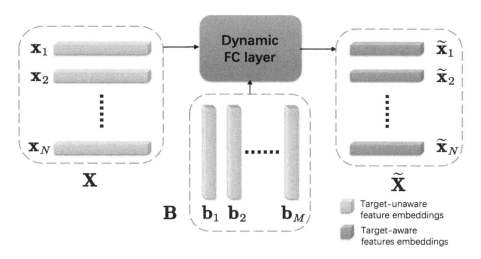

Fig. 3. Construction of target-aware features. It is performed by a dynamic filters layer, using the base sample matrix **B** as the weights of a fully-connected layer. For a single training/test sample, the dimensionality of features is transformed from D to M.

Then, given a D-dimensional sample **x**, the process of the target-aware feature construction is represented as follows,

$$\widetilde{\mathbf{x}} = [\langle \mathbf{x}, \mathbf{b}_1 \rangle, \langle \mathbf{x}, \mathbf{b}_2 \rangle, ..., \langle \mathbf{x}, \mathbf{b}_M \rangle]^\top = \mathbf{B}^\top \mathbf{x}, \tag{1}$$

where $\langle , \rangle$ represents dot product.

Thus, we are able to obtain the target-aware feature embeddings of training samples and test ones by

$$\begin{aligned} \widetilde{\mathbf{X}} &= [\widetilde{\mathbf{x}}_1, \widetilde{\mathbf{x}}_2, ..., \widetilde{\mathbf{x}}_N] = \mathbf{B}^\top \mathbf{X}, \\ \widetilde{\mathbf{Z}} &= [\widetilde{\mathbf{z}}_1, \widetilde{\mathbf{z}}_2, ..., \widetilde{\mathbf{z}}_N] = \mathbf{B}^\top \mathbf{Z}. \end{aligned} \tag{2}$$

Further, to obtain target-aware feature embeddings efficiently and enable learning target-unaware ones in an end-to-end way, we view the projection operation as a dynamic fully connected layer, whose weights are equal to the generated base sample matrix **B**, as shown in Fig. 3. As such, we are able to construct a specific feature space for each target object by changing the input base frame.

2.3 Ridge Regression with Target-Aware Feature Embeddings

Ridge Regression with Single Frame. As a classical discriminative model, ridge regression model has been confirmed to be simple yet effective in the field of visual tracking by many trackers [6,10,14,35], and thereby employed in our approach.

The optimization problem of ridge regression with target-aware feature embeddings is formulated as follows,

$$\min_{\alpha} \|\widetilde{\mathbf{X}}^{\top}\alpha - \mathbf{y}\|_2^2 + \lambda\|\alpha\|_2^2, \tag{3}$$

where $\mathbf{y} \in \mathbb{R}^{N \times 1}$ is the groundtruth labels and $\lambda \geq 0$ is a regularization parameter. The optimal α^* can be analytically solved by

$$\alpha^* = \left(\widetilde{\mathbf{X}}\widetilde{\mathbf{X}}^{\top} + \lambda\mathbf{I}\right)^{-1}\widetilde{\mathbf{X}}\mathbf{y}, \tag{4}$$

where $\mathbf{I} \in \mathbb{R}^{D \times D}$ is the identity matrix.

Note that in the optimization problem of ridge regression with the D-dimensional target-unaware feature embeddings , we achieve the optimal solution with Gauss elimination method and its time complexity is $O(D^3/2)$. It means the time-consuming grows cubically with the dimension of the feature embeddings. However, due to the common characteristic in CNNs that deep features are high-dimensional and $M < D$ in our approach ($M = 23 \times 23$ and $D = 31 \times 31$), a faster solution of the optimization problem can be achieved, $i.e.$, α is obtained with time complexity of $O(M^3/2)$ in our approach.

Ridge Regression with Historical Frames. Since the size of discriminative model is reduced by the target-aware feature construction, the influence of the online optimization on efficiency is weaken, and we are able to extend our model with historical frames for robust tracking without increasing heavy computational burden. To be specific, we draw training samples from p historical frames in the training set, $i.e.$, $\{\mathbf{X}_j\}_{j=1}^{p}$. We consider the optimization problem of multiple frames in our approach as follows,

$$\min_{\alpha_p} \frac{1}{p} \sum_{j=1}^{p} \|\widetilde{\mathbf{X}}_j^{\top}\alpha_p - \mathbf{y}\|_2^2 + \lambda\|\alpha_p\|_2^2, \tag{5}$$

where $\widetilde{\mathbf{X}}_j = \mathbf{B}^{\top}\mathbf{X}_j$.

The objective function in Eq. (5) is minimized by

$$\alpha_p^* = \left(\sum_{j=1}^{p} \widetilde{\mathbf{X}}_j\widetilde{\mathbf{X}}_j^{\top} + \lambda\mathbf{I}\right)^{-1} \sum_{j=1}^{p} \widetilde{\mathbf{X}}_j\mathbf{y}. \tag{6}$$

Similar to DCFST [35], the model solver, $i.e.$, solving for α_p by directly using Eq. (6), can be integrated into the networks as a layer with forward and backward processes during offline training, benefiting an end-to-end manner to learn the target-unaware feature embeddings coupled with our approach.

2.4   Offline Training

To take advantage of multiple frames from the sequence, during offline training, each input image tuple consists of 3 training frames, 3 test ones, and a base one sampled in a sequence. Then the training sample matrices $\{\mathbf{X}_j\}_{j=1}^3$, the test sample matrices $\{\mathbf{Z}_j\}_{j=1}^3$, and the base sample matrix $\mathbf{B}$ are obtained through the target-unaware feature extraction, respectively.

After obtaining $\boldsymbol{\alpha}_p^*$ over the three training frames by Eq. (6), we calculate the predicted labels $\hat{\mathbf{y}}_j \in \mathbb{R}^{N \times 1}$ by $\hat{\mathbf{y}}_j = \widetilde{\mathbf{Z}}_j^\top \boldsymbol{\alpha}_p^*$, where $\widetilde{\mathbf{Z}}_j = \mathbf{B}^\top \mathbf{Z}_j$. In the end, we adopt the modified shrinkage loss like in [21, 35] over the three test frames. The objective loss used for offline training is given below.

$$\mathcal{L} = \sum_{j=1}^q \frac{\exp(\mathbf{y}) \cdot \|\hat{\mathbf{y}}_j - \mathbf{y}\|^2}{1 + \exp(c_1 \cdot (c_2 - |\hat{\mathbf{y}}_j - \mathbf{y}|))}, \tag{7}$$

where $c_1 = 10$ and $c_2 = 0.2$ are set the same as in [21, 35].

It is worth noting that even though the shrinkage loss and our target-aware feature construction both aim at alleviating the problems caused by the imbalance of foreground and background samples, they focus on different aspects. The shrinkage loss focuses on the imbalance issue of samples from test frames in the offline training of the network and is used to prevent the vast number of easy background samples from overwhelming the learning of feature embeddings [35]. Different from that, the target-aware feature construction focuses on the problems caused by the imbalance of samples from training frames in model learning and enables the learned discriminative model fit training samples well in the online tracking.

2.5   Online Tracking

Updating. Several recent tracking approaches [29, 34, 35] use a moving average strategy to update the discriminative model. Despite its simplicity, this strategy limits the flexibility of the update mechanism due to synthetic appearance and a constant update rate across the whole sequence. To address the drawbacks of this strategy, our approach allows the discriminative model to be easily updated by adding new frames to the training set as in [2]. We ensure a maximum memory size of 50 for the training set. During online tracking, the model is updated every 15 frames. Besides, we update the base sample matrix in every frame as follows.

$$\begin{aligned} \overline{\mathbf{B}}_T &= \mathbf{B}, \quad T = 1, \\ \overline{\mathbf{B}}_T &= (1 - \gamma)\overline{\mathbf{B}}_{T-1} + \gamma \mathbf{B}, \quad T > 1, \end{aligned} \tag{8}$$

where γ is a weight parameter and T is the frame number.

Localization. In a new test frame, given the optimal solution to Problem (5), $\boldsymbol{\alpha}_p^*$, the updated base sample matrix $\overline{\mathbf{B}}$ and the test sample matrix $\mathbf{Z}$, we predict the target location by $\hat{\mathbf{y}} = \widetilde{\mathbf{Z}}^\top \boldsymbol{\alpha}_p^*$, where $\widetilde{\mathbf{Z}} = \overline{\mathbf{B}}^\top \mathbf{Z}$. The sample corresponding to the maximum element of $\hat{\mathbf{y}}$ is regarded as the target object.

Refinement. After locating the target in a new test frame, we refine its bounding box by ATOM [5] for more accurate tracking.

3   Experiments

Our tracker is implemented in Python using PyTorch. On a single TITAN RTX GPU, employing ResNet-50 [9] as the backbone network, TAFDT achieves a tracking speed of 30 FPS. We validate it by performing comprehensive experiments on four benchmarks: OTB-2015 [31], VOT-2018 [16], NfS [13], and GOT-10k [11].

Table 1. Comparisons of different feature constructions for TAFDT on the combined NfS and OTB-2015 benchmarks.

Method	Ours		Larger area		Smaller area		Baseline
	Target-aware	None-aware	Target-aware+	Target-aware−	Center-aware	Target-unaware
Sample area	3 × 3	5 × 5	4 × 4	2 × 2	1.5 × 1.5	−
Mean DP(%)	**81.2**	78.8	80.6	79.7	80.4	79.0
AUC (%)	**68.5**	64.9	66.2	66.1	64.0	64.7

3.1   Implementation Details

Training Data. We use the training splits of large-scale tracking datasets including TrackingNet [24], GOT-10k [11], and LaSOT [8]. The networks are trained with 20 image tuples sampled from a random segment of 50 frames in a sequence. For each training/test frame, the search area is 5 × 5 times the target area. Base samples are extracted from the region centered at the target, with an area of 3 × 3 times the target area. In addition, we use image flipping and color jittering for data augmentation.

Network Learning. We freeze the weights of the ResNet-50 [9] backbone network during training. The networks are trained for 50 epochs with 1500 iterations per epoch within 30 h. We use ADAM [15] with learning rate decay of 0.2 every 15 epochs, using a initial learning rate of 0.005.

3.2   Feature Comparisons

In order to confirm the effectiveness of the proposed target-aware feature construction, we compare it with other four feature constructions on the combined OTB-2015 and NfS benchmarks. To compare equitably, the only difference between these feature constructions is the area of the region where base samples are extracted. As shown in Table 1, applying none-aware features only obtains similar results to those of the baseline applying target-unaware features directly from the feature extraction network. Applying target-aware+ features obtains better results due to the reduction of redundant background information, with the mean AUC score of 66.2% and DP of 80.6%. Our tracker with target-aware features further improves the results, giving absolute gains of 2.3% and 0.6% in mean AUC scores and DP, respectively. Note that when more background information is contained in base samples as in target-aware+ features and none-aware ones, it is hard for the discriminative model to pay attention to foreground and fit training samples well. In contrast, when less foreground information is contained in base samples as in center-aware features and target-aware− ones, the results degenerate because the discriminative model is less robust to variation of target appearance. The results show that the area of 3×3 times the target size is the most suitable for our target-aware feature construction, which is the maximum area where each base sample contains part of the target object.

3.3   State-of-the-Art Comparisons

OTB-2015. Figure 4 and Table 2 show the comparisons of TAFDT with the state-of-the-art online discriminative trackers. TAFDT achieves the AUC score

Table 2. The mean overlap of TAFDT and other nine state-of-the-art online discriminative trackers on OTB-2015.

Tracker	TAFDT	ECO [6]	ATOM [5]	DiMP50 [2]	DCFST50 [35]
Venue	Ours	CVPR2017	CVPR2019	ICCV2019	ECCV2020
Mean OP(%)	88.9	84.2	83.4	84.6	87.3
FPS	30	6	30	30	25
Tracker	DCFST18 [35]	VITAL [26]	fdKCF [34]	MDNet [25]	DeepSTRCF [18]
Venue	ECCV2020	CVPR2018	ICCV2019	CVPR2016	CVPR2018
Mean OP(%)	87.2	85.7	82.7	84.7	84.5
FPS	40	2	24	1	5

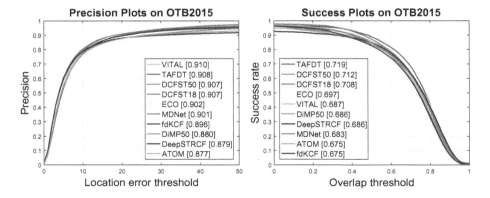

Fig. 4. The mean precision and success plots of the proposed TAFDT and other six state-of-the-art online discriminative trackers on OTB-2015. The mean distance precisions and AUC scores are reported in the legends.

of 71.9%, outperforming the latest DCFST50 [35] and DiMP50 [2] by 0.7% and 3.3%, respectively. In addition, TAFDT achieves the mean OP of 88.9% (see Table 2), leading the second best tracker DCFST50 with an absolute gain of 1.6%. The results show that the target-aware feature construction helps our new discriminative tracker TAFDT achieve new state of the art on this benchmark, while running at 30 FPS, comparable with other trackers.

Table 3 shows the comparison of TAFDT with the Siamese network based trackers including the latest target-aware feature-based trackers MLT [4] and TADT [19]. TAFDT achieves a gain of 1.8% in AUC scores compared with the state-of-the-art SiamRCNN [30] which employs stronger backbone network (ResNet-101) and more training datas (*e.g.*, COCO [20]), respectively. Moreover, TAFDT surpasses the previous 'target-aware' trackers TADT and MLT with significant gains of 5.9% and 10.8%, respectively, confirming the effectiveness and superiority of our target-aware feature construction.

Table 3. The mean AUC scores of TAFDT and other four state-of-the-art Siamese-based trackers on OTB-2015.

Tracker	TAFDT	SiamRCNN [30]	SiamRPN++ [17]	MLT [4]	TADT [19]
Venue	Ours	CVPR2020	CVPR2019	ICCV2019	CVPR2019
AUC(%)	71.9	70.1	69.6	61.1	66.0

Table 4. State-of-the-art comparisons on the VOT-2018 in terms of expected average overlap (EAO), accuracy and robustness.

Tracker	UPDT [3]	SiameseRPN++ [17]	ATOM [5]	DiMP50 [2]	DCFST50 [35]	SiamRCNN [30]	TAFDT
Venue	ECCV2018	CVPR2019	CVPR2019	ICCV2019	ECCV2020	CVPR2020	Ours
EAO	0.378	0.414	0.401	0.440	0.452	0.408	0.455
Robustness	0.184	0.234	0.204	0.153	0.136	0.220	0.180
Accuracy	0.536	0.600	0.590	0.597	0.607	0.609	0.611

VOT-2018. We evaluate TAFDT on VOT-2018, then compare its accuracy, robustness, and EAO score with six state-of-the-art trackers. As shown in Table 4, the proposed TAFDT achieves 45.5% and 61.1% in terms of EAO and accuracy, respectively, surpassing the second best DCFST50 by 0.3% and 0.4% in terms of EAO and accuracy, respectively.

NfS. We evaluate the proposed TAFDT on the 30 FPS version of NfS. Table 5 shows the comparison with six state-of-the-art trackers. Our approach achieves the AUC score of 65.5%. TAFDT consistently outperforms DCFST50 and SiamRCNN with gains of 1.4% and 1.6% in AUC scores, respectively.

GOT-10k. We evaluate the proposed TAFDT on the test set of GOT10k including 180 test videos. In contrast to other datasets, trackers are restricted to use only the training split of the dataset to ensure a fair comparison. Accordingly, we retrain our networks by using only the training split. Figure 5 shows the success plots of TAFDT and other nine state-of-the-art trackers. The success rates at overlap threshold 0.5 are shown in legend which represent the percentages of successfully tracked frames. TAFDT outperforms all previous trackers, achieving an absolute gain of 0.5% in success rates over the previous best method, DCFST50.

Table 5. State-of-the-art comparison on the 30 FPS version of NfS in terms of AUC scores.

Tracker	TAFDT	ECO [6]	UPDT [3]	ATOM [5]	DiMP50 [2]	DCFST50 [35]	SiamRCNN [30]
Venue	Ours	CVPR2017	ECCV2018	CVPR2019	ICCV2019	ECCV2020	CVPR2020
AUC(%)	65.5	46.6	53.7	58.4	62.0	64.1	63.9

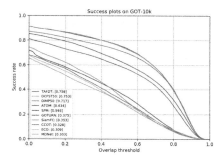

Fig. 5. Success plots of the proposed TAFDT and other nine state-of-the-art trackers on GOT-10k. The success rates at overlap threshold 0.5 are shown in legend.

4   Conclusion

In this paper, we propose a simple yet powerful construction scheme of target-aware features for online discriminative tracking and a target-aware features-based online discriminative tracker, TAFDT. Based on our target-aware feature embeddings, the learned discriminative model is capable of being less affected by the imbalance issue, and thus fitting training samples well in online tracking. The devised networks are able to learn the target-unaware feature embeddings that are tightly coupled with our novel target-aware feature construction and the discriminative tracking scheme. We extensively validate the proposed TAFDT on four public benchmarks. Experimental results verify that target-aware feature construction is effective and leads TAFDT to achieve state-of-the-art performance.

Acknowledgements. This work was supported by the Key-Areas Research and Development Program of Guangdong Province (No. 2020B010165001). This work was also supported by National Natural Science Foundation of China under Grants 61772527, 61976210, 62076235, and 62002356.

References

1. Bertinetto, L., Henriques, J.F., Torr, P., Vedaldi, A.: Meta-learning with differentiable closed-form solvers. In: ICLR (2019)
2. Bhat, G., Danelljan, M., Gool, L.V., Timofte, R.: Learning discriminative model prediction for tracking. In: ICCV (2019)
3. Bhat, G., Johnander, J., Danelljan, M., Shahbaz Khan, F., Felsberg, M.: Unveiling the power of deep tracking. In: ECCV (2018)
4. Choi, J., Kwon, J., Lee, K.M.: Deep meta learning for real-time target-aware visual tracking. In: ICCV (2019)
5. Danelljan, M., Bhat, G., Khan, F.S., Felsberg, M.: Atom: accurate tracking by overlap maximization. In: CVPR (2019)
6. Danelljan, M., Bhat, G., Shahbaz Khan, F., Felsberg, M.: Eco: efficient convolution operators for tracking. In: CVPR (2017)
7. Danelljan, M., Hager, G., Shahbaz Khan, F., Felsberg, M.: Learning spatially regularized correlation filters for visual tracking. In: ICCV (2015)
8. Fan, H., et al.: Lasot: a high-quality benchmark for large-scale single object tracking. In: CVPR (2019)
9. He, K., Zhang, X., Ren, S., Sun, J.: Deep residual learning for image recognition. In: CVPR (2016)
10. Henriques, J.F., Caseiro, R., Martins, P., Batista, J.: High-speed tracking with kernelized correlation filters. TPAMI **37**, 583–596 (2014)
11. Huang, L., Zhao, X., Huang, K.: Got-10k: a large high-diversity benchmark for generic object tracking in the wild. TPAMI **43**, 1562–1577 (2019)
12. Jiang, B., Luo, R., Mao, J., Xiao, T., Jiang, Y.: Acquisition of localization confidence for accurate object detection. In: ECCV (2018)
13. Galoogahi, H.K., Fagg, A., Huang, C., Ramanan, D., Lucey, S.: Need for speed: a benchmark for higher frame rate object tracking. In: ICCV (2017)
14. Galoogahi, H.K., Fagg, A., Lucey, S.: Learning background-aware correlation filters for visual tracking. In: ICCV (2017)
15. Kingma, D.P., Ba, J.: Adam: a method for stochastic optimization. arXiv preprint arXiv:1412.6980 (2014)
16. Kristan, M., Leonardis, A., et al.: The visual object tracking vot2018 challenge results. In: ECCV (2018)
17. Li, B., Wu, W., Wang, Q., Zhang, F., Xing, J., Yan, J.: Siamrpn++: evolution of siamese visual tracking with very deep networks. In: CVPR (2019)
18. Li, F., Tian, C., Zuo, W., Zhang, L., Yang, M.H.: Learning spatial-temporal regularized correlation filters for visual tracking. In: CVPR (2018)
19. Li, X., Ma, C., Wu, B., He, Z., Yang, M.H.: Target-aware deep tracking. In: CVPR (2019)
20. Lin, T.Y., et al.: Microsoft COCO: common objects in context. In: Fleet, D., Pajdla, T., Schiele, B., Tuytelaars, T. (eds.) ECCV 2014. LNCS, vol. 8693, pp. 740–755. Springer, Cham (2014). https://doi.org/10.1007/978-3-319-10602-1_48
21. Lu, X., Ma, C., Ni, B., Yang, X., Reid, I., Yang, M.H.: Deep regression tracking with shrinkage loss. In: ECCV (2018)

22. Lukezic, A., Vojir, T., Cehovin Zajc, L., Matas, J., Kristan, M.: Discriminative correlation filter with channel and spatial reliability. In: CVPR (2017)
23. Ma, C., Huang, J.B., Yang, X., Yang, M.H.: Hierarchical convolutional features for visual tracking. In: ICCV (2015)
24. Muller, M., Bibi, A., Giancola, S., Alsubaihi, S., Ghanem, B.: Trackingnet: a large-scale dataset and benchmark for object tracking in the wild. In: ECCV (2018)
25. Nam, H., Han, B.: Learning multi-domain convolutional neural networks for visual tracking. In: CVPR (2016)
26. Song, Y., et al.: Vital: visual tracking via adversarial learning. In: CVPR (2018)
27. Tang, M., Yu, B., Zhang, F., Wang, J.: High-speed tracking with multi-kernel correlation filters. In: CVPR (2018)
28. Tang, M., Zheng, L., Yu, B., Wang, J.: Fast kernelized correlation filter without boundary effect. In: WACV (2021)
29. Valmadre, J., Bertinetto, L., Henriques, J., Vedaldi, A., Torr, P.H.: End-to-end representation learning for correlation filter based tracking. In: CVPR (2017)
30. Voigtlaender, P., Luiten, J., Torr, P.H., Leibe, B.: Siam r-cnn: visual tracking by re-detection. In: CVPR (2020)
31. Wu, Y., Lim, J., Yang, M.H.: Object tracking benchmark. TPAMI **37**, 1834–1848 (2015)
32. Zheng, L., Chen, Y., Tang, M., Wang, J., Lu, H.: Siamese deformable cross-correlation network for real-time visual tracking. Neurocomputing **401**, 36–47 (2020)
33. Zheng, L., Tang, L., Wang, J.: Learning robust gaussian process regression for visual tracking. In: IJCAI (2018)
34. Zheng, L., Tang, M., Chen, Y., Wang, J., Lu, H.: Fast-deepkcf without boundary effect. In: ICCV (2019)
35. Zheng, L., Tang, M., Chen, Y., Wang, J., Lu, H.: Learning feature embeddings for discriminant model based tracking. In: ECCV (2020)

3D Multi-object Detection and Tracking with Sparse Stationary LiDAR

Meng Zhang, Zhiyu Pan, Jianjiang Feng$^{(\boxtimes)}$, and Jie Zhou

Department of Automation, Beijing National Research Center for Information
Science and Technology, Tsinghua University, Beijing 100084, China
{zhangm20,pzy20}@mails.tsinghua.edu.cn
{jfeng,jzhou}@tsinghua.edu.cn

Abstract. The advent of low-cost LiDAR in recent years makes it feasible for LiDAR to be used in visual surveillance applications such as detection and tracking of players in a football game. However, the extreme sparsity of point cloud acquired by such LiDAR is a challenge for object detection and tracking in large-scale scenes. To alleviate this problem, we propose a method of multi-object detection and tracking from sparse point clouds comprising a short-term tracklet regression stage and a 3D D-IoU data association stage. In the former stage, temporal information is aggregated by the proposed temporal fusion module to predict short-term tracklets formed by three bounding boxes. In the latter stage, the Distance-IoU scores of current tracklets and historical trajectories are computed to associate the data using Hungarian matching algorithm. To reduce the cost of manual annotations, we build a simulated point cloud dataset using Google Research Football for training. A real test dataset of football game is acquired by Livox Mid-100 LiDAR. Our experimental results on both datasets show that fusing multi-frames conduces to improving detection and tracking performance from sparse point clouds. Our 3D D-IoU tracking method also gets a promising performance on the nuScenes autonomous driving dataset.

Keywords: 3D detection · Multi-object tracking · Sparse Stationary LiDAR

1 Introduction

3D multi-object detection and tracking play an important role in the visual surveillance applications, such as football player tracking [11]. These visual surveillance systems mainly utilize RGB cameras for acquisition until now, which leads to some challenging problems: poor performance under non-ideal light and weather conditions, insufficient ability to distinguish foreground and background, and inaccurate object location estimations in 3D space. These challenges can be eliminated by using high-resolution LiDAR to capture point cloud of the scene, yet which is only used in the autonomous driving scene due to the high price.

H. Ma et al. (Eds.): PRCV 2021, LNCS 13019, pp. 16–28, 2021.
https://doi.org/10.1007/978-3-030-88004-0_2

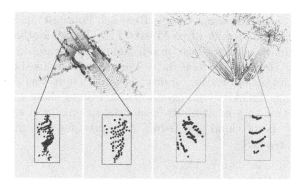

Fig. 1. Point clouds from Velodyne HDL-64E (left) and Livox Mid-100 (right). Point clouds of four persons with labeled distance from LiDAR are present in the second row.

In recent years, with the launch of low-cost LiDAR, such as Livox Mid-100, it is becoming feasible to apply LiDAR to visual surveillance applications. As shown in Fig. 1, the point cloud density of low-cost LiDAR is much lower than the normal LiDAR such as Velodyne HDL-64E laser scanner, making 3D object detection and tracking a very challenging problem.

Despite rapid progress, 3D MOT systems within tracking-by-detection paradigm still suffer a lot from the extreme sparsity of point clouds, which impairs both object detection stage and tracking stage. Most previous object detection works focus on single point cloud frame and leave out the temporal information between multi-frames. Subsequently, tracking-by-detection tracking methods struggle to find the association between current and past trajectories only using the detection bounding boxes of the current frame. To improve the suboptimal solutions based on single frame detection and tracking-by-detection methods, a novel architecture of detection and tracking for point clouds is introduced in this paper by considering the feature of stationary LiDAR. It is rational to aggregate point clouds over different timestamps and leverage temporal information more effectively when the LiDAR is fixed to the ground.

In this paper we propose a 3D multi-object detection and tracking method, with the application in football player tracking. Our method is composed of two stages: a short-term tracklet regression stage and a 3D Distance-IoU [25] (DIoU) data association stage. The first stage takes as input three successive point cloud frames and extracts features separately by the backbone of PointPillars [6]. The three feature maps are aggregated together utilizing 3D convolutional operations to predict short-term tracklets over three past frames. Since adjacent short-term tracklets are overlapping, the following data association stage calculates the DIoU scores of the overlapped pairs from the same object. Finally, a combination of split, birth and death module is applied to get the final tracking result.

As most of public point cloud datasets for 3D object detection and tracking are captured in autonomous driving scene such as KITTI [4], nuScenes [2] and Waymo [18], we build a simulated point cloud dataset for 3D detection and tracking with stationary LiDAR using Google Research Football [5]. We also acquire real point cloud data employing Livox Mid100 LiDAR on a football playground[1]. We train and evaluate our approach on our simulated dataset and test the generalization ability through the real data. We extend our method to nuScenes dataset to show its generalization ability despite that it was designed specifically for stationary LiDAR. Our experimental results show that aggregating temporal information to predict multi-frames detections improves the detection accuracy in sparse point cloud and D-IoU score used for matching conduces to a robust tracking result.

2   Related Work

2.1   3D Object Detection

Traditional football player detection and tracking algorithms are mostly based on RGB videos [11]. Despite rapid progress in CNN based object detection [1, 22], image-based methods face several challenges, such as processing of complex background and accurate estimation of 3D location.

3D object detection systems taking point clouds as input can output highly accurate 3D location, which can be divided into three categories [15]: view-based methods, voxel-grid based methods and raw point cloud-based methods. View-based [17] approaches always convert the point clouds into 2D image views such as the forward views, cylindrical views or bird's eye views. Voxel-grid based methods [7] firstly discretize the 3D point cloud into voxels, each of which is the binary occupancy of the voxel or the amount of points in the voxel. Point RCNN [16] takes as input unstructured point clouds without discretization, which is made up of two stages. Yang *et al.* [23] extended the raw point cloud-based method votenet [12] into the autonomous scenes with a novel fusion sampling strategy.

All of the detectors above process single frame without exploring the temporal information. However, Livox LiDAR can only capture one point on a human-sized object 100 m away in a scanning frame and five points of 50 m, which calls for the multi-frame fusion urgently. Though [24] leverages temporal information utilizing AST-GRU from consecutive frames, the detection results among each frames are lack of connection, which is not for tracking. Our temporal feature aggregation component instead outputs a short-term tracklet capturing temporal information, which contributes to robust tracking results.

2.2   3D Multi-Object Tracking

Most of the 3D multi-object tracking systems within the tracking-by-detection paradigm share the same structure as 2D MOT systems with changing the detection of 2D image plane to 3D space. Online Tracking methods [3,20,21] extract motion features or appearance features to compute the affinity matrix

---

[1] We plan to make both datasets publicly available.

and solve the bipartite graph matching problem by greedy or Hungarian algorithm. Kalman filters are commonly used to estimate the current state of motion features. Weng *et al.* [20] used 3D IoU to match detections with the previous frame and Chiu *et al.* [3] employed the Mahalanobis distance for data association between predicted and actual object detections instead. In [21], a novel joint feature extractor was proposed to learn appearance and motion features from 2D and 3D space simultaneously and a GNN-based architecture is used for data association.

Several methods [10,19] focus on the detection and tracking jointly from point cloud videos. Wang *et al.* [19] proposed an end-to-end 3D object detection and tracking network to generate point-wise tracking displacements for each object. Luo *et al.* [10] proposed a single network to do 3D detection, motion forecasting and tracking together with the spatio-temporal information. Qi *et al.* [14] proposed a end-to-end trained network only for 3D single object tracking in point clouds. Our network directly utilizes PointPillars [6] as backbone to predict 3D multi-object short-term tracklets over historical frames, which is beneficial to the tracking stage.

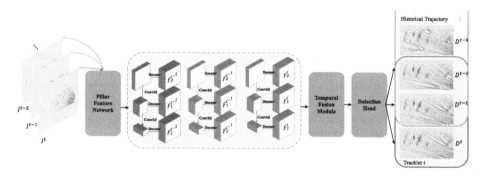

Fig. 2. System overview of our tracklet regression stage. We encode point cloud by a PFN layer and extract high-level features utilizing a 2D convolutional backbone. A temporal fusion module is applied to aggregate the temporal information of three frames. Finally, the detection head predicts a short-term tracklet (blue) as the result, which has two frames overlap with historical trajectory (orange) for tracking. (Color figure online)

3   Proposed Method

In this section, we elaborate on the framework of our 3D MOT method which consists of two main stages: tracklet regression stage and tracking stage. In the following, we first describe the tracklet regression stage in Sect. 3.1, which takes as input three point cloud frames and outputs the short-term tracklets. The data association stage is presented in Sect. 3.2 which links the tracklet regression results to the historical trajectories. Finally, the details of our simulated dataset and real data are provided in Sect. 3.3.

3.1   Tracklet Regression

Our data acquisition experiment showed that there is at most one point on a human-sized object 100 m away from the Livox Mid100 LiDAR in a scanning frame, which leads to a serious decline in the detection performance. A study on nuScenes dataset [2] has reported that accumulation of ten LiDAR sweeps by merging ten sweeps together in the coordinate system of key frame leads to a significant performance increase. However, the raw data accumulation makes it inefficient to leverage the temporal information. Our detector instead introduces a temporal fusion module of three frames in the stationary scenes to avoid the impairment of sparsity. The aggregated temporal information makes the network capable of capturing the motion features and giving a multi-frame tracklet as the result. The structure of our detector is shown in Fig. 2.

Feature Extracting Backbone. The tracklet detector first takes three consecutive point cloud frames $\{I^{t-2}, I^{t-1}, I^t\}$ as input, each of which is discretized uniformly into a set of pillars in the x-y plane after removing the surroundings. The non-empty pillars are fed into a Pillar Feature Network (PFN) [6] which is a simplified version of PointNet [13] here to extract the features and then scattered back to form a pseudo-image of size (C, H, W), where the H and W are decided by the grid size. The pseudo-images of three inputs are denoted by $\{F^{t-2}, F^{t-1}, F^t\}$. Then a top-down network composed of three 2D convolutional blocks is applied to each pseudo-image to extract high-level features. The structure of 2D backbone shown in Figs. 2 shares the same settings with the pedestrian backbone in PointPillars [6]. The features of each top-down block are upsampled to the input size by the following upsampling operation so that 3×3 final features in total are generated from three input frames.

Temporal Fusion Module. Towards the goal of predicting the tracklet, we take all of the output features from the 2D backbone to aggregate the temporal information. In contrast to the accumulation of nuScenes dataset [2] which is denoted by *early fusion*, our temporal fusion module applied after feature extracting is called *late fusion*. The structure of our temporal fusion module is illustrated in Fig. 3. With 3×3 final features of size (C, H, W), we utilize the features of the same level to form a group, which can be taken as a 4D tensor of size $(C, 3, H, W)$. Then a 3D convolution layer with $3 \times 3 \times 3$ kernel followed by BatchNorm and a ReLU is performed on each 4D tensor with no padding in the temporal dimension to aggregate the information of the same level. The aggregated features of three levels are squeezed and combined together through concatenation to form a tensor of size $(3C, H, W)$, which is fed into the detection head to obtain the final detection results.

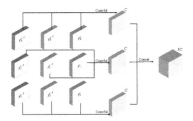 

Fig. 3. The structure of the temporal fusion module. The extracted features of three frames at the same level are formed to create a 4D tensor and three levels are fused through performing a 3D convolution and concatenation.

Fig. 4. The collision of the blue object and the red object leads to a undetected error. The black bounding box is the fusion of two objects actually so we obtain dotted bounding boxes M^t to split the black box D^t to regain the real bounding boxes. (Color figure online)

Detection Head. The fusion of temporal information allows the detection head to capture the motion features such as velocity or acceleration and thus to predict the tracklet of the object. Our detection head following SSD [9] uses two predefined anchor boxes at each location for short-term tracklet regression, which is denoted by three 3D bounding boxes $\{D^{t-2}, D^{t-1}, D^t\}$ for 7×3 regression values. The ground truth tracklet composed of 3D bounding boxes $\{G^{t-2}, G^{t-1}, G^t\}$ is assigned through the 2D IoU between current bounding box G^t and anchor boxes, and then all three boxes are encoded by the corresponding anchor with height and heading angle of the object as additional regression targets. And thus, in addition to the classification scores s_i and box regression of current frame $(x_i^t, y_i^t, z_i^t, l_i^t, w_i^t, h_i^t, \theta_i^t)$ at location i, we add two branches to predict the other two bounding boxes $(x_i^m, y_i^m, z_i^m, l_i^m, w_i^m, h_i^m, \theta_i^m)_{m \in \{t-1, t-2\}}$, which means that each location in the feature map obtains bounding boxes of three frames $\{D^{t-2}, D^{t-1}, D^t\}$ to make up of a short-term tracklet. The short-term tracklet non maximum suppression (NMS) is actually processed on the bounding boxes of current frame, and then tracklets are filtered by classification scores to get the final regression result.

3.2   Data Association

The data association stage takes as input the predicted short-term tracklets $\{D^i\}_{i=t-2}^t$ and the historical trajectories $\{M^i\}_{i=1}^{t-1}$, and finally outputs the updated trajectories $\{M^i\}_{i=1}^t$. We utilize 3D Distance-IoU (DIoU) [25] as the distance measurement since sparse point clouds contain no appearance information of objects. Due to the tracklet regression results, we compute the DIoU scores of overlapping frames directly without Kalman filters to predict the object state. Then we apply a Hungarian matching algorithm to link the trajectories.

We propose a trajectory management module to handle the collision, birth and death of trajectories to improve the tracking robustness.

3D DIoU. As we have mentioned in Sect. 3.1, the extreme sparsity of points in the large-scale scenes impairs the appearance features for that a few 3D points per object make it hard to distinguish the object from others through for even manual annotations. Hence we utilize the motion features represented by the short-term tracklets $\{D^{t-2}, D^{t-1}, D^t\}$ to compute the affinity of objects. Rather than using normal IoU in previous works [20], we use the 3D DIoU [25] instead to avoid the collapse of the normal IoU for non-overlapping or other special cases. Normal IoU only works when two bounding boxes have overlap, and would keep zero while providing no affinity information for non-overlapping cases. And thus, DIoU add a penalty term to consider the normalized distance between two central points in addition to the overlapped area. We extend the 3D DIoU following the penalty term of 2D DIoU while considering the additional dimension and heading angle, which is calculated as follows:

$$DIoU_{3D} = IoU_{3D} - \frac{\rho^2(c_1, c_2)}{\rho^2(a, b)} \tag{1}$$

where c_1, c_2 are the center of two bounding boxes and $\rho(a, b)$ is the farthest point pair of the two boxes. With the penalty term, two bounding boxes far apart will get a lower 3D DIoU score despite no overlap. The distance measurement for non-overlapping case is necessary for the tolerance of inevitable detection errors due to the movement of object in the large-scale scene.

Association Module. Given the short-term tracklet $\{D^i\}_{i=t-2}^t$ and the historical trajectories $\{M^i\}_{i=1}^{t-1}$, the association module utilizes two overlapped frames to compute the affinity matrix. Considering the p objects of the short-term tracklet D and q objects of the trajectory M, we apply the two overlapped pairs of $\{D^{t-2}, M^{t-2}\}$ and $\{D^{t-1}, M^{t-1}\}$ to compute two affinity matrices of $S_{(p \times q)}^{t-1}$ and $S_{(p \times q)}^{t-2}$ using 3D DIoU. The final affinity matrix $S_{(p \times q)}$ is given through the weighted summing of two DIoU matrices considering the precision of bounding boxes regression over three frames. Given the affinity matrix between the tracklets and historical trajectories, we adopt the Hungarian algorithm to solve the bipartite graph matching problem. Finally, the matched pairs of $\{D^i\}_{i=t-2}^t$ and $\{M^i\}_{i=1}^{t-1}$ are combined together to update the trajectories as $\{M^i\}_{i=1}^t$.

Trajectory Management Module. Considering the frequent collision of football players, we propose a split module to deal with the players joined together. As shown in Fig. 4, the collision of the blue object and the red object leads to an undetected error that the black detection bounding box is the fusion of two objects. In this way, the data association module might mistakenly switch the track IDs of these two objects after the collision. Based on the observation that the location and heading angle of the fusion bounding box are almost the average of two separate objects, we propose a split module to generate two bounding boxes instead of the false one to avoid the ID switch, which is fulfilled by taking the separate bounding boxes to split the fused bounding boxes.

Occlusion of objects or the sparsity of point clouds give rise to the birth and death of trajectories in the center of the scene. We retain all the unmatched historical trajectories as the dying trajectories for at most 6 frames. The dying trajectories are updated by the Kalman filter to keep in the scene, which matched in 6 frames are recovered to the normal ones to avoid the false negative, otherwise those unmatched for 6 continual frames are deleted from the scene. The association with a lower threshold of dying trajectories is performed after the matching between short-term tracklets and normal trajectories to reduce the trajectory breaks. At the same time, we accept the unlinked short-term tracklets as the initialized trajectories with new tracking identities. These initialized trajectories will be killed unless it has been consecutively detected for three frames to avoid false positives.

(a)   (b)

Fig. 5. The simulated dataset (left) and the real data (right). We present the RGB data in the top and point cloud in the bottom.

3.3   Football Game Dataset

Several 3D point cloud datasets have been released in the past decade for autonomous driving while point cloud data of surveilance scenes captured by LiDAR fixed to the ground is especially scarce. Hence we build a simulated point cloud dataset and capture the real data for the football game scenes.

The cost of 3D point cloud dataset annotations is extremely high. For this reason, we use the Google Research Football [5] to generate the simulated data.

To be consistent with the real world data from the Livox Mid LiDAR, we sample on the depth map of google research football with a non-repetitive scanning pattern as shown in Fig. 5a, which is implemented by a polar rose curve $\rho = 350 \sin{(3\pi\theta + \theta_0)}$. The origin of coordinates is set to the center of the football playground. Google Research Football provides 12 keypoints of each person in the playground without bounding box. Accordingly, we consider the midperpendicular of the line between left and right shoulder as the heading direction and use the keypoints to generate the enclosing bounding box modeled as x, y, z, width, length, height and yaw angle. The simulated dataset is captured 11 Hz and consists of four point cloud videos for 821 frames in total.

Moreover, we record a football game using Livox Mid100 LiDAR with frequencies 20 Hz. Subsequently, we merge every two sweeps to the frequency 10 Hz following the ground plane calibration. The real data is composed of 1,121 consecutive frames for 116 s. The scene and the point cloud are illustrated in the Fig. 5b. We have manually labelled 200 frames to test the generalization ability of our method.

4   Experiments

4.1   Settings

Datasets and Metrics. We evaluated the proposed algorithm both on the point cloud simulated dataset and the real data. We also conducted experiments on nuScenes [2] dataset to prove the effectiveness of the 3D DIoU tracking method. We merged ten LiDAR sweeps to the key frame as the raw input data due to 2 Hz annotations.

The detection results of simulated dataset are all reported through employing the similar metrics of KITTI [4] in both bird's eye view (BEV) and 3D over IoU threshold of 0.5. The mAP (mean average precision) is used to evaluate the performance of three bounding boxes regression. We followed the AB3DMOT [20] and used sAMOTA, AMOTA and AMOTP as main tracking metrics. In addition, MOTA, MOTP, IDS, FRAG from CLEAR metrics are also applied as secondary metrics.

Implementation Details. As for simulated dataset, we only considered the point cloud within the range of $[-40, 40] \times [-40, 40] \times [-0.1, 2.5]$ meters to remove the surroundings, which was discretized to the grids of 0.25^2 m^2. The details of our backbone follow the implementation of PointPillars [6].

The lower bound threshold of matching algorithm is set to -0.125 for normal trajectories and -0.5 for dying trajectories. The maximum duration of dying trajectories is 6 frames and the minimum duration of new trajectories is 3 frames.

We modified the loss function to add the loss of past and future frames, which is shown as follows:

$$\mathcal{L} = \alpha\mathcal{L}_{cls} + \sum_{i=t-2}^{t} \lambda_i \mathcal{L}_{loc}^i + \beta\mathcal{L}_{dir} \qquad (2)$$

where $\mathcal{L}_{cls}$ is the focal loss [8], $\mathcal{L}_{loc}^i$ is smooth L1-norm and $\mathcal{L}_{dir}$ is the classification loss of heading angles as PointPillars [6]. We set $\alpha = 1, \lambda_{t-2} = \lambda_{t-1} = \lambda_t = 2, \beta = 0.2$ to make a balance.

4.2 Experimental Results

Results on Football Dataset. We report our method's short-term tracklet regression results on the simulated dataset. Firstly, to verify the effectiveness of our temporal fusion module, we present the D^t bounding box detection performance comparison of our method and other approaches in Table 1. PointPillars [6] with a single frame was used as the baseline in the experiment. Early fusion was conducted by merging 3 successive LiDAR sweeps as raw input of PointPillars. We can see that both early fusion and our late fusion method aggregate the temporal information and thus outperform the baseline of the single-frame method as expected. Additionally, late fusion achieved a higher mAP compared with the early fusion and outperformed by a large margin (6.18% in BEV and 14.86% in 3D), which revealed that the fusion of extracted features leverages temporal information more effectively. We evaluated on the point cloud of twice density and we also found that our multi-frame fusion module suffers less from the sparsity of point cloud. The 3D performance of our late fusion method was reduced by 11.83% mAP with the half density while single frame method got 17.91% drop.

Table 1. The comparative results of methods with different temporal information on the simulated dataset

Method	BEV	3D	Δ of 3D
Single Frame ($\times 2$)	56.46	51.65	–
Late Fusion ($\times 2$)	86.26	83.30	–
Single Frame ($\times 1$)	44.51	33.74	–17.91
Early Fusion ($\times 1$)	69.03	56.61	–
Late Fusion ($\times 1$)	75.21	71.47	–11.83

Table 2. Tracking performance on the simulated dataset

Method	sAMOTA↑	AMOTA↑	AMOTP↑	MOTA↑	MOTP↑	IDS↓	FRAG↓
AB3DMOT [20]	76.43	34.68	41.59	82.11	51.03	**2**	167
3D DIoU	84.49	39.58	**46.96**	85.82	51.72	42	196
3D DIoU + Birth/Death	85.14	39.43	45.22	86.14	**51.73**	5	**163**
3D DIoU + Birth/Death/Split	**85.56**	**39.96**	45.01	**86.24**	51.69	4	**163**

The main results of 3D multi-object tracking are summarized in Table 2. We report the AB3DMOT [20] result in Table 2 as baseline. We can see that our method achieves higher AMOTA and AMOTP compared with the AB3DMOT baseline applying the public detection of the tracklet regression stage. We also conducted an experiment to evaluate the effect of the management modules in our method. As shown in the Table 2, adding the birth and death module reduces IDS and FRAG which enhances the robustness of trajectories. The birth module reduces the false positives caused by noise point cloud, and the death module alleviates the effect of occlusion which may lead to tracjectory breaks. Including the split module improves the tracking performance further.

We tested our model on 200 annotated frames of the real data. Our tracking results with a higher sAMOTA and AMOTA compared with AB3DMOT [20] are showed in Table 3. The qualitative 3D tracking results of the real data are shown in Fig. 6 with different color trajectories for different players.

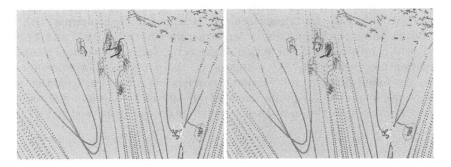

Fig. 6. 3D tracking visualization of real data. The annotated trajectories (left) and our tracking results (right) are drawn in different colors for different players.

Quantitative Results on nuScenes Benchmark. In addition to evaluation on football game dataset, we also report our preliminary results without fine adjustment on the car and pedestrian subset of nuScenes [2] dataset, which was captured using LiDAR mounted on cars. We should emphasize that the input point cloud of nuScenes is very dense since 10 sweeps are merged as input to be consistent with 2 Hz annotations and evaluation benchmark, which weakens the strength of our method. In addition, point cloud from moving LiDAR makes it hard to fuse the temporal information compared to the stationary LiDAR. In order to evaluate 3D trajectory more effectively, we proposed a metric of the 3D trajectory estimation error. The 3D trajectory estimation error of our method tested on nuScenes is 3.620 m for cars and 3.826 m for pedestrians, which is a promising result. Table 4 presents the AMOTA tracking results of our method on nuScenes validation set with the PointPillars detection results. We also report the official results of AB3DMOT [20] with a better detection results than PointPillars. We can see that our method outperforms by 4.3% compared with AB3DMOT in spite of the poorer detection due to the effectiveness of our 3D DIoU tracklet association method.

Table 3. Tracking performance on real data

Method	sAMOTA↑	AMOTA↑
AB3DMOT [20]	44.04	19.97
Ours	49.92	21.53

Table 4. Tracking results on nuScenes val set

Method	Car	Pedestrian
AB3DMOT [20]	69.4	58.7
Ours	73.7	63.0

5 Conclusion

We propose an online 3D multi-object detection and tracking method for sparse stationary LiDAR, which consists of two stages: tracklet regression stage and data association stage. We propose a temporal fusion module to give the short-term tracklet as output and a data association stage that utilizes 3D DIoU to link the tracklets and historical trajectories. Moreover, we build a simulated point cloud dataset for the football game scenes and capture the point cloud data of a real football game. The experimental results show that our tracklet regression and 3D DIoU data association method get a promising performance. We hope to reduce the computational cost to meet real time requirements later. We will also push on the collection and annotation of more real data and generate more realistic simulated data in the future.

Acknowledgments. The work was supported by the National Key Research and Development Program of China under Grant 2018AAA0102803.

References

1. Buric, M., Ivasic-Kos, M., Pobar, M.: Player tracking in sports videos. In: Cloud-Com (2019)
2. Caesar, H., Bankiti, V., Lang, A.H.: nuScenes: a multimodal dataset for autonomous driving. CoRR (2019)
3. Chiu, H., Prioletti, A., Li, J., Bohg, J.: Probabilistic 3D multi-object tracking for autonomous driving. CoRR (2020)
4. Geiger, A., Lenz, P., Urtasun, R.: Are we ready for autonomous driving? the KITTI vision benchmark suite. In: CVPR (2012)
5. Kurach, K., et al.: Google research football: a novel reinforcement learning environment. In: AAAI (2020)
6. Lang, A.H., Vora, S., Caesar, H., Zhou, L., Yang, J., Beijbom, O.: PointPillars: fast encoders for object detection from point clouds. In: CVPR (2019)
7. Li, B.: 3D fully convolutional network for vehicle detection in point cloud. In: IROS (2017)
8. Lin, T., Goyal, P., Girshick, R.B., He, K., Dollár, P.: Focal loss for dense object detection. IEEE Trans. Pattern Anal. Mach. Intell. **42**, 318–327 (2020)
9. Liu, W., et al.: SSD: single shot multibox detector. In: Leibe, B., Matas, J., Sebe, N., Welling, M. (eds.) ECCV 2016. LNCS, vol. 9905, pp. 21–37. Springer, Cham (2016). https://doi.org/10.1007/978-3-319-46448-0_2

10. Luo, W., Yang, B., Urtasun, R.: Fast and furious: real time end-to-end 3D detection, tracking and motion forecasting with a single convolutional net. In: CVPR (2018)
11. Manafifard, M., Ebadi, H., Moghaddam, H.A.: A survey on player tracking in soccer videos. Comput. Vis. Image Underst. **159**, 19–46 (2017)
12. Qi, C.R., Litany, O., He, K., Guibas, L.J.: Deep hough voting for 3D object detection in point clouds. In: ICCV (2019)
13. Qi, C.R., Su, H., Mo, K., Guibas, L.J.: PointNet: deep learning on point sets for 3D classification and segmentation. In: CVPR (2017)
14. Qi, H., Feng, C., Cao, Z., Zhao, F., Xiao, Y.: P2B: point-to-box network for 3D object tracking in point clouds. In: CVPR (2020)
15. Rahman, M.M., Tan, Y., Xue, J., Lu, K.: Recent advances in 3D object detection in the era of deep neural networks: a survey. IEEE Trans. Image Process. **29**, 2947–2962 (2020)
16. Shi, S., Wang, X., Li, H.: Pointrcnn: 3D object proposal generation and detection from point cloud. In: CVPR (2019)
17. Su, H., Maji, S., Kalogerakis, E., Learned-Miller, E.G.: Multi-view convolutional neural networks for 3D shape recognition. In: ICCV. IEEE Computer Society (2015)
18. Sun, P., et al.: Scalability in perception for autonomous driving: waymo open dataset. CoRR (2019)
19. Wang, S., Sun, Y., Liu, C., Liu, M.: Pointtracknet: an end-to-end network for 3-D object detection and tracking from point clouds. IEEE Rob. Autom. Lett. **5**, 3206–3212 (2020)
20. Weng, X., Wang, J., Held, D., Kitani, K.: 3D multi-object tracking: a baseline and new evaluation metrics. In: IROS (2020)
21. Weng, X., Wang, Y., Man, Y., Kitani, K.M.: GNN3DMOT: graph neural network for 3D multi-object tracking with 2D–3D multi-feature learning. In: CVPR (2020)
22. Yang, Y., Xu, M., Wu, W., Zhang, R., Peng, Y.: 3D multiview basketball players detection and localization based on probabilistic occupancy. In: DICTA (2018)
23. Yang, Z., Sun, Y., Liu, S., Jia, J.: 3DSSD: point-based 3D single stage object detector. In: CVPR (2020)
24. Yin, J., Shen, J., Guan, C., Zhou, D., Yang, R.: LiDAR-based online 3D video object detection with graph-based message passing and spatiotemporal transformer attention. In: CVPR (2020)
25. Zheng, Z., Wang, P., Liu, W., Li, J., Ye, R., Ren, D.: Distance-IoU loss: faster and better learning for bounding box regression. In: AAAI (2020)

CRNet: Centroid Radiation Network for Temporal Action Localization

Xinpeng Ding[1], Nannan Wang[2(✉)], Jie Li[1], and Xinbo Gao[1,3]

[1] State Key Laboratory of Integrated Services Networks,
School of Electronic Engineering, Xidian University, Xi'an 710071, China
xpding@stu.xidian.edu.cn, leejie@mail.xidian.edu.cn
[2] State Key Laboratory of Integrated Services Networks,
School of Telecommunications Engineering, Xidian University, Xi'an 710071, China
nnwang@xidian.edu.cn
[3] Chongqing Key Laboratory of Image Cognition, Chongqing University of Posts
and Telecommunications, Chongqing 400065, China
gaoxb@cqupt.edu.cn

Abstract. Temporal action localization aims to localize segments in an untrimmed video that contains different actions. Since contexts at boundaries between action instances and backgrounds are similar, how to separate the action instances from their surrounding is a challenge to be solved. In fact, the similar or dissimilar contents in actions play an important role in accomplishing the task. Intuitively, the instances with the same class label are affinitive while those with different labels are divergent. In this paper, we propose a novel method to model the relations between pairs of frames and generate precise action boundaries based on the relations, namely Centroid Radiation Network (CRNet). Specifically, we propose a Relation Network (RelNet) to represent the relations between sampled pairs of frames by employing an affinity matrix. To generate action boundaries, we use an Offset Network (OffNet) to estimate centroids of each action segments and their corresponding class labels. Based on the assumption that a centroid and its propagating areas have the same action label, we obtain action boundaries by adopting random walk to propagate a centroid to its related areas. Our proposed method is an one-stage method and can be trained in an end-to-end fashion. Experimental results show that our approach outperforms the state-of-the-art methods on THUMOS14 and ActivityNet.

This work was supported in part by the National Key Research and Development Program of China under Grant 2018AAA0103202; in part by the National Natural Science Foundation of China under Grant Grants 62036007, 61922066, 61876142, 61772402, and 62050175; in part by the Xidian University Intellifusion Joint Innovation Laboratory of Artificial Intelligence; in part by the Fundamental Research Funds for the Central Universities. **Student Paper.**

Electronic supplementary material The online version of this chapter (https://doi.org/10.1007/978-3-030-88004-0_3) contains supplementary material, which is available to authorized users.

© Springer Nature Switzerland AG 2021
H. Ma et al. (Eds.): PRCV 2021, LNCS 13019, pp. 29–41, 2021.
https://doi.org/10.1007/978-3-030-88004-0_3

Keywords: Temporal action localization · Random walk · Video understanding

1   Introduction

Video understanding has become an important and promising task in computer vision because of its wide applications in security surveillance, video retrieval and other areas [3–8,22,24,25]. Temporal action localization which aims to localize an action instance and classify it in an untrimmed video is more challenging compared with other video understanding problems, such as action recognition [22,25] and video object detection [26]. For example, given a basketball game video, the detector should not only determine whether slam dunk occurs, but also find the start and end frames of slam dunk in this long range video containing many background frames.

In recent years, many methods based on deep learning have achieved great performances for temporal action localization [11,12,14,21]. Existing methods can be roughly divided into two categories: two-stage methods and one-stage methods. In two-stage methods [11,14], firstly, probabilities for estimating start and end frames are generated. Then, the confidence scores for constructed proposals with starting and ending frames are evaluated. Finally, the non-maximum suppression (NMS) algorithm is used to suppress redundant proposals by decaying their scores. In one-stage methods [1,16,28], each action instance, which consists of boundary, action label, and ranking score, is predicted first. Then, the NMS algorithm is conducted to suppress the redundant action instances. However, these methods may fail to generate the precise boundaries when contexts at boundaries between actions and backgrounds are similar.

In fact, the relations of inter and intra action instances play an import role in accomplishing the temporal action localization task. For example, given a pair of frames A and B, the semantic information of A and B are similar if they belong to the actions with the same label. While those of A and B are dissimilar if they belong to the actions with different labels. In this paper, we introduce a Centroid Radiation Network (CRNet) to explicitly model the frame-wise relations for temporal action localization. To achieve our purpose, we firstly introduce a Relation Network (RelNet) to model the affinity of each pair of frames. We then devise two methods named as Offset Network (OffNet) and Center Network (CenNet) to find centroids (the most salient regions for each action segments) and learn a classifier to determine class labels of the centroids. Under the rationale that a centroid and its neighbour frames with high affinity are likely to belong to the same action class, we finally adopt a random walk algorithm to generate instance boundaries by exploiting the affinity of each pair of frames and centroids. Due to make full use of the relations between each pair of frames, our model can separate contexts at boundaries between action instances and backgrounds to generate more precis boundaries. To sum up, our method is an end-to-end and one-stage method. Our contributions are as follows: (a) A Relation Network (RelNet) is proposed to construct an affinity matrix which models the relations between the pair-wise frames to separate contexts at boundaries between action instances and backgrounds; (b) A Offset Network (OffNet) is proposed to find centroids of action instances which is then exploited as initial seeds to generate action instances by a random walk algorithm; (c) Experimental results conducted on THUMOS14 and ActivityNet v1.3 empirically prove that our proposed method has a better performance than the state-of-the-art temporal action localization methods.

2 Related Work

Early methods are based on hand-crafted features and sliding windows [23] for temporal action localization before the employment of deep learning. Recurrently, temporal action localization based on deep learning have achieved significant performances [11,14,21]. There are two main categories of approaches: two-stage and one-stage methods. Two-stage methods [11,14] firstly generate proposals and then conduct classification to evaluate proposals. For one-stage methods [1,16,28], they directly predict action confidence, boundaries, and action labels based on aggregated features.

However, proposals or action instances predicted by these methods may fail to generate precise boundaries when contexts are similar at boundaries between action instances and backgrounds. Because the relations of inter and intra action instances have not been fully exploited.

3 Our Approach

3.1 Notation and Preliminaries

In this section, we introduce the notation and preliminaries of our problem. Let us denote an untrimmed video as $V = \{\mathbf{v}_t\}_{t=1}^{T}$, where T and $\mathbf{v}_t$ denote the duration and the t-th frame, respectively. Within each video V, ground-truth action instances are denote as $\mathcal{G} = \{g_i | g_i = (c_i, (t_{i,s}, t_{i,e}))\}_{i=1}^{N}, c_i = 1 \ldots C$, where g_i is the i-the action instance in $\mathcal{G}$, with c_i, $t_{i,s}$, and $t_{i,e}$ being the action label, start, and end time of the action, respectively. N and C are the number of action instances and categories, respectively. Distinct from previous works, we also consider the background information. Specifically, let $\mathcal{B} = \{b_i | b_i = (0, (t_{i,s}, t_{i,e}))\}_{i=1}^{N}$ denote the background instances which we sample from the video V with the same size as $\mathcal{G}$.

3.2 Feature Extractor Network

We first need to extract features to get temporal visual contents of untrimmed videos. In our paper, we adopt TSN [25] which has achieved great performances in both action recognition [27] and temporal action localization [10,20,29]. We denote snippets sequence constructed from a video V as $X = \{\mathbf{x}_n\}_{n=1}^{l_x}$, where l_x is the length of snippets sequence. Following [14], we extract snippets with

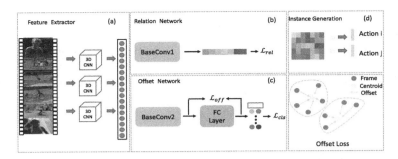

Fig. 1. The framework of the proposed Centroid Radiation Network (CRNet).

a frame interval σ to reduce the computation cost, i.e., $l_x = T/\sigma$. Given X, we extract appearance and motion features using TSN. We denote the fused features as $F = \{\mathbf{f}_{t_n}\}_{n=1}^{l_x}$. To model the relations between pair-wise frames, we first sample the pairs of neighboring frames as follows:

$$\mathcal{S} = \{(\mathbf{f}_{t_1}, \mathbf{f}_{t_2}) \mid |t_1 - t_2| < \delta, \forall t_1 \neq t_2\}, \tag{1}$$

where δ is a radius denoting the maximum distance of a pair of frames. Then we construct intra and inter frame pair information as follows:

$$\mathcal{S}^+ = \{(\mathbf{f}_{t_1}, \mathbf{f}_{t_2}) \mid L(\mathbf{f}_{t_1}) = L(\mathbf{f}_{t_2}), (\mathbf{f}_{t_1}, \mathbf{f}_{t_2}) \in \mathcal{S}\}, \tag{2}$$

$$\mathcal{S}^- = \{(\mathbf{f}_{t_1}, \mathbf{f}_{t_2}) \mid L(\mathbf{f}_{t_1}) \neq L(\mathbf{f}_{t_2}), (\mathbf{f}_{t_1}, \mathbf{f}_{t_2}) \in \mathcal{S}\}, \tag{3}$$

where $L(\mathbf{f})$ indicates the class label of $\mathbf{f}$. Note that $\mathcal{S}^+$ denotes the collection of pairs of features with the same action class; while $\mathcal{S}^-$ denotes the collection of pairs of features from different action classes.

3.3 Relation Network

In this section, we tends to model frame-wise relations based on the key assumption that features of frames in action instances with the same class label are similar while those in action instances with different class labels are dissimilar. Formally, let us define $\mathcal{A}(\mathbf{f}_{t_1}, \mathbf{f}_{t_2})$ as the affinity between $\mathbf{f}_{t_1}$ and $\mathbf{f}_{t_2}$. What we want to obtain is an affinity matrix that can indicate the similarity between $\mathbf{f}_{t_1}$ and $\mathbf{f}_{t_2}$ and should satisfy the following relations:

$$\mathcal{A}(\mathbf{f}_{t_1}, \mathbf{f}_{t_2}) \approx \mathcal{A}(\mathbf{f}_{t_2}, \mathbf{f}_{t_1}), \tag{4}$$

$$\mathcal{A}(\mathbf{f}_{t_1}, \mathbf{f}_{t_2}) \gg \mathcal{A}(\mathbf{f}_{t_1}, \mathbf{f}_{t_3}), \tag{5}$$

where $\mathbf{f}_{t_1}$, $\mathbf{f}_{t_2}$ and $\mathbf{f}_{t_3}$ are three temporal features that satisfy $(\mathbf{f}_{t_1}, \mathbf{f}_{t_2}) \in \mathcal{S}^+$ and $(\mathbf{f}_{t_1}, \mathbf{f}_{t_3}) \in \mathcal{S}^-$. For a pair of $(\mathbf{f}_{t_1}, \mathbf{f}_{t_2}) \in \mathcal{S}$, we define their affinity $a_{t_1 t_2}$ as:

$$a_{t_1 t_2} = \min_{k \in [t_1, t_2]} \mathcal{R}(\mathbf{f}_k), \tag{6}$$

where $[t_1, t_2]$ in Eq. 6 means a range from t_1 to t_2 and $\mathcal{R}$ is the proposed RelNet. If $\mathbf{f}_{t_1}$ and $\mathbf{f}_{t_2}$ are in the same instance, the minimal similarity of pairwise frames between them should still be high. While $\mathbf{f}_{t_1}$ and $\mathbf{f}_{t_2}$ are not in the same instance, the minimal similarity, which will occur at boundaries between action instances and backgrounds, should be nearly zero. Then the objective of RelNet is designed as follows:

$$\mathcal{L}_{rel} = -\sum_{(\mathbf{f}_{t_1}, \mathbf{f}_{t_2}) \in \mathcal{S}^+} \frac{\log a_{t_1 t_2}}{|\mathcal{S}^+|} - \sum_{(\mathbf{f}_{t_1}, \mathbf{f}_{t_2}) \in \mathcal{S}^-} \frac{\log (1 - a_{t_1 t_2})}{|\mathcal{S}^-|}. \tag{7}$$

Finally, the affinity matrix is constructed by $\mathcal{A} = [a_{t_1 t_2}]$. An example of the outputs of the RelNet is shown in Fig. 2(a). The relationship between the affinity matrix and the outputs of RelNet is shown in Eq. 6. In Fig. 2(a), darker color denotes higher output value. We can see that the output values at the boundaries are low while those within action instances are high, **clearly separating the context at boundaries between action instances and backgrounds.** It can be verified that Eq. 5 and Eq. 4 are satisfied. Due to $l_x = 100$ which is much smaller than T, the dimension of $\mathcal{A}$ is not too high and the computation cost can be acceptable. The exhaustive architecture of RelNet is illustrated in Fig. 1 (b). The details of BaseConv1 can be found in the **Supplementary Material**.

3.4   Centroids Prediction

In this section, we introduce the method to predict centroids and train a classifier to determine their labels which is then used to generate action instances. We define centroids as a set of the most discriminative parts of an action instance, since these parts are easily to predict compared with other parts [17]. An Intuitive approach is to set frames near the middle frame of an action instance as centroids. We define a center loss modified from [31] to regress this kind of centroids. For each ground-truth middle frame $p_i = \frac{t_{i,s} + t_{i,e}}{2}$, we splat it onto a heatmap $Y \in [0, 1]^{T \times C}$ which is defined as:

$$Y_{tc} = \exp \left(-\frac{(t - p_i)^2}{2\sigma_p^2} \right), \tag{8}$$

where σ_p is an instance-size-adaptive standard deviation. The training objective is a penalty-reduced frame-wise logistic regression with focal loss [15]:

$$\mathcal{L}_{cen} = \frac{-1}{N} \sum_t \begin{cases} \left(1 - \hat{Y}_{tc} \right)^\alpha \log \left(\hat{Y}_{tc} \right), & \text{if } Y_{tc} = 1 \\ (1 - Y_{tc})^\beta \left(\hat{Y}_{tc} \right)^\alpha \\ \log \left(1 - \hat{Y}_{tc} \right), & \text{otherwise} \end{cases}, \tag{9}$$

where $\hat{Y}_{tc}$ indicates the predicted confidence scores of action instances for label c, α and β are hyperparameters. Finally, we obtain centroids as follows:

$$\hat{\mathcal{C}}_{cen} = \left\{ (t, c) | \max_{t \in [t-l_c, t+l_c]} \hat{Y}_{tc} \geq \gamma \right\}, \tag{10}$$

where l_c is the average duration of each action instance whose label is c.

In fact, **the middle frames are not necessarily the most discrimina-tive parts of action instances.** For example, the most discriminative parts of 'LongJump' are frames where the athlete jumps over the bunker rather than the middle frames where the athlete are running on the track.

Motivated by class activation map (CAM) [30] which can roughly localize object areas by drawing attentions on discriminative parts of object classes. Following CAM, there are some works proposing class activation sequence (CAS) for weakly supervised action localization [17,19]. We define offset as the difference between the salient values of the centroid and other frames, e.g., as shown in Fig. 1, to help find the centroids. The offset is small when the salient frame and the centroid are not very far from each other. Then the frames whose offsets equal nearly to zero can be regarded as the centroids.

We devise an Offset Network (OffNet) $\mathcal{O}$ and denote its output as Q. Let $\hat{\mathbf{f}}_{t_1}$ and $\hat{\mathbf{f}}_{t_2}$ denote the salient value of $\mathbf{f}_{t_1}$ and $\mathbf{f}_{t_2}$ in the same action instance g_i. $\hat{\mathbf{f}}_{t_1}$ and $\hat{\mathbf{f}}_{t_2}$ are encouraged to be the same, though there may be several centroids in g_i. Then, we can get the following equation:

$$\hat{\mathbf{f}}_{t_1} + Q_{t_1 c} = \hat{\mathbf{f}}_{t_2} + Q_{t_2 c}, \quad \text{s.t. } \hat{\mathbf{f}}_{t_1}, \hat{\mathbf{f}}_{t_2} \in g_i, \tag{11}$$

where $Q \in \mathbb{R}^{T \times C}$ indicating the offset between $\hat{\mathbf{f}}_{t_1}$ and its centroid for class label c. We define the following equation:

$$J_{\mathcal{O}}(\hat{\mathbf{f}}_{t_1}, \hat{\mathbf{f}}_{t_2}) = \hat{\mathbf{f}}_{t_1} - \hat{\mathbf{f}}_{t_2} + Q_{t_1 c} - Q_{t_2 c}, \tag{12}$$

and we can obtain the following equation because of Eq. 11:

$$J_{\mathcal{O}}(\hat{\mathbf{f}}_{t_1}, \hat{\mathbf{f}}_{t_2}) = 0, \forall f_{t_1}, f_{t_2} \in g_i. \tag{13}$$

Then the offset to obtain can be changed to: $\min_{\mathcal{O}} J_{\mathcal{O}}(\hat{\mathbf{f}}_{t_1}, \hat{\mathbf{f}}_{t_2})$. Based on the above analysis, the objective of OffNet can be defined as follows:

$$\mathcal{L}_{off}^{\mathcal{G}} = \frac{1}{|\mathcal{G}|} \sum_{(\hat{\mathbf{f}}_{t_1}, \hat{\mathbf{f}}_{t_2}) \in g} |J_{\mathcal{O}}(\hat{\mathbf{f}}_{t_1}, \hat{\mathbf{f}}_{t_2})| + \lambda_1 \|\mathcal{O}\|, \tag{14}$$

where λ_1 is the trade-off parameters and g is an instance in $\mathcal{G}$. Since there are no significant and indefinite centroids for background frames, we define the following loss for background frames:

$$\mathcal{L}_{off}^{\mathcal{B}} = \frac{1}{|\mathcal{B}|} \sum_{(\hat{\mathbf{f}}_t) \in \mathcal{B}} |\hat{\mathbf{f}}_t|. \tag{15}$$

Finally, we obtain the offset loss by combining two losses:

$$\mathcal{L}_{off} = \mathcal{L}_{off}^{\mathcal{G}} + \mathcal{L}_{off}^{\mathcal{B}}. \tag{16}$$

We select the temporal frames as centroids $\mathcal{C}_{off}$ by a threshold μ as follows:

$$\mathcal{C}_{off} = \{(t,c)|Q_{tc} \le \mu\}. \tag{17}$$

The action classification scores for label c can be defined as follows:

$$y_c^a = 1 - \sum_{(t,c)\in\mathcal{C}_{off}} \frac{Q_{tc}}{|\mathcal{C}_{off}|}. \tag{18}$$

Based on Eq. 18, we can obtain $\mathbf{y^a} = [y_1^a, y_2^a, \ldots, y_C^a]$, which indicates the probabilities belonging to C action classes. We measure the classification loss via the softmax loss:

$$\mathcal{L}_{cls} = -\sum_{n=0}^{C} I_{n=c} \log(y_n^a), \tag{19}$$

where the indicator function $I_{n=c} = 1$ if n equals to the ground truth action label c, otherwise $I_{n=c} = 0$. Since the numbers of elements in sampled $\mathcal{S}^+$ and $\mathcal{S}^-$ are the same, imbalance training can be avoided. During prediction, the centroids can be obtained by:

$$\hat{\mathcal{C}}_{off} = \left\{ (t,c) | \min_{t\in[t-l_c,t+l_c]} Q_{tc} \le \mu \right\}, \tag{20}$$

where l_c is the average duration of each action instance whose label is c. As Fig. 1(c) shows, OffNet consists of a BaseConv2 and a fully connected layer. The details of BaseConv2 can be found in **Supplementary Material**.

3.5   Instance Generation

In this section, we describe instance generation based on the centroids and the affinity matrix. We use the random walk similar to [9] to generate instances. First of all, we define $\mathcal{N}(\mathbf{f}_t) = \mathbf{f}_t \pm 1$ as the neighbors of $\mathbf{f}_t$. With $\mathcal{N}$, we define the adjacency matrix for R as follows:

$$R_{t_1 t_2} = \begin{cases} 1, & if \ \mathbf{f}_{t_1} \in \mathcal{N}(\mathbf{f}_{t_2}) \\ 0, & otherwise \end{cases} \tag{21}$$

Laplacian matrix is defined as $L = D - R$, where D is the degree matrix. To obtain instances, we firstly construct initial action instance vectors as follows:

$$\mathcal{I}_c(t) = \begin{cases} 1, & if \ \mathbf{f}_t \ has \ a \ label \ c \\ 0, & otherwise \end{cases} \tag{22}$$

Combining all vector $\mathcal{I}_c, c = 0 \ldots C$, we can obtain the initial instance matrix $\mathcal{I}$. According to [9], the combinatorial Dirichlet problem has the same solution as the desired random walker probabilities. Therefore, instance generation can be solved as:

$$\nabla^2(\frac{1}{2}\mathcal{I}^T L \mathcal{I}) = 0. \tag{23}$$

We divide $\mathcal{I}$ into two parts $\mathcal{I}_M$ and $\mathcal{I}_U$, where $\mathcal{I}_M$ corresponds to the controid frames whose class labels are predicted and $\mathcal{I}_U$ corresponds to the non-controid frames whose class labels are uncertain and to be predicted by the random walk algorithm. Then $\frac{1}{2}\mathcal{I}^T L \mathcal{I}$ can be expressed as:

$$\frac{1}{2}\mathcal{I}^T L \mathcal{I} = \frac{1}{2} \begin{bmatrix} \mathcal{I}_M^T \ \mathcal{I}_U^T \end{bmatrix} \begin{bmatrix} L_M & B \\ B^T & L_U \end{bmatrix} \begin{bmatrix} \mathcal{I}_M \\ \mathcal{I}_U \end{bmatrix} \tag{24}$$
$$= \frac{1}{2}\left(\mathcal{I}_M^T L_M \mathcal{I}_M + 2\mathcal{I}_U^T B^T \mathcal{I}_M + \mathcal{I}_U^T L_U \mathcal{I}_U \right).$$

Instances can be located by the solution of following equation:

$$L_U \mathcal{I}_U = -B^T \mathcal{I}_M, \tag{25}$$

which can be solved by iterative methods [9]. Finally we recombine $\mathcal{I}_M$ and $\mathcal{I}_U^{(t)}$ to get the generated instance vectors $\mathcal{I}^{(t)}$, where t indicates the number of iterations.

3.6 Overall Objective Before Random Walk

Our CRNet is trained by minimizing the following objective, i.e.,

$$\mathcal{L} = \mathcal{L}_{cls} + \lambda_2 \mathcal{L}_{off} + \lambda_3 \mathcal{L}_{rel}, \tag{26}$$

where λ_2 and λ_3 and are the trade-off parameters.

3.7 Prediction and Post-processing

During the prediction of action localization, CRNet first generates an affinity matrix which presents frame-wise relations information by using RelNet. Then OffNet predicts centroids and determine their class labels in a video. Based on the random walk algorithm which is described in Sect. 3.5, action instances are generated. Our CRNet is an end-to-end and one-stage method.

4 Experiments

4.1 Datasets

The THUMOS14 [18] and ActivityNet v1.3 [2] are two popular benchmark datasets for action localization. The THUMOS14 dataset has over 13,000 temporally trimmed videos for training, 1,010 videos for validation and 1,574 videos for testing from 20 classes. ActivityNet v1.3 is a large dataset for general temporal action proposal generation and detection, which contains 19994 videos with 200 action classes collected from Youtube.

4.2 Implementation Details

We use TSN [25] to extract features, where spatial and temporal sub-networks adopt ResNet and BN-Inception network, respectively. We conduct experiments on THUMOS14 and ActivityNet v1.3, respectively. We do not fine tune the feature extractors. We set δ in Eq. 1 to 10. The trade-off parameters λ_1 in Eq. 14, λ_2 and λ_3 in Eq. 26 are set to 0.01, 10, and 10, respectively. In Eq. 9, we set $\alpha = 2$ and $\beta = 4$. The threshold γ in Eq. 10 and and μ in Eq. 17 are set to 0.45 and 0.06 respectively.

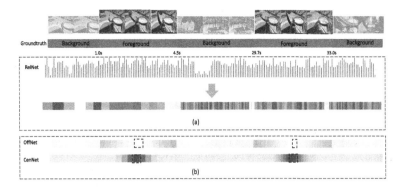

Fig. 2. Visualization of Instance Generation. For better presentation, we normalize the duration of each instance in temporal axis.

4.3 Evaluation of RelNet, CenNet and OffNet

We conduct experiments and visualize the training process of 'High Jump' on THUMOS14 dataset, as illustrated in Fig. 3. Affinity loss continuously reduces the output of RelNet near boundaries and increases the output of RelNet inside the instances. Specifically, when iterating 100 times, the probability values of frames inside the instances is already high, but the probability values of the frames near the boundaries is low as those of the frames at boundaries. This is because in the initial stage of training, the output of RelNet around the boundaries are reduced by the second loss of Eq. 7. As the number of training iterations increases, the first part loss of Eq. 7 makes the probability values of the frames inside instances larger and larger, and the second part loss of Eq. 7 makes the probability values of frames at boundaries lower and lower. After about 15,000 iterations, the probability values of the frames at the boundaries are already low enough to distinguish them from their nearby frames.

4.4 Performance with Fewer Data

We also discover that our method can maintain good performances with using fewer training data. We train our model with only ϕ percent of training videos, denoted as model@ϕ. The comparison of OffNet and CentNet are reported in Table 1.

4.5 Comparisons with State-of-the-Art

Table 2 presents the mAP performances with different IoU thresholds on THU-MOS14 dataset. 'Ours (CenNet)' and 'Ours (OffNet)' indicate the CRNet adopting CenNet and OffNet to predict centroids respectively. The results with different IoU values illustrate that Our CRNet manifestly outperforms others. We also

Table 1. Comparisons of OffNet and CenNet with fewer datasets on THUMOS14.

Method	0.3	0.4	0.5	0.6	0.7	AVG
OffNet@30%	46.3	35.1	24.8	17.2	11.9	27.1
CenNet@30%	43.9	29.7	21.3	12.5	6.3	22.7
OffNet@60%	54.3	44.1	35.5	26.6	17.3	35.6
CenNet@60%	52.6	39.4	30.9	22.8	15.5	32.2
OffNet@100%	58.1	48.3	39.7	31.4	22.0	39.9
CenNet@100%	57.6	47.8	38.9	29.9	20.7	38.9

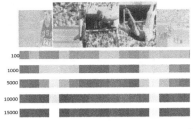

Fig. 3. Visualization of Training Rel-Net. With the iterations of training increases, the value of the frames at the boundaries will be lower and lower under the effect of affinity loss, and the value of the frames within instances will become higher.

Table 2. Comparisons of temporal action localization on THUMOS14.

Method	0.3	0.4	0.5	0.6	0.7	AVG
Two-stage methods						
BSN [14]	53.5	45.0	36.9	28.4	20.0	36.8
BMN [11]	56.0	47.4	38.8	29.7	20.5	38.5
One-stage methods						
SMC [28]	36.5	27.8	17.8	–	–	–
SS-TAD [1]	45.7	-	29.2	–	–	–
GTAN [16]	57.8	47.2	38.8	–	–	–
Ours (CenNet)	57.6	47.8	38.9	29.9	20.7	38.9
Ours (OffNet)	**58.1**	**48.3**	**39.7**	**31.4**	**22.0**	**39.9**

Table 3. Comparisons of temporal action localization on THUMOS14.

Method	0.5	0.75	0.95	AVG
Lin et al. [13]	49.0	32.9	7.9	32.3
BSN [14]	52.5	33.5	8.9	33.7
GTAN [16]	52.6	34.1	8.9	34.3
BMN [11]	52.1	34.8	8.3	33.9
Ours (CenNet)	52.7	34.5	9.0	34.4
Ours (OffNet)	**53.8**	**35.3**	**9.4**	**35.6**

carry experiments on ActivityNet v1.3 in Table 3 to show our method's superiority compared with other approaches. The performance of CRNet (OffNet) will be increase to 53.8%,35.3% and 9.4% on mAP@0.5, mAP@0.75, mAP@0.95 respectively, outperforming significantly over CRNet (CenNet). Qualitative results on THUMOS14 and ActivityNet v1.3 dataset are illustrated in the **Supplementary Material**.

5   Conclusion

In this paper, we introduce the Centroid Radiation Network (CRNet) for temporal action localization by exploiting the relations between pairs of frames to construct the affinity matrix. By this means, our model can separate contexts at the boundaries between action instances and backgrounds even they are similar. Meanwhile, we propose two methods named as OffNet and CenNet to predict centroids. Compared to previous regression based methods, our model propagates centroids to its related regions by a random walk algorithm to generate action instances, avoiding the context separation. Extensive experiments demonstrate that CRNet outperforms other state-of-the-art methods in the temporal action localization task. We also verify that our model used OffNet can achieve good performance with fewer training data.

References

1. Buch, S., Escorcia, V., Ghanem, B., Fei-Fei, L., Niebles, J.C.: End-to-end, single-stream temporal action detection in untrimmed videos. In: Proceedings of the British Machine Vision Conference, vol. 2, p. 7 (2017)
2. Caba Heilbron, F., Escorcia, V., Ghanem, B., Carlos Niebles, J.: Activitynet: a large-scale video benchmark for human activity understanding. In: Proceedings of the IEEE Conference on Computer Vision and Pattern Recognition, pp. 961–970 (2015)
3. Duan, X., Huang, W., Gan, C., Wang, J., Zhu, W., Huang, J.: Weakly supervised dense event captioning in videos. In: Proceedings of Advances in Neural Information Processing Systems, pp. 3059–3069 (2018)
4. Fan, L., Huang, W., Gan, C., Ermon, S., Gong, B., Huang, J.: End-to-end learning of motion representation for video understanding. In: Proceedings of the IEEE Conference on Computer Vision and Pattern Recognition, pp. 6016–6025 (2018)
5. Gan, C., Gong, B., Liu, K., Su, H., Guibas, L.J.: Geometry guided convolutional neural networks for self-supervised video representation learning. In: Proceedings of the IEEE Conference on Computer Vision and Pattern Recognition, pp. 5589–5597 (2018)
6. Gan, C., Wang, N., Yang, Y., Yeung, D.Y., Hauptmann, A.G.: Devnet: a deep event network for multimedia event detection and evidence recounting. In: Proceedings of the IEEE Conference on Computer Vision and Pattern Recognition, pp. 2568–2577 (2015)
7. Gan, C., Yang, Y., Zhu, L., Zhao, D., Zhuang, Y.: Recognizing an action using its name: a knowledge-based approach. Int. J. Comput. Vis. **120**(1), 61–77 (2016)
8. Gan, C., Yao, T., Yang, K., Yang, Y., Mei, T.: You lead, we exceed: labor-free video concept learning by jointly exploiting web videos and images. In: Proceedings of the IEEE Conference on Computer Vision and Pattern Recognition, pp. 923–932 (2016)
9. Grady, L.: Random walks for image segmentation. IEEE Trans. Pattern Anal. Mach. Intell. **11**, 1768–1783 (2006)
10. Ji, J., Cao, K., Niebles, J.C.: Learning temporal action proposals with fewer labels. arXiv preprint arXiv:1910.01286 (2019)
11. Lin, T., Liu, X., Li, X., Ding, E., Wen, S.: Bmn: boundary-matching network for temporal action proposal generation. arXiv preprint arXiv:1907.09702 (2019)
12. Lin, T., Zhao, X., Shou, Z.: Single shot temporal action detection. In: Proceedings of the 25th ACM International Conference on Multimedia, pp. 988–996. ACM (2017)
13. Lin, T., Zhao, X., Shou, Z.: Temporal convolution based action proposal: submission to activitynet 2017. arXiv preprint arXiv:1707.06750 (2017)
14. Lin, T., Zhao, X., Su, H., Wang, C., Yang, M.: Bsn: boundary sensitive network for temporal action proposal generation. In: Proceedings of the European Conference on Computer Vision, pp. 3–19 (2018)

15. Lin, T.Y., Goyal, P., Girshick, R., He, K., Dollár, P.: Focal loss for dense object detection. In: Proceedings of the IEEE International Conference on Computer Vision, pp. 2980–2988 (2017)
16. Long, F., Yao, T., Qiu, Z., Tian, X., Luo, J., Mei, T.: Gaussian temporal awareness networks for action localization. In: Proceedings of the IEEE Conference on Computer Vision and Pattern Recognition, pp. 344–353 (2019)
17. Nguyen, P., Liu, T., Prasad, G., Han, B.: Weakly supervised action localization by sparse temporal pooling network. In: Proceedings of the IEEE Conference on Computer Vision and Pattern Recognition. pp. 6752–6761 (2018)
18. Oneata, D., Verbeek, J., Schmid, C.: The lear submission at thumos 2014 (2014)
19. Paul, S., Roy, S., Roy-Chowdhury, A.K.: W-talc: weakly-supervised temporal activity localization and classification. In: Proceedings of the European Conference on Computer Vision, pp. 563–579 (2018)
20. Shou, Z., Chan, J., Zareian, A., Miyazawa, K., Chang, S.F.: Cdc: convolutional-de-convolutional networks for precise temporal action localization in untrimmed videos. In: Proceedings of the IEEE Conference on Computer Vision and Pattern Recognition, pp. 5734–5743 (2017)
21. Shou, Z., Wang, D., Chang, S.F.: Temporal action localization in untrimmed videos via multi-stage cnns. In: Proceedings of the IEEE Conference on Computer Vision and Pattern Recognition, pp. 1049–1058 (2016)
22. Simonyan, K., Zisserman, A.: Two-stream convolutional networks for action recognition in videos. In: Proceedings of Advances in Neural InforKFCmation Processing Systems, pp. 568–576 (2014)
23. Tang, K., Yao, B., Fei-Fei, L., Koller, D.: Combining the right features for complex event recognition. In: Proceedings of the IEEE International Conference on Computer Vision, pp. 2696–2703 (2013)
24. Tran, D., Bourdev, L., Fergus, R., Torresani, L., Paluri, M.: Learning spatiotemporal features with 3D convolutional networks. In: Proceedings of the IEEE International Conference on Computer Vision, pp. 4489–4497 (2015)
25. Wang, L., et al.: Temporal segment networks: towards good practices for deep action recognition. In: Leibe, B., Matas, J., Sebe, N., Welling, M. (eds.) ECCV 2016. LNCS, vol. 9912, pp. 20–36. Springer, Cham (2016). https://doi.org/10.1007/978-3-319-46484-8_2
26. Xiao, F., Jae Lee, Y.: Video object detection with an aligned spatial-temporal memory. In: Proceedings of the European Conference on Computer Vision, pp. 485–501 (2018)
27. Yan, S., Xiong, Y., Lin, D.: Spatial temporal graph convolutional networks for skeleton-based action recognition. In: Proceedings of Thirty-Second AAAI Conference on Artificial Intelligence (2018)
28. Yuan, Z., Stroud, J.C., Lu, T., Deng, J.: Temporal action localization by structured maximal sums. In: Proceedings of the IEEE Conference on Computer Vision and Pattern Recognition, pp. 3684–3692 (2017)
29. Zhao, Y., Xiong, Y., Wang, L., Wu, Z., Tang, X., Lin, D.: Temporal action detection with structured segment networks. In: Proceedings of the IEEE International Conference on Computer Vision, pp. 2914–2923 (2017)

30. Zhou, B., Khosla, A., Lapedriza, A., Oliva, A., Torralba, A.: Learning deep features for discriminative localization. In: Proceedings of the IEEE Conference on Computer Vision and Pattern Recognition, pp. 2921–2929 (2016)
31. Zhou, X., Wang, D., Krähenbühl, P.: Objects as points. In: arXiv preprint arXiv:1904.07850 (2019)

Weakly Supervised Temporal Action Localization with Segment-Level Labels

Xinpeng Ding[1], Nannan Wang[2(✉)], Jie Li[1], and Xinbo Gao[1,3]

[1] State Key Laboratory of Integrated Services Networks,
School of Electronic Engineering, Xidian University, Xi'an 710071, China
`xpding@stu.xidian.edu.cn, leejie@mail.xidian.edu.cn`
[2] State Key Laboratory of Integrated Services Networks,
School of Telecommunications Engineering, Xidian University, Xi'an 710071, China
`nnwang@xidian.edu.cn`
[3] Chongqing Key Laboratory of Image Cognition, Chongqing University of Posts
and Telecommunications, Chongqing 400065, China
`gaoxb@cqupt.edu.cn`

Abstract. Temporal action localization presents a trade-off between test performance and annotation-time cost. Fully supervised methods achieve good performance with time-consuming boundary annotations. Weakly supervised methods with cheaper video-level category label annotations result in worse performance. In this paper, we introduce a new segment-level supervision setting: segments are labeled when annotators observe actions happening here. We incorporate this segment-level supervision along with a novel localization module in the training. Specifically, we devise a partial segment loss regarded as a loss sampling to learn integral action parts from labeled segments. Since the labeled segments are only parts of actions, the model tends to overfit along with the training process. To tackle this problem, we first obtain a similarity matrix from discriminative features guided by a sphere loss. Then, a propagation loss is devised based on the matrix to act as a regularization term, allowing implicit unlabeled segments propagation during training. Experiments validate that our method can outperform the video-level supervision methods with almost same the annotation time.

Keywords: Temporal action localization · Weak supervision · Regularization

This work was supported in part by the National Key Research and Development Program of China under Grant 2018AAA0103202; in part by the National Natural Science Foundation of China under Grant Grants 62036007, 61922066, 61876142, 61772402, and 62050175; in part by the Xidian University Intellifusion Joint Innovation Laboratory of Articial Intelligence; in part by the Fundamental Research Funds for the Central Universities. **Student Paper.**

Electronic supplementary material The online version of this chapter (https://doi.org/10.1007/978-3-030-88004-0_4) contains supplementary material, which is available to authorized users.

H. Ma et al. (Eds.): PRCV 2021, LNCS 13019, pp. 42–54, 2021.
https://doi.org/10.1007/978-3-030-88004-0_4

1   Introduction

There are many works [1,4,8,10,21] in recent years to tackle temporal action localization which aims to localize and classify actions in untrimmed videos. These methods are introduced with **full supervision** setting: annotations of temporal boundaries (start time and end time) and action category labels are provided in the training procedure as shown in Fig. 1 (a). Although great improvement has been gained under this setting, obtaining such annotations is very time-consuming in long untrimmed videos [26]. To alleviate the requirement for temporal boundary annotations, weakly supervised methods [7,13–16,28] have been developed. The most common setting is **video-level supervision**: only category labels are provided for each video in training time as shown in Fig. 1 (c). In these methods, researchers aim to learn class activation sequences (CAS) for action localization using excitation back-propagation with video-level category label supervision guided by a classification loss, which are simple yet efficient for weakly supervised action localization. Along with the learning procedure, CAS will shrink to the discriminative parts of action due to the discriminative parts are capable of minimizing the action classification loss. Therefore, they are usually observed to active discriminative action parts instead of full action extent. Existing approaches [18,28] have explored erasing salient parts to expand temporal class activation maps and pursue full action extent. Nevertheless, these methods may result in decreased action classification accuracy and incomplete semantic information of actions, due to the lack of some action parts.

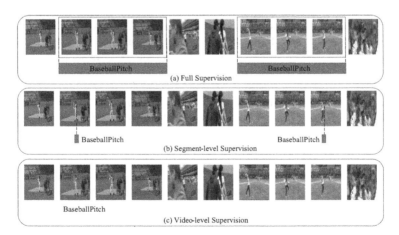

Fig. 1. A video annotated with (a) full supervision, (b) segment-level supervision and (c) video-level supervision.

In this paper, we first divide an video into non-overlap segments, each of which contains 16 frames. Then, we propose a new **segment-level supervision** setting: one or two segments and their corresponding action category labels are provided in training time as shown in Fig. 1(b). In this setting, annotators browse the video for an action instance and simultaneously label one or two interval seconds that belong to the action instance. The segments contain the labeled seconds are regard as the ground-truth labeled segments. Compared with boundary annotations, segment annotations do not require time consumption on finding precise start and end time which is sometimes accurate to 0.1 s. Furthermore, segment-level supervision can provide extra localization information compared with video-level supervision.

To make full use of this segment-level information, we propose a localization module which consists of three loss terms: a partial segment loss, a sphere loss and a propagation loss. Compared with video-level supervision methods, the partial segment loss uses the labeled segments to **learn more parts of action instances instead of just focusing on discriminative parts**. Since the labeled segments are only a part of an action instance rather than the full extent, the model will overfit along with the training process. To address the problem, we first define the segments that have high feature similarity with labeled segments as implicit segments, which is motivated by the intuition that the features of the segments belonging to the same action instance are similar. To measure the similarity between pairs of segments, we obtain a similarity matrix generated from the discriminative features. Guided by the sphere loss, the discriminative features have **smaller maximal intra-class distance than minimal inter-class distance**. Then, based on the obtained similarity matrix, the propagation loss is introduced to **act as a regularization term which propagates labeled segments to implicit ones**. The main contributions of this paper are as follows: (a) A new segment-level supervision setting is proposed for weakly supervised temporal action localization, costing almost the same annotation time as the video-level supervision; (b) A novel localization module guided by a sphere loss, a partial segment loss and a propagation loss is proposed to exploit both labeled and implicit segment to keep from focusing only on the discriminative parts; (c) Experimental results demonstrate that the proposed method outperforms the state-of-the-art weakly supervised temporal action localization methods with video-level supervision setting.

2   Related Work

Temporal Action Localization. Temporal action localization in full supervision has gained significant developments in recent years [4, 8–10, 13, 17]. However, obtaining precise temporal boundaries is very time-consuming in long untrimmed videos. To reduce the time-consumption of boundary annotations, weakly supervised temporal action localization in video-level category label supervision has attracted growing attentions. Given only category labels, most of methods [11, 15, 16, 18, 21, 28] tend to generate class activation sequences (CAS) from a classification loss. However, CAS guided by the classification loss is observed to shrink to salient parts instead of the full action extent. The reason behind the phenomenon lies in that the networks tend to learn the most compact features to

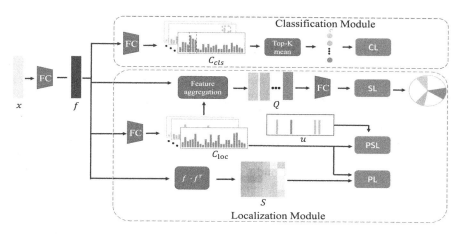

Fig. 2. Architecture of Our Approach. There are two main modules: a classification module and a localization module. The classification module is trained by a classification loss (CL) and the localization module is guided by a partial segment loss (PSL), a sphere loss (SL) and a propagation loss (PL).

distinguish different categories when optimizing the classification loss and ignore less discriminative ones [25].

Regularization for Neural Networks. Regularization is a set of techniques that can prevent overfitting in neural networks and has been widely used to improve the performance, e.g. norm regularization [5]. Motivated from the semi-supervised learning [23], our proposed propagation loss differs from these regularization in parameters.

3 Our Approach

3.1 Problem Statement and Notation

Let define untrimmed videos as $V = \{v_i\}_{i=1}^{M}$, where M denotes the number of videos. We divide each video into non-overlap segments $\{g_t\}_{t=1}^{l_i}$, where l_i denotes number of segments. Each segment consists of 16 frames. The extracted feature of v_i is denoted as $x_i \in \mathbb{R}^{l_i \times D}$, where D is the dimension. Let the action label be denoted as $Y = \{y_i\}_{i=1}^{M}$, where $y_i \in \{0,1\}^{N}$ is a multi-hot vector and N is the number of action classes. For the video v_i, we denote its segment label as $u_i \in \{0,1\}^{l_i \times N}$. $u_i(t, n) = 1$ when there is an action instance with category n occurring in t-th segment and $u_i(t, n) = 0$ when none action instance with category n occurs in the t-th segments. For simplicity, we use x, u and l instead of x_i, u_i and l_i when there is no confusion. We use $Q(i, j)$ to represent the elements in the i-th row and j-th column of matrix Q. Naturally, $Q(i, :)$ and $Q(:, j)$ indicate the i-th row vector and j-th column vector of matrix Q respectively.

3.2   Architecture

The architecture of our approach is shown in Fig. 2. The fused feature x described in Sect. 3.1 is fed into a fully connected (FC) layer to get the discriminative feature $f \in \mathbb{R}^{l \times D}$. Following is two main modules: a classification module to learn discriminative parts for distinguishing different action classes and a localization module to observe integral action regions.

The output of a fully connected layer in the classification module is the class activation sequence (CAS) which is a class-specific 1D temporal map, similar to the 2D class activation map in object detection [29]. We denote the CAS for classification as $C_{cls} \in \mathbb{R}^{l \times N}$. Conducting the top-k mean operation on C_{cls}, a probability mass function (PMF), denoted by p, is generated for a classification loss. Similar to other video-level supervision methods [15,16], the classification loss encourages the model to distinguish the different action categories.

In the localization module, we obtain a localization CAS, denoted by $C_{loc} \in \mathbb{R}^{l \times N}$, with a fully connected layer similar to the classification module. Guided by a partial segment loss, the model can pay attention to labeled segments rather than only the discriminative ones learned from the classification loss. Since the labeled segments are only part of action instances, the model are prone to overfit as the training proceeds. To solve this drawback, we define the segments having high similarity with labeled segments as the implicit segments. In order to measure the similarity of pairs of segments, a sphere loss is first adopted to ensure f has smaller maximal intra-class distance than minimal inter-class distance. Then, we measure the similarity between pairs of segments by the similarity matrix $S = f \cdot f^T$, where $\cdot$ and f^T indicates the matrix multiplication and the transpose of f respectively. Finally, we add a propagation loss to propagate C_{loc} over partially labeled segments to the entire action instances including unlabeled implicit segments. The objective of our framework is formulated as follows:

$$\mathcal{L} = \mathcal{L}_{cls} + \alpha\mathcal{L}_{segment} + \beta\mathcal{L}_{sphere} + \gamma\mathcal{L}_{prop}, \tag{1}$$

where $\mathcal{L}_{cls}$, $\mathcal{L}_{segment}$, $\mathcal{L}_{sphere}$ and $\mathcal{L}_{prop}$ indicate the classification loss, the partial segment loss, the sphere loss and the propagation loss respectively. α, β and γ are trade-off hyperparameters.

3.3   Classification Loss

Due to the variation temporal duration, we use the top-k mean to generate a single class score aggregated from C_{cls} described in Sect. 3.2, similar to [16]. The class score for the n-th category, denoted by s^n, is defined as follows:

$$s^n = \frac{1}{k} \max_{\mathcal{M} \subset C_{cls}(:,n)} \sum_{m \in \mathcal{M}} m, \tag{2}$$

where $|\mathcal{M}| = k$, $k = \lfloor \frac{T}{r} \rfloor$ and r is a hyperparameter to control the ratio of selected segments in a video. Then, a probability mass function (PMF), p^n, is computed by employing softmax:

$$p^n = \frac{\exp(s^n)}{\sum_{n=1}^{N} \exp(s^n)}. \tag{3}$$

Finally, the classification loss (CL) is defined as follows:

$$\mathcal{L}_{cls} = \frac{1}{N} \sum_{n=1}^{N} -y^n \log(p^n), \tag{4}$$

where y^n is the ground-truth label for n-th class which described in Sect. 3.1.

Along with the training process, the CAS guided by the classification loss will shrink to only the discriminative parts rather than the whole action instances. Specific activated action parts are capable of minimizing the action classification loss, but difficult to optimize action localization. *The only goal of optimizing CL is to capture the relevant action parts between f and y to distinguish action categories. Along with training, the relevant parts become more and more discriminative while the irrelevant parts with no contribution to the prediction of y are suppressed.*

3.4 Partial Segment Loss

In order to tackle the problem described in Sect. 3.3, we introduce a partial segment loss in segment-level supervision. An intuitive choice is ℓ_2 loss, denoted by $\ell_2 = \sum_{n=1}^{N} \sum_{t=1}^{l} \|C_{loc}(t, n) - u(t, n)\|^2$. The model guided by l_2 will urge C_{loc} to fit u. However, the ground-truth labels u are only the partial action instances rather than the entire ones. Therefore, We introduce a partial segment loss which only considers the cross entropy loss for labeled segments u and effectively ignores other parts. We first conduct softmax on C_{loc} to obtain the normalized CAS, defined as:

$$a(t, n) = \frac{\exp(C_{loc}(t, n))}{\sum_{i=1}^{l} \exp(C_{loc}(i, n))}. \tag{5}$$

Then, the partial segment loss can be defined as follows:

$$\mathcal{L}_{PSL} = \sum_{(t,n)\in\Omega} -u(t, n) \cdot \log(a(t, n)), \tag{6}$$

where $\Omega = \{(t, n)|u(t, n) = 1\}$. This partial segment loss can be seen as a sampling of the loss which is consistent with the annotation of the segment-level supervision. In segment-level supervision, the process of labeling segments can be regarded as a sampling of action instances.

3.5 Sphere Loss

As described in Sect. 3.2, we denote the similarity matrix between two segments as $S = f \cdot f^T$. To ensure features with the same category have higher similarity than those with different categories, f should have the property that maximal intra-class distance is smaller than minimal inter-class distance. The A-Softmax loss introduced in [12] learns the features by constructing a discriminative angular distance metric, making the decision boundary more stringent and separated. However, A-Softmax loss is used for face recognition, which is trained on examples containing single-label instances with no backgrounds.

In our task, we integrate the sphere loss adapted from A-Softmax loss [12] into our network for multi-label action instances. Since an untrimmed video contains many background clips, a feature aggregation is needed to obtain a class-specific feature without background regions. Specifically, let $\tau^n = median\left(a(:,n)\right)$ and we first compute the high attention along the temporal axis for class n as follows:

$$A\left(t,n\right) = \begin{cases} a\left(t,n\right), & \text{if } a\left(t,n\right) \geq \tau^n \\ 0, & \text{if } a\left(t,n\right) < \tau^n \end{cases}, \tag{7}$$

where $a\left(t,n\right)$ is the normalized CAS in Eq. 5. We refer to A as attention, as it attends to the portions of the video where an action of a certain category occurs. For example, if $A(t,n)$ equals $a(t,n)$ instead of 0, the t-th segments of the video may contain action instances of category n. Then as in [16], we obtain the high attention region aggregated class-wise feature vectors for category n as follows:

$$F^n = (f)^T \cdot A(:,n), \tag{8}$$

where $F^n \in \mathbb{R}^D$. Following [12], we define the fully connect layer as W and $\theta_{i,j}$ is the angle between $W(j)$ and $F^n(i)$. Then, the A-Softmox loss for category n is formulated as:

$$\mathcal{L}_{ang}^n = \frac{1}{D}\sum_{i=1}^{D} -\log\left(\frac{e^{\|F^n(i)\|\psi(\theta_{n,i})}}{e^{\|F^n(i)\|\psi(\theta_{n,i})} + \sum_{j\neq n} e^{\|F^n(i)\|\cos(\theta_{j,i})}}\right), \tag{9}$$

where $\psi\left(\theta_{n,i}\right) = (-1)^k \cos\left(m\theta_{n,i}\right) - 2, \theta_{n,i} \in \left[\frac{k\pi}{m}, \frac{(k+1)\pi}{m}\right]$ and $k \in [0, m-1]$. $m \geq 1$ is an integer that controls the size of angular margin. More detail explanation and provement can be found in [12]. Then, the sphere loss for multi-label action categories can be formulated as:

$$\mathcal{L}_{sphere} = \sum_{n=1}^{N} \mathcal{L}_{ang}^n. \tag{10}$$

3.6   Propagation Loss

In segment-level supervision, only a part of action instances is labeled, compared with entire regions in full supervision methods. This setting is similar to the semi-supervised learning which combines a small amount of labeled data with a large amount of unlabeled data during training. In many semi-supervised algorithms [23], a key assumption is the structure assumption: points within the same structure (such as a cluster or a manifold) are likely to have the same label. Under this assumption, the aim is to use this structure to propagate labeled data to unlabeled data. In [23], authors add a semi-supervised loss (regularizer) to the supervised loss on the entire network's output:

$$\sum_{i=1}^{L} \ell\left(E\left(x_i\right), y_i\right) + \gamma \sum_{i,j=1}^{L+U} H\left(E\left(x_i\right), E\left(x_j\right), W_{ij}\right), \tag{11}$$

where L and U indicate the number of the labeled and unlabeled examples respectively. E indicates the encoding function and W_{ij} specifies the similarity or dissimilarity between examples x_i and x_j. ℓ is the loss for labeled examples and H is the loss between pairs of examples. γ is the trade-off hyperparameter. In our approach, we rewrite the Eq. 11 as follows:

$$\underbrace{\sum_{(t,n)\in\Omega} -u(t,n)\cdot\log\left(a(t,n)\right)}_{\text{Partial segment Loss}} + \gamma \underbrace{\sum_{n=1}^{N} \mathcal{L}_{prop}\left(C_{loc}^n, S\right)}_{\text{Regularizer}}, \tag{12}$$

where the propagation loss is defined as follows:

$$\mathcal{L}_{prop}\left(C_{loc}^n, S\right) = \sum_{i,j=1}^{l} S\left\|C_{loc}^n(i) - C_{loc}^n(j)\right\|^2, \tag{13}$$

where S is the similarity matrix described in Sect. 3.2.

3.7 Classification and Localization

We first get the final CAS, denoted by $C_a = \frac{C_{cls}+C_{loc}}{2}$. Then, s_a and p_a are computed by Eq. 2 and 3. As in [16], we use the computed PMF p_a with a threshold to classify the video to contain one or more action categories. For localization, we discard the categories of which s_a are below a certain threshold (0 in our experiments). Thereafter, for each of the remaining categories, we apply a threshold to C_a along the temporal axis to obtain the action proposals.

4 Experiments

4.1 Experimental Setup

We evaluate our method on two popular action localization benchmark datasets: THUMOS14 [6] and ActivityNet [2]. The details about the datasets, evaluation metric and implementation can be found in the **Supplementary Material**.

4.2 Exploratory Experiments

In the following experiments, we take I3D [3] as the feature extractor.

Table 1. Action localization performance comparison (mAP) of our method with the state-of-the-art methods on the THUMOS14 dataset.

Methods	mAP @ IoU							
	0.1	0.2	0.3	0.4	0.5	0.6	0.7	AVG
Fully-supervised methods								
SSN [27]	60.3	56.2	50.6	40.8	29.1	–	–	47.4
Chao et al. [4]	59.8	57.1	53.2	48.5	42.8	33.8	20.8	52.3
GTAN [13]	**69.1**	**63.7**	**57.8**	**47.2**	**38.8**	–	–	**55.32**
Weakly-supervised methods								
Hide-and-Seek [18]	36.4	27.8	19.5	12.7	6.8	–	–	20.6
Zhong et al. [28]	45.8	39.0	31.1	22.5	15.9	–	–	30.9
STPN (UNT) [15]	45.3	38.8	31.1	23.5	16.2	9.8	5.1	31.0
Liu et al. (UNT) [11]	53.5	46.8	37.5	29.1	19.9	12.3	6.0	37.4
BaSNet (UNT) [7]	56.2	50.3	42.8	34.7	25.1	17.1	9.3	41.8
Ours (UNT)	**59.1**	**53.5**	**45.7**	**37.5**	**28.4**	**20.3**	**11.8**	**44.8**
W-TALC (I3D) [16]	55.2	49.6	40.1	31.1	22.8	–	7.6	39.8
Liu et al. (I3D) [11]	57.4	50.8	41.2	32.1	23.1	15.0	7.0	40.9
3C-Net (I3D) [14]	59.1	53.5	44.2	34.1	26.6	8.1	–	43.5
BaSNet (I3D) [7]	58.2	52.3	47.6	36.0	27.0	18.6	10.4	43.6
Ours (I3D)	**61.6**	**55.8**	**48.2**	**39.7**	**31.6**	**22.0**	**13.8**	**47.4**

Table 2. Results (mAP) with different loss terms on THUMOS14 at IoU = 0.5. 'one-segment' and 'two-segment' indicate labeling one segment and two segments for each action instance respectively.

CL (baseline)	✓	✓	✓	✓	✓	✓	✓	✓
PSL		✓			✓	✓		✓
SL			✓		✓		✓	✓
PL				✓		✓	✓	✓
One-segment	19.4	27.0	24.4	19.7	28.6	27.2	26.1	**29.3**
Two-segment	19.4	28.5	24.4	19.7	29.9	29.1	26.1	**31.6**

Ablation Study. We set the model guided by the classification loss (CL) alone as the baseline. The comparison of temporal action localization performance (mAP) with different loss terms on THUMOS14 at IoU = 0.5 are shown in Table 2.

Comparisons of the Trade-off Between Annotation Time and Performance. To evaluate the annotation time, we define a new metric named as

Table 3. Action localization performance comparison (mAP) of our method with the state-of-the-art methods on ActivityNet1.2 dataset.

Methods	mAP @ IoU			
	0.5	0.75	0.95	AVG
Fully-supervised methods				
SSN [27]	**41.3**	**27.0**	**6.1**	**26.6**
Weakly-supervised methods				
W-TALC [16]	37.0	–	–	18.0
Liu et al. [11]	36.8	22.0	5.6	22.4
3C-Net [14]	37.2	23.7	9.2	21.7
BaSNet [7]	38.5	24.2	5.6	24.3
Ours	**41.7**	**26.7**	**6.3**	**26.4**

Table 4. Action localization performance comparison (mAP) of our method with the state-of-the-art methods on ActivityNet1.3 dataset.

Methods	mAP @ IoU			
	0.5	0.75	0.95	AVG
Fully-supervised methods				
GTAN [13]	**52.6**	34.1	**8.9**	**34.3**
BMN [8]	50.1	**34.8**	8.3	33.9
Weakly-supervised methods				
STPN [15]	29.3	16.9	2.6	–
Liu et al. [11]	34.0	20.9	5.7	21.2
BaSNet [7]	34.5	22.5	4.9	22.2
Ours	**37.7**	**25.6**	**6.8**	**24.8**

Table 5. Action classification performance comparison of our method with the state-of-the-art methods on the THUMOS14 and ActivityNet 1.2 datasets.

Methods	THUMOS14	ActivityNet1.2
iDT+FV [20]	63.1	66.5
C3D [19]	–	74.1
TSN [22]	67.7	88.8
W-TALC [16]	85.6	93.2
3C-Net [14]	86.9	92.4
Ours	**87.6**	**93.2**

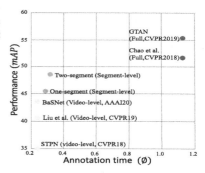

Fig. 3. Trade-off of annotation time and performance.

annotation-duration ratio which is denoted by $\phi = \frac{t_a}{l}$, here t_a and l indicate the annotation time and duration time of videos respectively. Using the COIN annotation tool [24] to label on the THUMOS14 dataset (details can be found in the **Supplementary Material**). We present the trade-off between annotation time and performance on the THUMOS14 dataset in Fig. 3. As Fig. 3 indicates, our approach in segment-level supervision can significantly improve the performance compared with video-level supervision methods.

4.3 Comparisons with the State-of-the-art

We conduct experiments on THUMOS14 and ActivityNet datasets to compare with several state-of-the-art techniques.

Action Localization. Table 1 reports the comparison of our method with existing approaches on the THUMOS14 dataset. We report mAP scores at different IoU thresholds (0.1:0.1:0.7). 'AVG' represents the average mAP at IoU thresholds form 0.1 to 0.5. Results show that our model can perform better than previous video-level weakly supervised methods at all IoU thresholds for both UNT and I3D feature extractors. Table 3 shows the state-of-the-art comparison on the ActivityNet1.2 dataset. Our approach achieves an average mAP of 26.4% which surpasses all existing video-level weakly-supervised methods. Table 4 illustrates the performance comparison of our method with the start-of-the-art on ActivityNet1.3 dataset.

Action Classification. Table 5 reports action classification performance comparison (mAP) of our method with the state-of-the-art methods on the THUMOS14 and ActivityNet 1.2 datasets. Since r varies with the number of labeled segments in the video, it can represent more appropriate sampling information compared with the fixed number.

5   Conclusion

In this work, we propose a new segment-level supervision setting for weakly supervised temporal action localization, which costs almost the same annotation time as video-level supervision. Based on the segment-level supervision, we devise a localization module guided by the partial segment loss, the sphere loss and the propagation loss. Compared with video-level supervision, our approach, exploiting the segment labels and propagating them to explicit segments based on the discriminative features, significantly improves the integrity of predicted segments.

References

1. Alwassel, H., Caba Heilbron, F., Ghanem, B.: Action search: spotting actions in videos and its application to temporal action localization. In: Proceedings of the European Conference on Computer Vision (ECCV), pp. 251–266 (2018)
2. Caba Heilbron, F., Escorcia, V., Ghanem, B., Carlos Niebles, J.: Activitynet: a large-scale video benchmark for human activity understanding. In: Proceedings of the IEEE Conference on Computer Vision and Pattern Recognition, pp. 961–970 (2015)
3. Carreira, J., Zisserman, A.: Quo vadis, action recognition? a new model and the kinetics dataset. In: Proceedings of the IEEE Conference on Computer Vision and Pattern Recognition, pp. 6299–6308 (2017)
4. Chao, Y.W., Vijayanarasimhan, S., Seybold, B., Ross, D.A., Deng, J., Sukthankar, R.: Rethinking the faster r-cnn architecture for temporal action localization. In: Proceedings of the IEEE Conference on Computer Vision and Pattern Recognition, pp. 1130–1139 (2018)
5. Goodfellow, I., Bengio, Y., Courville, A.: Deep Learning. MIT press, Cambridge (2016)

6. Idrees, H., et al.: The thumos challenge on action recognition for videos "in the wild.". Comput. Vis. Image Underst. **155**, 1–23 (2017)
7. Lee, P., Uh, Y., Byun, H.: Background suppression network for weakly-supervised temporal action localization. In: AAAI (2020)
8. Lin, T., Liu, X., Li, X., Ding, E., Wen, S.: Bmn: boundary-matching network for temporal action proposal generation. arXiv preprint arXiv:1907.09702 (2019)
9. Lin, T., Zhao, X., Shou, Z.: Single shot temporal action detection. In: Proceedings of the 25th ACM International Conference on Multimedia, pp. 988–996. ACM (2017)
10. Lin, T., Zhao, X., Su, H., Wang, C., Yang, M.: Bsn: boundary sensitive network for temporal action proposal generation. In: Proceedings of the European Conference on Computer Vision (ECCV), pp. 3–19 (2018)
11. Liu, D., Jiang, T., Wang, Y.: Completeness modeling and context separation for weakly supervised temporal action localization. In: Proceedings of the IEEE Conference on Computer Vision and Pattern Recognition, pp. 1298–1307 (2019)
12. Liu, W., Wen, Y., Yu, Z., Li, M., Raj, B., Song, L.: Sphereface: deep hypersphere embedding for face recognition. In: Proceedings of the IEEE Conference on Computer Vision and Pattern Recognition, pp. 212–220 (2017)
13. Long, F., Yao, T., Qiu, Z., Tian, X., Luo, J., Mei, T.: Gaussian temporal awareness networks for action localization. In: Proceedings of the IEEE Conference on Computer Vision and Pattern Recognition, pp. 344–353 (2019)
14. Narayan, S., Cholakkal, H., Khan, F.S., Shao, L.: 3c-net: category count and center loss for weakly-supervised action localization. In: Proceedings of the IEEE International Conference on Computer Vision, pp. 8679–8687 (2019)
15. Nguyen, P., Liu, T., Prasad, G., Han, B.: Weakly supervised action localization by sparse temporal pooling network. In: Proceedings of the IEEE Conference on Computer Vision and Pattern Recognition, pp. 6752–6761 (2018)
16. Paul, S., Roy, S., Roy-Chowdhury, A.K.: W-talc: weakly-supervised temporal activity localization and classification. In: Proceedings of the European Conference on Computer Vision (ECCV), pp. 563–579 (2018)
17. Shou, Z., Wang, D., Chang, S.F.: Temporal action localization in untrimmed videos via multi-stage CNNs. In: Proceedings of the IEEE Conference on Computer Vision and Pattern Recognition, pp. 1049–1058 (2016)
18. Singh, K.K., Lee, Y.J.: Hide-and-seek: forcing a network to be meticulous for weakly-supervised object and action localization. In: Proceedings of the IEEE International Conference on Computer Vision (ICCV), pp. 3544–3553. IEEE (2017)
19. Tran, D., Bourdev, L., Fergus, R., Torresani, L., Paluri, M.: Learning spatiotemporal features with 3d convolutional networks. In: Proceedings of the IEEE International Conference on Computer Vision, pp. 4489–4497 (2015)
20. Wang, H., Schmid, C.: Action recognition with improved trajectories. In: Proceedings of the IEEE International Conference on Computer Vision, pp. 3551–3558 (2013)
21. Wang, L., Xiong, Y., Lin, D., Van Gool, L.: Untrimmednets for weakly supervised action recognition and detection. In: Proceedings of the IEEE Conference on Computer Vision and Pattern Recognition, pp. 4325–4334 (2017)
22. Wang, L., et al.: Temporal segment networks: towards good practices for deep action recognition. In: Leibe, B., Matas, J., Sebe, N., Welling, M. (eds.) ECCV 2016. LNCS, vol. 9912, pp. 20–36. Springer, Cham (2016). https://doi.org/10.1007/978-3-319-46484-8_2

23. Weston, J., Ratle, F., Mobahi, H., Collobert, R.: Deep learning via semi-supervised embedding. In: Montavon, G., Orr, G.B., Müller, K.-R. (eds.) Neural Networks: Tricks of the Trade. LNCS, vol. 7700, pp. 639–655. Springer, Heidelberg (2012). https://doi.org/10.1007/978-3-642-35289-8_34

24. Tang, Y., et al.: Coin: a large-scale dataset for comprehensive instructional video analysis. In: IEEE Conference on Computer Vision and Pattern Recognition (CVPR) (2019)

25. Yuan, Y., Lyu, Y., Shen, X., Tsang, I.W., Yeung, D.Y.: Marginalized average attentional network for weakly-supervised learning. arXiv preprint arXiv:1905.08586 (2019)

26. Zhao, H., Yan, Z., Wang, H., Torresani, L., Torralba, A.: Slac: a sparsely labeled dataset for action classification and localization (2017)

27. Zhao, Y., Xiong, Y., Wang, L., Wu, Z., Tang, X., Lin, D.: Temporal action detection with structured segment networks. In: Proceedings of the IEEE International Conference on Computer Vision, pp. 2914–2923 (2017)

28. Zhong, J.X., Li, N., Kong, W., Zhang, T., Li, T.H., Li, G.: Step-by-step erasion, one-by-one collection: a weakly supervised temporal action detector. arXiv preprint arXiv:1807.02929 (2018)

29. Zhou, B., Khosla, A., Lapedriza, A., Oliva, A., Torralba, A.: Learning deep features for discriminative localization. In: Proceedings of the IEEE Conference on Computer Vision and Pattern Recognition, pp. 2921–2929 (2016)

Locality-Constrained Collaborative Representation with Multi-resolution Dictionary for Face Recognition

Zhen Liu[1,2], Xiao-Jun Wu[1(✉)], Hefeng Yin[1], Tianyang Xu[1], and Zhenqiu Shu[3]

[1] School of Artificial Intelligence and Computer Science, Jiangnan University,
Wuxi, China
wu_xiaojun@jiangnan.edu.cn
[2] School of Computer and Information Science, Hubei Engineering University,
Xiaogan, China
[3] Faculty of Information Engineering and Automation, Kunming University
of Science and Technology, Kunming, China

Abstract. Sparse learning methods have drawn considerable attention
in face recognition, and there are still some problems need to be further
studied. For example, most of the conventional sparse learning methods concentrate only on a single resolution, which neglects the fact that
the resolutions of real-world face images are variable when they are
captured by different cameras. Although the multi-resolution dictionary
learning (MRDL) method considers the problem of image resolution, it
takes a lot of training time to learn a concise and reliable dictionary
and neglects the local relationship of data. To overcome the above problems, we propose a locality-constrained collaborative representation with
multi-resolution dictionary (LCCR-MRD) method for face recognition.
First, we extend the traditional collaborative representation based classification (CRC) method to the multi-resolution dictionary case without
dictionary learning. Second, the locality relationship characterized by the
distance between test sample and training sample is used to learn weight
of representation coefficient, and the similar sample is forced to make
more contribution to representation. Last, LCCR-MRD has a closed-
form solution, which makes it simple. Experiments on five widely-used
face databases demonstrate that LCCR-MRD outperforms many state-
of-art sparse learning methods. The Matlab codes of LCCR-MRD are
publicly available at https://github.com/masterliuhzen/LCCR-MRD.

Keywords: Collaborative representation · Multi-resolution
dictionary · Locality constraint · Face recognition

1 Introduction

Sparse learning methods have drawn considerable attention in computer vision
and pattern recognition, such as object detection and tracking [3,12,25], image

---

Z. Liu—The first author is a graduate student.

© Springer Nature Switzerland AG 2021
H. Ma et al. (Eds.): PRCV 2021, LNCS 13019, pp. 55–66, 2021.
https://doi.org/10.1007/978-3-030-88004-0_5

super-resolution [4,18,30] and image classification [2,6,19,24,27], etc. Sparse learning methods can be roughly divided into two types:dictionary learning based methods and sparse representation based methods.

Sparse representation based methods mostly aim at gaining sparse representation of sample from the dictionary of all training samples. Sparse representation based classification (SRC) [27] learns the sparsest representation of the test sample over the training samples. In [31], a feature fusion approach based on sparse representation and local binary pattern was presented for face recognition. Collaborative representation based classification (CRC) [33] uses l_2 norm regularization to replace the time-consuming l_1 norm regularization of SRC for face recognition, which obtains competitive performance with lower time cost. In [5], the working mechanism of CRC/SRC are explained. Xu et al. [28] present a novel sparse representation method for face recognition, in which the label information of the training samples are used to enhance the discriminative ability of representation results.

Dictionary learning based methods mainly aim at obtaining reliable dictionary and brief representation of samples [32]. K-SVD [1] is one of the famous unsupervised dictionary learning methods. In [21], both unlabeled and labeled data are exploited for semi-supervised dictionary learning to deal with the shortcomings of deficient label samples. Discriminative K-SVD (D-KSVD) [17] and label consistent K-SVD(LC-KSVD) [11] effectively use the label information of samples for supervised dictionary learning.

Sparse learning methods have obtained good performance, however, there are still some shortcomings. For example, it is quite time-consuming to solve the l_1 norm and l_21 norm minimum for sparse learning methods [6,15,27]. Furthermore, most of them concentrate only on a single resolution yet neglect the fact that the resolutions of real-world face images are variable.

To address the above problems, we present a locality-constrained collaborative representation with multi-resolution dictionary method, named LCCR-MRD. The multiple dictionaries each being with a resolution are used to cope with the issue of face images with different resolutions, and the locality-constraint measured by the distance between test sample and training sample is used to improve the discrimination ability of representation coefficient. LCCR-MRD uses the multi-resolution dictionary directly, thus we do not need to spend a lot of time on dictionary learning. In addition, the proposed LCCR-MRD has a closed-form solution, thus it has global optimal solution, which could lead to better recognition result.

In summary, the main contributions of our proposed LCCR-MRD method are presented as follows:

1. We propose a locality-constrained collaborative representation based classification method with multi-resolution dictionary (LCCR-MRD) for face recognition when the resolution of the test image is uncertain.
2. We use the multi-resolution dictionary to deal with the issue of resolution of the test image is uncertain, and the local structure relation is used to enhance the discriminative representation coefficient.

3. Experimental results on five face databases show that our proposed method achieves excellent performance on small-scale datasets, even better than some typical deep learning methods.

2 Proposed Method

2.1 Notations

Given a training set $X = [X_1, X_2, \ldots, X_C] = [x_1, x_2, \ldots, x_n] \in R^{m \times n}$ with C classes and n samples, where m is the dimensionality of sample. $D = [D_1, D_2, \ldots, D_k] \in R^{m \times kn}$ is the multi-resolution dictionary corresponding to the training set X, k is the number of resolutions. $D_i = [D_i^1, D_i^2, \ldots, D_i^C] \in R^{m \times n}$ represents the ith sub-dictionary of D and $D_i^j = [d_i^1, d_i^2, \ldots, d_i^{n_j}] \in R^{m \times n_j}$ is the jth class images of the ith resolution dictionary, where n_j is the number of jth class training samples and $d_i^{n_j}$ denotes the n_jth training sample for the ith resolution, thus $n = \sum_{j=1}^{C} n_j$. $A = [A_1; A_2; \ldots; A_k] \in R^{kn \times 1}$ is the representation coefficient vector corresponding to the multi-resolution training images dictionary D. $A_i = [A_i^1; A_i^2; \ldots; A_i^C] \in R^{n \times 1}$ represents the coefficient vector corresponding to the ith resolution dictionary, and $A_i^j = [a_i^1, a_i^2, \ldots, a_i^{nj}]^T \in R^{n_j \times 1}$ denotes the representation coefficient vector corresponding to jth class training samples for the ith resolution. $Y = [y_1, y_2, \ldots, y_p]$ represents the test images with different resolutions, where p is the number of test samples. Furthermore, we presume the training samples of each resolution involve all categories.

2.2 Model of LCCR-MRD

In this section, we introduce the idea of LCCR-MRD in detail. The idea of multi-resolution dictionary [15] is used to cope with the issue of samples with different resolutions, the CRC framework [33] is employed to collaboratively represent the test sample with multi-resolution dictionary, which can obtain a closed-form solution. Thus ,we can get the followings model,

$$\min_{A_1, A_2, \ldots, A_k} \|y - \sum_{i=1}^{k} D_i A_i\|_2^2 + \gamma \|A_1; A_2; \ldots; A_k\|_2^2 \tag{1}$$

where γ is a positive regularization parameter. Let $D = [D_1, D_2, \ldots, D_k]$ and $A = [A_1; A_2; \ldots; A_k]$, the above model can be rewritten as

$$\min_{A} \|y - DA\|_2^2 + \gamma \|A\|_2^2 \tag{2}$$

In order to enhance the discriminative of representation vector, we consider the locality of data measured by the distance between the test sample and the training sample, and make it as weight to restrict corresponding representation coefficient. So the model of LCCR-MRD can be presented as follows,

$$\min_{A} \|y - DA\|_2^2 + \gamma \|WA\|_2^2 \tag{3}$$

Algorithm 1. Algorithm of LCCR-MRD

Input: The training images $X = [x_1, x_2, \ldots, x_n]$, the test images $Y = [y_1, y_2, \ldots, y_p]$ with different resolutions, the number of resolution type k, and the regularization parameter γ.
1.Construct multi-resolution training images dictionary $D = [D_1, D_2, \ldots, D_k]$ by resolution pyramid methods.
2.Calculate diagonal weighed matrix W by (4).
3.Calculate representation coefficient vector A by (5).
4.Calculate class-specific reconstruction residuals (6).
5.Classify the test image y using (7).
Output: The category of the test image y

---

where W is a diagonal weight matrix and defined as follows,

$$
\begin{aligned}
W = diag(&\|y - x_1^1\|_2, \|y - x_1^2\|_2, \ldots, \|\|y - x_1^n\|_2, \\
&\ldots, \|y - x_2^1\|_2, \|y - x_2^2\|_2, \ldots, \|\|y - x_2^n\|_2, \\
&\ldots, \|y - x_k^1\|_2, \|y - x_k^2\|_2, \ldots, \|\|y - x_k^n\|_2)
\end{aligned}
\tag{4}
$$

where $\|y - x_k^j\|_2$ denotes the similarity between the test sample y and the jth training sample of the kth resolution dictionary, the value of $\|y - x_k^j\|_2$ is used as weigh to restrict corresponding representation coefficient. Intuitively, if the test sample y and the jth training sample of the kth resolution dictionary is very similar, the value of $\|y - x_k^j\|_2$ will be small, which may promote the training sample and the test sample y to obtain a relatively large representation coefficient. Otherwise, we will obtain a relatively small representation coefficient. According to multi-resolution dictionary and the weighting strategy, our model can adaptively select similar atoms from the multi-resolution dictionary to represent the test sample. When $k = 1$ and W is an identity matrix, the model of LCCR-MRD degenerates into conventional CRC. That is, the traditional CRC is a special case of LCCR-MRD when the resolution of training sample is unique and does not consider the similarity between the test sample and the training sample.

2.3 Optimization

As the regularization term $\|WA\|_2^2$ is l_2 norm regularized, we can get the solution to the objective function very simply and efficiently as follows,

$$
A = (D^T D + \gamma W^T W)^{-1} D^T y
\tag{5}
$$

2.4 Classification

After obtaining the representation coefficient vector A, we can get the sub-vectors $A_1, A_2, \ldots, A_k$. The class-specific reconstruction residual can be computed by

$$\psi_j = \|y - \sum_{i=1}^{k} D_i^j A_i^j\|_2 \tag{6}$$

where $A_i^j (1 \le i \le k, 1 \le j \le C)$ is the representation coefficient vector corresponding to the jth class training samples for the ith resolution. If the test sample comes from the j-th class, the class-specific reconstruction residual of the j-th class will be minimal, and the training samples with the same resolution contribute the most to the class-specific reconstruction error. Therefore, the training samples with the same class, different resolution can be adaptively selected to represent the test sample. When all the class-specific reconstruction residuals are available, we can classify the test image y by the following criterion,

$$Classif(y) = arg \min_{j} \psi_j, j = 1, 2, \ldots, C. \tag{7}$$

The main steps of LCCR-MRD are given in Algorithm 1.

3 Experiments

3.1 Experimental Settings

Databases. We conduct experiments on five widely-used face databases (Extended YaleB [8], ORL [20], PIE [22], AR [16] and LFW [10]) to evaluate our model.

Extended YaleB (EYaleB) [8]. It contains 2414 face images of 38 persons with different light changes. We randomly select 32 images of each person as training set and treat the rest images as test set.

ORL [20] There are 400 face images of 40 individuals with facial expressions, lighting and facial details changes. We randomly select 5 images of each individual as training samples and treat the remaining images as test samples.

PIE [22] It includes face images of 68 persons that are captured in different illumination conditions, poses and facial expressions. We use face images from pose 05 as a subset, which contains 3332 face images of 68 persons. We randomly select 25 images of each person to form the training set and treat the rest images as the test set.

AR [16] There are 3276 face images of 126 individuals that varies in facial expression, dresses and illumination conditions. We select 3120 images of 120 individuals to form a subset. 13 images of each individuals are selected as the training set and the remaining images are chosen as the test set.

LFW [10] It contains more than 13,000 face images collected in an unconstrained environment. The same as reference [15,29], we exploit a subset(LFW crop) in the LFW database, which includes 158 individuals with 10 images of each person. We randomly select 5 images from each person as training set and treat the rest images as test set.

Some images with different resolutions on these database are shown in the Fig. 1. Following the experiment setting in [15], the images with different resolutions are obtained by resolution pyramid method [7]. Specifically, we first

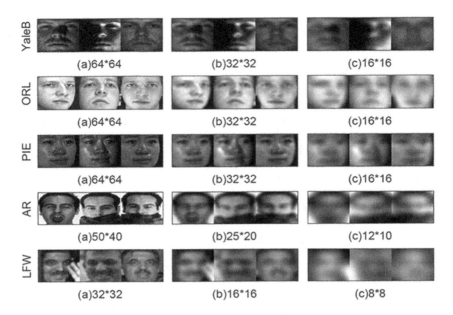

Fig. 1. The first to fifth line are some images from the Extended YaleB, ORL, PIE, AR, and LFW databases with different resolution, respectively.

divide the face database into original test set T and original training set X. For the images in the original test set T, we first divide images of each subject into k equal parts, and then use the resolution pyramid method to convert them into images with different resolutions. For the images in the original training set X, the resolution pyramid method is directly used to transform them into k-resolution training images $D_1, D_2, \ldots, D_k$. So the test set Y with different resolution has the same number of samples to the original test set T and the number of samples of the training set $D = [D_1, D_2, \ldots, D_k]$ with different resolutions is k times the number of samples of original training set X, both of them have k resolutions. The Fig. 2 shows the process of converting the original database to test set and training set with different resolutions. In our experiment, we have used the same databases used in [15].

Comparative Methods. In this paper, LCCR-MRD is compared with several state-of-art methods, i.e., NSC [14], SRC [27], CRC [33], ProCRC [5], DSRC [28], K-SVD [1],D-KSV [17], LC-KSVD [11], DLSPC [26] and MRDL [15]. To further show the effectiveness of LCCR-MRD, we also compare it with three CNN based models(AlexNet [13], VGG [23], ResNet [9]).

Evaluation Metrics. We employ average recognition accuracy of ten runs to evaluate the recognition performance, the best average recognition accuracy are shown in bold. We also compare the training time and average classification time (classify a test sample) to show the efficiency of LCCR-MRD and the baseline WRDL.

3.2   Results and Discussions

Comparison with Traditional Methods. We use the experimental results of SRC [27], CRC [33], K-SVD [1], D-KSV [17], LC-KSVD1 [11], LC-KSVD2 [11], DLSPC [26] and MRDL [15] that was reported in [15], and report the best result of CRC [33], ProCRC [5] and DSRC [28] on these databases from the candidate parameter values [0.00001 0.0001 0.001 0.01 0.1 1 10 100]. The experimental results are shown in Table 1. From these results, we have several observations. ① Our proposed LCCR-MRD method is appropriate for face recognition. Though the databases we utilised include diverse facial expressions, illumination conditions and poses, our proposed LCCR-MRD method exhibits satisfactory performance. ② We can observe that LCCR-MRD and MRDL achieve better average recognition accuracy than other methods, which indicates that when the resolution of the test image is uncertain, the recognition accuracy can be improved by using the multi-resolution dictionary reasonably. ③ The proposed LCCR-MRD outperforms all the compared algorithms on five face databases. The main reason is that our LCCR-MRD model can adaptively select similar atoms from the multi-resolution dictionary to represent the test sample when the resolution of the test sample is uncertain. In addition, the weigh restriction to the representation coefficient also helps to enhance the recognition performance. ④ LCCR-MRD gains at least 5.4%, 3.6%, 2.6%, 13.5% and 10.0% performance

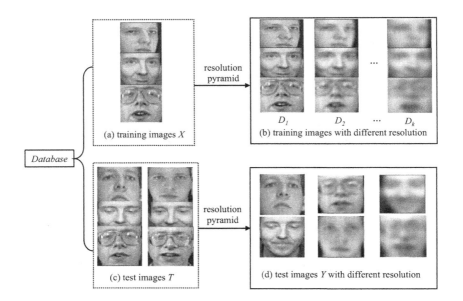

Fig. 2. Convert images to have different resolution.

improvement over the second best result on EYaleB, ORL, AR, PIE and LFW face databases respectively, which demonstrates that the proposed LCCR-MRD is suit for face recognition when the resolution of test images is uncertain. ⑤ We set $\gamma = 0.01$ on all face databases and achieve significantly better performance than other methods, which indicates that our proposed LCCR-MRD method is robust on these databases.

Comparison with Deep Learning Methods. Due to deep learning method needs a large number of training samples yet the five databases we used are not large enough. Thus, in order to objectively compare the performance of our proposed LCCR-MRD method with deep learning method, we exploit them in two ways: without pre-trained model VS with pre-trained model [15]. In addition to LCCR-MRD method, we use the experimental results that reported in [15]. The experimental results are exhibited in Table 2. From these results, we have several observations. ① The deep learning methods with pre-trained model obtain better results than the deep learning methods without pre-trained model, the reason is that the pre-trained process can learn more discriminative features for face recognition. ② The LCCR-MRD and MRDL achieve competitive average recognition accuracy with the deep learning methods, and the LCCR-MRD gains at least 2.8%, 3.7%, 26.6%, 5.3% and 14.3% performance improvement over the second best result on Extended YaleB, ORL, AR, PIE and LFW face databases respectively, which demonstrates that the traditional sparse learning method can achieve competitive or even better results than some deep learning methods in face recognition when the resolution of image is uncertain. Although deep learning methods have achieved much better recognition performance than traditional methods on large scale face datasets, our proposed LCCR-MRD method shows that the traditional method can achieve competitive or even better performance than the deep learning method on small scale face databases. Compared with

Table 1. The average recognition accuracy (%) and standard deviation (%) on five face databases.

Methods	EYaleB	ORL	PIE	AR	LFW
KSVD [1]	72.69 ± 0.67	89.15 ± 2.14	67.03 ± 1.02	69.99 ± 1.08	11.08 ± 1.43
D-KSVD [17]	74.88 ± 1.60	87.45 ± 2.73	60.52 ± 1.07	65.17 ± 1.11	7.70 ± 0.98
LC-KSVD1 [11]	74.95 ± 1.44	87.45 ± 1.83	67.02 ± 0.90	74.58 ± 1.40	9.86 ± 0.88
LC-KSVD2 [11]	76.70 ± 1.64	88.85 ± 2.61	67.74 ± 0.85	75.30 ± 1.24	11.54 ± 1.19
DLSPC(GC) [26]	77.59 ± 0.78	84.60 ± 2.98	63.31 ± 1.67	68.38 ± 1.15	9.47 ± 0.61
SRC [27]	81.85 ± 1.12	90.01 ± 1.96	70.66 ± 1.15	78.18 ± 1.45	14.79 ± 1.21
NSC [14]	85.08 ± 0.81	77.55 ± 3.80	79.08 ± 0.73	73.06 ± 1.13	11.29 ± 0.79
ProCRC [5]	87.97 ± 0.58	90.75 ± 1.42	84.78 ± 0.89	73.34 ± 0.76	23.84 ± 1.01
DSRC [28]	89.48 ± 0.80	90.00 ± 0.97	83.85 ± 0.71	74.84 ± 2.50	22.13 ± 1.27
CRC [33]	81.69 ± 1.38	86.60 ± 2.67	82.07 ± 0.62	73.51 ± 0.81	20.46 ± 0.81
MRDL [15]	88.58 ± 1.53	92.15 ± 1.51	95.81 ± 0.50	82.19 ± 1.54	16.63 ± 0.63
LCCR-MRD	**94.94** ± 0.69	**95.75** ± 1.55	**98.49** ± 0.22	**95.75** ± 0.70	**33.87** ± 1.75

the deep learning method, the traditional method usually has shorter training time, lower computing resources and more explanatory model.

Computing Time. In order to evaluate the efficiency of our proposed LCCR-MRD method. We compare the training time and the average classification time of classifying a test sample. The experimental results of LCCR-MRD, MRDL and CRC on ORL and LFW databases are shown in Table 3. From the results we can see that the average classification time of our proposed LCCR-MRD method is slightly higher than that of the benchmark MRDL method. The main reason is that our approach does not involve dictionary learning, thus the dictionary size of LCCR-MRD is big than that of MRDL, which results in a slight increase in the average classification time of LCCR-MRD compared with MRDL. It is worth noting that our proposed LCCR-MRD does not involve dictionary learning, and thus compared with MRDL, our method can save substantial training time.

Table 2. The average recognition accuracy (%) of different deep learning methods on five face databases.

Method	EYaleB	ORL	PIE	AR	LFW
AlexNet without pre-training [13]	7.85	2.50	6.74	5.51	0.63
AlexNet with pre-training [13]	51.84	64.50	46.08	53.46	6.84
VGG16 without pre-training [23]	78.55	5.50	66.73	77.12	1.90
VGG16 with pre-training [23]	82.97	79.50	69.00	86.35	8.61
ResNet18 without pre-training [9]	86.48	84.00	69.00	86.60	15.19
ResNet18 with pre-training [9]	92.07	92.00	71.81	90.38	19.49
LCCR-MRD	**94.94**	**95.75**	**98.49**	**95.75**	**33.87**

Table 3. The average computing time(second) of LCCR-MRD, CRC, and MRDL on the ORL and LFW face databases.

Database	Method	Training time	Average classification time
ORL	CRC	0	0.0019
	MRDL	0.4624	0.0238
	LCCR-MRD	0	0.0676
LFW	CRC	0	0.0026
	MRDL	2.4181	0.2099
	LCCR-MRD	0	0.6747

Parameter Setting and Sensitivity. In our model, there are two parameters γ and k that are needed to set. We choose the value of γ from the candidate parameter values [0.00001 0.0001 0.001 0.01 0.1 1 10 100]. The parameter sensitivity analysis of γ on ORL [20] and LFW [10]) databases are shown in Fig. 3. It can be found that the LCCR-MRD and MRDL methods can achieve a relatively stable performance with the parameter γ in [0.0001 0.001 0.01 0.1], and our proposed LCCR-MRD method achieve better result than MRDL [15] with the same value of γ. Moreover, our proposed LCCR-MRD method is more robust than MRDL. In order to reduce the time of tuning parameter, we set $\gamma = 0.01$ on all databases. For the parameter k, we follow the experiment setting in [15] that sets $k = 3$.

Ablation Experiment. In order to verify the effect of locality constrain on the accuracy of face recognition, we conduct ablation experiment on five face databases. If we do not consider the locality constraint, the model degenerates into model 2 and names it as CR-MRD. We use the same setting (as shown in the Sect. 3.1) and the experimental results on five databases shown in the Table 4. From the results, we can find that the locality constraint can effectively improve the accuracy of face recognition.

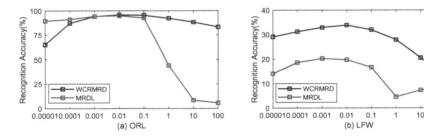

Fig. 3. The recognition accuracy of LCCR-MRD and MRDL with different value of γ on the (a) ORL and (b) LFW face databases, respectively.

Table 4. The average recognition accuracy (%) and standard deviation (%) of LCCR-MRD and CR-MRD on five face databases.

Methods	Extended YaleB	ORL	PIE	AR	LFW
CR-MRD	92.54 ± 0.50	94.40 ± 1.56	98.19 ± 0.27	94.31 ± 0.57	30.77 ± 1.39
LCCR-MRD	**94.94** ± 0.69	**95.75** ± 1.55	**98.49** ± 0.22	**95.75** ± 0.70	**33.87** ± 1.75

4   Conclusion

In this paper, a novel face recognition method based on multi-resolution dictionary is proposed to deal with the face recognition problem when the resolution of test sample is uncertain. The multi-resolution dictionary composed of training samples with different resolution is directly used to represent test samples collaboratively, and the data locality measured by the distance between the test sample and the training sample is used as the weight to further enhance the discriminative ability of the representation coefficient. Experimental results on five widely-used face databases demonstrate that the proposed LCCR-MRD method is effective.

Acknowledgments. This work was supported by the National Natural Science Foundation of China (62020106012, U1836218, 61672265), the 111 Project of Ministry of Education of China (B12018), and the Natural Science Foundation of Xiaogan, China (XGKJ2020010063).

References

1. Aharon, M., Elad, M., Bruckstein, A.M.: k-svd: an algorithm for designing overcomplete dictionaries for sparse representation. IEEE Trans. Signal Process. **54**(11), 4311–4322 (2006)
2. Abdi, A., Rahmati, M., Ebadzadeh, M.E.: Entropy based dictionary learning for image classification. Pattern Recogn. **110**, 107634 (2021)
3. Bai, T., Li, Y.F., Tang, Y.: Robust visual tracking with structured sparse representation appearance model. Pattern Recogn. **45**(6), 2390–2404 (2012)
4. Li, B., Yuan, Z., Yeda, Z., Aihua, W.: Depth image super-resolution based on joint sparse coding. Pattern Recogn. Lett. **130**, 21–29 (2020)
5. Cai, S., Zhang, L., Zuo, W., Feng, X.: A probabilistic collaborative representation based approach for pattern classification. In: IEEE Conference on Computer Vision and Pattern Recognition (CVPR), pp. 2950–2959 (2016)
6. Chen, Z., Wu, X., Kittler, J.: A sparse regularized nuclear norm based matrix regression for face recognition with contiguous occlusion. Pattern Recogn. Lett. **125**, 494–499 (2019)
7. Frucci, M., Ramella, G., Baja, G.S.D.: Using resolution pyramids for watershed image segmentation. Image Vis. Comput. **25**(6), 1021–1031 (2007)
8. Georghiades, A.S., Belhumeur, P.N., Kriegman, D.J.: From few to many: illumination cone models for face recognition under variable lighting and pose. IEEE Trans. Pattern Anal. Mach. Intell. **23**(6), 643–660 (2001)
9. He, K., Zhang, X., Ren, S., Sun, J.: Deep residual learning for image recognition. In: Computer Vision and Pattern Recognition, CVPR, pp. 770–778 (2016)
10. Huang, G.B., Ramesh, M., Berg, T., Learned-Miller, E.: Labeled faces in the wild: a database for studying face recognition in unconstrained environments. Technical Report 07–49, University of Massachusetts, Amherst (October 2007)
11. Jiang, Z., Lin, Z., Davis, L.S.: Label consistent k-svd: learning a discriminative dictionary for recognition. IEEE Trans. Pattern Anal. Mach. Intell. **35**(11), 2651–2664 (2013)
12. Kang, B., Zhu, W., Liang, D., Chen, M.: Robust visual tracking via nonlocal regularized multi-view sparse representation. Pattern Recogn. **88**, 75–89 (2019)
13. Krizhevsky, A., Sutskever, I., Hinton, G.E.: Imagenet classification with deep convolutional neural networks. Neural Inf. Process. Syst. **141**(5), 1097–1105 (2012)

14. Lee, K.C., Ho, J., Kriegman, D.J.: Acquiring linear subspaces for face recognition under variable lighting. IEEE Trans. Pattern Anal. Mach. Intell. **27**(5), 684–698 (2005)
15. Luo, X., Xu, Y., Yang, J.: Multi-resolution dictionary learning for face recognition. Pattern Recogn. **93**, 283–292 (2019)
16. Martínez, A., Benavente, R.: The AR face database: CVC Technical Report 24, pp. 1–8 (1998)
17. Zhang, Q., Li, B.: Discriminative k-svd for dictionary learning in face recognition. In: 2010 IEEE Computer Society Conference on Computer Vision and Pattern Recognition, pp. 2691–2698 (2010)
18. Dehkordi, R.A., Khosravi, H., Ahmadyfard, A.: Single image super resolution based on sparse representation using dictionaries trained with input image patches. IET Image Process. **14**(8), 1587–1593 (2020)
19. Rong, Y., Xiong, S., Gao, Y.: Double graph regularized double dictionary learning for image classification. IEEE Trans. Image Process. **29**, 7707–7721 (2020)
20. Samaria, F.S., Harter, A.C.: Parameterisation of a stochastic model for human face identification. In: Proceedings of 1994 IEEE Workshop on Applications of Computer Vision, pp. 138–142 (1994)
21. Shrivastava, A., Pillai, J.K., Patel, V.M., Chellappa, R.: Learning discriminative dictionaries with partially labeled data. In: 19th IEEE International Conference on Image Processing, pp. 3113–3116 (2012)
22. Sim, T., Baker, S., Bsat, M.: The CMU pose, illumination, and expression database. IEEE Trans. Pattern Anal. Mach. Intell. **25**(12), 1615–1618 (2003)
23. Simonyan, K., Zisserman, A.: Very deep convolutional networks for large-scale image recognition, pp. 1–14 (2015). arXiv:1409.1556v6
24. Song, X., Chen, Y., Feng, Z., Hu, G., Zhang, T., Wu, X.: Collaborative representation based face classification exploiting block weighted LBP and analysis dictionary learning. Pattern Recogn. **88**, 127–138 (2019)
25. Sun, J., Chen, Q., Sun, J., Zhang, T., Fang, W., Wu, X.: Graph-structured multi-task sparsity model for visual tracking. Inf. Sci. **486**, 133–147 (2019)
26. Wang, D., Kong, S.: A classification-oriented dictionary learning model: explicitly learning the particularity and commonality across categories. Pattern Recogn. **47**(2), 885–898 (2014)
27. Wright, J., Yang, A.Y., Ganesh, A., Sastry, S.S., Ma, Y.: Robust face recognition via sparse representation. IEEE Trans. Pattern Anal. Mach. Intell. **31**(2), 210–227 (2009)
28. Xu, Y., Zhong, Z., Yang, J., You, J., Zhang, D.: A new discriminative sparse representation method for robust face recognition via l2 regularization. IEEE Trans. Neural Netw. Learn. Syst. **28**(10), 2233–2242 (2017)
29. Xu, Y., Li, Z., Zhang, B., Yang, J., You, J.: Sample diversity, representation effectiveness and robust dictionary learning for face recognition. Inf. Sci. **375**, 171–182 (2017)
30. Yang, J., Wright, J., Huang, T.S., Ma, Y.: Image super-resolution via sparse representation. IEEE Trans. Image Process. **19**(11), 2861–2873 (2010)
31. Yin, H.F., Wu, X.J.: A new feature fusion approach based on LBP and sparse representation and its application to face recognition. In: Multiple Classifier Systems, pp. 364–373 (2013)
32. Xu, Y., Li, Z., Tian, C., Yang, J.: Multiple vector representations of images and robust dictionary learning. Pattern Recogn. Lett **128**, 131–136 (2019)
33. Zhang, L., Yang, M., Feng, X., Ma, Y., Zhang, D.: Collaborative representation based classification for face recognition. In: Computer Vision and Pattern Recognition, pp. 1–33 (2012)

Fast and Fusion: Real-Time Pedestrian Detector Boosted by Body-Head Fusion

Jie Huang[1], Xiaoling Gu[1(✉)], Xun Liu[2], and Pai Peng[3]

[1] Hangzhou Dianzi University, Hangzhou 310018, China
{huangjiejie,guxl}@hdu.edu.cn
[2] Shanghai Jiaotong University, Shanghai 200030, China
lxbcd94@alumni.sjtu.edu.cn
[3] YoutuLab, Tencent Technology, Shanghai 201103, China
popeyepeng@tencent.com

Abstract. As pedestrian detection plays a critical role in various real-world applications, real-time processing becomes a natural demand. However, the existing pedestrian detectors either are far from being real-time or fall behind in performance. From our observation, two-stage detectors yield stronger performance but their speed is limited by the propose-then-refine pipeline; Current single-stage detectors are significantly faster yet haven't addressed the occlusion problem in the pedestrian detection task, which leads to poor performance. Thus, we propose FastNFusion, which enjoys the merit of anchor-free single-stage detectors and mitigates the occlusion problem by fusing head features into the body features using body-head offset since heads suffer from occlusion more rarely. Particularly, our body-head fusion module introduces marginal computational overhead, which enables the real-time processing. In addition, we design an auxiliary training task to further boost the learning of full-body bounding box and body-head offset prediction, which is cost-free during inference. As a result, FastNFusion improves the MR^{-2} by 3.6% to 41.77% on the CrowdHuman dataset, which is state-of-the-art, while runs at 16.8 fps on single Tesla P40.

Keywords: Pedestrian detection · Real-time · Body-head fusion · Auxiliary training task

1 Introduction

Pedestrian detection plays a critical role in various real-world applications, such as intelligent surveillance, robotics, and automatic driving. Moreover, it is also fundamental to important research topics, such as human action recognition, face recognition, person re-identification, and so on. Hence, it continues to attract

Electronic supplementary material The online version of this chapter (https://doi.org/10.1007/978-3-030-88004-0_6) contains supplementary material, which is available to authorized users.

© Springer Nature Switzerland AG 2021
H. Ma et al. (Eds.): PRCV 2021, LNCS 13019, pp. 67–79, 2021.
https://doi.org/10.1007/978-3-030-88004-0_6

massive attention over the past few years. As the cornerstone of the pedestrian detection, the generic object detection has provided a lot of successful techniques, which can be divided into two categories, i.e., two-stage approaches and single-stage approaches.

Two-stage approaches [2, 6–8, 14, 26] leverage Region Proposal Network (RPN) to generate proposals in the first stage by classifying the pre-defined anchor boxes as foreground or background and regressing the foreground boxes. And in the second stage, a small fully convolutional network (FCN) is used to classify the proposals into different categories and refine their bounding box coefficients. Although the two-stage approaches consistently achieve top accuracy, it is hard to achieve real-time detection speed since the propose-then-refine pipeline is not computationally efficient enough.

However, most of the current pedestrian detectors [3–5, 11, 20, 22, 29, 33, 35, 39] are based on two-stage baselines because their goal is to optimize the performance by addressing the occlusion problem in pedestrian detection while does not take into account the speed. Meanwhile, pedestrian detection has great value in various applications thus the real-time processing becomes a natural demand. Hence, in this paper, we explore whether a single-stage detector can achieve comparable or even better performance while keeping running in real-time[1].

Different from two-stage detectors, single-stage approaches [17, 23–25] directly classify the pre-defined anchor boxes into categories and regress them into target boxes, which significantly improves the speed. Furthermore, anchor-free single-stage methods [13, 30, 41] are proposed to eliminate the hand-crafted anchor design and further boost the performance. Based on single-stage baselines, TLL [28], ALFNet [18], and CSP [19] are proposed with better performance-speed trade-off.

However, the current existing single-stage pedestrian detectors have not taken special consideration to the occlusion problem. On the other hand, several two-stage methods try to handle the occlusion by partitioning the pedestrian into parts [29, 33] or leveraging additional annotations such as visible boxes [11, 22, 39] and head boxes [3, 4, 20, 35]. Since occlusion is more frequent and severe in pedestrian detection than generic object detection, this further enlarges the performance gap between single-stage and two-stage pedestrian detectors. Thus, in this paper, we aim to boost the performance of single-stage detector to that of the two-stage counterparts by occlusion handling with minimal computational overhead to preserve the real-time processing speed.

Specifically, we propose FastNFusion, a novel body-head fusion approach based on a single-stage detector to increase the recall of the pedestrians whose bodies suffer from occlusion. The key idea is that heads are more rarely occluded and can serve as a strong cue to infer the full-body prediction. Our body-head fusion module is simple yet efficient. We first predict the offsets between head and body centers, then translate each pixel in the head feature map following the predicted corresponding offset, and finally, pixel-wise concatenation between body and head feature maps is performed to fuse the body and head feature. As

---

[1] In this paper, we consider real-time as over 15 fps.

the offset prediction branch adds only two convolutional layers and the translation operation can be efficiently implemented on GPU, the whole module costs marginal speed overhead. In addition, we design an auxiliary training task to further boost the learning of body box and body-head offset prediction, which is cost-free during inference, i.e. constructing the head center prediction map from body center prediction map and body-head center offsets then getting supervision from the ground-truth of the head center.

To prove the effictiveness of FastNFusion, we conduct comprehensive experiments on CrowdHuman [27] and CityPersons [37] datasets, compare the performance and speed with the current state-of-the-art methods, and visualize the behavior of the proposed body-head fusion module. As a result, FastNFusion improves the MR^{-2} by 3.6% to 41.77% on the CrowdHuman dataset, which is state-of-the-art, while runs at 16.8 fps on single Tesla P40.

Our contributions can be summarized as follows:(1)We propose a novel body-head fusion module that improves the detection of occluded pedestrians using head cues.(2)We design an auxiliary training task that enhances the full-body bounding box and body-head offset learning, which is cost free during inference.(3)The proposed FastNFusion achieves significant performance improvements (3.6% gain on CrowdHuman) while keeping running in real-time.

2   Related Work

Generic Object Detection. Mounts of research efforts have been made in generic object detection in recent years, which can be divided into two categories, i.e., the two-stage approaches and the single-stage approaches.

The dominant position of the two-stage approaches is established by Faster R-CNN [26], which generates proposals by Region Proposal Network (RPN) in the first stage and refines the proposals with a fully convolutional network (FCN) in the second stage. Based on the Faster R-CNN, lots of algorithms are proposed to improve the performance, including FPN [14], Mask R-CNN [8], Cascade R-CNN [2], etc. Generally speaking, two-stage approaches consistently obtain the top performance due to the refinement procedure, but meanwhile, the propose-then-refine pipeline is not computationally efficient, which prohibits the processing speed from being real-time.

On the other hand, single-stage approaches, such as YOLO [23] and SSD [17], directly classify and regress the anchor boxes into final predictions, which achieves significant speed improvement and comparable performance. And recently, anchor-free single-stage detectors [13,30,41] are attracting much attention as they get rid of the handcrafted anchor design by making predictions based on keypoints on feature map instead of anchor boxes, which not only avoids the anchor hyper-parameters finetuning but also improves the performance. Thus, to fulfill the real-time processing demand, our method is based on the Center-Net [41] as it delivers the best speed-performance trade-off.

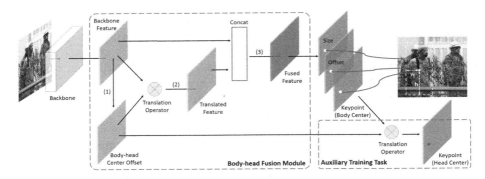

Fig. 1. Overview of FastNFusion. In the body-head fusion module, (1) we first predict the body-head center offset, (2) then translate the original backbone feature guided by the predicted offset, (3) finally concatenate the original feature and translated feature. Furthermore, the auxiliary training task refers to constructing head center predictions from both the body-head center offset and body center prediction then getting supervision from head center annotations. This module will not be used during inference.

Single-stage Pedestrian Detection. Prior works have made efforts on developing pedestrian detectors on single-stage baselines. TLL [28] proposes to better detect the small-scale pedestrians by replacing the box annotations with line annotations during training and introducing temporal feature aggregation for video inputs. It also proposes a post-processing scheme based on Markov Random Field (MRF) to compensate for the ambiguity introduced by the line predictions under occlusion. ALFNet [18] boosts the single-stage pipeline with a light-weight refinement module that translates the anchor boxes into the final predictions in multiple steps. CSP [19] is the first anchor-free pedestrian detector. Similar to the CenterNet, it simplifies the detection into a center and scale prediction task, but it has not made explicit efforts to address the special issues in pedestrian detection, such as occlusion and small-scale targets (Fig. 1).

Occulsion Handling by Pedestrian Parts. Occlusion remains one of the most significant challenges in pedestrian detection since it occurs much more frequently than generic object detection. To mitigate the problem, DeepParts [29] consists of extensive part detectors instead of a single detector so that the detections of occluded pedestrians can be reconstructed by the unoccluded parts. PSC-Net [33] extends the idea by modeling the intra and inter relationship between pedestrian parts with graph convolutional networks (GCNs). Another line of work seeks to leverage additional annotations, such as head boxes or visible boxes, as these parts are rarely occluded. MGAN [22] and PedHunter [3] introduce an attention module on visible boxes and head boxes respectively. PBM [11] and Bi-Box [39] explicitly predict the visible boxes and later fuse the prediction to improve the full-body detection. Double-Anchor R-CNN [35] generates head and full-body proposals with the same anchor. HBAN [20] uses the emperical knowledge of the body-head spatial relationship to create auxiliary supervision on full-body box training. JointDet [4] generates full-body proposals directly from head proposals with fixed ratio. Note that all these methods are two-stage and inspired by their success, we propose to efficiently fuse the body-head feature in a single-stage framework. Apart from boosting by pedestrian parts, there are also excellent prior works focusing on novel loss [31,38], better NMS [1,16,40], and data augmentation[3] to improve the detection performance, but please note that these works are orthogonal to our contributions.

3   Fast and Fusion

In this section, we first briefly recap our baseline framework named Center-Net [41]. Then we propose a novel body-head fusion module, which improves the full-body detection performance under the help of the pedestrian heads to address the occlusion problem with negligible computational overhead. Later, an auxiliary training task is introduced, which facilitates the learning of body-head offset and full-body box predictions with no cost during inference. Finally, we cover the implementation details of the proposed method.

3.1   Baseline

As a single-stage method, CenterNet consists of backbone network and detection branch. The backbone network can be any fully-convolutional encoder-decoder network, such as Hourglass [21], ResNet [9,32], and DLA [34], which projects the input images $I \in \mathcal{R}^{W \times H \times 3}$ into the feature space.

The detection branch takes the backbone feature as input and generates three prediction maps in parallel using an FCN, namely keypoint, offset, and size predictions.

Each point on the keypoint map $\hat{Y} \in [0,1]^{\frac{W}{R} \times \frac{H}{R} \times C}$ represents the confidence score of this location being an object center in a certain category, where H, W, R, C denote input height, width, output stride, and number of categories respectively. For each ground truth keypoint of category c, we denote the location in the input image as $p_c \in \mathcal{R}^2$ and the location in the low-resolution feature map as $\hat{p}_c = \lfloor \frac{p_c}{R} \rfloor$. We use a Gaussian kernel $Y_{xyc} = \exp{-\frac{(x_c - \hat{p}_{x,c})^2 + (y_c - \hat{p}_{y,c})^2}{2\sigma_p^2}}$ to splat all ground-truth keypoints, where σ_p is an object size-adaptive standard deviation. The training objective of keypoint heatmap is a penalty-reduced pixel-wise logistic regression with focal loss [15]:

$$L_k = \frac{-1}{N} \sum_{xyc} \begin{cases} (1 - \hat{Y}_{xyc})^\alpha \log(\hat{Y}_{xyc}) & \text{if } Y_{xyc} = 1 \\ (1 - Y_{xyc})^\beta (\hat{Y}_{xyc})^\alpha & \\ \quad \log(1 - \hat{Y}_{xyc}) & \text{otherwise} \end{cases} \tag{1}$$

where α and β are the hyper-parameters of focal loss and N denotes the total number of keypoints.

Offset prediction $\hat{O} \in \mathcal{R}^{\frac{W}{R} \times \frac{H}{R} \times 2}$ for each center point is used to recover the discretization error caused by the output stride, which is trained with L1 loss:

$$L_{off} = \frac{1}{N} \sum_p \|\hat{O}_{\hat{p}} - \left(\frac{p}{R} - \hat{p}\right)\|_1. \tag{2}$$

Finally, for each center point, the width and height are predicted for the corresponding bounding box, which is also supervised by L1 loss:

$$L_{size} = \frac{1}{N} \sum_p \|\hat{S}_p - s_p\|_1, \tag{3}$$

where $\hat{S} \in \mathcal{R}^{\frac{W}{R} \times \frac{H}{R} \times 2}$ denotes the size prediction and s_p represents the ground truth.

3.2   Body-Head Fusion

Occlusion occurs more frequently in pedestrian detection than generic object detection which causes a significant performance drop due to that (1) pedestrians occluded by background objects have partially visible feature, which degrades the confidence score and leads to potential miss; (2) pedestrians occluded with each other make it hard to distinguish instance boundaries, which results in inaccurate bounding box predictions and sometimes multiple instances are mistakenly predicted as one. Fortunately, pedestrian heads suffer from occlusion much rarely, which can serve as cues to mitigate the occlusion problem.

Our body-head fusion module consists of three steps: (1) take the backbone feature F as input and predict the body-head center offset $\hat{B}$; (2) guided by $\hat{B}$, we translate each pixel on F and get F'; (3) F and F' are fused then used as the input to the detection branch. We will cover each step in the following.

Body-Head Center Offset Prediction. Similar to the keypoint prediction in CenterNet, the Body-head Center Offset branch outputs a prediction map $\hat{B} \in \mathcal{R}^{\frac{W}{R} \times \frac{H}{R} \times 2}$ with the 2-d vector on each position representing the offset pointing from full-body center to head center. When constructing training targets, in order to increase the number of training samples, in addition to the sample of body center p_b pointing to head center p_h, we let all the positions around p_b within certain range δ point to p_h as well. Finally, we impose KL loss [10] for training:

$$L_{bh} = \frac{1}{N} \sum_p \frac{(KL(\hat{B}_p - (p_h - p)))}{\times \max(0, \delta - |p_h - p|)}, \tag{4}$$

where δ is size-adaptive to the size of full-body box. Since this branch only consists of two convolutional layers with one relu layer in between, the computational cost is negligible.

Feature Translation. The goal of the translation module is to re-locate each pixel on the original backbone feature map to a new position according to the predicted body-head center offset. We achieve this using the Differentiable Image Sampling (DIS) proposed by STN [12]. We encourage you to refer to their paper for details, but in brief, DIS generates the new feature map by sampling on the original feature map. And to allow backpropagation of the loss, we choose bilinear sampling kernel:

$$F'_p = \sum_n^H \sum_m^W \begin{matrix} (F_{nm} \max(0, 1 - |(p + \hat{B}_p)_w - m|) \\ \max(0, 1 - |(p + \hat{B}_p)_h - n|)). \end{matrix} \tag{5}$$

Note that, this operator can be implemented very efficiently on GPU.

Fig. 2. Qualitative results of baseline (1st row) and FastNFusion (2nd row). Blue boxes denote the detected pedestrian bounding boxes and red boxes denote the pedestrians detected by our method while miss detected by the baseline. (Color figure online)

Feature Fusion. After F' is generated, the feature fusion between F and F' is straightforward. We simply concatenate them along the channel dimension. Then we feed the fused feature into the aforementioned CenterNet detection branch to get the keypoint, offset, and size predictions.

3.3 Auxiliary Training Task

To strengthen the learning of the full-body bounding box prediction and body-head center offset prediction, we design an auxiliary training task, which will not be used during inference. The key idea is to construct the head keypoint prediction map $\hat{M}_h$ from the body keypoint prediction map $\hat{M}_b$ and the center offset $\hat{B}$ then supervise by the head box annotations so that if $\hat{M}_b$ and $\hat{B}$ are not well-learned, the loss will be large since there is no way to construct a good $\hat{M}_h$.

Table 1. Performance and speed on the CityPersons validation set. $1\times$ scale means train and test on original resolution (1024×2048) while $1.3\times$ is 1.3 times of that. Adaptive-NMS+ refers to Adaptive-NMS trained with AggLoss.

Stage(s)	Method	Scale	MR^{-2}	FPS
Two	Faster-RCNN	$1.3\times$	15.4	**5.2**
	Bi-Box	$1.3\times$	11.2	·
	JointDet	$1.3\times$	**10.2**	·
	CrowdDet	$1.3\times$	10.7	4.5
	PSC-Net	$1\times$	**10.6**	·
	MGAN	$1\times$	11.5	·
	PBM	$1\times$	11.1	·
	HBAN	$1\times$	12.5	·
	OR-RCNN	$1\times$	12.8	·
	RepLoss	$1\times$	13.2	·
	Adaptive-NMS+	$1\times$	11.9	·
Single	TLL	$1\times$	14.4	·
	ALFNet	$1\times$	12.0	5.5
	CSP	$1\times$	11.0	2.9
	FastNFusion (ours)	$1\times$	**10.3**	**9.3**

Specifically, we apply the same translation operator used in body-head feature fusion module on $\hat{M}_b$ and $\hat{B}$ to generate $\hat{M}_h$. Then we contruct the training targets using the same method as for full-body keypoint. Finally, focal loss is used for training and we denote the loss of the auxiliary task as L_{aux}.

4   Experiment

4.1   Datasets and Evaluation Metric

We conduct experiments on CrowdHuman [27] and CityPersons [37] datasets. We choose MR^{-2} [36] as the performance metric, which is calculated by the log-average miss rate over 9 points ranging from 10^{-2} to 10^0 false positive per image (FPPI). And a lower score indicates better performance. Besides, we also report the inference speed of various detectors as it is one of the focus of this paper. More detailed information can be referred in the supplementary material.

Table 2. Performance and speed on the CrowdHuman validation set. All the two-stage methods in this table are based on Faster-RCNN thus the FPS of Faster-RCNN is their upper bound. Our method achieves comparable performance to two-stage state-of-the-art while being significantly faster. Note that, (800,1400) refers to the short edge is at 800 pixels while the long edge is no more than 1400 pixels and * denotes implemented by us.

Stage(s)	Method	Scale	Backbone	MR^{-2}	FPS	Time (ms)
Two	Faster-RCNN	(800,1400)	ResNet-50	50.42	**11.8**	**84.7**
	Adaptive-NMS	(800,∞)	ResNet-50	49.73	·	·
	JointDet	(800,1333)	ResNet-50	46.50	·	·
	PBM	(800,1400)	ResNet-50	43.35	·	·
	CrowdDet	(800,∞)	ResNet-50	**41.40**	9.7	103.1
Single	CenterNet*	(800,1216)	DLA-34	45.37	**18.2**	**54.9**
	FastNFusion (ours)	(800,1216)	DLA-34	**41.77**	16.8	59.5

4.2 Evaluation on Extended CityPersons Dataset

As shown in Table 1, we compare FastNFusion with various state-of-the-art methods in terms of performance and speed on Citypersons validation set. We group the methods into three folds: two-stage methods with 1× and 1.3× input scales, and single-stage methods with 1× input scale. Compared to other single-stage methods, FastNFusion outperforms the second-best by 0.7% while being the fastest. Besides, under the same input scale, FastNFusion still surpasses all the two-stage methods in performance. Finally, when using the 1.3× input scale, only one two-stage method (JointDet) performs better than ours (by 0.1%) but ours is significantly faster since we are 1.8 times faster than the Faster-RCNN baseline.

4.3 Evaluation on CrowdHuman Dataset

Table 2 reports the performance and speed on the CrowdHuman validation set. Compared to our baseline (CenterNet), we improve the performance by 3.6% with merely 4.6 ms speed overhead. Furthermore, when comparing with the current state-of-the-art, FastNFusion is the only pedestrian detector processing at over 15 fps with comparable performance. Specifically, FastNFusion is both faster and more accurate than methods except for CrowdDet [5]. And compared to CrowdDet, ours is 43.6 ms faster and falls behind only by 0.37% in performance.

4.4 Ablation Study

To evaluate the effectiveness of different components in the proposed model, we conduct ablation studies for each component on the CrowdHuman dataset, as shown in Table 3. Moreover, we also provide the visualization of the behavior of each component in the supplementary material to assist better understanding.

Table 3. Ablation study on the CrowdHuman validation dataset.

Feature fusion	Auxiliary task	MR^{-2}
✗	✗	45.37
✓	✗	43.23 (+2.14)
✓	✓	41.77 (+1.46)

Body-Head Feature Fusion. The body-head fusion module is the key component of FastNFusion, which improves the baseline by 2.14% on the CrowdHuman dataset (See Table 3). Since CrowdHuman aims at highlighting the occlusion issue in pedestrian detection, this indicates that spatially transforming the head feature and fusing it into the full-body feature does mitigate the problem and recall more missed detections in the baseline. We list several examples in Fig. 2, where the missed pedestrian in the crowd with heavy occlusion can be recalled, especially when the head part is visible.

Moreover, another key contribution of ours is that we propose to pair the head and full-body feature with their center offsets. And to enable accurate feature fusion, the center offsets need to be accurate as well. Thus, the performance gain also implies that the center offset prediction branch works up to expected as shown in Fig. 3, where the predicted offset can accurately point to the head for pedestrians with various scales.

Fig. 3. Qualitative results of body-head center offset prediction. Ground-truth and prediction results under various scales are shown in the left column and right column respectively.

Auxiliary Training Task. As shown in Table 3, the auxiliary training task further boosts the FastNFusion by 1.46%, resulting in the state-of-the-art performance of 41.77%. Note that this module won't be used during inference and is supposed to help train the full-body center prediction and body-head center offset.

5   Conclusion

In this paper, we propose FastNFusion, which is the first real-time pedestrian detector that is capable of occlusion handling thus yielding state-of-the-art performance. We achieve this by fusing head features into full-body features since the head suffers less from occlusion. Besides, we design an auxiliary training task to facilitate the learning of full-body box and body-head offset prediction. Finally, comprehensive experiments and analyses are done on CityPersons and CrowdHuman to show the strength of FastNFusion.

Acknowledgements. This work was supported in part by the Zhejiang Provincial Natural Science Foundation of China (No. LY21F020019) and in part by the National Science Foundation of China under Grants 61802100,61971172 and 61971339. This work was also supported in part by the China Post-Doctoral Science Foundation under Grant 2019M653563.

References

1. Bodla, N., Singh, B., Chellappa, R., Davis, L.S.: Soft-nms-improving object detection with one line of code. In: Proceedings of the IEEE International Conference on Computer Vision, pp. 5561–5569 (2017)
2. Cai, Z., Vasconcelos, N.: Cascade r-cnn: delving into high quality object detection. In: CVPR, pp. 6154–6162 (2018)
3. Chi, C., Zhang, S., Xing, J., Lei, Z., Li, S.Z., Zou, X.: Pedhunter: occlusion robust pedestrian detector in crowded scenes. arXiv preprint arXiv:1909.06826 (2019)
4. Chi, C., Zhang, S., Xing, J., Lei, Z., Li, S.Z., Zou, X.: Relational learning for joint head and human detection. arXiv preprint arXiv:1909.10674 (2019)
5. Chu, X., Zheng, A., Zhang, X., Sun, J.: Detection in crowded scenes: one proposal, multiple predictions. In: Proceedings of the IEEE/CVF Conference on Computer Vision and Pattern Recognition (CVPR) (2020)
6. Dai, J., et al.: Deformable convolutional networks. In: ICCV, pp. 764–773 (2017)
7. Girshick, R.: Fast r-cnn. In: CVPR, pp. 1440–1448 (2015)
8. He, K., Gkioxari, G., Dollár, P., Girshick, R.: Mask r-cnn. In: ICCV, pp. 2961–2969 (2017)
9. He, K., Zhang, X., Ren, S., Sun, J.: Deep residual learning for image recognition. In: Proceedings of the IEEE Conference on Computer Vision and Pattern Recognition, pp. 770–778 (2016)
10. He, Y., Zhu, C., Wang, J., Savvides, M., Zhang, X.: Bounding box regression with uncertainty for accurate object detection. In: CVPR, pp. 2888–2897 (2019)
11. Huang, X., Ge, Z., Jie, Z., Yoshie, O.: Nms by representative region: towards crowded pedestrian detection by proposal pairing. In: Proceedings of the IEEE/CVF Conference on Computer Vision and Pattern Recognition, pp. 10750–10759 (2020)
12. Jaderberg, M., Simonyan, K., Zisserman, A., et al.: Spatial transformer networks. In: NIPS, pp. 2017–2025 (2015)
13. Law, H., Deng, J.: Cornernet: detecting objects as paired keypoints. In: ECCV, pp. 734–750 (2018)
14. Lin, T.Y., Dollár, P., Girshick, R., He, K., Hariharan, B., Belongie, S.: Feature pyramid networks for object detection. In: CVPR, pp. 2117–2125 (2017)

15. Lin, T.Y., Goyal, P., Girshick, R., He, K., Dollár, P.: Focal loss for dense object detection. In: CVPR, pp. 2980–2988 (2017)
16. Liu, S., Huang, D., Wang, Y.: Adaptive NMS: refining pedestrian detection in a crowd. In: CVPR, pp. 6459–6468 (2019)
17. Liu, W., et al.: SSD: single shot multibox detector. In: Leibe, B., Matas, J., Sebe, N., Welling, M. (eds.) ECCV 2016. LNCS, vol. 9905, pp. 21–37. Springer, Cham (2016). https://doi.org/10.1007/978-3-319-46448-0_2
18. Liu, W., Liao, S., Hu, W., Liang, X., Chen, X.: Learning efficient single-stage pedestrian detectors by asymptotic localization fitting. In: Proceedings of the European Conference on Computer Vision (ECCV), pp. 618–634 (2018)
19. Liu, W., Liao, S., Ren, W., Hu, W., Yu, Y.: High-level semantic feature detection: a new perspective for pedestrian detection. In: CVPR, pp. 5187–5196 (2019)
20. Lu, R., Ma, H.: Semantic head enhanced pedestrian detection in a crowd. arXiv preprint arXiv:1911.11985 (2019)
21. Newell, A., Yang, K., Deng, J.: Stacked hourglass networks for human pose estimation. In: Leibe, B., Matas, J., Sebe, N., Welling, M. (eds.) ECCV 2016. LNCS, vol. 9912, pp. 483–499. Springer, Cham (2016). https://doi.org/10.1007/978-3-319-46484-8_29
22. Pang, Y., Xie, J., Khan, M.H., Anwer, R.M., Khan, F.S., Shao, L.: Mask-guided attention network for occluded pedestrian detection. In: Proceedings of the IEEE International Conference on Computer Vision, pp. 4967–4975 (2019)
23. Redmon, J., Divvala, S., Girshick, R., Farhadi, A.: You only look once: unified, real-time object detection. In: CVPR, pp. 779–788 (2016)
24. Redmon, J., Farhadi, A.: Yolo9000: better, faster, stronger. In: CVPR, pp. 7263–7271 (2017)
25. Redmon, J., Farhadi, A.: Yolov3: an incremental improvement. arXiv preprint arXiv:1804.02767 (2018)
26. Ren, S., He, K., Girshick, R., Sun, J.: Faster r-cnn: towards real-time object detection with region proposal networks. In: NIPS, pp. 91–99 (2015)
27. Shao, S., et al.: Crowdhuman: a benchmark for detecting human in a crowd. arXiv preprint arXiv:1805.00123 (2018)
28. Song, T., Sun, L., Xie, D., Sun, H., Pu, S.: Small-scale pedestrian detection based on topological line localization and temporal feature aggregation. In: Proceedings of the European Conference on Computer Vision (ECCV), pp. 536–551 (2018)
29. Tian, Y., Luo, P., Wang, X., Tang, X.: Deep learning strong parts for pedestrian detection. In: CVPR, pp. 1904–1912 (2015)
30. Tian, Z., Shen, C., Chen, H., He, T.: Fcos: fully convolutional one-stage object detection. In: ICCV (2019)
31. Wang, X., Xiao, T., Jiang, Y., Shao, S., Sun, J., Shen, C.: Repulsion loss: detecting pedestrians in a crowd. In: CVPR, pp. 7774–7783 (2018)
32. Xiao, B., Wu, H., Wei, Y.: Simple baselines for human pose estimation and tracking. In: Proceedings of the European Conference on Computer Vision (ECCV), pp. 466–481 (2018)
33. Xie, J., Pang, Y., Cholakkal, H., Anwer, R.M., Khan, F.S., Shao, L.: Psc-net: learning part spatial co-occurence for occluded pedestrian detection. arXiv preprint arXiv:2001.09252 (2020)
34. Yu, F., Wang, D., Shelhamer, E., Darrell, T.: Deep layer aggregation. In: Proceedings of the IEEE Conference on Computer Vision and Pattern Recognition, pp. 2403–2412 (2018)
35. Zhang, K., Xiong, F., Sun, P., Hu, L., Li, B., Yu, G.: Double anchor r-cnn for human detection in a crowd. arXiv preprint arXiv:1909.09998 (2019)

36. Zhang, S., Benenson, R., Omran, M., Hosang, J., Schiele, B.: How far are we from solving pedestrian detection? In: CVPR, pp. 1259–1267 (2016)

37. Zhang, S., Benenson, R., Schiele, B.: Citypersons: a diverse dataset for pedestrian detection. In: CVPR, pp. 3213–3221 (2017)

38. Zhang, S., Wen, L., Bian, X., Lei, Z., Li, S.Z.: Occlusion-aware r-cnn: detecting pedestrians in a crowd. In: Proceedings of the European Conference on Computer Vision (ECCV), pp. 637–653 (2018)

39. Zhou, C., Yuan, J.: Bi-box regression for pedestrian detection and occlusion estimation. In: ECCV, pp. 135–151 (2018)

40. Zhou, P., et al.: Noh-nms: Improving pedestrian detection by nearby objects hallucination. arXiv preprint arXiv:2007.13376 (2020)

41. Zhou, X., Wang, D., Krähenbühl, P.: Objects as points. arXiv preprint arXiv:1904.07850 (2019)

STA-GCN: Spatio-Temporal AU Graph Convolution Network for Facial Micro-expression Recognition

Xinhui Zhao, Huimin Ma$^{(\boxtimes)}$, and Rongquan Wang

School of Computer and Communication Engineering, University of Science and Technology Beijing, Beijing 100083, China
mhmpub@ustb.edu.cn

Abstract. Facial micro-expression (FME) is a fast and subtle facial muscle movement that typically reflects person's real mental state. It is a huge challenge in the FME recognition task due to the low intensity and short duration. FME can be decomposed into a combination of facial muscle action units (AU), and analyzing the correlation between AUs is a solution for FME recognition. In this paper, we propose a framework called spatio-temporal AU graph convolutional network (STA-GCN) for FME recognition. Firstly, pre-divided AU-related regions are input into the 3D CNN, and inter-frame relations are encoded by inserting a Non-Local module for focusing on apex information. Moreover, to obtain the inter-AU dependencies, we construct separate graphs of their spatial relationships and activation probabilities. The relationship feature we obtain from the graph convolution network (GCN) are used to activate on the full-face features. Our proposed algorithm achieves state-of-the-art accuracy of 76.08% accuracy and F1-score of 70.96% on the CASME II dataset, which outperformance all baselines.

Keywords: Micro-expression · GCN · Emotion recognition · Non-local network · Facial action unit

1 Introduction

Facial micro-expression (FME) is a short-lived, unconscious facial muscle movement that usually lasts 40–200 ms with characterized by short duration and low intensity of movement. FME appear when people unconsciously or deliberately suppress the expression of emotions. It reflects authentic psychological state of an individual, so FME recognition is of great research value in areas such as emotion analysis and mental illness detection. In a FME sequence, the frame where the expression begins is called the onset frame, and the frame where the expression peaks is called the apex frame. In the facial action coding system (FACS) proposed by Ekman et al. [4], FME can be decomposed into a combination of different muscle action unit (AU) movements. Figure 1 shows an example

© Springer Nature Switzerland AG 2021
H. Ma et al. (Eds.): PRCV 2021, LNCS 13019, pp. 80–91, 2021.
https://doi.org/10.1007/978-3-030-88004-0_7

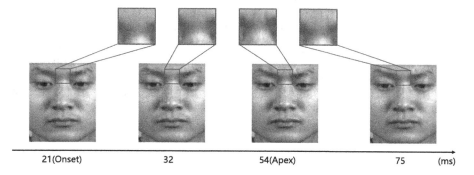

21(Onset)   32   54(Apex)   75   (ms)

Fig. 1. An example of a FME sequence containing AU4 from the CASME II dataset, the apex frame appears in frame 54.

of an FME sequence from CASME II dataset [24], the apex frame appears at about frame 54 and the AU is 4.

Traditional deep learning-based FME recognition utilizes video recognition methods. Methods based on 3D CNN were the first to be considered. Wang et al. [19] applies Eulerian motion magnification on micro-movement and utilizes 3D CNN to recognize FME. In recent years, more and more researchers work on the AU level based on FACS. AU-RCNN [10] is proposed to introduce expert prior knowledge in AU detection. Literature [27] performs AU intensity estimation by joint learning representation and estimation.

This paper reviews that past methods have two shortcomings. Firstly, FME recognition is different from traditional video recognition tasks. It is more important to capture apex information, and focus on the inter-frame relationships while describing the spatio-temporal features. Secondly, there is a potential correlation between facial muscle movements, which has not been exploited as much as possible in past AU-based FME recognition.

Therefore, we try to formulate two module to solve above shortcomings: 3D CNN with a non-local module (3DCNN-NL) and AU-attention graph convolution. We build our framework based on these two modules called spatio-temporal AU graph convolutional network (STA-GCN). In the 3DCNN-NL module, we adopt a non-local module [18] to extract spatio-temporal features from sequences of AU-related regions. The non-local network can capture long-distance dependencies well and build an attention mechanism based on the inter-frame relationships. In the AU-attention graph convolution module, since AUs are non-Euclidean structures, we consider further modeling the spatial position and activation probability among AUs through GCN. Then the AU-related attention is activated on the full-face feature, and classification of FME emotional attributes are achieved.

In summary, our contributions in this paper are as follows:

1. A novel GCN-based method were proposed to recognize FME. We build the attention mechanism based on AUs relation information by graph convolution, which is activated on the full-face features.

2. A framework called STA-GCN is proposed divided into 3DCNN-NL and AU attention graph convolution. Firstly, 3D CNN with a non-local module is used to extract the spatio-temporal features of FME sequence, so as to pay more attention to the peak information. Secondly, we model the relationship of AUs' space position and activation probability respectively, then the attention of AUs relationship imformation is estabilshed by GCN, which is activated on the full-face feature.
3. We verify the effectiveness of STA-GCN on the CASME II dataset and achieve performance improvement in terms of recognition accuracy and F1-score.

2   Related Work

2.1   Micro-Expression Recognition

In general, traditional FME feature extraction is mainly based on hand-crafted methods such as Local Binary Pattern (LBP) [1,2] and optical-flow methods, then features classification by SVM [11,12,17], etc. LBP-TOP [29] with three orthogonal planes of LBP operators is proposed to describe the local appearance and dynamic texture. The local binary pattern with six intersection points (LBP-SIP) [20] was proposed as an extension of LBP-TOP and achieved higher efficiency. The optical flow-based method uses optical-flow direction and optical strain as descriptors. Liu et al. [8] propose an effective Main Direction Mean Optical-flow method to obtain features by counting the optical-flow histogram on the ROI of the face and considering local motion and spatial information.

In recent years, deep learning is introduced in FME recognition to improve the accuracy and robustness limitations of traditional methods [13,15], and achieve better performance. Khor et al. [6] proposed a rich long-term recurrent convolutional network framework to predict FMEs. Literature [16] introduces LSTM [5] into micro-expression recognition. Model STRCN [21] introduced a recurrent convolutional network to extract visual features and use a classification layer for recognition. However, FME recognition based on deep learning is easily limited by dataset size and sample balance.

2.2   Graph Convolution Network

In real-world scenarios, many data structures such as graph cannot be modeled with traditional CNNs to extract features. Graph convolution network (GCN) [3,7,25] utilizes the edge information of nodes in the graph to aggregate node information, thus dynamically updating the node representation, where each node aggregates the information of neighboring nodes. Therefore we extract the relational features of nodes by GCN. In recent years, researchers start to introduce GCN into video or action classification tasks [23]. Literature [22] proposed a GCN model that fuses multi-level semantics into temporal features, using video clips as nodes to build graphs and update nodes dynamically. GCN also introduce into the AU detection task. Liu et al. [9] first model the AU relationship using GCN to detect the AU in a FME sample. In this paper, we further

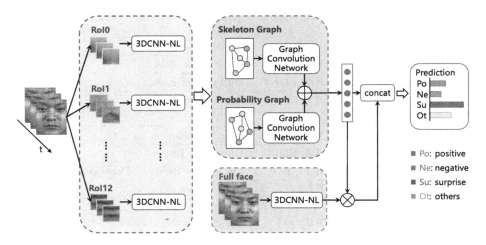

Fig. 2. The overall structure of STA-GCN.

implement the micro-expression emotion recognition using GCN, and achieve the best performance currently.

3   Method

In this section, we introduce the proposed STA-GCN architecture. The architecture consists of two module: 3DCNN-NL and AU-attention graph convolution. In the 3DCNN-NL module, a non-local module is used to extract AU-related region spatio-temporal features. In the AU-attention graph convolution module, we further model the AUs relationship by GCN, and perform activation on the full-face features. Finally, the activation features are used for emotion classification. The overall structure of STA-GCN is shown in Fig. 2.

3.1   ROI Division

To focus on the region where AUs occur to avoid the noise caused by useless parts of the face, we perform landmark detection on the face and select 13 key points as the center of ROIs [26,28]. We try to cover the region where AUs occur with ROIs and ensure that ROIs are independent of each other. Figure 4(a) shows the specific division of ROIs on the template face. Table 1 lists AUs, the corresponding ROIs, and the facial movements.

Table 1. ROIs, the corresponding AUs and the facial movements.

AUs	ROIs	Facial movements
AU1	ROI 2, ROI 3	Inner Brow Raiser
AU2	ROI 1, ROI 4	Outer Brow Raiser
AU4	ROI 0	Brow Lowerer
AU7	ROI 8, ROI 9	Lid Tightener
AU9	ROI 5	Nose Wrinkler
AU10	ROI 6, ROI 7	Upper Lip Raiser
AU12, Au15	ROI 10, ROI 11	Lip Corner Puller or Depressor
AU17	ROI 12	Chin Raiser

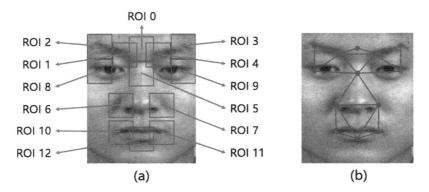

Fig. 3. (a) 13 ROIs on the template face. (b) Connect adjacent nodes to build a skeleton relationship graph

3.2   3D CNN with Non-Local Block

In general, as one of the video recognition tasks, deep learning-based FME recognition algorithms always adopt a 3D convolutional network to extract the spatio-temporal features [13,19]. Different from traditional video recognition tasks, expressions tend to change slowly when first appear, and the closer with the apex frame, the more drastic the change is. Therefore, we consider describing the spatio-temporal features with more attention to the inter-frame relationships, thus giving higher weights to the apex frames.

Non-local network [18] is a typical self-attention method, which is proposed to build a global attention mechanism for effectively capturing long-distance dependencies. It estabilshes attention by calculating the response to the current position at any location on the globe, and give a larger weight to important motion regions. For expressions, the motion amplitude near the apex frame is large and concentrated. We focus attention on the apex information in this way.

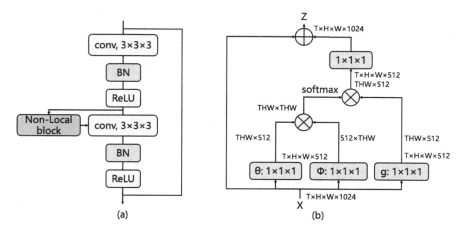

Fig. 4. (a) Insert a non-local block in the third block of 3DResnet18. (b) A spacetime non-local block.

The structure of a non-local block is shown in Fig. 4(a). We describe it by the following formula.

$$y_i = \frac{1}{C(x)} \sum_{\forall j} \eta(X_i, X_j)\gamma(X_j), \qquad (1)$$

where i represents the key of the pixel position whose response value is to be calculated and j represents the index of all the queried positions on the FME sequence. The global response at position X_i is obtained by calculating the similarity between X_i and X_j by the correlation function η. The representation of the feature map at position j is calculated by linear function γ. The final feature is normalized using the coefficient function $c(x)$. For each ROIs selected in the previous step, We adopt 3D CNN with a non-local module (3DCNN-NL) for spatio-temporal features extraction. For the specific implementation, we designed the model based on 3D-Resnet-18 architecture and a non-local module is inserted into the third block of 3D-Resnet-18 (shown in Fig. 4(b)). We denote the above spatio-temporal convolution module as $F(X, W)$.

3.3 AU-attention Graph Convolution

Graph Convolution Network. In the FME recognition task, muscle movement has hidden relevance. Since AUs are non-Euclidean structure, we utilize graph convolution network (GCN) to better extract the relationship information between AUs, thereby building a global attention mechanism for the full-face feature.

GCN takes the node feature matrix and the inter-node adjacency matrix as input, then uses the activation function to update the node representation of the deeper hidden layer. Based on the GCN proposed by Kipf et al. [7], we design the AU-attention graph convolution, and expressed the node feature matrix as $X = [X_{AU1}, X_{AU2}, ..., X_{AU17}]$. The adjacency matrix is A. The l-th graph convolution layer is defined as the following formula. We denote the weight matrix of the l-th layer by W_l, the nonlinear activation function by σ, and $\hat{A}$ is the normalization of the adjacency matrix A. The input H_0 of the first layer is the feature matrix X.

$$H_{l+1} = f(H_l, A, W_l) \tag{2}$$

$$H_0 = X \tag{3}$$

$$f(H_l, A, W_l) = \sigma(\hat{A} H_l W_l) \tag{4}$$

Build Graph. In this section, we build a skeleton graph for the AU spatical relationship, and a probability graph for the activation probability.

In the skeleton graph, we model the spatial adjacency relationship of AUs. Specifically, if two AUs are adjacent in the face structure, the weight of the adjacency matrix between them is 1, otherwise 0. Figure 3(b) shows the skeleton relationship we defined.

In the probability graph, we calculate the conditional probability of occurrence of AU_j under the condition of occurrence of AU_i based on CASME II dataset, and the activation probability AU_{ij} will be obtained as the adjacency of AU_i and AU_j. The calculation formula is as follows.

$$A_{ij} = P(Ai \mid Aj), \tag{5}$$

and the adjacency matrix we finally obtained is shown in Fig. 5.

For the two built graphs, we get the output through the two above-mentioned graph convolutional layers respectively. Denote the multi-layer graph convolution module as $g(X, A, W)$.

$$g(X, A, W) = f(X, A_1, W_1) + f(X, A_2, W_2) \tag{6}$$

AU-attention. To utilize the relationship information between AUs, we build attention in the full-face, and further introduce the global features of the face. The full FME sequence is first processed through the 3DCNN-NL module. Then we reduce the dimension of the relationship feature g obtained in the above to match the full-face representation, and activate g with sigmoid, distributed the feature probability between $[0, 1]$. The result is uesd to activate on the complete facial features. Finally, to further enrich the relationship information, we splice the results with the result of AUs relationship feature G, and make the final emotion classification through a simple full connection layer.

$$G = concat(g) \cdot w + b \tag{7}$$

$$Z = concat(G, G \cdot F(X_{full}, W_{full})) \tag{8}$$

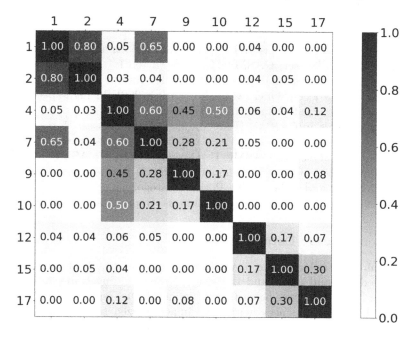

Fig. 5. Activation probability graph connection matrix on CASME II.

3.4 Loss Function

Considering the small size and unbalanced class distribution of the micro-expression dataset, we extend the cross-entropy function into the form of balanced sample loss with an inverse proportional weight.

$$L = -\frac{1}{N} \sum_i \lambda(1 - \beta_i)(y_i log \hat{y}_i + (1 - y_i)log(1 - \hat{y}_i)) \tag{9}$$

where β_i reflects the proportion of category i in all samples, and 1-β_i is the inverse weight of the sample. λ represents the scaling parameter.

In the first stage of 3DCNN-NL, we use the balanced cross-entropy loss function to constrain the RoIs separately. This loss function is still used to constrain the emotion classification in the AU-attention graph convolution in the second stage.

4   Experiment

In this section, we introduce the experimental details, including the dataset we used, implementation details, comparison methods, and experimental results.

4.1   Experimental Setting

The micro-expression dataset is divided into spontaneous and non-spontaneous. In the spontaneous dataset, CASME II [24] is currently the most widely used spontaneous micro-expression data set. CASME II contains 255 micro-expression samples from 26 subjects, which were shot with a 200FPS high-speed camera with a facial resolution of about 280 * 340 pixels.

Micro-expressions in CASME II are divided into 7 classes: Happiness, Disgust, Surprise, Sadness, Repression, Fear, and Others. Moreover, the AUs of each FME samples are labeled in CASME II. Considering that CASME II's sample has serious imbalances, we unify the emotional labels into four classes: Positive, Negative, Surprise, and Others. The Positive category includes happiness, and the Negative category includes Disgust, Sadness, Repression, and Fear.

4.2   Implementation Details

Before training, to enhance the diversity of training data, the images are randomly flipped left and right, and the sequences are sliding frame fetched [14]. We crop the FME sequences according to defined ROIs, and each of FME samples obtain 13 ROIs sequences, with each image resized to 112×112 and pixel normalized. The sequences were used as input the first stage of 3DCNN-NL. After obtaining the spatio-temporal features of ROIs, the 13 ROIs features on the same FME sample are used as nodes to build two relationship graphs, which are convolved through two graph convolution layers, and the results are summed to obtain AU relational features. Next, the spatio-temporal features of the full FME sequences are extracted by 3DCNN-NL, fuse the full-face features with AU relationship by dot product operation. Furthermore, we add the original AUs relational features to enhance the features, finally use the fully connected layer for emotion classification.

In the validation, Leave-One-Subject-Out (LOSO) Cross-Validation was adopted to evaluate experimental performance, in which all samples of a single subject in CASME II are used for validation and the samples of remaining subjects are used for training in each training process. Since the distribution of subjects in the training and test is independent, the LOSO validation better reflects the level of the model.

4.3   Experimental Result

We evaluate the proposed method on CASME II, accuracy and F1-score were selected as indicators for evaluating performance. To prove the effectiveness of the algorithm, we subject the structural decomposition to ablation experiments to verify the performance of 3DCNN-NL and STA-GCN respectively.

Table 2 summarizes the experimental results of our method compared to other FME recognition algorithms, the proposed algorithm achieves an excellent performance of 74.03% accuracy and 70.96% F1-score on the CASME II. Figure 6 shows the confusion matrix of 3D CNN, 3DCNN-NL and our proposed method.

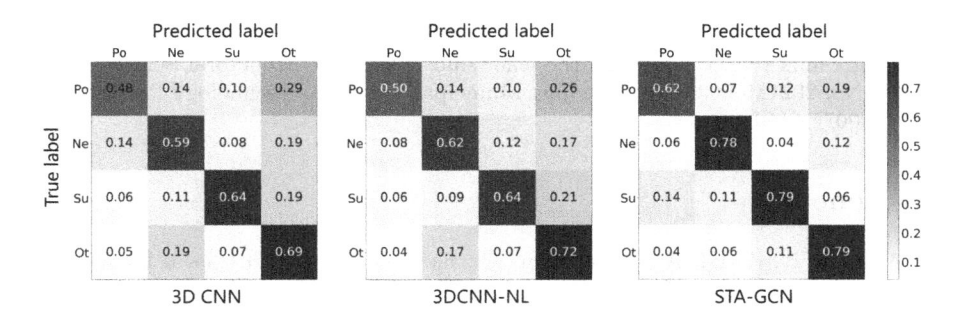

Fig. 6. The normalized confusion matrix of 3D CNN, 3DCNN-NL and STA-GCN on CASME II.

It can be seen that the 3DCNN-NL module improve the performance, but a large amount of data is still misclassified as Others due to the imbalance of the dataset. STA-GCN improve this by considering the relationship imformation between AUs.

Table 2. Performance of proposed methods vs. other methods on CASMEII dataset

Methods	Acc	F1-score
LBP-TOP [29]	0.4581	0.3813
LBP-SIP [20]	0.5172	0.4737
MDMO [8]	0.6310	0.5412
3D CNN	0.6240	0.5813
3DCNN-NL	0.6492	0.6027
STA-GCN (Ours)	**0.7608**	**0.7096**

5   Conclusion

In this paper, we propose a framework for facial micro-expression (FME) recognition based on graph convolutional networks, containing two modules: 3DCNN-NL and AU-attention graph convolution. The spatio-temporal features of AUs regions are extracted by 3D CNN, while a non-local module is used to focus on apex information. Then the AUs relationships are modeled using GCN, activated with relational features on full-face features. The proposed method achieves a competitive recognition rate of 76.08% and F1-score of 70.96% on the CASME II dataset, which outperformance all baselines. Furthermore, we demonstrate the effectiveness of the 3DCNN-NL module and the AU-attention graph convolution module by ablation experiments.

The current FME recognition algorithms are limited by the size and balance of dataset. In future research, we will focus on resolving the limitations.

Acknowledgments. This work was supported by the National Natural Science Foundation of China (No. U20B2062), the fellowship of China Postdoctoral Science Foundation (No. 2021M690354), the Beijing Municipal Science & Technology Project (No. Z191100007419001).

References

1. Ahonen, T., Hadid, A., Pietikäinen, M.: Face recognition with local binary patterns. In: Pajdla, T., Matas, J. (eds.) ECCV 2004. LNCS, vol. 3021, pp. 469–481. Springer, Heidelberg (2004). https://doi.org/10.1007/978-3-540-24670-1_36
2. Ahonen, T., Hadid, A., Pietikainen, M.: Face description with local binary patterns: application to face recognition. IEEE Trans. Pattern Anal. Mach. Intell. **28**(12), 2037–2041 (2006)
3. Defferrard, M., Bresson, X., Vandergheynst, P.: Convolutional neural networks on graphs with fast localized spectral filtering. arXiv preprint arXiv:1606.09375 (2016)
4. Ekman, R.: What the Face Reveals: Basic and Applied Studies of Spontaneous Expression using the Facial Action Coding System (FACS). Oxford University Press, USA (1997)
5. Hochreiter, S., Schmidhuber, J.: Long short-term memory. Neural Comput. **9**(8), 1735–1780 (1997)
6. Khor, H.Q., See, J., Phan, R.C.W., Lin, W.: Enriched long-term recurrent convolutional network for facial micro-expression recognition. In: 2018 13th IEEE International Conference on Automatic Face & Gesture Recognition (FG 2018), pp. 667–674. IEEE (2018)
7. Kipf, T.N., Welling, M.: Semi-supervised classification with graph convolutional networks. arXiv preprint arXiv:1609.02907 (2016)
8. Liu, Y.J., Zhang, J.K., Yan, W.J., Wang, S.J., Zhao, G., Fu, X.: A main directional mean optical flow feature for spontaneous micro-expression recognition. IEEE Trans. Affect. Comput. **7**(4), 299–310 (2015)
9. Liu, Z., Dong, J., Zhang, C., Wang, L., Dang, J.: Relation modeling with graph convolutional networks for facial action unit detection. In: Ro, Y.M., et al. (eds.) MMM 2020. LNCS, vol. 11962, pp. 489–501. Springer, Cham (2020). https://doi.org/10.1007/978-3-030-37734-2_40
10. Ma, C., Chen, L., Yong, J.: Au R-CNN: encoding expert prior knowledge into R-CNN for action unit detection. Neurocomputing **355**, 35–47 (2019)
11. Pfister, T., Li, X., Zhao, G., Pietikäinen, M.: Recognising spontaneous facial micro-expressions. In: 2011 International Conference on Computer Vision, pp. 1449–1456. IEEE (2011)
12. Platt, J.: Sequential minimal optimization: a fast algorithm for training support vector machines (1998)
13. Reddy, S.P.T., Karri, S.T., Dubey, S.R., Mukherjee, S.: Spontaneous facial micro-expression recognition using 3D spatiotemporal convolutional neural networks. In: 2019 International Joint Conference on Neural Networks (IJCNN), pp. 1–8. IEEE (2019)
14. Shi, X., Yang, C., Xia, X., Chai, X.: Deep cross-species feature learning for animal face recognition via residual interspecies equivariant network. In: Vedaldi, A., Bischof, H., Brox, T., Frahm, J.-M. (eds.) ECCV 2020. LNCS, vol. 12372, pp. 667–682. Springer, Cham (2020). https://doi.org/10.1007/978-3-030-58583-9_40

15. Verma, M., Vipparthi, S.K., Singh, G., Murala, S.: Learnet: dynamic imaging network for micro expression recognition. IEEE Trans. Image Process. **29**, 1618–1627 (2019)
16. Wang, S.J., et al.: Micro-expression recognition with small sample size by transferring long-term convolutional neural network. Neurocomputing **312**, 251–262 (2018)
17. Wang, S.J., et al.: Micro-expression recognition using color spaces. IEEE Trans. Image Process. **24**(12), 6034–6047 (2015)
18. Wang, X., Girshick, R., Gupta, A., He, K.: Non-local neural networks. In: Proceedings of the IEEE Conference on Computer Vision and Pattern Recognition, pp. 7794–7803 (2018)
19. Shi, X., Yang, C., Xia, X., Chai, X.: Deep cross-species feature learning for animal face recognition via residual interspecies equivariant network. In: Vedaldi, A., Bischof, H., Brox, T., Frahm, J.-M. (eds.) ECCV 2020. LNCS, vol. 12372, pp. 667–682. Springer, Cham (2020). https://doi.org/10.1007/978-3-030-58583-9_40
20. Wang, Y., See, J., Phan, R.C.W., Oh, Y.H.: Lbp with six intersection points: reducing redundant information in lbp-top for micro-expression recognition. In: Asian Conference on Computer Vision, pp. 525–537. Springer (2014)
21. Xia, Z., Hong, X., Gao, X., Feng, X., Zhao, G.: Spatiotemporal recurrent convolutional networks for recognizing spontaneous micro-expressions. IEEE Trans. Multimed. **22**(3), 626–640 (2019)
22. Xu, M., Zhao, C., Rojas, D.S., Thabet, A., Ghanem, B.: G-tad: Sub-graph localization for temporal action detection. In: Proceedings of the IEEE/CVF Conference on Computer Vision and Pattern Recognition, pp. 10156–10165 (2020)
23. Yan, S., Xiong, Y., Lin, D.: Spatial temporal graph convolutional networks for skeleton-based action recognition. In: Proceedings of the AAAI Conference on Artificial Intelligence, vol. 32 (2018)
24. Yan, W.J., et al.: Casme ii: an improved spontaneous micro-expression database and the baseline evaluation. PLoS ONE **9**(1), e86041 (2014)
25. Ying, R., He, R., Chen, K., Eksombatchai, P., Hamilton, W.L., Leskovec, J.: Graph convolutional neural networks for web-scale recommender systems. In: Proceedings of the 24th ACM SIGKDD International Conference on Knowledge Discovery & Data Mining, pp. 974–983 (2018)
26. Zhang, K., Zhang, Z., Li, Z., Qiao, Y.: Joint face detection and alignment using multitask cascaded convolutional networks. IEEE Signal Process. Lett. **23**(10), 1499–1503 (2016)
27. Zhang, Y., et al.: Joint representation and estimator learning for facial action unit intensity estimation. In: Proceedings of the IEEE/CVF Conference on Computer Vision and Pattern Recognition, pp. 3457–3466 (2019)
28. Zhang, Z., Luo, P., Loy, C.C., Tang, X.: Facial landmark detection by deep multitask learning. In: European Conference on Computer Vision, pp. 94–108. Springer (2014)
29. Zhao, G., Pietikainen, M.: Dynamic texture recognition using local binary patterns with an application to facial expressions. IEEE Trans. Pattern Anal. Mach. Intell. **29**(6), 915–928 (2007)

Attentive Contrast Learning Network for Fine-Grained Classification

Fangrui Liu[1], Zihao Liu[2,3], and Zheng Liu[1(✉)]

[1] Faculty of Applied Science, University of British Columbia, Vancouver, BC, Canada
{fangrui.liu,zheng.liu}@ubc.ca
[2] NELVT, Department of Computer Science, Peking University, Beijing, China
[3] Advanced Institute of Information Technology, Peking University, Beijing, China

Abstract. Fine-grained visual classification is challenging due to subtle differences between sub-categories. Current popular methods usually leverage a single image and are designed by two main perspectives: feature representation learning and discriminative parts localization, while a few methods utilize pairwise images as input. However, it is difficult to learn representations discriminatively both across the images and across the categories, as well as to guarantee for accurate location of discriminative parts. In this paper, different from the existing methods, we argue to solve these difficulties from the perspective of contrastive learning and propose a novel Attentive Contrast Learning Network (ACLN). The network aims to attract the representation of positive pairs, which are from the same category, and repulse the representation of negative pairs, which are from different categories. A contrastive learning module, equipped with two contrastive losses, is proposed to achieve this. Specifically, the attention maps, generated by the attention generator, are bounded with the original CNN feature as positive pair, while the attention maps of different images form the negative pairs. Besides, the final classification results are obtained by a synergic learning module, utilizing both the original feature and the attention maps. Comprehensive experiments are conducted on four benchmark datasets, on which our ACLN outperforms all the existing SOTA approaches. For reproducible scientific research https://github.com/mpskex/AttentiveContrastiveLearningNetwork.

1 Introduction

Going beyond normal image classification that recognizes basic classes, fine-grained visual classification (FGVC) aims at distinguishing between children categories of the main parent class (e.g., birds [14], dogs [8]), which has become a significant topic with a broad application value. Although CNN has achieved excellent progress in tackling this task, it still remains challenging, due to the close resemblance between sub-categories.

Existing methods can be divided into several categories, due to the difference in their perspectives. The first category is dedicated to learning fine-grained

© Springer Nature Switzerland AG 2021
H. Ma et al. (Eds.): PRCV 2021, LNCS 13019, pp. 92–104, 2021.
https://doi.org/10.1007/978-3-030-88004-0_8

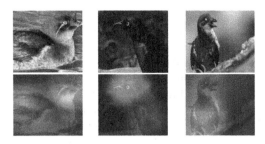

Fig. 1. Examples of three visually similar bird species. (a) is crested auklet, (b) is Parakeet auklet and (c) is least auklet. For each category, there are discriminative parts for the distinction, which are highlighted in the second row.

feature representation. Among these methods, one stream aims to obtain higher-order information. A two-branch network equipped with bilinear pooling is introduced in [10] to obtain second-order information from patches, unlike just utilizing first-order [20]. Follow this direction, high-order information is modelled in [3] via polynomial kernel formulation, and in [19] by cross-layer pooling. Nevertheless, it is not enough for mining deeper information among different regions of an image. Regarding this, another stream targets region-level representation learning. A destruction and construction method is introduced in [2] to learn comprehensive features of random different regions of a single image. The CDL [15] is proposed to exploit the discriminative potentials of correlations for different regions globally and locally. However, these methods mainly leverage information from a single image to enhance feature learning, thus the distinction between representations is limited, which is one of the most critical and difficult points to discriminate between different categories. This is called **"Representation Discrimination Difficulty"**. Besides, they ignored locating the class-discriminative parts in objects among images.

Another category of methods focuses on localizing discriminative parts in the object, which is crucial for the classification, as illustrated in Fig. 1. Early methods achieve this by utilizing part annotations for more visual details [9,20]. Despite the effectiveness, such annotations are costly. Hence, an increasing number of weakly supervised approaches have been designed. In [17], a self-supervision mechanism is introduced to effectively localize informative regions. Besides, by collecting peaks from class response maps, informative receptive fields and parts are estimated in [4], leading to better classification performance. Despite the effectiveness of these methods, they still could not utilize the information from multiple images. More importantly, the localization of parts is still not accurate enough, e.g., sometimes even deviate from the main object. Thus how to accurately locate these parts poses a difficulty, which is **"Discriminative Parts Difficulty"**. Although some recent methods use pairwise images to obtain contrastive clues to tackle this task [5,6,21], the two main difficulties still remain unsolved.

In this paper, different from the existing methods, we argue to view this task from a whole new perspective: contrastive learning, and propose a novel Attentive Contrast Learning Network (ACLN), to tackle these two difficulties. The original core idea of contrastive learning is to bring the representation of different views of the same image closer ('positive pairs'), and spread representations of views from different images ('negative pairs') apart [1,7,13,16]. We redefine the concept of 'pairs' in this task. The attention maps of each image, which is generated by a designed attention generator, and the original CNN feature of this image, are formed as the 'positive pair'. On the other hand, the attention maps among different images from different categories, are formed as the 'negative pairs'. On this basis, a novel contrastive learning module, as well as two contrastive losses, are proposed. It is divided into two blocks. The first block is the **positive learning block**, to deal with the "Discriminative Parts Difficulty". As known, the original CNN feature contains abundant information, and this information is guaranteed by strong supervision, i.e., the class label. But the attention maps, which are supposed to mine deeper information about discriminative parts, have no supervision. Thus by attracting the 'positive pair', the representation of attention maps could take more advantage from which of the original feature, mining more useful and meaningful information of distinctive parts and preventing over-divergent learning. The other block is the **negative learning block**, which tackles the "Representation Discrimination Difficulty". Images of different sub-categories suffer from similar features, and they are hard to discriminate. By repulsing these large number 'negative pairs', the network could be benefited by acquiring more discriminative feature representations. Note that all of the attention maps are stored in attentive feature memory. Besides, a novel synergic learning module is introduced to collaboratively utilize the original feature and the corresponding attention maps and fuse the messages, to achieve a more comprehensive recognition result.

Our major contributions are summarized as follows: (1) We propose an attentive contrast learning architecture to tackle the "Representation Discrimination Difficulty" and the "Discriminative Parts Difficulty", which fully take advantage of the abundant information within and across images. (2) We propose a synergic learning module to enhance the discernment ability by fusion of original feature information and attention information. (3) We conduct experiments on four challenging benchmark datasets: CUB-200-2011, Stanford Dogs, Flowers-102 and Oxford-IIIT Pets, and achieve state-of-the-art performance on all of them, which demonstrate the superiority of the proposed method.

2   Method

In this section, we present our Attentive Contrast Learning Network (ACLN), as shown in Fig. 2. Given an image I, it's first fed into the CNN backbone and generates the original feature F. The following attention generator takes F as input and obtains multiple attention maps M, in each of which discriminative parts and meaningful information are contained. The generator is regularized

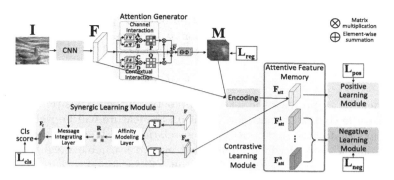

Fig. 2. An overview of the proposed ACLN: it consists of three modules: Attention Generator, Synergic Learning Module, and Contrastive Learning Module.

by an extra loss function L_{reg}. Then, F, M are fed to the contrastive learning module and encoded to obtain the attentive feature representation F_{att}. The positive learning block of this module takes F_{att} and outputs a positive learning score, reflecting the compact affinity between the information learned from the generator and the original feature. In parallel, the attentive feature of other N images from different categories, F_{att_1} to F_{att_n}, which are stored in attentive feature memory, are fed into the negative learning part and mixed with the original attentive feature F_{att}. Multiple negative learning scores will be given, denoting the spread degree of discriminative information learned from different categories. After that, the synergic learning module takes F and F_{att}, obtains the final recognition results.

2.1   Attention Generator

Overview. As illustrated in Fig. 2, the attention generator takes $F \in \mathbb{R}^{H \times W \times C}$ as input and outputs multiple attention maps $M \in \mathbb{R}^{L \times H \times W \times C}$, indicating different discriminative parts. Two blocks are designed, the channel interaction block and the contextual interaction block. They both take F as input and output an attention map P and Q respectively, to take advantage of both channel and contextual information. On this basis, an extra regularization loss L_{reg} is proposed for the diversity between different attention maps.

Specific Details. For the channel interaction block, F will first go through two learning functions ψ and Ψ to generate $A \in \mathbb{R}^{C \times HW}$ and $B \in \mathbb{R}^{HW \times C}$, then an activation function ρ respectively for further channel representation learning, and produce a product $P \in \mathbb{R}^{C \times C}$, which is the channel attention map derived in Eq. 1.

$$P = [\rho(\psi(F))] * [\rho(\Psi(F))] \tag{1}$$

where "*" means matrix multiplication. In P, each element P_{ij} represents the relationship between the i_{th} channel and j_{th} channel. So that important channels, reflecting to corresponding parts [18] and their relationships will be highlighted.

Meanwhile, for the contextual interaction block, F passes through two functions ϕ and ψ and an activation function ϑ to obtain $C \in \mathbb{R}^{HW \times C}$ and $D \in \mathbb{R}^{C \times HW}$, then outputs another matrix product $Q \in \mathbb{R}^{HW \times HW}$, as shown in Eq. 2:

$$Q = [\vartheta(\phi(F))] * [\vartheta(\varphi(F))] \tag{2}$$

Each element Q_{ij} represents the relationship between the i_{th} position and the j_{th} position, so the contextual correlation within the feature will be well modelled, which means that those parts with high correlation will be emphasized.

After that, P and Q will be multiplied by two learnable weights α and β, then multiply with F respectively. The result will be added to F to generate $\overline{F} \in \mathbb{R}^{H \times W \times C}$, as illustrated in Eq. 3:

$$\overline{F} = F + \alpha \cdot (F * P) + \beta \cdot (Q * F) \tag{3}$$

where "$\cdot$" means element-wise multiplication and "$*$" means matrix multiplication.

The contextually and channel-wisely interacted feature $\overline{F}$ successfully integrates the inter-dependencies between different parts semantically from both channel and contextual perspectives, resulting in a strengthened representation. Finally, $\overline{F}$ passes through an attention learner, including a learning function Θ and an activation function Φ. The learned information and knowledge are reflected attention maps $M \in \mathbb{R}^{L \times W \times C}$, representing different part visual patterns and emphasizing semantically valuable ones. L denotes the number of attention maps.

At the same time, an extra loss function L_{reg} is designed to regularize the position of the activation locations in different attention maps. We aim to encourage a diverse distribution of high-response activation parts between different attention maps from the same image, thus avoid compact and repeated attention parts. Hence, the coordinates of the maximum element in each attention map is calculated first, denoted as $m_i(x, y)$. Then a constraint between each two of them is conducted, illustrated as Eq. 4:

$$L_{reg} = -\frac{1}{L} \sum_{i=1}^{L-1} \sum_{j=i+1}^{L} max(0, \mid m_i(x, y) - m_j(x, y) \mid -mrg) \tag{4}$$

where mrg is a margin which controls the distance between different coordinates.

2.2   Contrastive Learning Module

Overview. Illustrated as Fig. 3, the contrastive learning will construct positive pairs, which are points with the same color, e.g., green points or red points; and negative samples, which are yellow points. Positive pairs aim to minimize the representation distance between them, for a more compact and continuous feature representation. Negative samples will encourage maximizing the distance between each other, supporting the decision boundary and pulling samples from different classes apart. That will ease the over-fitting problem as the feature distribution will be more diverged and more distinctive representations between different images are obtained. This module takes F and M of different images, and output a positive learning score and multiple negative scores by two blocks, the positive learning block and the negative learning block respectively, and supervised by two losses.

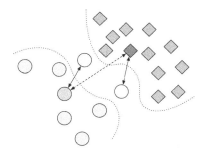

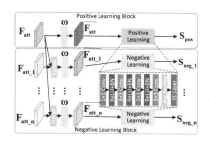

Fig. 3. Contrastive learning on attentive feature (Color figure online)

Fig. 4. The contrastive learning module.

Specific Details. First, a non-linear encoder g maps F and M, which are the 'positive pair', to an aggregated representation $F_{att} = g(F, M)$. Specifically, F will be masked by M and fed to the backbone CNN to generate attentive feature $F_{att} \in \mathbb{R}^{L \times H \times W \times C}$, which is the input of the positive learning block.

Shown in Fig. 4, F_{att} is applied by an average calculation through L and then passed through a learning layer ω, obtaining $\overline{F_{att}} \in \mathbb{R}^{H \times W \times C}$. F_{att} aggregates messages from original feature and feature of discriminative parts, as illustrated in Eq. 5:

$$\overline{F}_{att} = \omega(F_{att}) = \omega(\frac{1}{L} \sum_{i=1}^{L} M_i) \tag{5}$$

$\overline{F_{att}}$ is fed to positive learning block. It consists of four convolution layers, with a ReLU layer and a batch normalization layer after each convolution layer. Then, it goes through a fully connected layer to get the positive learning score S_{pos}, indicating the compact degree. S_c is expected be as close to pos_degree as possible, which will be supervised by L_{con}, in Eq. 6:

$$L_{con} = -log(1 + (S_{pos} - pos\_degree)) \tag{6}$$

Through such extra supervision, while maintaining discriminative and meaningful, the information provided by attention maps are a constraint to not deviate too much from the original feature, since it will confuse the final classification.

The other module is the negative learning block. As illustrated in Fig. 4, it takes F_{att}, F_{att_1} to F_{att_n} as input and obtain multiple negative learning scores S_{neg_1} to S_{neg_n}.

Each F_{att_n} will first fused with F_{att}. A hyper-parameter ρ is defined to selectively replace $\lfloor \rho L \rfloor$ channels from F_{att} with the counterpart in F_{att_n}. More specifically, the selected $\lfloor \rho L \rfloor$ indexes are randomly generated, indicating the exact channels from F_{att} being replaced by the corresponding index of channels from F_{att_n}. Then the result will go through ω and obtains $\overline{F}_{att\_n}$. Each F_{att}^n undergoes negative learning sub-block and obtain the negative learning scores S_{neg_1} to S_{neg_n}, indicating the spread degree. Each S_{neg_i} should be as close to neg_degree as possible, which is supervised L_{neg}, in Eq. 7:

$$L_{neg} = -\frac{1}{n} \sum_{i=1}^{n} (log(1 - (S_{neg\_i} - neg\_degree))) \tag{7}$$

2.3   Synergic Learning Module

Overview. As illustrated in Fig. 2, this module utilizes original feature $F \in \mathbb{R}^{H \times W \times C}$ and the attentive feature $F_{att} \in \mathbb{R}^{H \times W \times C}$ for the same image collaboratively to learn a better feature distribution and predict the category.

Specific Details. F and F_{att} will go through two learnable functions ζ and ξ respectively, and then fed to affinity modeling layer, which aims to learn the correlation between these two features. This layer outputs $R \in \mathbb{R}^{HW \times HW}$, each element of which reflects the affinity between these two features, illustrated as Eq. 8:

$$R_{jk} = (\zeta(F))_j \cdot (\xi(F_{att}))_k^T \tag{8}$$

Then F, F_{att} and R will be the input of message integrating layer. First, the concatenation of F and F_{att} go through a learnable function δ to learn features together, output $F_{concat} \in \mathbb{R}^{H \times W \times C}$.

$$F_{concat} = \delta(concat(F, F_{att})) \tag{9}$$

After that, the affinity from R is deeply merged to F_{concat} to obtain the final feature F_f, illustrated as Eq. 10:

$$F_f = F_{concat} + \gamma \cdot (R * F_{concat}) \tag{10}$$

where "*" indicates matrix multiplication, "·" indicates element-wise multiplication, γ is a learnable weighting factor. F_{att} will then flattened and fed into two fully connected layers for classification, supervised by cross-entropy loss L_{cls}.

2.4   Learning Attentive Contrast Learning Networks

The final loss is computed as Eq. 11:

$$L = L_{cls} + \lambda_1 L_{pos} + \lambda_2 L_{neg} + \lambda_3 L_{reg} \tag{11}$$

where λ_1, λ_2, λ_3 are hyper-parameters to balance the loss.

For network parameters, θ_{ag} denotes Attention Generator, θ_{sl} represents Synergic Learning module, θ_c denotes the Classifier and θ_{cl} denotes the Contrastive Learning Module.

3   Experiments

3.1   Datasets

We evaluate our approach on four challenging benchmark datasets, including Stanford Dogs [8] consists of 20580 images from 120 categories, CUB-200-2011 [14] with 11788 images from 200 classes, Flowers-102 [11] including 8209 images from 102 classes and Oxford-IIIT Pets [12] including 7049 images from 37 classes. We use the official training set and test set for evaluation and do not utilize any extra annotations (e.g., part annotations or object bounding boxes).

Algorithm 1: Learning Attentive Contrast Learning Network

Result: Network Parameters θ_{ag}, θ_{sl} and θ_c
Initialize Attentive Memory $\mathcal{M}$, θ_{ag}, θ_{sl}, θ_c and θ_{cl};
for *epochs* **do**
    foreach *sample in dataset* **do**
        Update θ_{ag}, θ_c, θ_{cl} and θ_{sl} according to $L_{cls} + \lambda_3 L_{reg}$;
        Put current attentive feature F_{att} to $\mathcal{M}$;
        Update θ_{cl}, θ_{ag} according to $\lambda_1 L_{pos}$;
        Pick F_{att}^n from $\mathcal{M}$ and mix with F_{att};
        Update θ_{cl}, θ_{ag} according to $\lambda_2 L_{neg}$;
    end
end

3.2   Implementation Details

ResNet-101 is chosen as our backbone. In our experiments, each image is resized to 512×512 and then cropped a 448×448 patches as the input of the network. ρ and ϑ are implemented by ReLU functions, ζ, ξ and Θ are 1×1 convolution layers, ω, ψ, Ψ, Φ and ϕ are all conducted by 3×3 convolution layers. δ is the combination of two 3×3 conv layers, with a ReLU layer between them. The number of the attentive feature memory is set to 8 on each class. *mrg*, *pos_degree* and *neg_degree* are set to 2, 0.9, 0.1, respectively. Both random rotation and random horizontal flip are applied for data augmentation. λ_1, λ_2 and λ_3 are all set to 0.1 in our implementation. The batch size is set to 128 and we trained the network with the standard SGD optimizer with momentum 0.9 and weight decay of $5e-4$ for all datasets. The learning rate is initialized to 0.001 and is scheduled with the cosine annealing strategy. The network is trained for 100 epochs to converge on the Standford Dogs dataset and 200 epochs on the other datasets. The training process takes 48 h on 16 NVIDIA V-100 GPUs.

3.3   Ablation Study

Impact of each component. The ablation study is conducted on two datasets, the Flowers-102 dataset, and the Oxford-IIIT Pets dataset, to illustrate the effectiveness of all the components in the network. The results are reported in Table 1, where "AG" denotes Attention Generator, "CL" stands for Contrastive Learning Module and "SL" represents Synergic Learning Module.

First, we extract features from the whole image through Resnet101 without any object or partial annotation for fine-grained classification and set it as the baseline. The accuracy is 94.2%. Then the attention generator is adopted, represented as "ResNet101 + AG". Note that since the synergic learning module has not been adopted, F and M are just simply concatenated and fed to two fully connected layers for the final classification. The classification accuracy exceeds by 1.2%, yielding 95.4%. After that, when we integrate a contrastive learning module, denoted as "ResNet101 + AG + CL", the performance achieves 97.4%, further improves by 2.0%. Then, the synergic learning module is adopted, represented as "ResNet101 + AG + SL", yielding 97.1% accuracy. It can be seen that "CL" improves the performance more than "SL". Finally, the full model is used, represented as "ResNet101 + AG + CL + SL ", achieving an accuracy of 98.0%. The whole network improves performance over the baseline by 3.8%. Thus, the discrimination ability of the network, indicating that discriminative representation and meaningful parts both within and between different images are successfully learned.

Table 1. Ablation study on Flowers-102 dataset

Method	Accuracy
ResNet101	94.2
ResNet101 + AG	95.4
ResNet101 + AG + CL	97.4
ResNet101 + AG + SL	97.1
ResNet101 + AG + CL + SL	98.0

Table 2. Ablation study on Oxford-IIIT pets dataset

Method	Accuracy
ResNet101	93.1
ResNet101 + AG	94.2
ResNet101 + AG + CL	96.0
ResNet101 + AG + SL	95.7
ResNet101 + AG + CL + SL	96.7

The evaluation result on the Oxford-IIIT Pets dataset is similar, which is shown in Table 2. The attention generator, denoted as "AG", brings 1.1% performance improvement to the baseline. On this basis, the contrastive learning module and synergic learning module further improves the accuracy by 1.8% and 1.5%. Finally, the full model yields an accuracy of 96.7%.

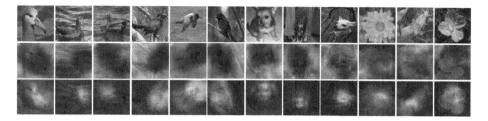

Fig. 5. The visualization results of CUB-100-2011, Stanford Dogs and Oxford Flowers Dataset. From top to below: the original image, the visualization of feature for baseline model, the visualization of feature for the proposed method. Comparing to the baseline, the discriminative regions between multiple images can be accurately discovered.

3.4   Comparison with Other Methods

We compare the proposed ACLN with several state-of-the-art methods on four public benchmark datasets.

Table 3. Comparison results on CUB-200-2011 and stanford dogs.

Method	Backbone	ACC (CUB)	ACC (Dogs)
RA-CNN	VGG19	85.3	87.3
MAMC	ResNet50	86.3	85.2
PC	ResNet50	86.9	83.8
DFL-CNN	ResNet50	87.4	–
MaxEnt	ResNet50	86.5	83.6
TASN	ResNet50	87.9	–
DCL	ResNet50	87.8	–
ACNet	ResNet50	88.1	–
DF-GMM	ResNet50	88.8	–
Subtle-CNN	ResNet50	88.6	87.7
GCL	ResNet50	88.3	–
MGE-CNN	ResNet101	89.4	–
S3N	ResNet101	88.5	–
CIN	ResNet101	88.1	–
NTS-Net	DenseNet161	87.5	–
FDL	DenseNet161	89.09	85.0
Our method	ResNet101	**89.8**	**89.7**
Our method	DenseNet161	**90.7**	**90.9**

Table 4. Comparison results on Oxford-IIIT pets dataset.

Method	Backbone	ACC
Deep standard	AlexNet	78.5
MsML	CNN	80.5
MsML+	CNN	81.2
NAC	AlexNet	85.2
Deep Optimized	AlexNet	88.1
ONE+SVM	VGGNet	90.3
FOAF	VGGNet	91.4
NAC	VGGNet	91.6
TL Atten	VGGNet	92.5
InterActive	VGGNet	93.5
OPAM	VGGNet	93.8
Our method	ResNet101	**96.0**
Our method	DenseNet161	**96.7**

Table 5. Comparison results on Flowers-102 dataset.

Method	Backbone	ACC
Det+Seg	–	80.7
Overfeat	–	86.8
PC	DenseNet161	91.4
NAC-CNN	VGG19	95.3
MGE-CNN	ResNet101	95.9
Our method	ResNet101	**97.5**
Our method	DenseNet161	**98.0**

CUB-200-2011. The classification performance of the CUB-200-2011 dataset is shown in Table 3. The accuracy of the proposed method is better than all the existing approaches. Choosing ResNet101 as our backbone, the proposed method achieves an accuracy of 89.8%. Besides, ACLN yields the best accuracy 90.7% when using DenseNet161 as the backbone, exceeding the MGE-CNN by 1.3%.

Stanford Dogs. Table 3 presents the classification results on Stanford Dogs Dataset. the proposed ACLN achieves the best performance on both backbones, with an accuracy of 90.9% with the backbone DenseNet161, and 89.7%with the backbone of ResNet101. The proposed ACLN exceeds the second-highest result Subtle-CNN by 3.2%.

Oxford Flowers-102. Table 5 reports the results on Oxford Flowers-102 Dataset. We yield an accuracy of 98.0% utilizing ResNet101 as the backbone, which exceeds the second place MGE-CNN by 2.1%.

Oxford-IIIT Pets. To verify the generalization ability of the proposed ACLN, we evaluate the performance on Oxford-IIIT Pets Dataset. The results are illustrated in Table 4, which is consistent with the former three datasets. the proposed ACLN has higher accuracy than all the existing methods, with an accuracy of 96.7%, which boosts OPAM by 2.9%. Using ResNet101 as the backbone, the proposed method also achieves the best accuracy of 96.0%.

3.5 Visualization Results

We qualitatively illustrate the effectiveness of the proposed ACLN in Fig. 5. Each Figure contains three rows, including the original image, the visualization of features for the baseline model, and the visualization of features for the proposed method. The six leftmost examples show the visualization on CUB-200-2011 dataset. Note that the head, the beak and the colour of the feathers are all very important to distinguish birds. It can be seen that compared to the baseline, the activations near the above regions become stronger through contrastive learning, indicating that the network has successfully noticed the key parts. The visualization results of Stanford Dogs and Oxford Flowers are presented in the other examples. For dogs, the head, eyes, and torso are significant regions for recognition, no matter the colour or the texture. And for flowers, the petals and the stamens are crucial. the proposed ACLN can effectively discover and distinguish discriminative regions between multiple images.

4 Conclusion

In this paper, we introduce a novel Attentive Contrast Learning Network (ACLN), containing an attention generator, a contrastive learning module and a synergic learning module. The attention generator automatically generates attention maps, localizing discriminative parts within an image. The contrastive learning module helps learn better feature representations and localize discriminative parts in objects. Besides, the synergic learning module utilizes both the original feature and the attention maps, learning synergically and obtain the final results. Extensive experiments on four public benchmark datasets demonstrate the superiority of our ACLN, comparing to other state-of-the-art approaches.

References

1. Chen, T., Kornblith, S., Norouzi, M., Hinton, G.: A simple framework for contrastive learning of visual representations. arXiv preprint arXiv:2002.05709 (2020)
2. Chen, Y., Bai, Y., Zhang, W., Mei, T.: Destruction and construction learning for fine-grained image recognition. In: Proceedings of the IEEE Conference on Computer Vision and Pattern Recognition, pp. 5157–5166 (2019)
3. Cui, Y., Zhou, F., Wang, J., Liu, X., Lin, Y., Belongie, S.: Kernel pooling for convolutional neural networks. In: Proceedings of the IEEE Conference on Computer Vision and Pattern Recognition, pp. 2921–2930 (2017)
4. Ding, Y., Zhou, Y., Zhu, Y., Ye, Q., Jiao, J.: Selective sparse sampling for fine-grained image recognition. In: Proceedings of the IEEE International Conference on Computer Vision, pp. 6599–6608 (2019)
5. Dubey, A., Gupta, O., Guo, P., Raskar, R., Farrell, R., Naik, N.: Pairwise confusion for fine-grained visual classification. In: Proceedings of the European Conference on Computer Vision (ECCV), pp. 70–86 (2018)
6. Gao, Y., Han, X., Wang, X., Huang, W., Scott, M.: Channel interaction networks for fine-grained image categorization. In: AAAI, pp. 10818–10825 (2020)
7. He, K., Fan, H., Wu, Y., Xie, S., Girshick, R.: Momentum contrast for unsupervised visual representation learning. In: Proceedings of the IEEE/CVF Conference on Computer Vision and Pattern Recognition, pp. 9729–9738 (2020)
8. Khosla, A., Jayadevaprakash, N., Yao, B., Li, F.F.: Novel dataset for fine-grained image categorization: stanford dogs. In: Proceedings of the CVPR Workshop on Fine-Grained Visual Categorization (FGVC), vol. 2 (2011)
9. Lin, D., Shen, X., Lu, C., Jia, J.: Deep lac: Deep localization, alignment and classification for fine-grained recognition. In: Proceedings of the IEEE Conference on Computer Vision and Pattern Recognition, pp. 1666–1674 (2015)
10. Lin, T.Y., RoyChowdhury, A., Maji, S.: Bilinear CNN models for fine-grained visual recognition. In: Proceedings of the IEEE International Conference on Computer Vision, pp. 1449–1457 (2015)

11. Nilsback, M.E., Zisserman, A.: Automated flower classification over a large number of classes. In: 2008 Sixth Indian Conference on Computer Vision, Graphics and Image Processing, pp. 722–729. IEEE (2008)
12. Parkhi, O.M., Vedaldi, A., Zisserman, A., Jawahar, C.: Cats and dogs. In: 2012 IEEE Conference on Computer Vision and Pattern Recognition, pp. 3498–3505. IEEE (2012)
13. Tian, Y., Krishnan, D., Isola, P.: Contrastive multiview coding. arXiv preprint arXiv:1906.05849 (2019)
14. Wah, C., Branson, S., Welinder, P., Perona, P., Belongie, S.: The caltech-ucsd birds-200-2011 dataset (2011)
15. Wang, Z., Wang, S., Zhang, P., Li, H., Zhong, W., Li, J.: Weakly supervised fine-grained image classification via correlation-guided discriminative learning. In: Proceedings of the 27th ACM International Conference on Multimedia, pp. 1851–1860 (2019)
16. Wu, Z., Xiong, Y., Yu, S.X., Lin, D.: Unsupervised feature learning via non-parametric instance discrimination. In: Proceedings of the IEEE Conference on Computer Vision and Pattern Recognition, pp. 3733–3742 (2018)
17. Yang, Z., Luo, T., Wang, D., Hu, Z., Gao, J., Wang, L.: Learning to navigate for fine-grained classification. In: Proceedings of the European Conference on Computer Vision (ECCV), pp. 420–435 (2018)
18. Yosinski, J., Clune, J., Nguyen, A., Fuchs, T., Lipson, H.: Understanding neural networks through deep visualization. arXiv preprint arXiv:1506.06579 (2015)
19. Yu, C., Zhao, X., Zheng, Q., Zhang, P., You, X.: Hierarchical bilinear pooling for fine-grained visual recognition. In: Proceedings of the European Conference on Computer Vision (ECCV), pp. 574–589 (2018)
20. Zhang, N., Donahue, J., Girshick, R., Darrell, T.: Part-based R-CNNs for fine-grained category detection. In: European Conference on Computer Vision, pp. 834–849. Springer (2014)
21. Zhuang, P., Wang, Y., Qiao, Y.: Learning attentive pairwise interaction for fine-grained classification. In: AAAI, pp. 13130–13137 (2020)

Relation-Based Knowledge Distillation
for Anomaly Detection

Hekai Cheng, Lu Yang$^{(\boxtimes)}$, and Zulong Liu

School of Automation Engineering, University of Electronic Science and Technology of China
(UESTC), Chengdu 611731, Sichuan, People's Republic of China
yanglu@uestc.edu.cn

Abstract. Anomaly detection is a binary classification task, which is to judge
whether the input image contains an anomaly or not and the difficulty is that only
normal samples are available at training. Due to this unsupervised nature, the
classic supervised classification methods will fail. Knowledge distillation-based
methods for unsupervised anomaly detection have recently drawn attention as it
has shown the outstanding performance. In this paper, we present a novel knowledge distillation-based approach for anomaly detection (RKDAD). We propose
to use the "distillation" of the "FSP matrix" from adjacent layers of a teacher
network, pre-trained on ImageNet, into a student network which has the same
structure as the teacher network to solve the anomaly detection problem, we show
that the "FSP matrix" are more discriminative features for normal and abnormal
samples than other traditional features like the latent vectors in autoencoders. The
"FSP matrix" is defined as the inner product between features from two layers
and we detect anomalies using the discrepancy between teacher's and student's
corresponding "FSP matrix". To the best of our knowledge, it is the first work to
use the relation-based knowledge distillation framework to solve the unsupervised
anomaly detection task. We show that our method can achieve competitive results
compared to the state-of-the-art methods on MNIST, F-MNIST and surpass the
state-of-the-art results on the object images in MVTecAD.

Keywords: Anomaly detection · Knowledge distillation · "FSP matrix"

1 Introduction

Anomaly detection (AD) is a binary classification problem and it has been approached
in a one-class learning setting, i.e., the task of AD is to identity abnormal samples during
testing while only normal samples are available at training. It has been an increasingly
important and demanding task in many domains of computer vision, like in the field of
visual industrial inspection [1, 2], the anomalies are rare events so it is usually required
that we only train machine learning models on normal product images and detect abnormal images during inference. Moreover, anomaly detection is widely used in health
monitoring [3], video surveillance [4] and other fields.

In recent years, a lot of studies have been done to improve the performance of anomaly
detection [5–13] in images. Among them, some methods especially the methods based

This research was supported by NSFC (No. 61871074).

H. Ma et al. (Eds.): PRCV 2021, LNCS 13019, pp. 105–116, 2021.
https://doi.org/10.1007/978-3-030-88004-0_9

on deep learning have achieved great success. Most methods [5–7] model the normal data abstraction by extracting discriminative latent features, which are used to determine whether the input is normal or abnormal. Some others [8] detect anomalies using per-pixel reconstruction errors or by evaluating the density obtained from the model's probability distribution. Some recent studies [5, 6] have shown that the knowledge distillation framework is effective for anomaly detection tasks. The main idea is to use a student network to distill knowledge from a pre-trained expert teacher network, i.e., to make the feature maps of certain layers of the two networks as equal as possible for the same input at training and anomalies are detected when the feature maps of certain layers of the two networks are very different at testing.

In this paper, we present a novel knowledge distillation-based approach (RKDAD) for anomaly detection. We use the "FSP matrix" as the distilled knowledge instead of the direct activation values of critical layers, the "FSP matrix" is defined as the inner product between features from two layers, i.e., the Gram matrix between two feature maps. By minimizing the L2 distance between the teacher's and the student's "FSP matrix" at training, the student network can learn the flow process of the normal sample features in the teacher network. We detect anomalies using the discrepancy between teacher's and student's corresponding "FSP matrix" at testing.

2   Related Work

Many previous studies have explored the anomaly detection tasks for images. In this section, we will provide a brief overview of the related works on anomaly detection tasks for images. We mainly introduce methods based on Convolutional Autoencoders (CAE), Generative Adversarial Networks (GAN) and methods based on knowledge distillation (KD).

2.1   CAE-Based Methods

The methods based on AE mainly use the idea that by learning the latent features of the normal samples, the reconstruction of abnormal inputs is not as precise as the normal ones, which results in larger reconstruction errors for abnormal samples, i.e., training on the normal data, the AE is expected to produce higher reconstruction error for the abnormal inputs than the normal ones and the reconstruction error can be defined as the L2 distance between the input and the reconstructed image. Abati et al. [7] proposed LSA [7] to train an autoregressive model at its latent space to better learn the normal latent features. MemAE [8] proposed the memory module to force the reconstructed image to look like the normal image, this mechanism increases the reconstruction error of abnormal images. CAVGA [9] proposed an attention-guided convolution adversarial variational autoencoder, which combines VAE with GAN to learn normal attention and abnormal attention in an end-to-end manner.

2.2   GAN-Based Methods

GAN-based methods attempt to find a specific latent feature space by training on normal samples. Then during testing, anomalies are detected based on the reconstruction or the feature error. AnoGAN [10] is the first research to use GAN for anomaly detection. Its main idea is to let the generator learn the distribution of normal images through adversarial training. When testing, the L1 distance between the generated image and the input image and the feature error will be combined to detect anomalies. GANomaly [11] proposed the Encoder1-Decoder-Encoder2 architecture with a discriminator. What is used to detect anomalies is not the difference between the input image and the reconstructed image but the difference between the features of the two encoders. Skip-GANomaly [12] improves the generator part and uses the U-net [13] architecture that is with stronger reconstruction capabilities.

2.3   KD-Based Methods

Recently, KD-based methods for anomaly detection have drawn attention as it has shown the outstanding performance. Bergmann et al. [5] proposed Uniformed Students [5] which is the first anomaly detection method based on knowledge distillation. In this method, several student networks are trained to regress the output of a descriptive teacher network that was pretrained on a large dataset. Anomalies are detected when the outputs of the student networks differ from that of the teacher network, and the intrinsic uncertainty in the student networks is used as an additional scoring function that indicates anomalies. Salehi et al. [6] proposed to use the "distillation" of features at various layers of the pre-trained teacher network into a simpler student network to tackle anomaly detection problem. The Uniformed Students solely utilizes the last layer activation values in distillation, the second method mentioned above shows that considering multiple layers' activation values leads to better exploiting the teacher network's knowledge and more distinctive discrepancy. However, the methods mentioned above only consider the direct activation values as the knowledge of distillation without considering the relations between layers, which is more representative of the essential characteristics of the normal samples.

3   Method

In this section, we will first introduce the details of the gram matrix and show how to use the Gram matrix to define the "FSP matrix". Then, we will introduce our approach to use the "FSP matrix" from two adjacent layers as the "distilled knowledge" to solve unsupervised anomaly detection tasks.

3.1   Gram Matrix and the "FSP Matrix"

As Eq. (1) shows, the matrix composed of the inner product of any k vectors in n-dimensional Euclidean space is defined as the Gram matrix of the k vectors. Obviously, Gram matrix of k vectors is a symmetric matrix. Gram matrix is often used in style transfer tasks, specifically, the feature map of the content image in a certain layer will be flattened into a one-dimensional feature vector according to the channel, and then a gram matrix composed of C vectors can be obtained, C is the number of channels in the feature map. Use the same operation to calculate the Gram matrix of the style image, then minimize the distance of the Gram matrix of the two images. The Gram matrix is used to measure the difference in the style of two images. If the distance between the Gram matrix of the feature vectors of the two images is very small, it can be determined that the styles of the two images are similar. Essentially, the Gram matrix can be regarded as the eccentric covariance matrix between feature vectors, i.e., the diagonal elements reflect the information of the different feature vectors themselves, that is, the intensity information of the feature vectors, and the off-diagonal elements provide correlation information between different feature vectors.

$$\Delta(\alpha_1, \alpha_2, \ldots, \alpha_k) = \begin{pmatrix} (\alpha_1, \alpha_1)(\alpha_1, \alpha_2) \ldots (\alpha_1, \alpha_k) \\ (\alpha_2, \alpha_1)(\alpha_2, \alpha_2) \ldots (\alpha_2, \alpha_k) \\ \cdots\cdots\cdots\cdots \\ (\alpha_k, \alpha_1)(\alpha_k, \alpha_2) \ldots (\alpha_k, \alpha_k) \end{pmatrix} \tag{1}$$

The "FSP matrix" [14] proposed by Yim [14] et al. As Fig. 1 shows, the FSP matrix is generated by the features from two layers instead of being generated by the feature map of a single layer like the standard Gram matrix. By computing the inner product which represents the direction, to generate the FSP matrix, the flow between two layers can be represented by the FSP matrix.

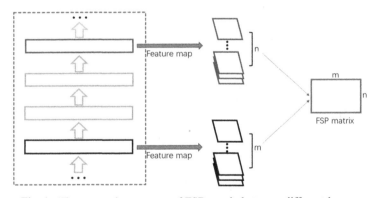

Fig. 1. The generation process of FSP matrix between different layers.

The calculation process of the FSP matrix is shown in Eq. (2), where F^1 and F^2 are two feature maps from different layers, h and w are the height and width of the feature map respectively, i and j are the channel indexes of the two feature maps, x and W are the input image and network parameters. It can be seen from Eq. (2) that the premise of calculating the FSP matrix is that the height and width of the two feature maps are equal.

$$FSP_{i,j}(x; W) = \sum_{s=1}^{h} \sum_{t=1}^{w} \frac{F_{s,t,i}^1(x; W) \times F_{s,t,j}^2(x; W)}{h \times w} \tag{2}$$

By letting the student network learn the flow of normal samples' knowledge in the teacher network during training, if the input is normal sample during testing, the flow of the teacher and the student network will be similar, while for the abnormal input, the flow will be very different.

3.2 The Proposed Approach

Given a training dataset $D_{train} = \{x_1, \ldots, x_n\}$ containing only normal images, we train a student network with the help of a pre-trained teacher network, and the teacher network remains the same throughout the training process. Given a test dataset D_{test}, we utilize the discrepancy of the "FSP matrix" between the teacher and the student network to detect anomalies during the test, therefore, the student network must be trained to mimic the behavior of the teacher network, i.e., the student network should learn the "FSP matrix" of the teacher network during training. Earlier KD-based works for anomaly detection such as [5], which strive to teach just the activation values of the final layer of the teacher to the student, and in [6] they encourage the student to learn the teacher's knowledge on normal samples through conforming its intermediate representations in a number of critical layers to the teacher's representations. Because the feature maps of different layers of the neural network correspond to different levels of abstraction, mimicking different layers leads to a more thorough understanding of normal data than using only the final layer. In the methods mentioned above, the knowledge taught by the teacher to the student is the direct activation values of the critical layers, considering that a real teacher teaches a student the flow for how to solve a problem, in [14], Yim et al. proposed to define high-level distilled knowledge as the flow for solving a problem. Because the Gram matrix can be generated by computing the inner product of feature vectors, it contains the directionality between features, the flow can be represented by using Gram matrix consisting of the inner products between features from two layers. As Fig. 1 shows, the Gram matrix across layers is defined as the "FSP matrix", i.e., the "relational knowledge" in our approach RKDAD, which is the higher-level knowledge for the normal images than the activation values of critical layers. We encourage the student to learn this higher-level knowledge of the teacher when training with normal image samples, during the test, if the input is an abnormal image the "FSP matrix" of the teacher network and the student network will be very different. Compared with the activation values of critical layers, the "FSP matrix" is more difficult to learn but it is also more discriminative for normal and abnormal samples.

In what follows, we refer to the "FSP matrix" between the i-th and the $(i + 1)$-th layers in the networks as $F_{i,i+1}$, and the "FSP matrix" between the two adjacent layers of the teacher network as $F^t_{i,i+1}$ and the student's one as $F^s_{i,i+1}$. The feature maps of the i-th layer and the $(i + 1)$-th layer should have the same resolution. The lost function of our approach can be defined as Eq. (3)

$$L_{FSP} = \frac{1}{N} \sum_{i=1}^{N} \lambda_i \times \left|\left|F^t_{i,i+1} - F^s_{i,i+1}\right|\right|_2^2 \tag{3}$$

where N represents the number of all convolutional layers in the network (The convolutional layer refers to a module with a convolution operator, an activation function and an optional pooling operation), and λ_i is used to control the number of "FSP matrix" in the loss function, i.e., when λ_i is equal to 0, the two adjacent convolutional layers starting from the i-th convolutional layer are not used.

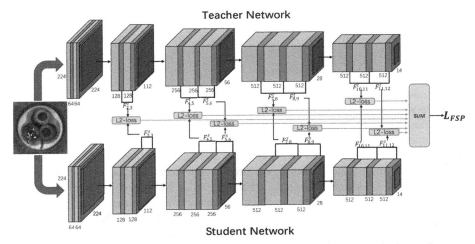

Fig. 2. Complete architecture of our proposed method. The student network shares the same structure with the teacher network pre-trained on a large dataset and learns the "FSP matrix" between two adjacent layers on normal data from the teacher. The discrepancy of their "FSP matrix" is formulated by a loss function and used to detect anomalies test time.

It should be noted that the teacher network and the student network share the same network architecture, While the teacher network should be deep and wide enough to learn all necessary features to perform well on a large-scale dataset, like ImageNet [15], and the teacher network should be pre-trained on a large-scale dataset. The goal of the student is to acquire the teacher's knowledge of "FSP matrix" of the normal data.

Anomaly Detection: To detect anomalous images, each image in D_{test} is fed to both the teacher and the student network, i.e., we need two forward passes for anomaly detection. L_{FSP}, the loss function of RKDAD, is also the final anomaly score. As the student only learned the knowledge of "FSP matrix" of the normal data from the teacher, when the input is an abnormal sample, the "FSP matrix" of the teacher and the student will be very different, the anomaly score can be thresholded for anomaly detection.

4 Experiments

In this section, we have done extensive experiments to verify the effectiveness of our method. We will first introduce the implementation details of our approach, and then we introduce the datasets used. At last, we will show the anomaly detection results on the datasets introduced. Specially, we report an average result sampled every 10 consecutive epochs instead of reporting the maximum achieved result in many methods. The average result is a better measure of a model's performance.

4.1 Implementation Details

VGG [16] has shown outstanding performance on classification tasks. In our approach, we choose the VGG-16 pre-trained on ImageNet as the teacher network, and a randomly initialized VGG-16 as the student network. Of course, there are many other excellent network structures that can be used. However, it is required that the feature map resolution of some adjacent layers in the network structure used is the same, so that the FSP matrix can be calculated. Similar to [6], we avoid using bias terms in the student network. The model architecture of our approach is described in Fig. 2. We add 7 pairs of FSP matrix to the loss function in total, the loss function is also the anomaly score that is ultimately used to detect anomalies.

In all experiments, we use Adam [17] for optimization. The learning rate is set to be 0.001 and the batch size is 64. We only use normal images which are fed to both the teacher and the student network, and the parameter weights of the teacher network remain unchanged while that of the student network are updated during the training process, i.e., there are only forward propagation in the teacher network, and both forward and back propagation in the student network. Because the FSP matrix is more difficult to learn during the training process than the direct activation value of feature maps for the student network, it is required to train many epochs, such as 1000 or more.

Fig. 3. Object (cable, toothbrush, capsule) and texture (wood, grid, leather) images in MVTecAD [2]. The images in the upper row are normal samples and the ones in the lower row are abnormal samples.

4.2 Datasets

We verified the effectiveness of our method on three datasets as follows.

MNIST [18]: a handwritten digit images dataset, which consists of 60k training and 10k test 28 × 28 Gy-scale images, includes numbers 0 to 9.

Fashion-MNIST [19]: a more complex image dataset proposed to replace MNIST, which covers images of 70 k different products in 10 categories, such as T-shirt, dress and coat etc.

MVTecAD [2]: a dataset dedicated to anomaly detection with more than 5 k images, which includes 5 categories of texture images and 10 categories of object images. For each category, the dataset train contains only normal images, and the test sets contain a variety of abnormal images and some normal images. In our experiment, the images will be scaled to the size 128 × 128. Some object and texture samples are shown in Fig. 3.

Note that for MNIST and Fashion-MNIST, we regard one class as normal and others as anomaly at training while at testing the whole test set is used. For MVTecAD, the datasets train and test are used.

4.3 Results

We use the area under the receiver operating characteristic curve (AUROC) for evaluation. The results are shown in Table 1, 2 for MNIST, Fashion-MNIST and Table3, 4, 5 for MVTecAD. Note that the AUROC values in the tables are the average of 10 consecutive epochs, not the maximum value. We have compared our method with many approaches, and the Tables show that our method can achieve competitive results with to state-of-the-art method on the datasets we used.

The Results on MNIST and Fashion-MNIST. As Table 1 and Table 2 show, our method RKDAD can achieve competitive results on mnist and fashion-mnist compared to the state-of-the-art methods with using only the sum of l2 distances of the "FSP matrix" as the loss function and the anomaly score, which verifies the effectiveness of our relation-based knowledge distillation framework for anomaly detection task.

Table 1. AUROC in % on MNIST.

Method	0	1	2	3	4	5	6	7	8	9	Mean
ARAE [20]	99.8	99.9	96.0	97.2	97.0	97.4	99.5	96.9	92.4	98.5	97.5
OCSVM [21]	99.5	99.9	92.6	93.6	96.7	95.5	98.7	96.6	90.3	96.2	96.0
AnoGAN [10]	96.6	99.2	85.0	88.7	89.4	88.3	94.7	93.5	84.9	92.4	91.3
DSVDD [22]	98.0	99.7	91.7	91.9	94.9	88.5	98.3	94.6	93.9	96.5	94.8
CapsNetpp [23]	99.8	99.0	98.4	97.6	93.5	97.0	94.2	98.7	99.3	99.0	97.7
OCGAN [24]	99.8	99.9	94.2	96.3	97.5	98.0	99.1	98.1	93.9	98.1	97.5
LSA [7]	99.3	99.9	95.9	96.6	95.6	96.4	99.4	98.0	95.3	98.1	97.5
CAVGA [9]	99.4	99.7	98.9	98.3	97.7	96.8	98.8	98.6	98.8	99.1	98.6
U-Std [5]	99.9	99.9	99.0	99.3	99.2	99.3	99.7	99.5	98.6	99.1	**99.35**
Mul-KD [6]	99.8	99.8	97.8	98.7	98.4	98.2	99.4	98.4	98.4	98.1	98.71
OURS	99.91	99.96	98.74	99.41	98.32	98.18	99.12	98.97	97.99	98.34	98.89

The Results on MVTecAD. Note that Table 3 shows the AUROC results of 10 categories of object images in MVTecAD, and it can be seen that the performance of the proposed approach RKDAD can surpass other methods to obtain the state-of-the-art results. Table 4 shows the results of 5 categories of texture images and Table 5 shows the average performance across all categories in MVTecAD. It can be seen from Table 4 that our method is not very good for the texture images but it still exceeds the performance of many methods. Table 5 shows that the performance of our method on MVTecAD is second only to the state-of-the-art method, and is significantly better than other methods.

Table 2. AUROC in % on Fashion-MNIST.

Method	0	1	2	3	4	5	6	7	8	9	Mean
ARAE [20]	93.7	99.1	91.1	94.4	92.3	91.4	83.6	98.9	93.9	97.9	93.6
OCSVM [21]	91.9	99.0	89.4	94.2	90.7	91.8	83.3	98.8	90.3	98.2	92.8
DAGMM [25]	30.3	31.1	47.5	48.1	49.9	41.3	42.0	37.4	51.8	37.8	41.7
DSEBM [26]	89.1	56.0	86.1	90.3	88.4	85.9	78.2	98.1	86.5	96.7	85.5
DSVDD [22]	98.2	90.3	90.7	94.2	89.4	91.8	83.4	98.8	91.9	99.0	92.8
LSA [7]	91.6	98.3	87.8	92.3	89.7	90.7	84.1	97.7	91.0	98.4	92.2
Mul-KD [6]	92.5	99.2	92.5	93.8	92.9	98.2	84.9	99.0	94.3	97.5	**94.5**
OURS	90.72	99.35	90.79	94.91	91.84	94.78	85.81	99.16	90.57	96.14	93.41

All experimental results show that although the model architecture of RKDAD is very simple, it still achieves excellent results, which proves that the proposed relation-based knowledge distillation framework for anomaly detection has great potential.

Table 3. AUROC in % on object images in MVTecAD. We surpass the SOTA method.

Method	Bottle	Hazelnut	Capsule	Metal Nut	Pill	Cable	Transistor	Toothbrush	Screw	Zipper	Mean
AVID [27]	88	86	85	63	86	64	58	73	66	84	75.3
AESSIM [28]	88	54	61	54	60	61	52	74	51	80	63.5
AEL2 [28]	80	88	62	73	62	56	71	98	69	80	73.9
AnoGAN [10]	69	50	58	50	62	53	67	57	35	59	56.0
LSA [7]	86	80	71	67	85	61	50	89	75	88	75.2
CAVGA [9]	89	84	83	67	88	63	73	91	77	87	80.2
DSVDD [22]	86	71	69	75	77	71	65	70	64	74	72.2
VAE-grad [29]	86	74	86	78	80	56	70	89	71	67	75.7

(continued)

Table 3. (*continued*)

Method	Bottle	Hazelnut	Capsule	Metal Nut	Pill	Cable	Transistor	Toothbrush	Screw	Zipper	Mean
GT [30]	74.29	33.32	67.79	82.37	65.16	84.70	79.79	94.00	44.58	87.44	71.34
Mul-KD [6]	99.39	98.37	80.46	73.58	82.70	89.19	85.55	92.17	83.31	93.24	87.80
OURS	99.05	96.82	73.08	82.79	78.70	92.71	87.75	88.33	94.67	89.97	**88.39**

Table 4. AUROC in % on texture images in MVTecAD.

Method	Leather	Wood	Carpet	Tile	Grid	Mean
AVID [27]	58	83	73	66	59	67.8
AESSIM [28]	46	83	67	52	69	63.4
AEL2 [28]	44	74	50	77	78	64.6
AnoGAN [10]	52	68	49	51	51	54.2
LSA [7]	70	75	74	70	54	68.6
CAVGA [9]	71	85	73	70	75	74.8
DSVDD [22]	73	87	54	81	59	70.8
VAE-grad [29]	71	89	67	81	83	78.2
GT [30]	82.51	48.24	45.90	53.86	61.91	58.49
Mul-KD [6]	73.58	94.29	79.25	91.57	78.01	**83.34**
OURS	60.90	92.37	77.71	78.79	66.42	75.24

Table 5. Mean AUROC in % on MVTecAD.

Method	AVID	AESSIM	AEL2	AnoGAN	LSA	CAVGA	DSVDD	VAE-grad	GT	Mul-KD	OURS
Mean	73	63	71	55	73	78	72	77	67.06	**87.87**	84.00

5   Conclusion

In this paper, we have presented a novel knowledge distillation-based approach for anomaly detection (RKDAD). We have further explored the possibility of anomaly detection using the "relational knowledge" between different layers when the knowledge of normal samples flows in the network, and we show that using the "distillation" of the "FSP matrix" from adjacent layers of a teacher network, pre-trained on ImageNet, into a student network which has the same structure as the teacher network, and then using the discrepancy between teacher's and student's corresponding "FSP matrix" at testing can achieve competitive results compared to the state-of-the-art methods. We have verified the effectiveness of our method on many datasets. In this paper, we only consider the

"FSP matrix" from adjacent layers as the "relational knowledge", more forms of "relational knowledge" can be explored to improve the performance of anomaly detection task in the future.

References

1. Shuang, M., Yudan, W., Guojun, W.: Automatic fabric defect detection with a multi-scale convolutional denoising autoencoder network model. Sensors **18**(4), 1064 (2018)
2. Bergmann P., Fauser M., Sattlegger D., et al.: MVTec AD — a Comprehensive Real-World dataset for unsupervised anomaly detection. In: 2019 IEEE Conference on Computer Vision and Pattern Recognition (CVPR), pp. 9592–9600 (2019)
3. Zhe L., et al.: Thoracic disease identification and localization with limited supervision. In: Proceedings of the IEEE Conference on Computer Vision and Pattern Recognition, pp. 8290–8299 (2018)
4. Zhou, J.T., Du, J., Zhu, H., et al.: AnomalyNet: an anomaly detection network for video surveillance. IEEE Trans. Inf. Forensics Secur. **14**(10), 2537–2550 (2019)
5. Bergmann P., Fauser M., Sattlegger D., et al.: Uninformed students: student-teacher anomaly detection with discriminative latent embeddings. In: IEEE Conference on Computer Vision and Pattern Recognition (CVPR), pp. 4182–4191 (2020)
6. Salehi M., Sadjadi N., Baselizadeh S., et al.: Multiresolution knowledge distillation for anomaly detection. arXiv preprint arXiv: 2011.11108 (2020)
7. Abati D., Porrello A., Calderara S., Cucchiara R.: Latent space autoregression for novelty detection. In: Proceedings of the IEEE Conference on Computer Vision and Pattern Recognition, pp. 481–490 (2019)
8. Gong D., Liu L., Le V., et al.: Memorizing normality to detect anomaly: memory-augmented deep autoencoder for unsupervised anomaly detection. In: 2019 IEEE/CVF International Conference on Computer Vision (ICCV), pp. 1705–1714 (2019)
9. Venkataramanan, S., Peng, K.-C., Singh, R.V., Mahalanobis, A.: Attention guided anomaly localization in images. In: Vedaldi, A., Bischof, H., Brox, T., Frahm, J.-M. (eds.) ECCV 2020. LNCS, vol. 12362, pp. 485–503. Springer, Cham (2020). https://doi.org/10.1007/978-3-030-58520-4_29
10. Schlegl, T., Seeböck, P., Waldstein, S.M., Schmidt-Erfurth, U., Langs, G.: Unsupervised anomaly detection with generative adversarial networks to guide marker discovery. In: Niethammer, M., et al. (eds.) IPMI 2017. LNCS, vol. 10265, pp. 146–157. Springer, Cham (2017). https://doi.org/10.1007/978-3-319-59050-9_12
11. Akcay, S., Atapour-Abarghouei, A., Breckon, T.P.: GANomaly: semi-supervised anomaly detection via adversarial training. In: Jawahar, C.V., Li, H., Mori, G., Schindler, K. (eds.) ACCV 2018. LNCS, vol. 11363, pp. 622–637. Springer, Cham (2019). https://doi.org/10.1007/978-3-030-20893-6_39
12. Akay, S., Atapour-Abarghouei, A., Breckon, T.P.: Skip-GANomaly: skip connected and adversarially trained encoder-decoder anomaly detection. In: IEEE International Joint Conference on Neural Networks (IJCNN), pp. 1–8 (2019)
13. Ronneberger, O., Fischer, P., Brox, T.: U-Net: convolutional networks for biomedical image segmentation. In: Navab, N., Hornegger, J., Wells, W.M., Frangi, A.F. (eds.) MICCAI 2015. LNCS, vol. 9351, pp. 234–241. Springer, Cham (2015). https://doi.org/10.1007/978-3-319-24574-4_28
14. Yim, J., Joo, D., Bae, J., et al.: A Gift from knowledge distillation: fast optimization, network minimization and transfer learning. In: 2017 IEEE Conference on Computer Vision and Pattern Recognition (CVPR), pp. 7130–7138 (2017)

15. Jia, D., Wei, D., Richard, S., Li-Jia, L., Kai, L., Fei-Fei L.: Imagenet: a large-scale hierarchical imagedatabase. In: 2009 IEEE Conference on Computer Vision and Pattern Recognition, pp. 248–255 (2009)
16. Simonyan, K., Zisserman, A.: Very Deep Convolutional Networks For Large-Scale Image Recognition. arXiv preprint arXiv:1409.1556 (2014)
17. Diederik, P.K., Jimmy, B.: Adam: a method forstochastic optimization. arXiv preprint arXiv: 1412.6980 (2014)
18. LeCun, Y., Cortes, C., et al. http://yann.lecun.com/exdb/mnist. Accessed 05 April 2021
19. Xiao, H., Rasul, K., Vollgraf, R.: Fashion-MNIST: a Novel Image Dataset For Benchmarking Machine Learning Algorithms. arXiv preprint arXiv:1708.07747 (2017)
20. Salehi, M., Arya, A., Pajoum, B., et al.: Arae: Adversarially robust training of autoencoders improves novelty detection. arXiv preprint arXiv:2003.05669 (2020)
21. Chen, Y., Xiang, S.Z., Huang, T.S.: One-class svm for learning in image retrieval. In: Proceedings 2001 International Conference on Image Processing, pp. 34–37 (2001)
22. Ruff, L., Vandermeulen, R.A., et al.: Deep one-class classification. In: International Conference on Machine Learning, pp. 4393–4402 (2018)
23. Li, X., Kiringa, I., Yeap, T., et al.: Exploring deep anomaly detection methods based on capsule net. In: ICML 2019 Workshop on Uncertainty and Robustness in Deep Learning, pp. 375–387 (2020)
24. Perera, P., Nallapati, R., Bing, X.: Ocgan: one-class novelty detection using gans with constrained latent representations. In: Proceedings of the IEEE Conference on Computer Vision and Pattern Recognition, pp. 2898–2906 (2019)
25. Zong, B., Song, Q., Martin Renqiang, M., et al.: Deep autoencoding gaussian mixture model for unsupervised anomaly detection. In: International Conference on Learning Representations (2018)
26. Shuangfei, Z., Yu, C., Weining, L., Zhongfei, Z.: Deep structured energy-based models for anomaly detection. In: International Conference on Machine Learning, pp. 1100–1109 (2016)
27. Sabokrou, M., Pourreza, M., Fayyaz, M., et al.: Avid: adversarial visual irregularity detection. In: Asian Conference on Computer Vision, pp. 488–505 (2018)
28. Bergmann, P., Lwe, S., Fauser, M., et al.: Improving unsupervised defect segmentation by applying structural similarity to autoencoders. In: International Joint Conference on Computer Vision, Imaging and Computer Graphics Theory and Applications (VISIGRAPP), pp. 372–180 (2019)
29. Dehaene, D., Frigo, O., Combrexelle, S., et al.: Iterative energy-based projection on a normal data manifold for anomaly localization. In: International Conference on Learning Representations (2020)
30. Golan, I., El-Yaniv, R.: Deep anomaly detection using geometric transformations. In: Advances in Neural Information Processing Systems, pp. 9758–9769 (2018)

High Power-Efficient and Performance-Density FPGA Accelerator for CNN-Based Object Detection

Gang Zhang[1], Chaofan Zhang[2]([✉]), Fan Wang[2,3], Fulin Tang[4], Yihong Wu[4], Xuezhi Yang[1], and Yong Liu[2]

[1] Hefei University of Technology, Hefei, Anhui, China
[2] Anhui Institute of Optics and Fine Mechanics, Hefei Institutes of Physical Science, Chinese Academy of Sciences, Hefei, Anhui, China
zcfan@aiofm.ac.cn
[3] University of Science and Technology of China, Hefei, Anhui, China
wangfan8@mail.ustc.edu.cn
[4] The National Laboratory of Pattern Recognition, Institute of Automation, Chinese Academy of Sciences, Beijing, China
{fulin.tang,yhwu}@nlpr.ia.ac.cn

Abstract. The Field Programmable Gate Array (FPGA) accelerator for CNN-based object detection has been attracting widespread attention in computer vision. For most existing FPGA accelerators, the inference accuracy and speed are affected negatively by the low power-efficient and performance-density. To address this problem, we propose a software and hardware co-designed FPGA accelerator for accurate and fast object detection with high power-efficient and performance-density. To develop the FPGA accelerator on CPU+FPGA heterogeneous platforms, a resource sensitive and energy aware FPGA accelerator framework is designed. In hardware, a hardware sensitive neural network quantization called Dynamic Fixed-point Data Quantization (DFDQ) is proposed to improve the power-efficient. In software, an algorithm-level convolution (CONV) optimization scheme is further proposed to improve the performance-density by paralleling block execution of CONV cores. To validate the proposed FPGA accelerator, a Zynq FPGA is used to build the acceleration platform of You Only Look Once (YOLO) network. Results demonstrate that the proposed FPGA accelerator outperforms the state-of-the-art methods in power-efficient and performance-density. Besides, the speed of object detection is increased by at most 16.5 times along with less than 1.5% accuracy degradation.

This work was supported by the Science and Technology Major Program of Anhui Province of China under Grants 202003a05020020, Joint fund of Science & Technology Department of Liaoning Province and State Key Laboratory of Robotics, China under Grant 2020-KF-22-16, Special Foundation of President of the Hefei Institutes of Physical Science under Grant YZJJ2020QN36, Anhui Provincial Key R&D Program under Grant 202104a05020043, the University Synergy Innovation Program of Anhui Province under Grant GXXT-2019-003.

H. Ma et al. (Eds.): PRCV 2021, LNCS 13019, pp. 117–128, 2021.
https://doi.org/10.1007/978-3-030-88004-0_10

Keywords: Object detection · Field Programmable Gate Array · Convolution optimization · Dynamic Fixed-Point Data Quantization

1 Introduction

Convolutional neural network (CNN) based Object detection has been widely used in diverse scenarios, such as robotics, intelligent traffic and drone [1–4]. Usually, CNN-based object detection is performed on Graphics Processing Unit (GPU) due to the vast computation cost of CNN. In order to apply it to lightweight mobile platforms, Field Programmable Gate Array (FPGA), a reconfigurable and low-power computing platform, is suitable. Therefore, FPGA based object detection has been a significant research topic [5–11].

However, there are still some challenges in FPGA based object detection. On one hand, as described in [12–14], CNN based object detection is extremely time-consuming. On the other hand, FPGA has limited on-chip resources [15,16]. These challenges make it hard to achieve high power-efficient and performance-density FPGA accelerator, further leading to poor speed and accuracy of object detection.

To address these problems, we propose a high power-efficient and performance-density FPGA accelerator to boost the performance of CNN-based object detection in speed while maintaining the accuracy. On CPU+FPGA heterogeneous platforms, we design a resource sensitive and energy aware FPGA accelerator framework. In hardware, we propose a Dynamic Fixed-point Data Quantization (DFDQ) method to quantify the convolutional parameters of YOLOv2 algorithm, which improves the power-efficient of FPGA accelerator. In software, we optimize the convolution (CONV) computation of FPGA accelerator to improve the performance-density, where a CONV optimization scheme is proposed by paralleling block execution of CONV cores.

The contributions are summarized as follows:

1) A resource sensitive and energy aware FPGA accelerator framework is developed on the most prospective heterogeneous framework (CPU+FPGA).
2) A Dynamic Fixed-point Data Quantization (DFDQ) method is proposed to improve the power-efficient of the designed FPGA accelerator, by sparse CNN's weights and bias.
3) A novel CONV optimization algorithm in accelerating the YOLOv2 algorithm is proposed to improve the performance-density of FPGA accelerator, by the parallel block CONV operations.

The rest of this letter is organized as follows. The related work is introduced in Sect. 2. The detail of the proposed FPGA accelerator is presented in Sect. 3. In Sect. 4, we provide the experimental results. Finally, in Sect. 5, we describe conclusions and future work.

2   Related Work

There are lots of works to improve the performance of FPGA accelerator, which can be divided into two categories: reducing the computation of CNN [17–19] and improving the utilization of limited on-chip resources [20–22].

For reducing the amount of computation, Nakahara et al. [17] proposed the lightweight YOLO-v2, by using a hybrid low-precision CNN. However, the ignored external memory access takes additional power consumption and latency. Nguyen et al. [18] used binary weights to represent the entire network and store them in on-chip memory, but this work is prone to waste resources since only 9.7% DSP resources are used. In [19], a heterogeneous weight quantization method was developed, by incorporating ADMM with FFT/IFFT. However, the FFT/ IFFT results in large redundant transform operations and memory requirements.

Considering the constraints on FPGA resources, Tiny-Yolo-v2 was presented in [20] to reduce the hardware resource consumption. However, it is hard to achieve more performance optimization based on specific on-chip resources. In [21], a "Multi-CLPs" approach that dynamically utilizes resource was presented to increase resource utilization, but it causes higher memory bandwidth and more power consumption. Li et al. [22] used dynamic parallelism to divide resources to achieve high resource utilization. This method requires the FPGA accelerator to have many computing resources.

In summary, most recent works in solving the problems of limited resources and huge computation amount achieve good results but with some unsatisfactory results like high power consumption and under-utilized resources. Different from these methods, our method considers power-efficient and performance-density at the same time.

3   Method

3.1   System Framework

To develop the FPGA accelerator, we design a resource sensitive and energy aware FPGA accelerator framework on the CPU+FPGA heterogeneous platforms. As shown in Fig. 1, the framework is split into Image Pre-processor, Neural Network Accelerator and Image Post-processor. In the Image Pre-processor, we mainly get the resized image by image size transformation algorithm on the CPU side. The Neural Network Accelerator is the core of the framework responsible for hardware acceleration. On the CPU side, the operating system is responsible for scheduling the operations of the FPGA, allowing FPGA accelerators and CPU to be tightly coupled through on-chip buses. The external DRAM stores the weight parameters, images and instructions. On the FPGA side, the main purpose is to accelerate the network and get an array containing detection information. Finally, we get the detection result including bounding box and predicted probability by the Image Post-processor on the CPU side.

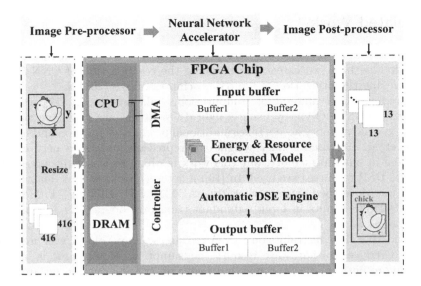

Fig. 1. The proposed FPGA accelerator framework consists of three parts: Image Pre-processor for data preparation, FPGA Accelerator for YOLOv2 network acceleration and Image Post-processor for detection results.

3.2   Neural Network Accelerator

The acceleration process of the neural network is carried out on the Neural Network Accelerator, in which the most important part is the energy and resource concerned model. We obtain the model by optimizing the original neural network with the proposed DFDQ method and CONV optimization. Besides, to reduce the transmission delay, the double buffering mechanism [23,24] operated in a ping-pong mode is applied in the input and output parts of each layer.

1) *DFDQ method*
The weights and bias after training are generally stored in 32-bit floating-point data (32-BFPD) types. Previous researches [25–28] have shown that shorter fixed-point data will reduce the total power consumption and improve power-efficient. However, shorter data representation will generate truncation error and cause certain accuracy lost in the inference results. Thus, it is of the utmost importance to strike a balance between data bit width and inference accuracy to achieve the best system performance. To convert 32-BFPD into fixed-point ones while achieving the highest accuracy, the DFDQ method is proposed to choose the best balance and improve the power-efficient.

The fixed-point number can be expressed:

$$N_{fixed} = \sum_{i=0}^{bw-1} B_i \times 2^{-Q} \times 2^i, \tag{1}$$

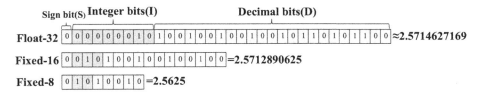

Fig. 2. 32-BFPD and fixed-point data. Suitable shorter data representation maintains data accuracy and reduces data redundancy.

where bw is the bit width, Q is the fractional length that can be negative and B_i is the binary number at i-th bit. The goal of the DFDQ is to find the optimal Q in a layer:

$$Q_d = \arg\min_Q \sum |N_{float} - N(bw, Q)| \qquad (2)$$

N_{float} denotes the original floating-point data after training in a layer. $N(bw, Q)$ is the fixed-point number transformed by (1). We dynamically adjust the Q value in each CONV layer to get the optimal Q_d by minimizing the truncation error of the DFDQ.

Further, as shown in Fig. 2, we analyze the quantization error theoretically because the detection accuracy is negatively correlated with the quantization error. The 32-BFPD consists of three parts: sign bit (S,1 bit), integer bits (I, 8 bits) and decimal bits (D, 23 bits). The error of the 16-bit fixed-point data type containing 1 sign bit, 3 integer bits, and 12 decimal bits is only 0.00017 compared with the 32-BFPD. However, the error is about 0.009 when using 8-bit data representation, nearly 51.6 times of 16 bits data representation.

2) *Convolutional Optimization*

To improve the performance-density, we take consideration into the algorithm-level convolution computational optimization of the CONV layer. The FPGA accelerator performance-density is derived from the Roofline Model [12] under giving a certain hardware resource condition, so we have:

$$Performance\ Bound = \frac{Total\ Number\ Of\ Operations}{Execution\ Cycles}$$

$$= \frac{f_{CONV}^{Time} + f_{Pooling}^{Time} + f_{others}^{Time}}{\frac{M}{T_m} \times \frac{N}{T_n} \times \frac{R}{T_r} \times \frac{C}{T_c} \times (T_r \times T_c \times K \times K + p)}$$

$$\approx \frac{f_{CONV}^{Time}}{\frac{M}{T_m} \times \frac{N}{T_n} \times R \times C \times K \times K}$$

$$= \frac{g(R \times C \times K^2 \times C_{in} \times C_{out})}{\frac{M}{T_m} \times \frac{N}{T_n} \times R \times C \times K \times K}$$

$$0 < T_m \times T_n \times (DSP_{Mul} + DSP_{Add}) < (\#of DSP), \qquad (3)$$

where Tm/Tn is the parallel degree of the M/N dimension of the output/input feature maps and f_{CONV}^{Time} and $f_{Pooling}^{Time}$ are the time complexities of convolutional and pooling layer, respectively. The time complexity of a certain layer in the CNN can be evaluated by the number of multiply-add operations. C_{in} denotes the number of input channels. C_{out} is the number of convolution kernels in the convolution layer.

Since convolution operations account for more than 90% of the operations in CNN [26, 28, 29], f_{CONV}^{Time} is much larger than the sum of $f_{Pooling}^{Time}$ and f_{others}^{Time}. Since only two-dimensional expansion of Tm and Tn, $Tr = Tc = 1$. Therefore, for a specific network, the parameters are constant except that T_m, T_n are variable. From (3), we get that the performance bottleneck is only determined by T_m and T_n, which are strictly restricted by on-chip DSP resources.

Through the above analysis, we focus on the optimization of the CONV layer and consider the actual hardware resources. We propose a novel convolution optimization algorithm by parallel block operation on on-chip resources to improve the performance-density of FPGA accelerator. Specifically, we use the following two methods.

a) *Loop Unrolling*
Loop Unrolling utilizes the parallelism between CONV kernels to perform parallel processing of multiple CONV operations. In this letter, the two dimensions M and N are partially expanded to perform parallel pipeline multiplication and addition computation in the CONV. Each layer has N input feature maps. After loop unrolling, for each time, the pixel blocks (Tn) at the same position are read from Tn independent input feature maps and the corresponding Tn weights are also read. Therefore, the Tn input feature maps require N/Tn times to read and calculate. In order to avoid waste of resources, N/Tn is an integer. $Tm \times Tn$ parallel multiplication units multiplex Tn input pixel blocks to perform multiplication operations, and Tm addition trees write the product addition result into the output buffer. As shown in the Fig. 3, the use of loop unrolling realizes the local parallel structure, which increases the execution speed and simultaneously improves the performance of each computing resource, that is, the performance-density [20].

b) *Loop Tiling*
Loop Tiling utilizes the locality of data calculated by convolution, that is, partitioning the entire data into multiple smaller blocks that can be accommodated in on-chip buffers. The blocks designed in this letter is shown in Fig. 3. The $T_{row} \times T_{cow}$ pixel blocks and corresponding $T_m \times T_n \times K^2$ weights of T_n input feature maps are read from the DRAM, then the feature map pixel blocks and weights parameters are reused on the chip. Further, the intermediate results are kept in the on-chip cache to reduce external memory access, reducing latency and improving performance. Only after the final output pixel blocks are obtained, the $T_r \times T_c$ pixel blocks of T_m output feature maps are written out.

After DFDQ and convolutional optimization methods, we get the energy and resource model.

Input Feature Maps   **Output Feature Maps**

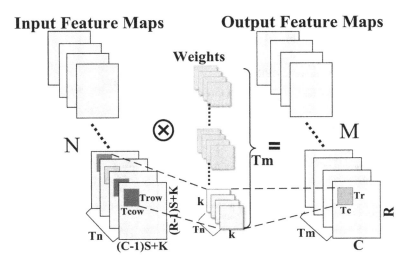

Fig. 3. Parallel block calculation after loop unrolling and loop tiling.

4   Experiments

In this section, we evaluate the proposed FPGA accelerator on a Xilinx PYNQ-z2 board, which contains a 630 KB block RAM, an external 512 MB DDR3, 220 DSP slices, and an ARM dual-core Cortex-A9 processor. We first choose the most suitable bit width through DFDQ experiments, then we compare our FPGA accelerator with the CPU, GPU implementations and recent FPGA accelerators on performance-density and power-efficient. The operating system is Ubuntu 16.04 and the deep learning software framework is Caffe [30] with the COCO dataset [31].

We get the FPGA accelerator performance and detection results through the design space exploration flow as shown in Fig. 4. Firstly, we get the energy and resource concerned model by using proposed DFDQ and CONV optimization methods. Then, the model automatically compiles the designed CONV code, collects and analyzes the hardware resource utilization extracted from the compilation report. Next, the flow generates the related peripheral circuits of the entire FPGA accelerator and outputs hardware design files. The above two steps will execute corresponding simulations to choose the best solutions. Finally, the hardware design bitstream file (.BIT) and the design instruction file (.TCL) are passed to the Linux OS layer to complete inference output using python language.

A) *DFDQ Results*
Through the analysis of the Sect.3, we conduct 16/8/4 bits DFDQ experiments on the weights and bias for each convolutional layer of YOLOv2 algorithm to make the best balance between bit width and accuracy. As shown in Fig. 5, the quantization errors of 16/8/4 bits DFDQ experiments are below 0.001, average

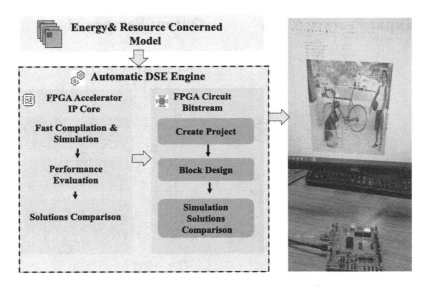

Fig. 4. FPGA accelerator design space exploration flow and a demo of the object detection on the PYNQ-z2 FPGA board.

above 0.01 and 0.1 to 1, respectively. The detection accuracy, which is negatively correlated with quantization errors, is 48.1% when using 32-BFPN. When applying 16-bit quantization strategy, there is only 1.2% acceptable accuracy lost close to the GPU's implementation. However, when adopting 8/4-bit quantization, the accuracy drops sharply. Therefore, we propose dynamic 16 bits precision, the best bit precision of DFDQ, as the FPGA-accelerated network parameter type.

B) *FPGA Accelerator Performance*
To validate the performance-density and energy-efficient of the proposed FPGA accelerator, we adopt the design flow shown in Fig. 4. The results in Table 1 show the proposed accelerator achieves the best power-efficient and almost the best performance-density. For [35], it is slightly higher than ours in terms of performance-density, since [35] has a slightly higher inference speed (44.97 vs 44.4) due to the simplified YOLOv2 network (our network time complexity is 5.3 times of [35] (29.46 vs 5.58)). What's worth mentioning is that the power-efficient of the proposed FPGA accelerator is much better than CPU (629.8 times), is 2.8 times of GPU, and stands out in all previous FPGA accelerators, which is due to the DFDQ's contribution to "cut" calculation and memory access power consumption.

C) *Object Detection Results*
For the designed FPGA accelerator's object detection accuracy, we have discussed in previous experiments that its accuracy can reach almost the same level as the GPU. So, we mainly discuss inference speed in this section. In Table 1,

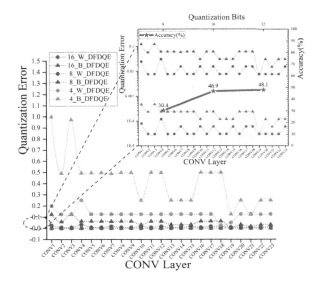

Fig. 5. 16/8/4 bits fixed-point quantization experiments on the weight and bias for each convolutional layer of YOLOv2 and the final detection accuracy in the upper right corner. "DFDQE" means the Dynamic Fixed-point Data Quantization Error. "B","W" denote Bias and weights, respectively.

Table 1. Comparison of the proposed design with previous works.

Experiment	CPU	GPU	[32]	[26]	[33]	[34]	[35]	Ours
Precision (Bit)	32	32	16	16	16	32	16	16
DSP Used (Slice)	N/A	N/A	N/A	780	68	800	153	155
Power (W)	22.6	100.6	8	9.63	2.1	1.17	2.7	0.63
Performance	2.54	2517.6	23.18	136.97	15.8	18.82	44.97	44.4
(GOP/S)	(1x)	(991.2x)	(9.13x)	(54x)	(6.2x)	(7.4x)	(17.7x)	(17.5x)
Performance-Density	N/A	N/A	N/A	0.176	0.232	0.024	0.294	**0.287**
(GOP/S/Slice)				(7.3x)	(9.7x)	(1x)	(12.3x)	(12x)
Power-Efficient	0.112	25.03	2.90	14.22	7.53	16.09	16.66	**70.54**
(GOP/S/W)	(1x)	(223.5x)	(25.9x)	(127x)	(67.2x)	(143.7x)	(148.8x)	(629.8x)

the proposed FPGA accelerator basically reaches the best performance. For [26], the design has achieved 136.97GOP/S performance, but its DSP consumption is 5 times higher than our implementation, that is, trading "area" for speed, which is not friendly for resource-constrained systems. For [35] using a simplified YOLOv2 network for inferring, we can achieve almost the same detection speed due to the CONV optimization strategies to speed up the calculation.

5   Conclusion

This letter proposes a high power-efficient and performance-density FPGA accelerator for CNN-based object detection on CPU+FPGA heterogeneous platform. The results demonstrate that the proposed FPGA accelerator achieves impressive power-efficient and performance-density, and the object detection results outperform current FPGA schemes. As a future work, we plan to explore the possibility of high accuracy and higher throughput FPGA accelerator for real-time object detection in mobile embedded platforms.

References

1. Guan, W., Wang, T., Qi, J., Zhang, L., Lu, H.: Edge-aware convolution neural network based salient object detection. IEEE Signal Process. Lett. **26**(1), 114–118 (2018)
2. Siagian, C., Itti, L.: Rapid biologically-inspired scene classification using features shared with visual attention. IEEE Trans. Pattern Anal. Mach. Intell. **29**(2), 300–312 (2007)
3. Fang, H., et al.: From captions to visual concepts and back. In: 2015 IEEE Conference on Computer Vision and Pattern Recognition (CVPR) (2015)
4. Borji, A., Frintrop, S., Sihite, D.N., Itti, L.: Adaptive object tracking by learning background context. In: Computer Vision & Pattern Recognition Workshops (2012)
5. Huang, H., Liu, Z., Chen, T., Hu, X., Zhang, Q., Xiong, X.: Design space exploration for yolo neural network accelerator. Electronics **9**(11), 1921 (2020)
6. Xu, K., Wang, X., Liu, X., Cao, C., Wang, D.: A dedicated hardware accelerator for real-time acceleration of yolov2. J. Real-Time Image Process. **18**(1), 481–492 (2021)
7. Wang, Z., Xu, K., Wu, S., Liu, L., Wang, D.: Sparse-yolo: Hardware/software co-design of an fpga accelerator for yolov2. IEEE Access **PP**(99), 1–1 (2020)
8. Bourrasset, C., Maggiani, L., Sérot, J., Berry, F.: Dataflow object detection system for fpga-based smart camera. Circuits Dev. Syst. Iet **10**(4), 280–291 (2016)
9. Kyrkou, C., Theocharides, T.: A parallel hardware architecture for real-time object detection with support vector machines. IEEE Trans. Comput. **61**(6), 831–842 (2012)
10. Ma, X., Najjar, W., Roy-Chowdhury, A.: High-throughput fixed-point object detection on fpgas. In: 2014 IEEE 22nd Annual International Symposium on Field-Programmable Custom Computing Machines, pp. 107–107 (2014). https://doi.org/10.1109/FCCM.2014.40
11. Ma, X., Najjar, W.A., Roy-Chowdhury, A.K.: Evaluation and acceleration of high-throughput fixed-point object detection on fpgas. IEEE Trans. Circuits Syst. Video Technol. **25**(6), 1051–1062 (2015)
12. Zhang, C., Li, P., Sun, G., Guan, Y., Xiao, B., Cong, J.: Optimizing fpga-based accelerator design for deep convolutional neural networks. In: Proceedings of the 2015 ACM/SIGDA International Symposium on Field-Programmable Gate Arrays, FPGA 2015, pp. 161–170. Association for Computing Machinery, New York (2015). https://doi.org/10.1145/2684746.2689060

13. Sharma, H., Park, J., Mahajan, D., Amaro, E., Esmaeilzadeh, H.: From high-level deep neural models to fpgas. In: IEEE/ACM International Symposium on Microarchitecture (2016)
14. Alwani, M., Chen, H., Ferdman, M., Milder, P.: Fused-layer cnn accelerators. In: 2016 49th Annual IEEE/ACM International Symposium on Microarchitecture (MICRO), pp. 1–12 (2016). https://doi.org/10.1109/MICRO.2016.7783725
15. Zhu, C., Huang, K., Yang, S., Zhu, Z., Zhang, H., Shen, H.: An efficient hardware accelerator for structured sparse convolutional neural networks on fpgas (2020)
16. Moini, S., Alizadeh, B., Emad, M., Ebrahimpour, R.: A resource-limited hardware accelerator for convolutional neural networks in embedded vision applications. IEEE Trans. Circuits Syst. I Express Briefs **64**, 1217–1221 (2017)
17. Nakahara, H., Yonekawa, H., Fujii, T., Sato, S.: A lightweight yolov2: a binarized cnn with a parallel support vector regression for an fpga. In: The 2018 ACM/SIGDA International Symposium (2018)
18. Nguyen, D.T., Nguyen, T.N., Kim, H., Lee, H.J.: A high-throughput and power-efficient fpga implementation of yolo CNN for object detection. IEEE Trans. Very Large Scale Integr. (VLSI) Syst. **27**, 1–13 (2019)
19. Ding, C., Wang, S., Liu, N., Xu, K., Wang, Y., Liang, Y.: Req-yolo: a resource-aware, efficient quantization framework for object detection on fpgas. In: the 2019 ACM/SIGDA International Symposium (2019)
20. Wai, Y.J., Yussof, Z., Salim, S., Chuan, L.K.: Fixed point implementation of tiny-yolo-v2 using opencl on fpga. Int. J. Adv. Comput. Sci. Appl. **9**(10) (2018)
21. Shen, Y., Ferdman, M., Milder, P.: Maximizing cnn accelerator efficiency through resource partitioning. Comput. Architecture News **45**(2), 535–547 (2017)
22. Li, H., Fan, X., Li, J., Wei, C., Wang, L.: A high performance fpga-based accelerator for large-scale convolutional neural networks. In: 2016 26th International Conference on Field Programmable Logic and Applications (FPL) (2016)
23. Fan, S., Chao, W., Lei, G., Xu, C., Zhou, X.: A high-performance accelerator for large-scale convolutional neural networks. In: 2017 IEEE International Symposium on Parallel and Distributed Processing with Applications and 2017 IEEE International Conference on Ubiquitous Computing and Communications (ISPA/IUCC) (2017)
24. Nguyen, D.T., Kim, H., Lee, H.J., Chang, I.J.: An approximate memory architecture for a reduction of refresh power consumption in deep learning applications. In: 2018 IEEE International Symposium on Circuits and Systems (ISCAS) (2018)
25. Simonyan, K., Zisserman, A.: Very deep convolutional networks for large-scale image recognition. Computer Science (2014)
26. Qiu, J., et al.: Going deeper with embedded fpga platform for convolutional neural network. In: Proceedings of the 2016 ACM/SIGDA International Symposium on Field-Programmable Gate Arrays, FPGA 2016, pp. 26–35. Association for Computing Machinery, New York (2016). https://doi.org/10.1145/2847263.2847265
27. He, K., Zhang, X., Ren, S., Sun, J.: Deep residual learning for image recognition (2015)
28. Lei, S., Zhang, M., Lin, D., Gong, G.: A dynamic multi-precision fixed-point data quantization strategy for convolutional neural network. In: CCF National Conference on Compujter Engineering and Technology (2016)
29. Li, S., Luo, Y., Sun, K., Yadav, N., Choi, K.: A novel fpga accelerator design for real-time and ultra-low power deep convolutional neural networks compared with titan x gpu. IEEE Access **PP**(99), 1 (2020)

30. Jia, Y., Shelhamer, E., Donahue, J., Karayev, S., Long, J., Girshick, R., Guadarrama, S., Darrell, T.: Caffe: Convolutional architecture for fast feature embedding. ACM (2014)
31. Lin, T.Y., Maire, M., Belongie, S., Hays, J., Zitnick, C.L.: Microsoft coco: Common objects in context. Springer International Publishing (2014)
32. Gokhale, V., Jin, J., Dundar, A., Martini, B., Culurciello, E.: A 240 g-ops/s mobile coprocessor for deep neural networks. In: IEEE (2014)
33. Dong, W., Ke, X., Jiang, D.: Pipecnn: An opencl-based open-source fpga accelerator for convolution neural networks. In: 2017 International Conference on Field Programmable Technology (ICFPT) (2017)
34. Zhao, R., Niu, X., Wu, Y., Luk, W., Qiang, L.: Optimizing cnn-based object detection algorithms on embedded fpga platforms. In: International Symposium on Applied Reconfigurable Computing (2017)
35. Bao, C., Xie, T., Feng, W., Chang, L., Yu, C.: A power-efficient optimizing framework fpga accelerator based on winograd for yolo. IEEE Access $\mathbf{PP}$(99), 1 (2020)

Relation-Guided Actor Attention for Group Activity Recognition

Lifang Wu, Qi Wang, Zeyu Li, Ye Xiang$^{(\boxtimes)}$, and Xianglong Lang

Faculty of Information Technology, Beijing University of Technology, Beijing, China
xiangye@bjut.edu.cn

Abstract. Group activity recognition has received significant interest due to its widely practical applications in sports analysis, intelligent surveillance and abnormal behavior detection. In a complex multi-person scenario, only a few key actors participate in the overall group activity and others may bring irrelevant information for recognition. However, most previous approaches model all the actors' actions in the scene equivalently. To this end, we propose a relation-guided actor attention (RGAA) module to learn reinforced feature representations for effective group activity recognition. First, a location-aware relation module (LARM) is designed to explore the relation among pairwise actors' features in which appearance and position information are both considered. We propose to stack all the pairwise relation features and the features themselves of an actor to learn actor attention which determines the importance degree from local and global information. Extensive experiments on two publicly benchmarks demonstrate the effectiveness of our method and the state-of-the-art performance is achieved.

Keywords: Group activity recognition · Attention mechanism · Relation representation · Video analysis

1 Introduction

Group activity recognition aims at figuring out what a group of people is performing in the video. It can be applied to many practical applications such as sports video analysis, video surveillance and social role understanding. This task is very challenging which requires understanding the individual actions and inferring the high-level relationships among individuals in the multi-person scenario. Hence, an effective model needs to exploit contextual information among actors and aggregate the actor-level information.

In the earlier period, a series of methods [1,3] designed the possibility graphical models based on handcrafted features to capture the contextual information. However, these methods rely on prior knowledge and are limited to the expressive power of features. Recently, several deep learning methods [6,14,15,18,19] have been proposed and made great progress for group activity recognition. Most of these methods [6,15] treat different actors in the scene with the same importance

© Springer Nature Switzerland AG 2021
H. Ma et al. (Eds.): PRCV 2021, LNCS 13019, pp. 129–141, 2021.
https://doi.org/10.1007/978-3-030-88004-0_11

and adopt max/average pooling to aggregate actor-level features. Yet, different actors have unequal contributions to overall group activity. In other words, only a few key actors play a vital role and others are irrelevant in the group activity. For instance, for the "left spike" activity in the volleyball game, more attention should be assigned to the spiking actor and blocking actors than the other actors. Therefore, it is significant to discover key actors for group activity recognition. To tackle down this problem, several attention based approaches [11,14,19] have been proposed. However, they essentially belong to the local attention which the attention of an actor is only determined by its own feature without considering the global relation information. Intuitively, to decide whether an actor is important or not, one ought to compare features with the global view. We argue that the global relation structural information is important for inferring attention and it is not dependable enough for these solutions to discover the key actors for group activity recognition.

In this paper, we first design a location-aware relation module (LARM) to compute the compact pairwise relation vector, which is able to represent the relation between two actors. Besides visual information, we further incorporate the spatial location information into the relation vector, letting the relation vector be aware of the actor locations. Then a relation-guided actor attention (RGAA) module is proposed to discover key actors and depress irrelevant actors for group activity automatically. The attention for each actor is determined by its appearance feature and the relation features with all other actors, which can exploit local and global information. Such a solution in which a global scope comparison is conducted is similar to the way humans determine the importance. Besides, we also combine the global relation feature with the original appearance feature to provide contextual information for the actor-level feature. Finally, we adopt adaptive weighted pooling to aggregate actor-level features into the scene-level feature, leading to a robust representation for group activity recognition.

In summary, the main contributions of this paper are as follows: (1) We propose a location-aware relation module to obtain compact relation features between pairwise actors. (2) We design a relation-guided actor attention module which learns attention for each actor by local appearance feature and the global view of relations. (3) We conduct extensive experiments on two widely-used datasets [3,6], and our method is competitive with the state-of-the-art approaches.

2   Related Works

Group Activity Recognition. Group activity recognition has attracted many researchers in the past years. At the early stage, many methods fed traditional features into the structured models [1,3]. With the recent revival of deep learning, various deep learning based methods have sprung up. Ibrahim *et al.* [6] proposed a hierarchical model consisting of two LSTMs to capture temporal dynamics of individual actions and group activity. Wang *et al.* [15] extended this work by another LSTM network to capture the interaction context of inter-group. Later some works proposed to model the relationships between actors. Azar *et al.* [2] explored spatial relations among persons using intermediate representations of CNN and refined group activity predictions iteratively. Wu *et al.* [17] proposed to construct a relation graph by appearance similarity and spatial position of actors. However, these methods treat all actors with equal importance. Yan *et al.* [19] modeled the dynamics of key actors with long motions and flash motions. Tang *et al.* [14] used attention learned from semantics information to guide the learning process in the vision domain. Qi *et al.* [10] designed a spatio-temporal attention model to attend to key persons and frames via the LSTM network. However, these methods essentially belong to the local attention mechanism and lack the global scope information in the attention learning process. In contrast, our actor attention considers the structural information in the global relation representation which is helpful for inferring attention.

Attention-Based Models. Attention mechanisms have been successfully leveraged in the field of vision and language. It aims to select vital features and suppress irrelevant features. Many works learn the attention using local feature [7,11]. To introduce contextual information, Woo *et al.* [16] utilized a filter size of 7×7 to convolve the spatial map to generate an attention map. Song *et al.* [13] designed spatial and temporal attention using the LSTM network to pay different degrees of attention to different joints and frames. Our method is similar to [8,20] in which explore global scope relations for attention learning. Different from these two works [8,20], we exploit the attention model to assign different weights to different actors in a multi-person scene for activity recognition. Moreover, we incorporate spatial position information into relation representations and further refine actor-level representations using global relation features.

3   Method

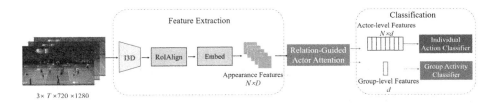

Fig. 1. Overview of our method.

The goal of our work is to model the participation degrees of actors and enhance and aggregate actor-level features for group activity recognition.

Our method consists of three main parts presented in Fig. 1: actor feature extractor, relation-guided actor attention module and classification layer. In the feature extraction stage, given a T frames video sequence we use the I3D as the backbone to extract the spatio-temporal context of the video clip. Then we apply

RoIAlign to extract a feature map for each actor bounding box in the middle frame. After that, a FC layer is adopted to embed the feature maps to the D dimensional actors appearance feature vectors $X = \{x_i \in \mathbb{R}^D\}_{i=1}^N$. Afterward, the location-aware relation module is used to explore global relation features for each actor. Upon the global relation features, the relation-guided actor attention module is exploited to model the importance degrees of actors for group activity and refine the actor-level representations. Finally, two classifiers are applied on the actor-level and group-level features for recognizing individual actions and group activity, respectively.

3.1   Location-Aware Relation Module

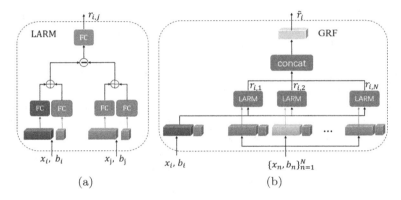

(a)   (b)

Fig. 2. Illustration of (a) the location-aware relation module and (b) the relation-guided actor attention module.

To explore the relationship between two features, we employ element-wise differences which can capture the similarity or difference of each element in the two features. However, the commonly element-wise difference is not compact and includes redundant information when the dimension of the feature is high. Therefore, we extend the element-wise difference operation to compute relations in a low-dimensional embedding space. Moreover, the spatial location of actors is important information to infer relationships between actors in the group activity recognition. Thus, we also consider the position information for the relation feature to make it more effective.

As shown in Fig. 2, the input of the location-aware relation module (LARM) is the pair actor features x_i, x_j and their spatial location $b_i, b_j \in \mathbb{R}^4$ representing the top-left and right-bottom coordinates of each bounding box. We first use two embedding function θ and ϕ to embed x_i, x_j into a subspace: $\theta(x_i) = \sigma(W_\theta x_i)$, $\phi(x_j) = \sigma(W_\phi x_j)$, where θ and ϕ are composed of fully-connected layer and rectified linear units, σ is the ReLU function, transformation function W_θ, $W_\phi \in$

$\mathbb{R}^{\frac{D}{c_1} \times D}$ are used to reduce the dimension of input feature, and c_1 is the dimension reduction factor. We also use a shared embedding function φ to encode spatial location features: $\varphi(b_i) = \sigma(W_\varphi b_i)$, where φ is composed of fully-connected layer $W_\varphi \in \mathbb{R}^{\frac{D}{c_1} \times D}$ and ReLU activation function and the embedding dimensions are the same as θ and ϕ. We used the element-wise sum to fuse the embedded actor features and spatial location features. We attempt the concatenation fusion in the experiment and the results indicate element-wise sum is better and fewer parameters. The difference of pair actor features is computed as follows:

$$x_{diff} = \left(\theta \left(x_i \right) + \varphi \left(b_i \right) \right) - \left(\phi \left(x_j \right) + \varphi \left(b_j \right) \right) \qquad (1)$$

Then we used a fully-connected layer and rectified linear unit to generate final relational vector:

$$r_{i,j} = LARM \left(x_i, x_j, b_i, b_j \right) = ReLU \left(W x_{diff} \right) \qquad (2)$$

where $W \in \mathbb{R}^{\frac{D}{c_2} \times \frac{D}{c_1}}$ and the dimension of relation feature is $\frac{1}{c_2}$ of input feature.

3.2   Relation-Guided Actor Attention Module

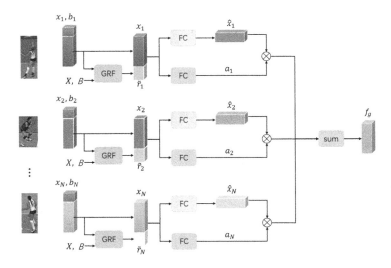

Fig. 3. The framework of the relation-guided actor attention (RGAA) module. The GRF indicates the global relation feature as shown in Fig. 2.

Most of the attention mechanisms used in the current group activity recognition methods are based on local self-attention, without considering the context information between actors and lack of global scope information. To deal with this problem, we proposed a relation-guided actor attention (RGAA) module in which the attention score of each actor is determined by its own feature and the relation vectors between all actors in the scene, simultaneously considering local and global information.

As demonstrated in the Fig. 3, given the N actors appearance features $X = \{x_i \in \mathbb{R}^D\}_{i=1}^N$ and their position coordinates $B = \{b_i \in \mathbb{R}^4\}_{i=1}^N$, we use the RGAA to refine the actor-level features and generate discriminative group-level features. Specifically, for each actor, we use the LARM to compute the relation features between other actors:

$$r_{i,j} = LARM\left(x_i, x_j, b_i, b_j\right), \quad 1 \leq j \leq N \tag{3}$$

All relation features related to actor i are concatenated to generate global relation feature:

$$\tilde{r}_i = concat\left([r_{i,1}, \ r_{i,2}, \ldots, r_{i,N}]\right) \tag{4}$$

where $\tilde{r}_i \in \mathbb{R}^{\frac{ND}{c_2}}$ is the global relation feature for actor i. Compared with original appearance, it contains global structural information which can better guide attention learning. In addition, the individual actions are related to each other in the multi-person scenario. The global relation features can provide contextual information between actors, and we exploit them to update original appearance features. The refined actor-level features and attention score of each actor can be computed as follows:

$$\hat{x}_i = ReLU\left(W_{ind}\left[x_i; \tilde{r}_i\right]\right) \tag{5}$$

$$s_i = tanh\left(W_{att}\left[x_i; \tilde{r}_i\right]\right) \tag{6}$$

where we use tanh activation function to normalize the scores between -1 and 1, $W_{ind} \in \mathbb{R}^{d \times (\frac{D}{c_2}+1)D}$, $W_{att} \in \mathbb{R}^{1 \times (\frac{D}{c_2}+1)D}$ are the weight matrices for refining features and attention learning, $\hat{x}_i \in \mathbb{R}^d$ is the final actor-level feature vector, and the attention score s_i is a scalar which conveys the participation degrees of actors. Softmax function is used to get the normalized score a_i:

$$a_i = \frac{\exp\left(s_i\right)}{\sum_{j=1}^N \exp\left(s_j\right)} \tag{7}$$

Instead of conventional aggregation approaches like max-pooling or average-pooling, we fuse the actor-level features $\hat{X} = \{\hat{x}_i \in \mathbb{R}^d\}_{i=1}^N$ into group-level feature F_g as:

$$F_g = \sum_{n=1}^N a_i \hat{x}_i \tag{8}$$

3.3   Classification Layer

The classification layer aims to define the objective function to optimize the whole network which is consisted of two branches for action recognition and group activity recognition. For both tasks we apply a standard cross-entropy loss and combine two losses in a weighted sum:

$$\mathcal{L} = \mathcal{L}_g\left(y_g, \tilde{y}_g\right) + \lambda\mathcal{L}_i\left(y_i, \tilde{y}_i\right) \tag{9}$$

where $\mathcal{L}_g$, $\mathcal{L}_i$ are the cross entropy losses, y_g and y_i are the ground truth labels, $\tilde{y}_g$ and $\tilde{y}_i$ are the model predictions for group activity and individual actions. λ are the weights to balance these two losses. We find the equal weights for two tasks perform best so we set $\lambda = 1$ in our experiments.

4   Experiments

4.1   Datasets and Implementation Details

Datasets. The Volleyball dataset [6] contains 4830 clips gathered from 55 volleyball games, with 3493 training clips and 1337 for testing. There are 9 individual action labels and 8 group activity labels. Each clip is only fully annotated in the middle frame with bounding boxes of each actor and their action labels. We use multi-class classification accuracy (MCA) and mean per-class accuracy (MPCA) of group activity as evaluation metrics.

The Collective Activity dataset [3] contains 44 video sequences of 5 collective activities and 6 individual actions. Following the same settings in [10], we select 1/3 of the video clips for testing and the rest for training. We use multi-class classification accuracy (MCA) of group activity as a metric following the same approach as the related work.

Implementation Details. Following recent methods [4,17], we use 10 frames to train our model on both datasets which corresponding to 5 frames before the middle frame and 4 frames after. For the VD we resize each frame to 720×1280 resolution, for the CAD to 480×720. We track $N = 12$ persons in both datasets. If the number of detected persons is less than 12, we fill zeros for the missing persons. We exploit I3D as backbone which is initialized with the pre-trained model on Kinetics-400 and use feature maps from Mixed-4f layer of I3D. Then we resize the feature maps to 90×160 and 60×90 respectively for VD and CAD. RoIAlign is adopted to crop feature maps of size 5×5 for each actor bounding box in the middle frame of the input sequence. We then embed the feature maps to the $D=1024$ dimension feature vector for each actor and the dimension of the final group-level feature d is also set to 1024. We utilize ADAM optimizer with $\beta_1 = 0.9$, $\beta_2 = 0.999$, $\epsilon = 10^{-8}$ following [17]. For the VD, we train the network in 70 epochs with a mini-batch size of 2 and a learning rate ranging from 0.0002 to 0.0000. For the CAD, we train the network in 50 epochs with a mini-batch size of 16 and the learning rate is set to 0.0002.

4.2   Ablation Study

In this subsection, we perform ablation studies on the Volleyball dataset to investigate the effect of our proposed relation-guided actor attention module and the influence of factors c_1, c_2.

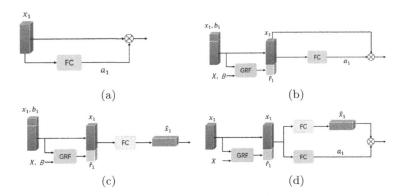

Fig. 4. Baseline models. (a) B2, (b) B3, (c) B4, (d) B5.

Effectiveness of the RGAA Module. We conduct experiments on five baselines to show the impact of each part of the RGAA module. The evaluated baselines are as follows.

B1: This is a base model which performs group activity and individual action classification on original appearance features.

B2: As shown in Fig. 4(a), this baseline adopts a local attention mechanism in which the attention score is only inferred by its own features. This baseline aims to show the importance of global relation features in attention learning.

B3: This baseline introduces a global relation feature to compute attention scores. The original appearance features are used to generate group-level features, as shown in Fig. 4(b). This baseline aims to illustrate the global relation feature can complement appearance features and refine actor-level features.

B4: As shown in Fig. 4(c), this baseline uses global relation features to refine original appearance features and the actor-level features are maxpooled to obtain group-level features without considering the contributions of different actors for group activity. This baseline can indicate the importance of attention modeling for key actors.

B5: As shown in Fig. 4(d), when computing pairwise relation features, this baseline ignores spatial position information and only explores appearance relations. This baseline can illustrate the importance of position information in relation modeling.

Table 1 shows the accuracy of our proposed model compared with the baselines. As is shown in this table, the proposed method achieves the best MCA and MPCA at the same time. The performance of B2 is worse than B1, indicating that simply local attention cannot effectively model the importance of actors. The results of B3 and B4 illustrate the actor-level features refined by global relation features are more discriminative than original appearance features. Compared to B4 and B5, our final model considers the importance of actors and spatial position information and achieves better performance.

Table 1. Comparison of different model on the Volleyball dataset

Model	MCA	MPCA
B1	89.8	88.9
B2	88.9	87.6
B3	90.2	89.7
B4	91.9	92.1
B5	92.6	93.1
Ours	**93.0**	**93.5**

Impact of Factors in LARM. In Table 2, we investigate the impact of factors c_1, c_2 in LARM on the Volleyball dataset. c_1 controls the dimension of features at embedding subspace. If c_1 is too large, it will lead to information loss, while if c_1 is too small, it will increase computational cost. c_2 affects the dimension of the global relation feature. The results of the model with different factors are all better than the base model and the best result achieves when c_1 and c_2 are set at 8 and 16, respectively. In our final model, we adopt this setting.

Table 2. Performance Comparison with difference factors on the Volleyball dataset.

c_1	c_2	MCA	MPCA
4	16	92.2	92.7
4	32	91.7	92.0
4	64	92.0	92.4
8	16	**93.0**	**93.5**
8	32	92.1	92.4
8	64	92.2	92.3

4.3 Comparison with the State-of-the-Arts

We compare our methods with the state-of-the-art methods on VD and CAD in Table 3. For a fair comparison, we only report the results of some methods [2,4,14] with RGB inputs.

138   L. Wu et al.

Table 3. Comparison with the state-of-the-art methods. Prefix '−' denotes results for the VD and prefix '−' for the CAD. The backbone used in SPTS [14] is not clarified in their paper. '−' denotes the result is not provided.

Method	Backbone	V-MCA	V-MPCA	C-MCA
HDTM [6]	AlexNet	81.9	82.9	81.5
PCTDM [19]	AlexNet	87.7	88.1	−
SPTS [14]	VGG*	89.3	89.0	−
CERN [12]	VGG16	83.3	83.6	87.2
stagNet [10]	VGG16	89.3	−	89.1
PRL [5]	VGG16	91.4	91.8	−
ARG [17]	Inception-v3	92.5	−	91.0
GAIM [9]	Inception-v3	92.1	92.6	90.6
CRM [2]	I3D	92.1	−	83.4
Actor-Transformer [4]	I3D	91.4	−	90.8
Ours	I3D	**93.0**	**93.5**	**91.5**

For the VD, the proposed method achieves the best performance and improves the performance to 93.0% and 93.5%, respectively, for MCA and MPCA metrics. ARG [17] attains an impressive performance by constructing an actor relation graph. Our method outperforms it by 0.5% for MCA by a much simpler framework. It is worth noting that with the same backbone I3D network, our results surpass the CRM [2] and Actor-Transformer [4] with 0.9% and 1.6% for MCA of group activity, since our model can refine actor-level features by global relation features and the group-level feature can be aggregated according to the contributions of different actors. Meanwhile, our method outperforms the recent two-stage RNN based methods [6,14], mostly because of the discriminative representation of spatio-temporal features extracted by 3D CNN. Figure 6(a) shows the confusion matrix of our model on the VD. Our method can achieve accuracy over 93% in terms of the majority of group activities. Nevertheless, it is easily confused by the set and pass activities as the two kinds of activities have similar appearance features and interaction relationships.

For CAD, our model again outperforms the previously published methods and achieves the state-of-the-art performance with 91.5%, as these methods neglect the unequal contributions of each actor for the overall group activity. This excellent performance demonstrates the effectiveness and generality of the proposed method. Additionally, we draw the confusion matrices based on our approach in Fig. 6(b). As shown in the figure, our model struggles to distinguish between crossing and walking. The reason may be that these two activities share high similarities and some works argued that their only difference is the relation between person and street.

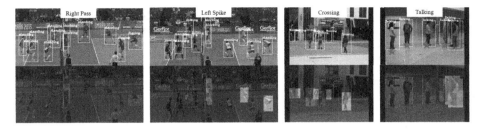

Fig. 5. Attention visualization on the VD and CAD. Individual action and group activity labels are shown in the first row. The key actors are denoted with red stars. Attention heat maps based on ground-truth bounding boxes are shown in the second row. The attention weights change from large to small along with the colors varying from red to blue. (Color figure online)

In Fig. 5, we visualize the attention learned by the RGAA module. We can see from the figure that some individual actions play vital roles in distinguishing group activity which indicates that the attention model is necessary and beneficial for recognizing the overall group activity. What's more, Our proposed RGAA module can focus on the key actors and ignore the outlier person.

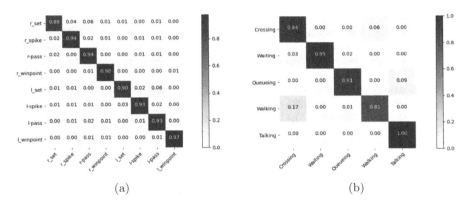

Fig. 6. Confusion matrices of our model for (a) VD and (b) CAD.

5   Conclusion

In this paper, we propose a simple yet effective framework for group activity recognition in video with relation-guided actor attention (RGAA) module. In the RGAA module, the global scope comparison information is exploited to guide the attention learning process and helps to discover key actors. Specifically, for an actor, we stack the pairwise relations between its feature and all other features together to infer the attention. The experiment results show that our method improves group activity recognition performance on two benchmarks.

Acknowledgement. This work was supported in part by the National Natural Science Foundation of China (61976010, 61802011), Beijing Municipal Education Committee Science Foundation (KM201910005024) and Postdoctoral Research Foundation (Q6042001202101).

References

1. Amer, M.R., Xie, D., Zhao, M., Todorovic, S., Zhu, S.C.: Cost-sensitive top-down/bottom-up inference for multiscale activity recognition. In: European Conference on Computer Vision, pp. 187–200. Springer (2012)
2. Azar, S.M., Atigh, M.G., Nickabadi, A., Alahi, A.: Convolutional relational machine for group activity recognition. In: Proceedings of the IEEE/CVF Conference on Computer Vision and Pattern Recognition, pp. 7892–7901 (2019)
3. Choi, W., Shahid, K., Savarese, S.: What are they doing?: Collective activity classification using spatio-temporal relationship among people. In: 2009 IEEE 12th international conference on computer vision workshops, ICCV Workshops, pp. 1282–1289. IEEE (2009)
4. Gavrilyuk, K., Sanford, R., Javan, M., Snoek, C.G.: Actor-transformers for group activity recognition. In: Proceedings of the IEEE/CVF Conference on Computer Vision and Pattern Recognition, pp. 839–848 (2020)
5. Hu, G., Cui, B., He, Y., Yu, S.: Progressive relation learning for group activity recognition. In: Proceedings of the IEEE/CVF Conference on Computer Vision and Pattern Recognition, pp. 980–989 (2020)
6. Ibrahim, M.S., Muralidharan, S., Deng, Z., Vahdat, A., Mori, G.: A hierarchical deep temporal model for group activity recognition. In: Proceedings of the IEEE Conference on Computer Vision and Pattern Recognition, pp. 1971–1980 (2016)
7. Li, W., Zhu, X., Gong, S.: Harmonious attention network for person re-identification. In: Proceedings of the IEEE Conference on Computer Vision and Pattern Recognition, pp. 2285–2294 (2018)
8. Li, X., Zhou, W., Zhou, Y., Li, H.: Relation-guided spatial attention and temporal refinement for video-based person re-identification. In: Proceedings of the AAAI Conference on Artificial Intelligence, vol. 34, pp. 11434–11441 (2020)
9. Lu, L., Lu, Y., Yu, R., Di, H., Zhang, L., Wang, S.: Gaim: graph attention interaction model for collective activity recognition. IEEE Trans. Multimed. **22**(2), 524–539 (2019)
10. Qi, M., Qin, J., Li, A., Wang, Y., Luo, J., Van Gool, L.: stagnet: An attentive semantic rnn for group activity recognition. In: Proceedings of the European Conference on Computer Vision (ECCV), pp. 101–117 (2018)
11. Ramanathan, V., Huang, J., Abu-El-Haija, S., Gorban, A., Murphy, K., Fei-Fei, L.: Detecting events and key actors in multi-person videos. In: Proceedings of the IEEE Conference on Computer Vision and Pattern Recognition, pp. 3043–3053 (2016)
12. Shu, T., Todorovic, S., Zhu, S.C.: Cern: confidence-energy recurrent network for group activity recognition. In: Proceedings of the IEEE conference on computer vision and pattern recognition. pp. 5523–5531 (2017)
13. Song, S., Lan, C., Xing, J., Zeng, W., Liu, J.: An end-to-end spatio-temporal attention model for human action recognition from skeleton data. In: Proceedings of the AAAI Conference on Artificial Intelligence. vol. 31 (2017)

14. Tang, Y., Wang, Z., Li, P., Lu, J., Yang, M., Zhou, J.: Mining semantics-preserving attention for group activity recognition. In: Proceedings of the 26th ACM international conference on Multimedia, pp. 1283–1291 (2018)
15. Wang, M., Ni, B., Yang, X.: Recurrent modeling of interaction context for collective activity recognition. In: Proceedings of the IEEE Conference on Computer Vision and Pattern Recognition, pp. 3048–3056 (2017)
16. Woo, S., Park, J., Lee, J.Y., Kweon, I.S.: Cbam: Convolutional block attention module. In: Proceedings of the European conference on computer vision (ECCV), pp. 3–19 (2018)
17. Wu, J., Wang, L., Wang, L., Guo, J., Wu, G.: Learning actor relation graphs for group activity recognition. In: Proceedings of the IEEE/CVF Conference on Computer Vision and Pattern Recognition, pp. 9964–9974 (2019)
18. Wu, L.F., Wang, Q., Jian, M., Qiao, Y., Zhao, B.X.: A comprehensive review of group activity recognition in videos. Int. J. Autom. Comput. **18**(3), 334–350 (2021)
19. Yan, R., Tang, J., Shu, X., Li, Z., Tian, Q.: Participation-contributed temporal dynamic model for group activity recognition. In: Proceedings of the 26th ACM International Conference on Multimedia, pp. 1292–1300 (2018)
20. Zhang, Z., Lan, C., Zeng, W., Jin, X., Chen, Z.: Relation-aware global attention for person re-identification. In: Proceedings of the IEEE/CVF Conference on Computer Vision and Pattern Recognition, pp. 3186–3195 (2020)

MVAD-Net: Learning View-Aware and Domain-Invariant Representation for Baggage Re-identification

Qing Zhao[1], Huimin Ma[1(✉)], Ruiqi Lu[2], Yanxian Chen[1], and Dong Li[3]

[1] University of Science and Technology Beijing, Beijing, China
s20200688@xs.ustb.edu.cn, mhmpub@ustb.edu.cn, 13001250622@163.com
[2] ByteDance Ltd., Beijing, China
luruiqi.v@bytedance.com
[3] Nuctech Co., Ltd., Beijing, China

Abstract. Baggage re-identification (ReID) is a particular and crucial object ReID task. It aims to only use the baggage image data captured by the camera to complete the cross-camera recognition of baggage, which is of great value to security inspection. Two significant challenges in the baggage ReID task are broad view differences and distinct cross-domain characteristics between probe and gallery images. To overcome these two challenges, we propose MVAD-Net, which aims to learn view-aware and domain-invariant representation for baggage ReID by multi-view attention and domain-invariant learning. The experiment shows that our network has achieved good results and reached an advanced level while consuming minimal extra cost.

Keywords: Baggage Re-Identification · Multi-view · Attention · Domain-invariant learning · Metric learning

1 Introduction

For security and customs declaration reasons, baggage from flights often needs to go through security checks before it can be delivered to passengers, and suspicious baggage needs to be tracked for a secondary inspection. Typically, once the flight arrives, the checked baggage is scanned by security equipment on the Baggage Handling System (BHS) conveyor. When suspicious items are found in baggage, it will be labeled with a Radio Frequency Identification (RFID) tag to indicate that the baggage needs to be further opened and checked manually. An alert will be triggered when the passenger picks up the baggage and takes it to the RFID detection area. However, this traditional baggage tracking method has a high cost and low efficiency. Besides, RFID tags may fall off, and passengers with ulterior motives may tear the tags off to avoid inspection.

Object re-identification (ReID) is mainly used to solve the cross-camera tracking task of the target, mainly focusing on person and vehicle re-identification. The baggage re-identification was proposed by Nuctech Company in [1], aiming to realize the tracking of

H. Ma et al. (Eds.): PRCV 2021, LNCS 13019, pp. 142–153, 2021.
https://doi.org/10.1007/978-3-030-88004-0_12

baggage only relying on its appearance, thus effectively replacing the traditional RFID-based method. The setting of baggage re-identification task is relatively fixed. First, three fixed cameras on the BHS conveyor belt capture the passing baggage to obtain images from three views. Images of these views constitute the galley image set. Then, when the passenger takes the baggage and passes through the checkpoint, the checkpoint camera will capture the baggage images again. The image acquired at checkpoint is a probe image. The task of baggage ReID is to obtain a baggage identity corresponding to a given probe by searching the gallery image, to achieve the purpose of tracking the baggage.

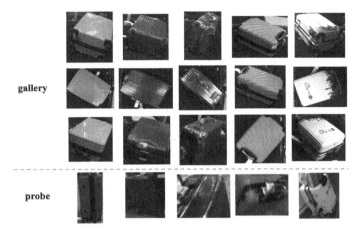

Fig. 1. Gallery and probe image display of 5 baggage. The gallery image was captured by three fixed cameras on BHS, and the probe image was captured by checkpoint. Intuitively, the view angle and image style of the baggage in probe image are quite different from those of gallery image.

Different from person and vehicle ReID, baggage ReID task has its unique characteristics and challenges: (a) The gallery images are fixed multi-view images. In many application scenarios, the single view image cannot accurately and completely show the whole picture of the object, and the use of multi-view image is a very intuitive way. As shown in Fig. 1. To get the features of the different surfaces of the baggage as much as possible, and to comprehensively represent them, the gallery image is obtained by shooting three different views with fixed cameras. This multi-view shooting method is intuitive, but there are few methods specifically to solve the problem of how to rationally use the multi-view images obtained by shooting. Our method explores this problem. (b) The gallery and probe images are clearly distinguishable. In the other ReID tasks, all images are randomly captured by cameras in different locations, and the probe and gallery are randomly divided. However, in the baggage ReID, we know for sure whether each image is a gallery image or a probe image because they are obtained from fixed cameras in different locations. As shown in Fig. 1, there are some differences between gallery and probe images of the same baggage in terms of brightness, color, blur degree, etc., which are mainly caused by different shooting scenes and carriers of the two types of images. This brings the distinct cross-domain problem but enables us to consider several gallery images of the same baggage as a whole and make full use of the domain information to design targeted strategies.

Our contributions are the following: (a) A multi-view attention method combined with metric learning, which is used to reasonably process and solve multi-view image problems and learn the view-aware representation. (b) A domain-invariant learning method to improve the baggage ReID performance by solving the cross-domain problem between probe and gallery images.

2 Related Works

2.1 Representation Learning and Metric Learning in ReID

In recent years, the ReID method based on deep learning can be divided into two ideas, one is to use the representation learning method, and the other is to use the metric learning method. The method based on representation learning does not directly obtain the similarity information of the image in the network, instead, it often transforms the ReID problem into a classification (such as [2–4], etc.) or a verification problem (such as [5, 6], etc.) to solve. Deep metric learning methods are often used in image retrieval tasks. Because they fit the characteristics of ReID tasks, they are a common method in the field today. [7] use triplet loss to build an end-to-end person ReID model to achieve the effect of reducing the distance of positive samples to the image and increasing the distance of negative samples to the image. Compared with representation learning, the metric learning method has a better effect and wider applicability, but the sample selection is difficult, and the training process is difficult to converge, so some studies combine the metric learning method with the representation learning method and optimize for the problem of difficult convergence. For example, [8] proposed the BNNeck structure to solve the problem of asynchronous convergence when the ID loss and triplet loss are used in combination. Our baseline draws the person ReID model proposed in [8], which uses both representation learning and metric learning and aims to obtain excellent feature embedding for baggage images.

2.2 View-Based Methods for ReID

Many methods start from the view of solving the ReID problem, which gives us some inspiration. In person ReID, [9] introduces people's coarse-grained view information, uses a view predictor to predict the view of the person relative to the camera, and derives three-view prediction scores to weight the output of each unit in the view unit. In vehicle ReID, like baggage, the same vehicle may have large differences in visual information from different views. [10] proposed VANet with view perception, which uses a view predictor to determine whether the view of the vehicles in the two images are the same, and then to deal with the images of the same view and different views separately, to overcome the ReID deviation caused by different views. [11] calculate the similarity between the angle of view of the vehicle and the similarity of the camera, combine it with the similarity of the vehicle to get the ReID result, and have achieved good results.

In the baggage ReID task, its gallery image has clear and relatively fixed three-view information. Therefore, when designing the algorithm, we make full use of this feature, referring to the self-attention [12], and propose a Multi-View Attention method to learn the view-aware representation for baggage ReID.

2.3 Domain Adaptation

Domain adaptation refers to mapping data distributed in different domains into the same feature domain to make the distance in the feature space as close as possible. [13] proposes the Maximum Mean Discrepancy, which maps features of different domains into the same Hilbert space. [14] uses Gradient Reversal Layer (GRL) to maximize the error of domain classification, so that features of different domains are closer in feature space and difficult to distinguish. [15] also uses a similar idea of adversarial learning, narrowing in the RGB image domain and the infrared image domain for person ReID. In this work, referring to GRL, we propose a domain-invariant learning method for baggage ReID, which effectively shortens the distribution distance of gallery image and probe image in the feature domain, and improves the performance.

3 The Proposed Method

MVAD-Net architectures we propose are shown in Fig. 2 and Fig. 3. It is an end-to-end baggage ReID model. The main methods are Multi-View Attention (MVA) and Domain-Invariant Learning (DIL), which are suitable for baggage ReID task.

3.1 Baggage ReID Baseline

Before introducing MVA and DIL, we first describe our baseline method and a method of uniformly utilizing images from different views of the same object (baggage) in the validation or testing phase.

In the verification or testing phase, all gallery images need to be input into the trained feature extractor to obtain their features f_G. After that, the probe image to be judged is input into the network to obtain its feature f_P. Finally, the cosine distance is calculated between f_P and all f_G. The smaller the distance, the greater the similarity, and the greater the possibility that they are the same baggage. The baggage category of the gallery image with the smallest distance from the probe image features can be used as the probe image category, which is the *original mode* of distance measurement in the testing and verification phase.

Nevertheless, there are 1 to 3 gallery images of each baggage, so our considerations are as following: calculate the distance between the feature f_P of a probe image and the feature $f_{G_i}(i = 1, 2, 3)$ of all gallery images of a baggage to get the distance $d\left(f_P, f_{G_i}\right)$, and the minimum, maximum and average values of $d\left(f_P, f_{G_i}\right)$. They can be used as the ranking basis of probe similar images, which are called *min-mode*, *max-mode*, and *mean-mode* respectively.

3.2 Multi-view Attention Model

Although the method described above comprehensively considers three-view gallery images of the same baggage, they are still rough. Therefore, we have designed MVA to use the multi-view images more reasonably.

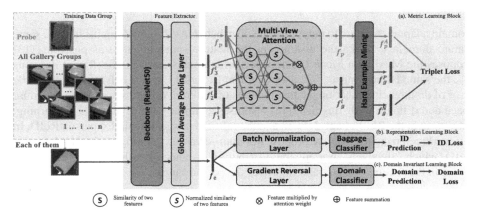

Fig. 2. Illustration of the architecture of MVAD-Net training. It is mainly composed of three parts in parallel: (a). metric learning block, (b). representation learning block and (c). domain-invariant learning block.

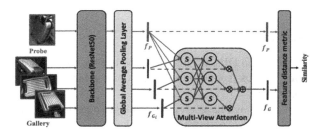

Fig. 3. Illustration of the architecture during MVAD-Net inference.

Multi-View Attention. In the baggage ReID, its gallery image has clear and relatively fixed three-view information. However, because the probe image may only be similar to one part of the baggage images from three views, the rest of the images may interfere with the judgment. We design the MVA to make the network have the view-aware ability. According to the current probe image, the MVA dynamically adjust the feature weight of each view of each baggage's multi-view gallery image, to use the multi-view information more reasonably and improve the ReID performance. As shown in Fig. 2, during training our MVAD-Net, the image data input to the neural network is divided into several training data groups. Each group contains a probe and all gallery images, and the gallery images of the same baggage are a group, a total of n groups. First, the input image is extracted by the feature extractor to obtain the feature vector. f_p is the feature vector of the probe, f_1^i, f_2^i and f_3^i are the features of three gallery images in group i, and f_e is the feature of each image in training data group. Then in block (a), the probe features and the gallery feature of each baggage's three views are input into the MVA to obtain the fused feature f_g^i. The calculation formula of MVA as:

$$Multi - View\ Attention\left(f_p, f_g^i\right) = \sum_{j=1}^{n} \alpha_j f_j^i \tag{1}$$

$$\alpha_j = softmax(s_j | s_1, s_2, \ldots, s_n) \tag{2}$$

$$s_j = sim\left(f_p, f_j^i\right) = f_p^T f_j^i \tag{3}$$

where f_p represents the probe feature as a query, $f_g = \left\{f_j^i \middle| f_1^i, f_2^i, f_3^i \ldots f_n^i\right\}$ is the gallery feature set of the baggage i, n is generally taken as 3. α_j represents the calculated attention weight of f_j^i relative to f_p. s_j is the similarity between f_j^i and f_p. There are many ways to calculate the similarity between the two features. We use the dot product in the MVA model. In the inference stage, as shown in Fig. 3, the similarity between f_G fused by MVA and probe feature f_P is calculated as the sorting basis of ReID results.

Hard-Negative Sample Triplet Training. We skillfully combine MVA training with triplet loss [7], and we call this hard-negative sample triplet training. As shown in block (a), we use the probe feature f_p, the query in MVA, as the anchor f_P^A in the triplet loss, and the fused gallery features f_g^P and f_g^N generated by MVA as the positive and negative sample of the triplet loss:

$$L_{HNS-Triplet} = [d_p - d_{hn} + \gamma]_+ \tag{4}$$

where d_p is the cosine distance between positive sample pairs, d_{hn} is the distance between negative sample pairs, γ is the super parameter — triplet margin, and $[z]_+$ is the maximum value between z and 0. Besides, to improve metric learning performance, the negative samples we selected are the fusion image features of the gallery image of some other baggage with the smallest distance from the probe feature, that is, the hard-negative samples.

3.3 Domain-Invariant Learning

In this task, the distinct cross-domain problem between the gallery and probe image will bring a certain impact on ReID. The key to ReID is to embed the image to the feature space to get the corresponding feature vector and measure the similarity between the probe feature and gallery feature. If there is a large gap between the two domains, the similarity measurement will inevitably be disturbed. It can be vividly assumed that the extracted features of the gallery and probe are shown in the left of Fig. 4 when in the same domain, and in the right of Fig. 4 when in different domains. It is observed that when they are in different domains, the difference between the gallery domain and

148   Q. Zhao et al.

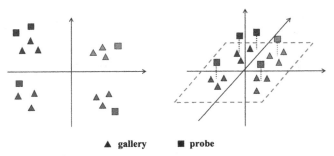

▲ gallery   ■ probe

Fig. 4. Image features of the gallery and probe are in the same feature domain (left) and different domains (right).

probe domain will also interfere with the similarity measurement (calculation of feature distance), even if the image features of certain baggage are clustered in a certain region.

We propose domain-invariant learning to solve the problem. Our domain-invariant learning is mainly realized by the gradient reverse layer (GRL) [14]. Gradient descent is the minimum loss function. GRL maximizes the loss function by reversing the gradient returned when the gradient is backpropagation, to update the parameters to the negative gradient direction.

As shown in Fig. 2 block (c), to reduce the difference between probe and gallery image feature space, the feature is input into a domain classifier with GRL after getting the feature vector of the image by using the feature extractor composed of convolution structure and global average pooling. The domain classifier is a binary classifier, which divides the image into probe or gallery, and then calculates the domain loss according to the domain prediction score calculated by the classifier. Domain loss is essentially a binary cross-entropy loss function, see Eq. (5). Where N represents the number of samples; t_i represents the domain category label of the image i; t_i equals 1 represents the gallery image and equals 0 represents the probe image; y_i represents the probability (domain prediction score) that image i is a gallery image.

$$L_{Domain} = -\sum_{i=1}^{N} t_i \log(y_i) + (1 - t_i)\log(1 - y_i) \tag{5}$$

The stronger the classification ability of domain classifier is, the greater the difference between gallery and probe image feature domain is. Weakening its classification ability realized by GRL can close the two domains. In forward propagation, the features of GRL remain unchanged, while in backward propagation, GRL reverses the gradient returned, which makes the neural network's ability to distinguish probe and gallery images worse. After domain-invariant learning, the accuracy of domain classifier for domain classification of baggage images can be effectively reduced, so that the probe and gallery images of the same baggage can be mapped into the same feature domain as far as possible. At the same time, ReID performance has been improved.

The combination of multi-view attention and domain-invariant learning makes up the final MVAD-Net, which can learn view-aware and domain-invariant representation for baggage ReID.

4 Experiments

4.1 Dataset and Protocols

At present, the only dataset for the baggage ReID task is the MVB dataset [1] released in 2019. It is also the first large baggage ReID dataset released to the public. It currently contains 4519 baggage identities and 22660 annotated baggage images as well as their surface material labels. Each baggage contains 1 to 3 gallery images and 1 to 4 probe images. In the divided training set, 20176 images of 4019 baggage are included. In the verification and test set, there are 2484 images of 500 baggage.

Cumulative matching characteristics (CMC) and mean average precision (mAP) are two commonly used measures of ReID model accuracy. For the case of single-gallery-shot, we generally use CMC top-k accuracy. For the case of multi-gallery-shot, CMC cannot conduct a comprehensive evaluation and we often introduce mAP [18]. In the baggage ReID task, there are multiple gallery images for each baggage, so if we use the general method and do not consider the gallery images of the same baggage uniformly, we should introduce mAP comprehensive evaluation while using CMC. However, once the gallery images of the same baggage are considered together, such as the MVA method we proposed, there is only one position for each baggage in the similar gallery sample sequence of the probe. In this case, the mAP is actually mean precision (mP), so we focus on using top-k accuracy such as Rank-1 to measure the performance of our algorithm, with mAP (mP) as an auxiliary.

4.2 Implementation Details

We use the ReaNet50 [19] model pre-trained on ImageNet [20] as our backbone. During training, before the image is input to the feature extractor, we first perform a series of data augmentation operations, including random rotation, random horizontal flip, random crop, and random erasing [16]. The image size of the input feature extractor is uniformly 224×224 pixels. The input images of each baggage include three gallery images and three probe images (when the number is insufficient, randomly select copies of other images of the same type and baggage as substitutes). We use Adam optimizer [21] to update the parameters. The initial learning rate is set to 0.0003. A total of 150 epochs are trained. In the first 10 epochs, the warmup learning rate [22] strategy is adopted, which gradually increases from 0.00003 to 0.0003. Then the learning rate is reduced to one-tenth of the original at the 40th and 70th epoch. In metric learning, the margin of the triple loss is set to 0.3, and the center loss weight is set to 0.0005. In the domain-invariant learning, the magnification of the gradient inversion of GRL is 2.0, and the learning rate of the domain classifier is individually set to 0.003. In representation learning, the weight of ID loss is set to 0.2. In the verification and testing phase, we used the re-ranking [24].

4.3 Effectiveness of Multi-view Attention

This section shows the effect of our proposed MVA method. As shown in Table 1, the first method is the baggage ReID baseline. It contains four different modes. The origin-mode directly calculates the similarity between the probe feature and all gallery features, and then uses this distance as the ranking basis. Min-mode, max-mode, and mean-mode are based on taking the minimum, maximum, and average distances between multiple gallery features and probe features of the same baggage as the sorting basis. It is a simple method of comprehensively considering gallery multi-view images. Mean-mode treats the gallery features of multiple viewing angles uniformly, and min-mode takes the minimum distance to mitigate the interference in extreme cases. It can be seen from the table that, compared to origin-mode, they all bring ReID performance improvement.

The second method is to add MVA based on the baseline, to perform attention weighted fusion on the gallery images of three views. As an upgraded version of min-mode and mean-mode, MVA has achieved better performance.

Table 1. The effect of multi-view attention model.

Method		Rank-1	mAP
Baseline	origin-mode [8]	83.2%	85.5%
	min-mode	83.2%	88.8%
	max-mode	82.4%	88.3%
	mean-mode	83.3%	89.0%
Baseline + MVA		**85.6%**	**90.3%**

4.4 Effectiveness of Domain-Invariant Learning

This section shows the effect of domain-invariant learning. We first added only the domain classifier based on the model using MVA, but not the GRL for domain-invariant learning. This is equivalent to pulling the gallery and probe image features that already have domain differences farther away, increasing their domain distance. As shown in the italic row of Table 2, after adding domain loss, the domain classification accuracy reaches 99.8%, and the ReID performance has significantly decreased. This indicates that the domain difference of probe and gallery features will have an adverse effect on baggage ReID, and it proves that our conjecture is reasonable.

Table 2. The effect of domain-invariant learning.

Method	Domain classification accuracy	Rank-1	mAP
Baseline (origin) [8]	-	83.2%	85.5%
+ MVA	-	85.6%	90.3%
+ Domian Loss	*99.8%*	*78.4%*	*85.8%*
+ GRL (MVAD-Net)	**76.0%**	**86.7%**	**91.2%**

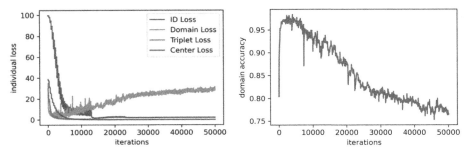

Fig. 5. Variation curves of various loss functions (left), and the change curve of domain classification accuracy during MVAD-Net training (right).

Next, we add the GRL before the feature is input to the domain classifier and introduce domain-invariant learning (this is our final MVAD-Net). As shown in Fig. 5, domain loss presents an upward trend with increased iteration times. Meanwhile, the domain classification accuracy also presents a downward trend, indicating that domain-invariant learning plays a role in pulling the probe's feature domain away from the gallery's feature domain. As shown in the last row of Table 2, our MVAD-Net finally achieved the excellent performance of Rank−1 86.7% and mAP 91.2%.

4.5 Comparison with Other Methods

We have also tested the performance of some person ReID methods on the MVB dataset. The comparison results are shown in Table 3, which shows that our method is very competitive.

Table 3. The performance of several ReID methods on MVB dataset.

Method	Rank-1	mAP
IDE-Resnet50 [2]	68.8%	64.6%
Verification-IDE [26]	68.8%	63.8%
PCB [25]	66.3%	63.6%
PCB + MHN4 [27]	70.0%	67.8%
DGNet [28]	67.1%	60.9%
Strong Baseline [8]	83.2%	85.5%
MVAD-Net (ours)	**86.7%**	**91.2%**

5 Conclusion

In this paper, we propose the MVAD-Net for baggage ReID with excellent performance. The key to it is multi-view attention and domain-invariant learning, which help the network learn view-aware and domain-invariant representation. The baggage ReID is a novel ReID task with significant application value and research value. It is different from other ReID tasks such as person ReID. We hope our work can serve as a starting point for more people to see and devote themselves to the research of baggage ReID. It is worth mentioning that the processing of multi-view image is a common problem, and our work can also provide a new reference solution for the task involving this kind of problem.

Acknowledgments. This work was supported by the National Natural Science Foundation of China (No. U20B2062), the fellowship of China Postdoctoral Science Foundation (No. 2021M690354), the Beijing Municipal Science & Technology Project (No. Z191100007419001).

References

1. Zhang, Z., Li, D., Wu, J., Sun, Y., Zhang, L.: MVB: a large-scale dataset for baggage re-identification and merged siamese networks. In: Lin, Z., et al. (eds.) PRCV 2019. LNCS, vol. 11859, pp. 84–96. Springer, Cham (2019). https://doi.org/10.1007/978-3-030-31726-3_8
2. Zheng, L., Yang, Y., Hauptmann, A.G.: Person re-identification: past, present and future. *arXiv preprint* arXiv:1610.02984 (2016)
3. Zheng, L., Zhang, H., Sun, S., Chandraker, M., Yang, Y., Tian, Q.: Person re-identification in the wild. In: Proceedings of the IEEE Conference on Computer Vision and Pattern Recognition, pp. 1367–1376 (2017)
4. Zheng, L., et al.: MARS: a video benchmark for large-scale person re-identification. In: Leibe, B., Matas, J., Sebe, N., Welling, M. (eds.) ECCV 2016. LNCS, vol. 9910, pp. 868–884. Springer, Cham (2016). https://doi.org/10.1007/978-3-319-46466-4_52
5. Yi, D., Lei, Z., Liao, S., Li, S.Z.: Deep metric learning for person re-identification. In: 2014 22nd International Conference on Pattern Recognition, pp. 34–39. IEEE, August 2014
6. Li, W., Zhao, R., Xiao, T., Wang, X.: Deepreid: deep filter pairing neural network for person re-identification. In: Proceedings of the IEEE Conference on Computer Vision and Pattern Recognition, pp. 152–159 (2014)
7. Hermans, A., Beyer, L., Leibe, B.: In defense of the triplet loss for person re-identification. arXiv preprint arXiv:1703.07737 (2017)
8. Luo, H., Gu, Y., Liao, X., Lai, S., Jiang, W.: Bag of tricks and a strong baseline for deep person re-identification. In: Proceedings of the IEEE Conference on Computer Vision and Pattern Recognition Workshops (2019)
9. Saquib Sarfraz, M., Schumann, A., Eberle, A., Stiefelhagen, R.: A pose-sensitive embedding for person re-identification with expanded cross neighborhood re-ranking. In: Proceedings of the IEEE Conference on Computer Vision and Pattern Recognition, pp. 420–429 (2018)
10. Chu, R., Sun, Y., Li, Y., Liu, Z., Zhang, C., Wei, Y.: Vehicle re-identification with viewpoint-aware metric learning. In: Proceedings of the IEEE International Conference on Computer Vision, pp. 8282–8291 (2019)
11. Zhu, X., Luo, Z., Fu, P., Ji, X.: VOC-ReID: vehicle re-identification based on vehicle-orientation-camera. In: Proceedings of the IEEE/CVF Conference on Computer Vision and Pattern Recognition Workshops, pp. 602–603 (2020)
12. Vaswani, A., et al.: Attention is all you need. In: Advances in Neural Information Processing Systems, pp. 5998–6008 (2017)
13. Long, M., Cao, Y., Wang, J., Jordan, M.I.: Learning transferable features with deep adaptation networks. In: Bach, F.R., Blei, D.M. (eds.) ICML, vol. 37, pp 97–105 (2015)

14. Ganin, Y., Lempitsky, V.: Unsupervised domain adaptation by backpropagation. In: International Conference on Machine Learning, pp. 1180–1189, June 2015
15. Dai, P., Ji, R., Wang, H., Wu, Q., Huang, Y.: Cross-modality person re-identification with generative adversarial training. In: IJCAI, vol. 1, pp. 2, July 2018)
16. Zhong, Z., Zheng, L., Kang, G., Li, S., Yang, Y.: Random erasing data augmentation. In: AAAI, pp. 13001–13008 (2020)
17. Szegedy, C., Vanhoucke, V., Ioffe, S., Shlens, J., Wojna, Z.: Rethinking the inception architecture for computer vision. In: Proceedings of the IEEE Conference on Computer Vision and Pattern Recognition, pp. 2818–2826 (2016)
18. Li, W., Sun, Y., Wang, J., Xu, H., Yang, X., Cui, L.: Collaborative Attention Network for Person Re-identification. arXiv preprint arXiv:1911.13008 (2019)
19. He, K., Zhang, X., Ren, S., Sun, J.: Deep residual learning for image recognition. In: Proceedings of the IEEE Conference on Computer Vision and Pattern Recognition, pp. 770–778 (2016)
20. Russakovsky, O., et al.: Imagenet large scale visual recognition challenge. Int. J. Comput. Vision **115**(3), 211–252 (2015)
21. Kingma, D.P., Ba, J.: Adam: A method for stochastic optimization. arXiv preprint arXiv: 1412.6980 (2014)
22. Fan, X., Jiang, W., Luo, H., Fei, M.: Spherereid: Deep hypersphere manifold embedding for person re-identification. J. Vis. Commun. Image Represent. **60**, 51–58 (2019)
23. Wen, Y., Zhang, K., Li, Z., Qiao, Y.: A discriminative feature learning approach for deep face recognition. In: Leibe, B., Matas, J., Sebe, N., Welling, M. (eds.) ECCV 2016. LNCS, vol. 9911, pp. 499–515. Springer, Cham (2016). https://doi.org/10.1007/978-3-319-46478-7_31
24. Zhong, Z., Zheng, L., Cao, D., Li, S.: Re-ranking person re-identification with k-reciprocal encoding. In: Proceedings of the IEEE Conference on Computer Vision and Pattern Recognition, pp. 1318–1327 (2017)
25. Sun, Y., Zheng, L., Yang, Y., Tian, Q., Wang, S.: Beyond part models: person retrieval with refined part pooling (and a strong convolutional baseline). In: Ferrari, V., Hebert, M., Sminchisescu, C., Weiss, Y. (eds.) ECCV 2018. LNCS, vol. 11208, pp. 501–518. Springer, Cham (2018). https://doi.org/10.1007/978-3-030-01225-0_30
26. Zheng, Z., Zheng, L., Yang, Y.: A discriminatively learned cnn embedding for person reidentification. ACM Trans. Multimedia Comput. Commun. Appl. (TOMM) **14**(1), 1–20 (2017)
27. Chen, B., Deng, W., Hu, J.: Mixed high-order attention network for person re-identification. In: Proceedings of the IEEE International Conference on Computer Vision, pp. 371–381 (2019)
28. Zheng, Z., Yang, X., Yu, Z., Zheng, L., Yang, Y., Kautz, J.: Joint discriminative and generative learning for person re-identification. In: Proceedings of the IEEE Conference on Computer Vision and Pattern Recognition, pp. 2138–2147 (2019)

Joint Attention Mechanism for Unsupervised Video Object Segmentation

Rui Yao[✉], Xin Xu, Yong Zhou, Jiaqi Zhao, and Liang Fang

School of Computer Science and Technology, China University of Mining and Technology,
Xuzhou 221116, China
ruiyao@cumt.edu.cn

Abstract. In this work, we propose an unsupervised video object segmentation framework based on a joint attention mechanism. Based on the feature extraction of video frames, this method constructs a joint attention module to mine the correlation information between different frames of the same video, and uses the global consistency information of the video to guide the segmentation. The joint attention module includes a soft attention unit and a co-attention unit. The former emphasizes important information in the feature embedding of a frame, and the latter enhances the features of the current frame by calculating the correlation between features from different frames. Furthermore, in order to exchange information more comprehensively and deeply in different frames, superimposing the joint attention module can achieve better performance. We conducted experiments on several benchmark datasets to verify the effectiveness of our algorithm, experimental results show that the joint attention module can capture global consistency information significantly and improves the accuracy of segmentation.

Keywords: Video object segmentation · Attention mechanism · Unsupervised learning

1 Introduction

Video object segmentation is a pixel-level task, which aims to segment the moving objects in each frame of the video. As a basic task of computer vision, video object segmentation has important research significance in application scenarios such as video surveillance, autonomous driving and motion recognition. The current unsupervised video object segmentation models based on deep learning mainly focus on the intra-frame property of the moving object that needs to be segmented in appearance or motion, while ignoring the valuable global consistency between multiple frames. These methods calculate the optical flow between several consecutive frames [1], which leads to the mining of temporal information only in a short temporal window.

The main objective in the unsupervised video object segmentation setting should satisfy the characteristics of local distinguishability and frequent occurrence in the entire video sequence. The two attributes are essential for determining the primary object. For example, it is difficult to determine the main object through only a short video clip. On

© Springer Nature Switzerland AG 2021
H. Ma et al. (Eds.): PRCV 2021, LNCS 13019, pp. 154–165, 2021.
https://doi.org/10.1007/978-3-030-88004-0_13

the contrary, if we view the entire video (or a sufficiently long video sequence), it is easy to find the primary object. Although the primary objects are highly correlated at the macro level (*i.e.*, in the entire video), at the micro level (*i.e.*, in a shorter video sequence) they often appear different due to human movement, occlusion, camera movement, *etc*. Microscopic changes are the main challenge of video object segmentation, so it is crucial to use information from other frames to obtain global consistency. Based on this view, the idea of our algorithm is to mine the rich correlation between different frames in the video when performing video object segmentation, and uses the global consistency information to guide the segmentation of current frame. In order to achieve this, our algorithm proposes a joint attention module that can fully capture the correlations between different frames. Furthermore, by cascading joint attention modules, the algorithm achieves deep information processing and obtains good performance. Comparing to state-of-the-art approaches, our method is much more robust and achieves the best performances on the DAVIS-16 and DAVIS-17 benchmarks.

Our main contributions are summarized as follows:

- We propose an end-to-end video object segmentation framework, which segments moving objects by mining the correlation information between frames.
- We construct a joint attention module to mine the correlation information between different frames of the same video, and use the global consistency information of the video to guide the segmentation.
- We conduct ablation experiments and comparison experiments with the state-of-the-art models on the DAVIS-16 [2] and DAVIS-17 [3] to verify the effectiveness of our algorithm. The experimental results show the effectiveness and robustness of our model.

2   Related Work

Unsupervised Video Object Segmentation (UVOS). Video object segmentation before the advent of deep learning uses hand-made features to model video information. After deep learning has demonstrated its powerful representation ability in the field of computer vision, it also provides new solutions for the field of VOS. In recent years, some methods have been proposed convolutional neural networks for the task of VOS. Early VOS methods used a CNN to generate salient objects, and then propagated moving objects in the video. Later, some methods build two-branch convolutional neural networks to segment objects. MP-Net [4] takes the optical flow fields of two adjacent frames in the video sequence as input, and generates motion labels for each pixel. Li *et al.* [5] transfer the knowledge encapsulated in the image-based instance embedding network and adjust the instance network for VOS. Wang *et al.* [6] propose an UVOS model driven by visual attention.

Attention Mechanism in VOS. The Attention mechanism, which are inspired by human perception [7], plays a significant role in improving sequential learning tasks [8]. The Attention mechanism help Convolutional Neural Network focus on the local information of the image to generate the corresponding sequence. For example, Chu *et al.* [9] estimate human pose through the capture of multi-context attention. In non-seq2seq tasks, such as text classification, or other classification problems, the attention mechanism will be used through self-attention. More recently, co-attention mechanisms have been studied in visual and language tasks. In [10], co-attention mechanisms are used to explore potential associations between different patterns. In this work, we use both self-attention and co-attention in order to fully capture the useful information in different frames.

3   Method

Our method introduces the attention mechanism into UVOS, and proposes a joint attention module, which explicitly encodes the correlations between video frames. So that the method can focus on frequent coherent regions, thereby further helping to discover foreground object and produce reasonable UVOS results. The overall framework of our method is shown in Fig. 1.

3.1   Joint Attention Mechanism

The joint attention mechanism consists of two soft attention units and a co-attention unit, as shown in Fig. 2. The soft attention unit can process and emphasize the most informative area of the appearance feature map of the video frame. The co-attention unit transfers the emphasis information of the reference frame captured by the soft attention unit to the current frame, which is convenient for the learning of spatio-temporal features.

Soft Attention Unit. The soft attention unit performs soft weight on each spatial position of the spatial feature map N_a and N_b of the input frames [11]. Taking the feature map N_a as an example, the soft attention unit outputs feature map $M_a \in \mathbb{R}^{W \times H \times C}$ with enhanced attention as follows:

$$G_a = softmax(W_a(N_a)), \tag{1}$$

$$M_a^c = G_a \odot N_a^c. \tag{2}$$

W_a is a convolution, which converts the feature map N_a into a probability map that characterizes the importance, and then normalizes it with *softmax* function to generate an attention map $G_a \in \mathbb{R}^{W \times H}$, where $\sum_{i=1}^{W \times H} G_a^i = 1$, each value of G_a^i is the importance of the position that the model predicts, and the larger the value, the more important the information of that position. N_a^c and M_a^c is the two-dimensional feature slice of N_a and M_a on the c-*th* channel, $\odot$ representing the Hadamard product, *i.e.*, multiplying the corresponding element with each slice of G_a and N_a on the c-*th* channel. Similarly, given N_b, the feature map M_b can be obtained by Eq. 1 and Eq. 2.

Co-attention Unit. In order to transfer the information of the feature map of the reference frame enhanced by soft attention unit, the non-local way is used to obtain the correlations between M_a and M_b as follows:

$$A = M_a^T W M_b \in \mathrm{R}^{(WH) \times (WH)}. \tag{3}$$

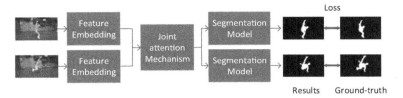

Fig. 1. Overall framework of the proposed algorithm during training.

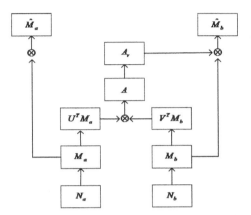

Fig. 2. Joint attention module.

$W \in \mathbb{R}^{C \times C}$ is a trainable weight matrix. The relationship matrix A can effectively capture the relationship between any two elements in the two feature spaces. However, this calculation introduces a large number of parameters, and increases the computation cost, and generates the risk of overfitting. To overcome this problem, W can be approximately decomposed into two low-rank matrices $U \in \mathbb{R}^{C \times \frac{C}{d}}$ and $V \in \mathbb{R}^{C \times \frac{C}{d}}$, where d ($d < 1$) represents the reduction ratio. This not only reduces the number of parameters by a factor of $2/d$, but also generates a compact channel feature representation for each channel. Therefore, Eq. 3 can be expressed as follows:

$$A = M_a^T W M_b = \left(U^T M_a \right)^T \left(V^T M_b \right). \tag{4}$$

After that, the relationship matrix A is normalized by rows to obtain the attention map A_r, and the enhanced feature map $\hat{M}_b \in \mathbb{R}^{WHC}$ of the current frame is obtained by multiplying it with the feature map M_b of the current frame, as follows:

$$A_r = softmax(A), \tag{5}$$

$$\hat{M}_b = M_b A_r. \tag{6}$$

Iterative Computation of Joint Attention Mechanism. The deep network structure has achieved great success due to its powerful representation ability. Therefore, we extend the joint attention module to a deeper structure to obtain more powerful temporal and spatial features. The deep joint attention module is composed of K cascaded joint attention modules $L^{(1)}, L^{(2)}, \ldots, L^{(K)}$. Let $\hat{M}_b^{(k-1)}$ and $\hat{M}_a^{(k-1)}$ represent the input features of the k-*th* layer joint attention module $L^{(k)}$. The output of the k-*th* layer joint attention module is $\hat{M}_a^{(k)}$ and $\hat{M}_b^{(k)}$ as follows:

$$\hat{M}_b^{(k)}, \hat{M}_a^{(k)} = L^{(k)}(\hat{M}_b^{(k-1)}, \hat{M}_a^{(k-1)}) \tag{7}$$

where $\hat{M}_b^{(k)}$ is calculated by Eq. 6 and set $\hat{M}_b^{(0)} = N_b$ and $\hat{M}_b^{(0)} = N_b$.

Since directly stacking the joint attention modules will lead to performance degradation, inspired by the residual network [12], we stack the joint attention modules together in a residual way as follows:

$$\hat{M}_b^{(k)} = \hat{M}_b^{(k-1)} + M_b^{(k-1)} A_r, \tag{8}$$

$$\hat{M}_a^{(k)} = \hat{M}_a^{(k-1)} + M_a^{(k-1)}. \tag{9}$$

3.2 Network Architecture

Our method is a Siamese network composed of three cascaded parts, including a feature embedding module based on DeepLabv3 [13], a joint attention module and a segmentation module. The feature embedding module encodes the video frame and extracts the appearance feature map of the video frame. The joint attention module captures the correlations between the two frames, enhances the feature map of the current frame, and supplements the appearance features of a single frame with the temporal information of the video. The segmentation module is used to obtain the segmentation results of the foreground objects. In this section, we will explain the architecture of our method from the scenarios of training and testing.

Network Architecture during Training
In the training process, our method randomly extracts a pair of video frames from the same video as input. The network architecture during training is shown in Fig. 1.

Our method alternately uses static image data and dynamic VOS data for training. The salient object segmentation dataset is used to train the backbone feature embedding module. So that the learned feature embedding module can capture and distinguish the most interesting objects. At the same time, in order to ensure that our method can capture the global consistency of the primary object, we use the VOS dataset to train the entire network, where the joint attention module captures the correlations between the video frames. Specifically, our method randomly selects two frames in a video sequence to construct a training pair.

Our method uses weighted binary cross-entropy loss to train the network as follows:

$$L_C(Y, O) = -\sum_x (1 - \eta) o_x \log(y_x) + \eta(1 - o_x) \log(1 - y_x), \tag{10}$$

where O represents the ground-truth, and Y represents the prediction results, o_x or y_x is the corresponding value of pixel x, and η is the ratio of foreground-background pixel numbers.

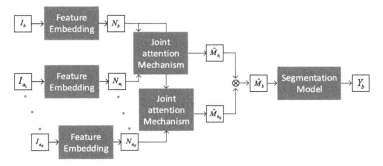

Fig. 3. Network architecture in the testing phase.

Network Architecture during Testing. After the network is trained, the testing video can be tested in an unsupervised manner. A variety of ways can be chosen to construct the network architecture during testing. The most intuitive way is to input each frame that needs to be segmented and the reference frame sampled in the same video into the network, and performs this operation frame-by-frame until all the segmentation results are obtained. However, the strategy generates too much noise, since global information in the video is not fully mined. Therefore, our method introduces multiple reference frames in the testing process to obtain more accurate segmentation results. When introducing multiple reference frames, a feasible way is to input N different reference frames uniformly sampled from the same video into the inference branch, and average all predictions. Specifically, for a query frame, a reference frame set containing K reference frames is used, and the enhanced features of the current frame from different reference frames are combined as follows: $\hat{M}_b \leftarrow \frac{1}{K} \sum_{k=1}^{K} \hat{M}_{b_k}$. The network architecture of the test phase is shown in Fig. 3. In the testing phase, the features enhanced by the joint attention module can effectively capture foreground information from the global view by considering more reference frames. Then we input into the segmentation module to generate the final output. Finally, CRF [14] is used as a post-processing step to get the final result.

4 Experiments

4.1 Datasets and Evaluation

DAVIS-16. DAVIS-16 [2] is currently one of the most popular and challenging video object segmentation benchmarks. The dataset contains 50 high-resolution video sequences, a total of 3455 densely labeled frames. DAVIS-16 is divided into 30 training videos and 20 test videos, covering multiple types of object categories.

DAVIS-17. DAVIS-17 [15] expands the training set and test set of DAVIS-16, and recalibrates all videos with multi-instance ground-truth, making it a multi-instance video segmentation dataset.

FBMS-59. FBMS-59 [16] is also a widely used unsupervised dataset. It contains 59 videos, of which 29 are training video sequences and 30 are test video sequences. Each video sequence of the FBMS-59 dataset contains multiple moving objects.

Evaluation. We adopt three standard evaluation metrics: (i) the spatial accuracy of segmentation, namely Region Similarity J, (ii) the consistency of similar contour, namely Contour Accuracy F, and (iii) Time Stability T.

4.2 Ablation Study

In this section, we verify the effectiveness of the joint attention module on the DAVIS-16 and the performance of different backbone networks.

The Effectiveness of the Joint Attention Module. In order to verify the effectiveness of the joint attention module, we compare the performance of the full model and the model with the same architecture but without the joint attention module. The full model represents the complete model that we proposed, the backbone network used is ResNet-101, and includes a joint attention module. Model_1 represents a model with the same architecture as the full model but without a joint attention module.

When excluding the joint attention module, the encoder in the network is equivalent to a standard two-stream model. As shown in Table 1, Model_1 encounters a huge performance degradation (1.7% in Mean J and 3.2% in Mean F), which demonstrates the effectiveness of the joint attention module.

In addition, we also evaluate the impact of the numbers of joint attention modules. Table 2 reports the results. As N increases, the performance of the model gradually improves, reaching the best performance at $N = 5$. At this situation, the Mean J is 80.9%, and the Mean F is 80.3%. When the number of cascaded joint attention module continues to increase, the performance of the model decreases. Based on the experimental results, our model uses $N = 5$ as the number of joint attention modules.

The Impact of Different Backbone Networks. In order to verify that the high performance of our model is not mainly due to the powerful backbone network, this section uses ResNet-50 instead of ResNet-101 to construct another network, which is represented as Model_2 in Table 1. The performance differences of different backbone networks on

Table 1. The ablation experiment on DAVIS-16, measured by the Mean J and Mean F.

	Full model	Model_1	Model_2
CAM	√	-	√
Res50	-	-	√
Res101	√	√	-
Mean J	80.9	79.1	79.7
Mean F	80.3	77.1	78.9

Table 2. Impact of different number of joint attention modules on DAVIS-16, measured by the Mean J and Mean F. N refers to the number of cascaded joint attention modules.

Metric	$N = 0$	$N = 1$	$N = 3$	$N = 5$	$N = 7$
Mean J	79.1	79.8	80.6	80.9	80.5
Mean F	77.1	79.2	79.9	80.3	80.3

Table 3. Experimental comparison of the algorithms on DAVIS-16, measured by the region similarity J, contour accuracy F and time stability T. The best result for each metric is bold-faced.

Method	J			F			T
	Mean	Recall	Decay	Mean	Recall	Decay	Mean
TRC[17]	47.3	49.3	8.3	44.1	43.6	12.9	39.1
CVOS[18]	48.2	54	10.5	44.7	52.6	11.7	25
KEY[19]	49.8	59.1	14.1	42.7	37.5	10.6	26.9
CUT[20]	55.2	57.5	2.2	55.2	51.9	3.4	27.7
SFL[21]	67.4	81.4	6.2	66.7	77.1	5.1	28.2
FSEG[1]	70.7	83.5	1.5	65.3	73.8	1.8	32.8
LVO[22]	75.9	89.1	**0**	72.1	83.4	1.3	26.5
ARP[23]	76.2	91.1	7	70.6	83.5	7.9	39.3
PDB[24]	77.2	90.1	0.9	74.5	84.4	**-0.2**	29.1
LSMO[25]	78.2	89.1	4.1	75.9	84.7	3.5	**21.2**
MotAdapt[26]	77.2	87.8	5	77.4	84.4	3.3	27.9
AGS[6]	79.7	91.1	1.9	77.4	**85.8**	1.6	26.7
Ours	**80.9**	**92.3**	2.3	**80.3**	83.5	1.9	30.2

DAVIS-16 are shown in Table 1. The two metrics of Model_2 in Mean J and Mean F are 1.2% and 1.4% lower than full model, but still have good performance. This further confirms the effectiveness of our model.

4.3   Comparison to the State-Of-The-Arts

Evaluation on DAVIS-16. Table 3 reports the performance comparison between our method and other state-of-the-art methods. It can be seen that our method is superior to all the reported methods. Compared with the second-best method AGS, our method achieves gains of 1.2% and 2.9% on Mean J and Mean F. The qualitative results of our method on DAVIS-16 are shown in Fig. 4.

In Table 3, several UVOS methods (*e.g.*, CUT, LMP, FSEG, and LVO) all use appearance information and additional motion information to improve segmentation performance. In contrast, our method only uses appearance information, but achieves leading performance. This is because our method fully considers the temporal information and

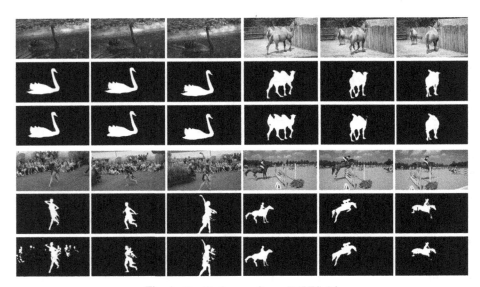

Fig. 4. Qualitative results on DAVIS-16.

Table 4. Experimental comparison on DAVIS-17, measured by the region similarity J and contour accuracy F. The best result for each metric is bold-faced.

Metrics		RVOS[27]	PDB[24]	AGS[6]	ours
J	Mean	36.8	53.2	55.5	**56.3**
	Recall	40.2	58.9	**61.6**	61.3
	Decay	**0.5**	4.9	7.0	6.2
F	Mean	45.7	57.0	59.5	**60.7**
	Recall	46.4	60.2	62.8	**63.7**
	Decay	**1.7**	6.8	9.0	8.8

captures the relationships between frames in the same video through the proposed joint attention module. Compared with the method of using optical flow to capture continuous temporal information, our method explores the correlations from a global perspective, has a greater advantage in dealing with the interference of similar objects.

Evaluation on DAVIS-17. We also perform experiments on DAVIS-17, and the qualitative results is shown in Fig. 5. The overall results are shown in Table 4. Our method has reached best performance among multiple metrics. Our method reached the best Mean J, which is 56.3%, it is 0.8% higher than the second-ranked AGS. We also reached the best Mean F, which is 60.7%. The experimental results verify that the joint attention module we proposed performs good in complex videos containing multiple moving instances.

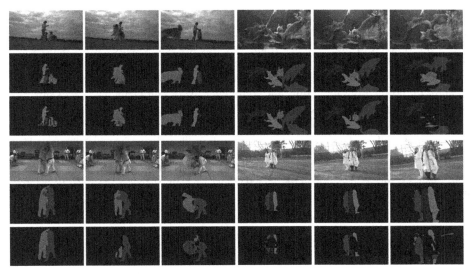

Fig. 5. Qualitative results on DAVIS-17.

Table 5. Experimental comparison on FBMS-59, measured by the mean region similarity (Mean J). The best result is bold-faced.

Metric	IET[5]	OBN[28]	FSEG[1]	PDB[24]	MSTP[29]	ARP[23]	Ours
Mean J	71.9	73.9	68.4	74.0	60.8	59.8	**75.8**

Evaluation on FBMS-59. We conduct experiments on FBMS-59 in order to complete the experiments. Table 5 records the quantitative results on the test set of FBMS-59. As shown in Table 5, our method has reached 75.8% on the Mean J metric, leading the advanced methods MSTP, ARP, IET, OBN, PDB.

Some methods in Table 5 use optical flow to predict the segmentation mask in addition to RGB image input. Considering that the appearance of many foreground objects in FBMS-59 is similar to the background, but the motion patterns of the two are different, the optical flow information has obvious advantages over this dataset, so such methods usually perform better. In contrast, our method only uses the original RGB frames, but achieves better performance, demonstrating its high performance and robustness.

5 Conclusion

In this work, we analyze the task of UVOS, and believe that the primary objects need to meet the characteristics of local saliency and global consistency. We propose an UVOS method based on joint attention mechanism. The main contribution of our method is the joint attention module, which is composed of two soft attention units and a co-attention unit. The soft attention unit enhances the features of the important positions in the feature space, and highlights the local saliency of the targets. The co-attention unit calculates the similarity between different frames in the same video, and uses the object information of the reference frame to guide the segmentation of current frame. In order to encourage deeper information exchange between video frames, we stack the joint attention modules together to form a deep joint attention mechanism, which captures the correlations more fully and comprehensively. Our method achieves superior performance on several representative VOS benchmarks.

Acknowledgements. This work was supported in part by the National Natural Science Foundation of China under Grant 62172417, 61772530 and 61806206, in part by the Six Talent Peaks Project in Jiangsu Province under Grant 2018-XYDXX-044, in part by the Natural Science Foundation of Jiangsu Province under Grant BK20180639, and in part by Xuzhou Science and Technology Plan Funds (No. KC19005).

References

1. Jain, S.D., Xiong, B., Grauman, K.: Fusionseg: learning to combine motion and appearance for fully automatic segmentation of generic objects in videos. In: 2017 IEEE Conference on Computer Vision and Pattern Recognition (CVPR). IEEE, pp. 2117–2126 (2017)
2. Perazzi, F., Pont-Tuset, J., McWilliams, B., Van Gool, L., Gross, M., Sorkine-Hornung, A.: A benchmark dataset and evaluation methodology for video object segmentation. In: Proceedings of the IEEE Conference on Computer Vision and Pattern Recognition, pp. 724–732 (2016)
3. Caelles, S., Montes, A., Maninis, K.-K., Chen, Y., Van Gool, L., Perazzi, F.: Pont-Tuset JJapa: The 2018 DAVIS challenge on video object segmentation (2018)
4. Tokmakov, P., Alahari, K., Schmid, C.: Learning motion patterns in videos. In: Proceedings of the IEEE Conference on Computer Vision and Pattern Recognition, pp. 3386–3394 (2017)
5. Li, S., Seybold, B., Vorobyov, A., Fathi, A., Huang, Q., Jay Kuo, C.-C.: Instance embedding transfer to unsupervised video object segmentation. In: Proceedings of the IEEE Conference on Computer Vision and Pattern Recognition, pp. 6526–6535 (2018)
6. Wang, W., et al.: Learning unsupervised video object segmentation through visual attention. In: Proceedings of the IEEE/CVF Conference on Computer Vision and Pattern Recognition, pp. 3064–3074 (2019)
7. Denil, M., Bazzani, L., Larochelle, H., de Freitas, N.J.Nc.: Learning where to attend with deep architectures for image tracking. **24**, 2151-2184 (2012)
8. Jetley, S., Lord, N.A., Lee, N., Torr, P.H.J.: Learn to pay attention (2018)
9. Chu, X., Yang, W., Ouyang, W., Ma, C., Yuille, A.L., Wang, X.: Multi-context attention for human pose estimation. In: Proceedings of the IEEE Conference on Computer Vision and Pattern Recognition, pp. 1831–1840 (2017)
10. Lu, J., Yang, J., Batra, D., Parikh, D.J.: Hierarchical question-image co-attention for visual question answering (2016)
11. Zhou, T., Wang, S., Zhou, Y., Yao, Y, Li, J., Shao, L.: Motion-attentive transition for zero-shot video object segmentation. In: Proceedings of the AAAI Conference on Artificial Intelligence, vol. 07, pp. 13066–13073 (2020)
12. Wang F et al.: Residual attention network for image classification. In: Proceedings of the IEEE Conference on Computer Vision and Pattern Recognition, pp. 3156–3164 (2017)
13. Chen, L.-C., Papandreou, G., Schroff, F., Adam, H.J.: Rethinking atrous convolution for semantic image segmentation (2017)

14. Perazzi, F., Khoreva, A., Benenson, R., Schiele, B., Sorkine-Hornung, A.: Learning video object segmentation from static images. In: Proceedings of the IEEE Conference on Computer Vision and Pattern Recognition, pp. 2663–2672 (2017)
15. Pont-Tuset, J., Perazzi, F., Caelles, S., Arbeláez, P., Sorkine-Hornung, A., Van Gool, L.J.: The 2017 davis challenge on video object segmentation (2017)
16. Ochs, P., Malik, J., Brox, T.J.: Intelligence m: Segmentation of moving objects by long term video analysis. **36**, 1187-1200 (2013)
17. Fragkiadaki, K., Zhang, G., Shi, J.: Video segmentation by tracing discontinuities in a trajectory embedding. In: 2012 IEEE Conference on Computer Vision and Pattern Recognition. IEEE, pp. 1846–1853 (2012)
18. Taylor, B., Karasev, V., Soatto, S.: Causal video object segmentation from persistence of occlusions. In: Proceedings of the IEEE Conference on Computer Vision and Pattern Recognition, pp. 4268–4276 (2015)
19. Lee, Y.J., Kim, J., Grauman, K.: Key-segments for video object segmentation. In: 2011 International Conference on Computer Vision. IEEE, pp. 1995–2002 (2011)
20. Cheng, J., Tsai, Y.-H., Wang, S., Yang, M.-H.: Segflow: Joint learning for video object segmentation and optical flow. In: Proceedings of the IEEE International Conference on Computer Vision, pp. 686–695 (2017)
21. Keuper, M., Andres, B., Brox, T.: Motion trajectory segmentation via minimum cost multicuts. In: Proceedings of the IEEE International Conference on Computer Vision, pp. 3271–3279 (2015)
22. Tokmakov, P., Alahari, K., Schmid, C.: Learning video object segmentation with visual memory. In: Proceedings of the IEEE International Conference on Computer Vision, pp. 4481–4490 (2017)
23. Koh, Y.J., Kim, C.-S.: Primary object segmentation in videos based on region augmentation and reduction. In: 2017 IEEE Conference on Computer Vision and Pattern Recognition (CVPR). IEEE, pp. 7417–7425 (2017)
24. Song, H., Wang, W., Zhao, S., Shen, J., Lam, K.-M.: Pyramid dilated deeper ConvLSTM for video salient object detection. In: Ferrari, V., Hebert, M., Sminchisescu, C., Weiss, Y. (eds.) ECCV 2018. LNCS, vol. 11215, pp. 744–760. Springer, Cham (2018). https://doi.org/10.1007/978-3-030-01252-6_44
25. Fragkiadaki, K., Arbelaez, P., Felsen, P., Malik, J.: Learning to segment moving objects in videos. In: Proceedings of the IEEE Conference on Computer Vision and Pattern Recognition, pp. 4083–4090 (2015)
26. Siam, M., et al.: Video object segmentation using teacher-student adaptation in a human robot interaction (hri) setting. In: 2019 International Conference on Robotics and Automation (ICRA). IEEE, pp. 50–56 (2019)
27. Ventura, C., Bellver, M., Girbau, A., Salvador, A., Marques, F., Giro-i-Nieto, X.: Rvos: End-to-end recurrent network for video object segmentation. In: Proceedings of the IEEE Conference on Computer Vision and Pattern Recognition, pp. 5277–5286 (2019)
28. Li, S., Seybold, B., Vorobyov, A., Lei, X., Kuo, C.-C.: Unsupervised video object segmentation with motion-based bilateral networks. In: Ferrari, V., Hebert, M., Sminchisescu, C., Weiss, Y. (eds.) ECCV 2018. LNCS, vol. 11207, pp. 215–231. Springer, Cham (2018). https://doi.org/10.1007/978-3-030-01219-9_13
29. Hu, Y.-T., Huang, J.-B., Schwing, A.G.: Unsupervised video object segmentation using motion saliency-guided spatio-temporal propagation. In: Ferrari, V., Hebert, M., Sminchisescu, C., Weiss, Y. (eds.) ECCV 2018. LNCS, vol. 11205, pp. 813–830. Springer, Cham (2018). https://doi.org/10.1007/978-3-030-01246-5_48

Foreground Feature Selection and Alignment for Adaptive Object Detection

Zixuan Huang[1,2,3], Huicheng Zheng[1,2,3(✉)], and Manwei Chen[1,2,3]

[1] School of Computer Science and Engineering, Sun Yat-sen University,
Guangzhou, China
zhenghch@mail.sysu.edu.cn
[2] Key Laboratory of Machine Intelligence and Advanced Computing,
Ministry of Education, Guangzhou, China
[3] Guangdong Key Laboratory of Information Security Technology,
Guangzhou, China

Abstract. Recently, remarkable progress has been witnessed in adaptive object detection, which aims to mitigate the distributional shifts between source domain and target domain. Domain-adversarial learning methods align the features of different levels to minimize the domain discrepancy, which have been proven effective for adapting object detectors. Most domain adaptation methods align whole-image features. Therefore, foreground alignment may be interfered by the backgrounds. In this work, we propose Foreground Feature Alignment Framework (FFAF) that strengthens the foreground alignment. One of our key contributions is the Foreground Selection Module (FSM), which captures the foreground features that are crucial for object detection and helpful for subsequent feature alignment. Additionally, we align the foreground features by integrating multi-level domain classifiers. Multi-level Domain adaptation (MDA) can simultaneously bridge the domain gap at various representation levels. We evaluate our method with multiple experiments, whose results demonstrate that our method achieves significant improvements in different cross-domain object detection tasks.

Keywords: Object detection · Domain adaptation · Foreground feature alignment

1 Introduction

Object detection is a significant computer vision task which aims to localize and identify all objects of interest in an image. In recent years, a number of state-of-the-art (SOTA) CNN-based detectors have been proposed, such as Faster R-CNN [21], SSD [18] and YOLO [20]. However, when these traditional detectors face a new domain where the background, viewpoint, illumination differ from those in the source domain, their performances drop significantly. As shown in

H. Ma et al. (Eds.): PRCV 2021, LNCS 13019, pp. 166–178, 2021.
https://doi.org/10.1007/978-3-030-88004-0_14

(a) Normal weather   (b) Foggy weather

Fig. 1. Examples of cross-domain object detection. The green boxes and the yellow boxes indicate the detected and the missing instances, respectively. The object detector is trained on Cityscapes [4]. When the detector faces different weather conditions, its performance drops significantly. (a) Detection result on the Cityscapes. (b) Detection result on the Foggy Cityscapes [24], which is generated by applying fog synthesis on the Cityscapes.

Fig. 1, the model trained in normal images can accurately detect objects in an image of the same style, but misses many objects in a foggy image.

In order to address the domain shift problem, unsupervised domain adaptation offers a promising solution to l earn a robust object detector by utilizing the labeled data of source domain and the unlabeled data of the target domain.

Recently, there have been a lot of significant advances in adaptive object detection. We observe that most of the previous methods learn the domain-confused features by aligning features of different levels across the two domains. The early work DA Faster R-CNN [3] proposed image-level and instance-level feature alignment. When performing the instance-level feature alignment, all instances need to be aligned without considering if they are foregrounds or backgrounds. Another representative work Strong-Weak Distribution Alignment (SWDA) [23] further proposed local alignment to align low-level features. Whether it is local or global alignment, foregrounds and backgrounds are aligned simultaneously, which means backgrounds may interfere the foreground alignment. Some researchers [30, 32] pointed out that foregrounds share more common features than backgrounds, so foreground alignment is vital in the adaptive object detection.

In order to align the regions that contain objects of interest, Zhu et al. [32] proposed region mining method that applies K-means clustering to find the centroids of foreground regions. Then a fixed-size region is assigned to each centroid for feature alignment. However, these regions still contain backgrounds inevitably, because each object has a different size. Zheng et al. [30] proposed Attention-based Region Transfer (ART) to raise more attention to foreground region alignment according to the activation of the feature maps. However, it is uncertain that the high activation of features is related to foreground instances.

Motivated by the importance of foreground features and the shortcomings of the above methods, we propose a novel end-to-end Foreground Feature Alignment Framework (FFAF), which can boost the foreground alignment. Our main idea of the framework is that we should focus more on aligning foreground features, since they are especially crucial in the detection task. Concretely, we introduce a Foreground Selection Module to select foreground features of different sizes rather than a fixed size. After selecting foreground features, we perform domain-adversarial learning by Multi-level Domain Adaptation, since more feature alignment modules can enhance the discrimination ability of model in different domains. Due to the precise multi-level alignment of transferable foregrounds, the object detector can achieve better performance.

In summary, the contributions of this work can be summarized as follows:

- We design a Foreground Selection Module to search for foreground regions. It can be utilized as a plug-and-play component before the domain classifiers for better subsequent feature alignment.
- We apply Multi-level Domain Adaptation to align the features from multiple scales, which can enhance the effect of feature alignment. Besides, unlike some previous methods, we do not add any additional component to our framework during inference, which means that the inference time of our model is the same as the original detector.
- Extensive experiments on various datasets show superior performance of our framework.

2   Related Work

2.1   Object Detection

In recent years, with the rapid development of convolutional neural network, object detection has made great progress. Many SOTA methods are based on two-stage detectors. The most representative two-stage detectors are the R-CNN series detectors [7,8,21]. R-CNN [8] is one of the earliest deep networks that generate region proposals by low-level vision techniques, which feeds each region of interest (RoI) into a network for classification and bounding box regression. Fast R-CNN [7] introduces a shared convolutional neural network to extract the RoI features. Faster R-CNN [21] introduces a Region Proposal Network (RPN) to generate a sparse set of candidate proposals, which can achieve better performance. R-CNN series detectors achieve dominant performance on benchmark datasets, such as PASCAL VOC [5] and MS COCO [17]. Moreover, FPN [15], Cascade R-CNN [1] and Mask R-CNN [9] further develop the two-stage detectors and improve the performance.

However, these detectors face great challenges when there is a domain gap between source domain and target domain.

2.2   Adaptive Object Detection

Recently, adaptive object detection has advanced rapidly. DA Faster R-CNN [3] is the first work for adaptive object detection, which introduces image-level and instance-level feature alignment to align features of different levels by adversarial learning. Besides, it exploits the consistency check to strengthen the robustness of the detector. Saito et al. [23] proposed Strong-Weak Distribution Alignment (SWDA) which consists of strong local feature alignment and weak global feature alignment. Xie et al. [27] presented a multi-level domain adaptive model (MAD) to align the distributions of features, which demonstrates that more domain classifiers can better bridge the domain gap. Zhu et al. [32] adjusted region mining method and region-level alignment to mine the discriminative regions. Zheng et al. [30] proposed Attention-based Region Transfer (ART) and Prototype-based Semantic Alignment (PSA). ART aims to focus more on foreground regions alignment by attention mechanism. PSA aligns different categories of feature distributions by prototype-based adversarial learning.

Nevertheless, these methods cannot align the transferable foregrounds properly. In this work, we focus on enhancing the foreground alignment.

3   Method

In the task of adaptive object detection, we can access an annotated source domain dataset $\mathcal{D}_s = \{(x_i^s, y_i^s)\}_{i=1}^{N^s}$ of N^s samples and an unlabeled target domain dataset $\mathcal{D}_t = \{(x_i^t)\}_{i=1}^{N^t}$ of N^t samples. x_i^s and x_i^t denote the i-th image from source domain and target domain, respectively. y_i^s denotes the corresponding labels which consist of bounding boxes and the associated categories. The source domain dataset and target domain dataset are sampled from different data distributions, but they share a same label space. The domain label of source domain and target domain is 1 and 0, respectively. Our goal is to learn an adaptive object detector trained in $\mathcal{D}_s$ and $\mathcal{D}_t$, which can perform well in $\mathcal{D}_t$.

3.1   Framework Overview

The overview of our proposed FFAF is shown in Fig. 2, which is based on Faster R-CNN [21]. We propose a novel Foreground Selection Module to select the foreground features for feature alignment, which is based on the prediction of the detector. Moreover, we exploit Multi-level Domain Adaptation to simultaneously align the feature distributions of different levels.

3.2   Foreground Selection Module

In order to boost the foreground alignment, we propose a Foreground Selection Module (FSM) to select the foreground features. For the source domain images, we can simply exploit their ground-truth bounding boxes since the source domain dataset is annotated. For the unlabeled target domain images, we can exploit the bounding box and category predictions.

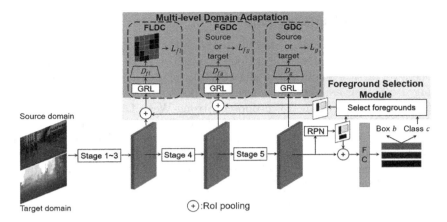

Fig. 2. Overview of our Foreground Feature Alignment Framework, which consists of two parts, i.e., Foreground Selection Module (FSM) and Multi-level Domain Adaptation (MDA). FSM is a plug-and-play component to select foreground features according to the predictions of instances. MDA aims to align features of different levels, which consists of three modules, i.e., Foreground Local Domain Classifier (FLDC), Foreground Global Domain Classifier (FGDC) and Global Domain Classifier (GDC).

More specifically, for the target domain images, we pick out the region proposals from the output of the RPN where category predictions c are not background, since the goal of RPN is to predict the foreground regions. For the source images, to stay consistent with the target images, we also select foreground regions from RPN rather than using the ground truth directly. For the annotated images, it is worth noting that some instances from RPN are not positive samples due to the RoI sampling strategy [21], so we only choose the positive samples as foreground instances.

As aforementioned, the sizes of the selected features are different, so it is hard to train and design the domain classifiers. In order to obtain same sized foreground features, we resize them by a RoI pooling layer. By this way, the resized foregrounds contain less background information than the fixed-size regions [32]. After resizing these features, we can put these RoI-pooled features together as a batch to train, which can improve the training speed and reduce the difficulty of designing the domain classifiers. The selected foregrounds can be represented as $\{R(F(x_i), b_{ij}^f)\}_{j=1}^{N_i}$, where x_i denotes the i-th image, N_i denotes the number of foreground regions selected by FSM, b_{ij}^f denotes the corresponding bounding box of j-th selected foreground region, R denotes the RoI pooling operation, F denotes the feature extractor.

3.3 Multi-level Domain Adaptation

Domain adaptation usually utilizes the domain classifier and the gradient reverse layer (GRL) [6] to align the features. The domain classifier aims to identify which domain the input features belong to. The GRL reverses the sign of the gradient, which can help the feature extractor confuse the domain classifier and extract the domain-invariant features. To bridge the domain gap in different levels, we exploit multi-level domain classifiers to align the features.

The Foreground Local Domain Classifier (FLDC) takes the foreground features selected by FSM as input. FLDC aims to distinguish which domain each patch of the foreground comes from. N_i^s and N_i^t denote the number of selected foreground regions from the i-th source and target image, respectively. F_l is part of the backbone and extracts the feature maps whose width and height is H and W, respectively. b_{ij}^{tf} and b_{ij}^{sf} denote j-th selected bounding box of i-th image from target domain and source domain, respectively. $R(F_l(x_i^t), b_{ij}^{tf})$ and $R(F_l(x_i^s), b_{ij}^{sf})$ denote the pooled foreground features from source domain and target domain, respectively, $D_{fl}(\cdot)_{wh}$ denotes the domain prediction of the foreground features at (w, h). We utilize the least-squares loss to train FLDC, which is the same as the loss function of local alignment in [23]. The loss $\mathcal{L}_{fl}$ is summarized as,

$$
\begin{aligned}
\mathcal{L}_{fl} = {} & \frac{1}{N^t N_i^t W H} \sum_{i=1}^{N^t} \sum_{j=1}^{N_i^t} \sum_{w=1}^{W} \sum_{h=1}^{H} D_{fl}(R(F_l(x_i^t), b_{ij}^{tf}))_{wh}^2 \\
& + \frac{1}{N^s N_i^s W H} \sum_{i=1}^{N^s} \sum_{j=1}^{N_i^s} \sum_{w=1}^{W} \sum_{h=1}^{H} (1 - D_{fl}(R(F_l(x_i^s), b_{ij}^{sf})))_{wh}^2
\end{aligned}
\tag{1}
$$

Similar to FLDC, we also feed the selected foreground features into the Foreground Global Domain Classifier (FGDC). FGDC aims to identify the domain of global foreground features. F_{fg} is part of the backbone that contains 4 stages. $D_{fg}(\cdot)$ denotes the domain prediction of the foreground features. We adopt cross-entropy loss, which is always used in the classification task, as the loss function of FGDC. We denote the loss of FGDC as $\mathcal{L}_{fg}$ as follows,

$$
\begin{aligned}
\mathcal{L}_{fg} = {} & -\frac{1}{N^t N_i^t} \sum_{i=1}^{N^t} \sum_{j=1}^{N_i^t} log(1 - D_{fg}(R(F_{fg}(x_i^t), b_{ij}^{tf}))) \\
& - \frac{1}{N^s N_i^s} \sum_{i=1}^{N^s} \sum_{j=1}^{N_i^s} log(D_{fg}(R(F_{fg}(x_i^s), b_{ij}^{sf})))
\end{aligned}
\tag{2}
$$

Although foregrounds play an important role in object detection, some backgrounds are also crucial because they may contain context information, which means image-level alignment is helpful. We utilize Global Domain Classifier (GDC) to align entire feature maps. The global features are extracted by F_g. $D_g(\cdot)$ denotes the domain prediction of global features. We follow [23] to train GDC by Focal loss [16]. The loss $\mathcal{L}_g$ is summarized as,

$$\mathcal{L}_g = -\frac{1}{N^t} \sum_{i=1}^{N^t} [D_g(F_g(x_i^t))^\gamma log(1 - D_g(F_g(x_i^t)))]$$
$$-\frac{1}{N^s} \sum_{i=1}^{N^s} [(1 - D_g(F_g(x_i^s)))^\gamma log(D_g(F_g(x_i^s)))] \tag{3}$$

3.4   Overall Objective

As aforementioned, we adopt adversarial loss $\mathcal{L}_{adv}$ to train a domain adaptive object detector, which can be summarized as,

$$\mathcal{L}_{adv} = \mathcal{L}_{fl} + \mathcal{L}_{fg} + \mathcal{L}_g \tag{4}$$

We denote the loss of the detection as $\mathcal{L}_{det}$, which is the same as Faster R-CNN [21]. The detector only utilizes the detection loss when it is trained in the source domain images. The overall objective of FFAF is,

$$\max_{D} \min_{F} (\mathcal{L}_{det} - \mathcal{L}_{adv}) \tag{5}$$

4   Experiments

We evaluate our framework on three cross-domain object detection tasks, including PASCAL [5] to Clipart [13], Cityscapes [4] to Foggy Cityscapes [24] and Cityscapes to BDD100k [29]. The results demonstrate the effectiveness of FFAF.

4.1   Implementation Details

We adopt VGG-16 [26] or ResNet-101 [10] pre-trained in ImageNet [22] as backbone, which follows other methods [12, 23, 28]. The initial learning rate is 1×10^{-3}, which is divided by 10 after every 50k iterations out of the total 120k iterations. We set γ to 5.0. For different cross-domain detection tasks, the losses of domain classifiers and the detection loss should balance in different ways. In practice, we balance the loss by modifying the gradient scale in the GRL. More specifically, we multiply the gradients from FLDC, FGDC and GDC by hyper parameters β_1, β_2 and β_3 in GRL during backpropagation respectively.

4.2   Adaptation Results

Adaptation to Dissimilar Domains. PASCAL VOC [5] is a well-known benchmark dataset, which contains 20 categories real world objects. We use the training set and validation set of VOC 07 and 12 as source domain, which contains 16, 551 images. Clipart [13] contains 1, 000 images and has the same 20 categories objects as PASCAL. Following [23], we utilize Clipart as target domain and use all images for both training (without labels) and test. In this experiment, we use ResNet-101 [10] as the backbone of Faster R-CNN, which is the same as all comparison methods. β_1, β_2 and β_3 are all set to -1.

As shown in Table 1, our proposed method outperforms all other existing methods. The source only model trained on the source domain can be regarded as the baseline. Our model can outperform the most competitive method by 0.8% mAP (42.1% → 42.9%).

In order to prove the effectiveness of proposed modules, we further conduct ablation studies. The mAP of the model without FSM (w/o-FSM) is 40.7%. With the help of FSM, our model can further improve mAP by 2.2% (40.7% → 42.9%), which validates the benefit of FSM. Without FLDC (w/o-FLDC) or FGDC (w/o-FGDC), the mAP drops to 41.9% or 40.0% respectively, which indicates that both FLDC and FGDC are vital for our proposed framework.

Table 1. Results on adaptation from PASCAL to Clipart. "w/o-FLDC-FGDC" indicates dropping both FLDC and FGDC, which means that we only exploit GDC. "w/o-FGDC" indicates dropping FGDC. "w/o-FSM" indicates dropping FSM, which means that we take global features as the input of FLDC and FGDC. "w/o-FLDC" indicates dropping FLDC.

Methods	Aero	Bcycle	Bird	Boat	Bottle	Bus	Car	Cat	Chair	Cow
Source only	19.8	43.0	15.9	24.3	25.2	32.5	32.0	15.4	20.9	29.3
SWDA [23]	26.2	48.5	32.6	33.7	38.5	54.3	37.1	18.6	34.8	58.3
HTCN [2]	33.6	58.9	34.0	23.4	45.6	57.0	39.8	12.0	39.7	51.3
SCL [25]	44.7	50.0	33.6	27.4	42.2	55.6	38.3	19.2	37.9	69.0
DD-MRL [14]	25.8	63.2	24.5	42.4	47.9	43.1	37.5	9.1	47.0	46.7
ATF [12]	41.9	67.0	27.4	36.4	41.0	48.5	42.0	13.1	39.2	75.1
Ours-w/o-FLDC-FGDC	31.4	46.7	31.2	29.0	36.9	51.5	41.3	15.2	33.1	46.5
Ours-w/o-FGDC	40.7	48.8	35.3	24.9	45.3	57.2	40.3	17.5	40.6	48.4
Ours-w/o-FSM	29.9	60.3	36.2	28.9	42.0	54.5	40.6	19.2	38.8	54.6
Ours-w/o-FLDC	32.9	53.8	36.9	32.9	40.7	48.9	43.0	23.9	41.2	54.6
Ours	38.2	54.5	34.6	30.0	36.5	60.4	43.5	22.0	40.2	60.8

Methods	table	dog	horse	mbike	prsn	plant	sheep	sofa	train	tv	mAP
Source only	13.4	6.2	31.0	54.7	35.9	30.6	2.6	19.9	34.2	27.4	25.7
SWDA [23]	17.0	12.5	33.8	65.5	61.6	52.0	9.3	24.9	54.1	49.1	38.1
HTCN [2]	21.1	20.1	39.1	72.8	63.0	43.1	19.3	30.1	50.2	51.8	40.3
SCL [25]	30.1	26.3	34.4	67.3	61.0	47.9	21.4	26.3	50.1	47.3	41.5
DD-MRL [14]	26.8	24.9	48.1	78.7	63.0	45.0	21.3	36.1	52.3	53.4	41.8
ATF [12]	33.4	7.9	41.2	56.2	61.4	50.6	42.0	25.0	53.1	39.1	42.1
Ours-w/o-FLDC-FGDC	15.5	17.9	29.0	58.7	54.0	45.5	22.1	28.0	51.0	49.1	36.7
Ours-w/o-FGDC	11.9	21.9	36.8	73.6	59.8	53.6	20.2	30.0	45.9	48.1	40.0
Ours-w/o-FSM	17.0	27.3	31.3	81.1	58.2	49.8	22.5	26.0	50.1	46.3	40.7
Ours-w/o-FLDC	20.2	29.7	43.2	65.6	64.6	46.5	26.9	25.0	52.9	54.1	41.9
Ours	18.6	31.1	32.0	83.1	65.3	46.3	27.2	32.3	50.7	51.8	42.9

Adaptation to Different Scenes. In this section, we evaluate our model on different scenes. We choose Cityscapes as source domain and BDD100k [29] as target domain. BDD100k contains 100k images from daytime or nighttime. We follow [28] and only choose the daytime subset to train and test, including 36, 728 training images and 5, 258 test images. It is worth noting that we do not report the detection result of the "train" category, because only 8 instances fall into the "train" category in the target domain. We follow [28] and utilize VGG-16 as the backbone. We set $\beta_1 = -0.01$, $\beta_2 = -0.1$ and $\beta_3 = -0.01$.

As shown in Table 2, the scene adaptation is more difficult than the weather adaptation. Our method outperforms the SOTA method [28] with a considerable margin of 2.7% mAP (26.9% → 29.6%).

Table 2. Results on adaptation from Cityscapes to BDD100k.

Methods	Person	Rider	Car	Truck	Bus	Train	Mcycle	Bicycle	mAP
Source only	27.3	27.8	42.2	18.2	18.5	-	18.9	20.1	24.7
SW-Faster-ICR-CCR [28]	31.4	31.3	46.3	19.5	18.9	-	17.3	23.8	26.9
Ours	34.0	34.0	46.3	21.2	21.6	-	22.2	27.5	29.6

Adaptation to Different Weather Conditions. We further evaluate our method on adaptation from normal weather to foggy weather. We use Cityscapes [4] as the source domain and Foggy Cityscapes [24] as the target domain. Foggy Cityscapes is generated by applying fog synthesis on the Cityscapes, so both datasets have 2, 975 images in the training set and 500 images in the test set. We use VGG-16 as the backbone of Faster R-CNN, which is the same as other comparison methods. β_1, β_2 and β_3 are set to -0.01, -0.1 and -0.01, respectively.

As shown in Table 3, we can see that our method also outperforms the SOTA method ATF [12] by 0.5% mAP (38.7% → 39.2%). Although HTCN [2] surpasses our method in this task, they exploit CycleGAN [31] to generate synthetic samples for training, which means their training set differs from ours. Besides, as shown in Table 1, our method can outperform HTCN by 2.6% mAP (40.3% → 42.9%) on adaptation from PASCAL to Clipart.

4.3 Visualization and Discussion

Detection Examples. Some of the detection results are shown in Fig. 3. We can observe that SCL [25] produces less true positives and more false positives. Thanks to the precise foreground feature alignment, our model detects objects more accurately.

Feature Visualization. We visualize the instance features on adaptation from PASCAL to Clipart. It is worth noting that we only select 6 categories of instances for a clear visualization. As shown in Fig. 4, the instance representations of the same category from two domains obtained by our method stay closer than those obtained by the source only model, which means that the instance features extracted by our model are domain-invariant.

Table 3. Results on adaptation from Cityscapes to Foggy Cityscapes.

Methods	Person	Rider	Car	Truck	Bus	Train	Mcycle	Bicycle	mAP
Source only	23.7	31.4	27.2	11.4	25.8	9.1	15.2	25.8	21.2
DA-Faster [3]	25.0	31.0	40.5	22.1	35.3	20.2	20.0	27.1	27.6
SCDA [32]	33.5	38.0	48.5	26.5	39.0	23.3	28.0	33.6	33.8
MAF [11]	28.2	39.5	43.9	23.8	39.9	33.3	29.2	33.9	34.0
SWDA [23]	29.9	42.3	43.5	24.5	36.2	32.6	30.0	35.3	34.3
iFAN [33]	32.7	40.0	48.5	27.9	45.5	31.7	22.8	33.0	35.3
MAD [27]	33.2	44.2	44.8	28.2	41.8	28.7	30.5	36.5	36.0
SW-Faster-ICR-CCR [28]	32.9	43.8	49.2	27.2	45.1	36.4	30.3	34.6	37.4
SCL [25]	31.6	44.0	44.8	30.4	41.8	40.7	33.6	36.2	37.9
ART-PSA [30]	34.0	46.9	52.1	30.8	43.2	29.9	34.7	37.4	38.6
ATF [12]	34.6	47.0	50.0	23.7	43.3	38.7	33.4	38.8	38.7
HTCN [2]	33.2	47.5	47.9	31.6	47.4	40.9	32.3	37.1	39.8
Ours	32.7	47.8	48.0	29.9	47.1	38.3	32.5	37.4	39.2

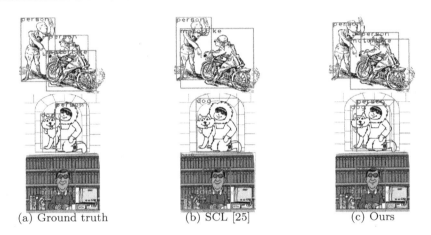

(a) Ground truth   (b) SCL [25]   (c) Ours

Fig. 3. Qualitative results on the PASCAL to Clipart adaptation.

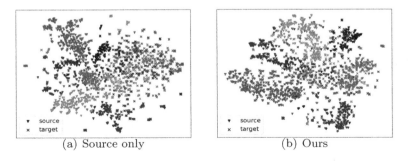

(a) Source only   (b) Ours

Fig. 4. Visualization of instance features with t-SNE [19]. Different color markers represent the features of different categories. We apply RoI pooling on the ground-truth bounding boxes to obtain the instance features for visualization.

5   Conclusions

In this work, we present FFAF for adaptive object detection. Specifically, we introduce a novel FSM to select the foreground regions for subsequent feature alignment, which is helpful for foreground alignment. Furthermore, we employ MDA to align features of different levels. In experiments, our model surpasses most of the existing domain adaptive detectors on three cross-domain object detection tasks, which shows the superior performance of our proposed framework.

Acknowledgments. This work was supported in part by the National Natural Science Foundation of China under Grant 61976231, Grant U1611461, Grant 61573387, and Grant 61172141, in part by the Guangdong Basic and Applied Basic Research Foundation under Grant 2019A1515011869, and in part by the Science and Technology Program of Guangzhou under Grant 201803030029.

References

1. Cai, Z., Vasconcelos, N.: Cascade R-CNN: delving into high quality object detection. In: Proceedings of the IEEE Conference on Computer Vision and Pattern Recognition, pp. 6154–6162 (2018)
2. Chen, C., Zheng, Z., Ding, X., Huang, Y., Dou, Q.: Harmonizing transferability and discriminability for adapting object detectors. In: Proceedings of the IEEE Conference on Computer Vision and Pattern Recognition, pp. 8869–8878 (2020)
3. Chen, Y., Li, W., Sakaridis, C., Dai, D., Van Gool, L.: Domain adaptive faster R-CNN for object detection in the wild. In: Proceedings of the IEEE Conference on Computer Vision and Pattern Recognition, pp. 3339–3348 (2018)
4. Cordts, M., et al.: The Cityscapes dataset for semantic urban scene understanding. In: Proceedings of the IEEE Conference on Computer Vision and Pattern Recognition, pp. 3213–3223 (2016)
5. Everingham, M., Van Gool, L., Williams, C.K., Winn, J., Zisserman, A.: The PASCAL visual object classes (VOC) challenge. Int. J. Comput. Vision **88**(2), 303–338 (2010)
6. Ganin, Y., Ustinova, E., Ajakan, H., Germain, P., Larochelle, H., Laviolette, F., Marchand, M., Lempitsky, V.: Domain-adversarial training of neural networks. J. Mach. Learn. Res. **17**(1), 2096–2030 (2016)
7. Girshick, R.: Fast R-CNN. In: Proceedings of the IEEE International Conference on Computer Vision, pp. 1440–1448 (2015)
8. Girshick, R., Donahue, J., Darrell, T., Malik, J.: Rich feature hierarchies for accurate object detection and semantic segmentation. In: Proceedings of the IEEE Conference on Computer Vision and Pattern Recognition, pp. 580–587 (2014)
9. He, K., Gkioxari, G., Dollár, P., Girshick, R.: Mask R-CNN. In: Proceedings of the IEEE International Conference on Computer Vision, pp. 2961–2969 (2017)
10. He, K., Zhang, X., Ren, S., Sun, J.: Deep residual learning for image recognition. In: Proceedings of the IEEE Conference on Computer Vision and Pattern Recognition, pp. 770–778 (2016)
11. He, Z., Zhang, L.: Multi-adversarial faster R-CNN for unrestricted object detection. In: Proceedings of the IEEE International Conference on Computer Vision, pp. 6668–6677 (2019)

12. He, Z., Zhang, L.: Domain adaptive object detection via asymmetric tri-Way Faster R-CNN. In: Proceedings of the European Conference on Computer Vision, pp. 309–324 (2020)
13. Inoue, N., Furuta, R., Yamasaki, T., Aizawa, K.: Cross-domain weakly-supervised object detection through progressive domain adaptation. In: Proceedings of the IEEE Conference on Computer Vision and Pattern Recognition, pp. 5001–5009 (2018)
14. Kim, T., Jeong, M., Kim, S., Choi, S., Kim, C.: Diversify and match: a domain adaptive representation learning paradigm for object detection. In: Proceedings of the IEEE Conference on Computer Vision and Pattern Recognition, pp. 12456–12465 (2019)
15. Lin, T.Y., Dollár, P., Girshick, R., He, K., Hariharan, B., Belongie, S.: Feature pyramid networks for object detection. In: Proceedings of the IEEE Conference on Computer Vision and Pattern Recognition, pp. 2117–2125 (2017)
16. Lin, T.Y., Goyal, P., Girshick, R., He, K., Dollár, P.: Focal loss for dense object detection. In: Proceedings of the IEEE International Conference on Computer Vision, pp. 2980–2988 (2017)
17. Lin, T.Y., et al.: Microsoft COCO: common objects in context. In: Proceedings of the European Conference on Computer Vision, pp. 740–755 (2014)
18. Liu, W., et al.: SSD: single shot multibox detector. In: Proceedings of the European Conference on Computer Vision, pp. 21–37 (2016)
19. Van der Maaten, L., Hinton, G.: Visualizing data using t-SNE. J. Mach. Learn. Res. 9(11), 2579–2605 (2008)
20. Redmon, J., Divvala, S., Girshick, R., Farhadi, A.: You only look once: unified, real-time object detection. In: Proceedings of the IEEE Conference on Computer Vision and Pattern Recognition, pp. 779–788 (2016)
21. Ren, S., He, K., Girshick, R., Sun, J.: Faster R-CNN: towards real-time object detection with region proposal networks. IEEE Trans. Pattern Anal. Mach. Intell. 39(6), 1137–1149 (2016)
22. Russakovsky, O., Deng, J., Su, H., Krause, J., Satheesh, S., Ma, S., Huang, Z., Karpathy, A., Khosla, A., Bernstein, M., et al.: Imagenet large scale visual recognition challenge. Int. J. Comput. Vision 115(3), 211–252 (2015)
23. Saito, K., Ushiku, Y., Harada, T., Saenko, K.: Strong-weak distribution alignment for adaptive object detection. In: Proceedings of the IEEE Conference on Computer Vision and Pattern Recognition, pp. 6956–6965 (2019)
24. Sakaridis, C., Dai, D., Van Gool, L.: Semantic foggy scene understanding with synthetic data. Int. J. Comput. Vision 126(9), 973–992 (2018)
25. Shen, Z., Maheshwari, H., Yao, W., Savvides, M.: SCL: towards accurate domain adaptive object detection via gradient detach based stacked complementary losses. arXiv preprint arXiv:1911.02559 (2019)
26. Simonyan, K., Zisserman, A.: Very deep convolutional networks for large-scale image recognition. In: International Conference on Learning Representations, pp. 1–14 (2015)
27. Xie, R., Yu, F., Wang, J., Wang, Y., Zhang, L.: Multi-level domain adaptive learning for cross-domain detection. In: Proceedings of the IEEE International Conference on Computer Vision, pp. 3213–3219 (2019)
28. Xu, C.D., Zhao, X.R., Jin, X., Wei, X.S.: Exploring categorical regularization for domain adaptive object detection. In: Proceedings of the IEEE Conference on Computer Vision and Pattern Recognition, pp. 11724–11733 (2020)

29. Yu, F., et al.: BDD100k: a diverse driving dataset for heterogeneous multitask learning. In: Proceedings of the IEEE Conference on Computer Vision and Pattern Recognition, pp. 2636–2645 (2020)
30. Zheng, Y., Huang, D., Liu, S., Wang, Y.: Cross-domain object detection through coarse-to-fine feature adaptation. In: Proceedings of the IEEE Conference on Computer Vision and Pattern Recognition, pp. 13766–13775 (2020)
31. Zhu, J., Park, T., Isola, P., Efros, A.A.: Unpaired image-to-image translation using cycle-consistent adversarial networks. In: Proceedings of the IEEE International Conference on Computer Vision, pp. 2242–2251 (2017)
32. Zhu, X., Pang, J., Yang, C., Shi, J., Lin, D.: Adapting object detectors via selective cross-domain alignment. In: Proceedings of the IEEE Conference on Computer Vision and Pattern Recognition, pp. 687–696 (2019)
33. Zhuang, C., Han, X., Huang, W., Scott, M.: iFAN: image-instance full alignment networks for adaptive object detection. In: Proceedings of the AAAI Conference on Artificial Intelligence, pp. 13122–13129 (2020)

Exploring Category-Shared and Category-Specific Features for Fine-Grained Image Classification

Haoyu Wang[1], DongLiang Chang[1], Weidong Liu[3], Bo Xiao[1(✉)], Zhanyu Ma[1,2], Jun Guo[1], and Yaning Chang[1]

[1] Beijing University of Posts and Telecommunications, Beijing 100876, People's Republic of China
{wanghaoyumartin,changdongliang,xiaobo,mazhanyu, guojun,cyaning}@bupt.edu.cn
[2] Beijing Academy of Artificial Intelligence, Beijing 100876, People's Republic of China
[3] China Mobile Research Institute, Beijing 100876, People's Republic of China
Liuweidong@chinamobile.com

Abstract. The attention mechanism is one of the most vital branches to solve fine-grained image classification (FGIC) tasks, while most existing attention-based methods only focus on inter-class variance and barely model the intra-class similarity. They perform the classification tasks by enhancing inter-class variance, which narrows down the intra-class similarity indirectly. In this paper, we intend to utilize the intra-class similarity as assistance to improve the classification performance of the obtained attention feature maps. To obtain and utilize the intra-class information, a novel attention mechanism, named category-shared and category-specific feature extraction module (CSS-FEM) is proposed in this paper. CSS-FEM firstly extracts the category-shared features based on the intra-class semantic relationship, then focuses on the discriminative parts. CSS-FEM is assembled by two parts: 1) The category-shared feature extraction module extracts category-shared features that contain high intra-class semantic similarity, to reduce the large intra-class variances. 2) The category-specific feature extraction module performs spatial-attention mechanism in category-shared features to find the discriminative information as category-specific features to decrease the high inter-class similarity. Compared with the state-of-the-art methods, the experimental results on three commonly used FGIC datasets show that the effectiveness and competitiveness of the proposed CSS-FEM. Ablation experiments and visualizations are also provided for further demonstrations.

Keywords: Fine-grained image classification · Semantic intra-class similarity · Channel-wise attention · Spatial-wise attention

© Springer Nature Switzerland AG 2021
H. Ma et al. (Eds.): PRCV 2021, LNCS 13019, pp. 179–190, 2021.
https://doi.org/10.1007/978-3-030-88004-0_15

1   Introduction

The image classification has been redefined with the rapid development in the amount of the training datasets. The computer vision community has been dominated by the deep convolutional neural networks [2]. Training CNNs requires massive amounts of labeled data [20]; but in fine-grained image collections [8], where the categories are visually very similar, the data population decreases significantly. It drives us to focus on the features that contain more category information [18]. As for the fine grained image classification task, it is much more challenging than the normal image classification task. Aiming to recognize hundreds of subcategories under the same basic-level category [6], the fine-grained image classification task is even difficult for the human to recognize hundreds of subcategories, such as 200 bird subcategories or 196 car subcategories. Due to the exiguity variance in object appearances, subtle and local differences are the key points for fine-grained image classification, such as the color of the back, the shape of the bill and the texture of the feather for the bird. Since these subtle and local differences locate at the discriminative objects [1], fine-grained image classification is a highly important task with wide applications, such as automatic driving, biological conservation.

Fine-grained image classification (FGIC) aims to achieve a classifier to distinguish different subordinate categories (e.g., birds species [22], cars [15] and aircraft models [19], etc.) within super-ordinate categories. Due to significant intra-class variance and inter-class similarity [10] in the FGIC datasets, FGIC remains to be a challenging task. How to reduce the intra-class variance and the inter-class similarity is the main task of the FGIC.

Recently, FGIC task is dominated by the deep convolutional neural network [11,14] due to its outstanding performance of extracting channel and spatial-wise features, according to previous works [11,17,26]. Moreover, the channel-wise features contain semantic information and spatial-wise ones contain morphological and location information [12]. In order to take advantage of this feature information, the former studies come up with the attention mechanism to find discriminative details of the feature information [5,14]. Complicated convolutional neural network has been proposed to perform part learning with the attention mechanism, to ensure that features learned are maximally discriminative.

The Fig. 1 (a) shows the normal attention mechanism, the predominant point of the normal attention mechanism is to make the model focus on the key parts of the classification task object [24]. So far, this method has shown an outstanding performance in finding the discriminative parts of the image. It helps to decrease the inter-class similarity, and reduce the intra-class variance indirectly. However, performing in one image, previous attention mechanisms may be susceptible to particular backgrounds. Some features of background may be more discriminative than the task-relevant object, leading to being given higher confidence, which leaves the intra-class variance unconsidered. Therefore, with the category-relevant information in the image pair (bird species, etc.), we intend to extract the features of the image pair that contain high intra-class similarity to

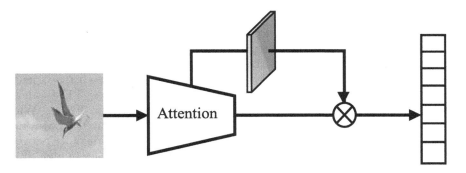

(a) normal attention mechanism

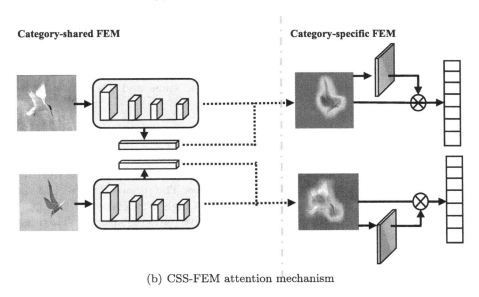

(b) CSS-FEM attention mechanism

Fig. 1. The motivation of CSS-FEM. Most existing attention methods mainly focus on the discriminative parts. The CSS-FEM first establishes the intra-class relationship with category-shared features to reduce intra-class variance. Then, it applies spatial-attention mechanism to obtain category-specific features to achieve a better FGIC performance

achieve a better attention mechanism, then focus on the discriminative parts in them to reduce the false rate.

The concept of extracting similar or common objects is widely used in co-segmentation tasks [3,21], which seems to be helpful in FGIC tasks. The co-segmentation method uses co-information from the image collection to predict the foreground pixels of the common specific objects in them. Most existing works consider the whole image collection to achieve a maximization in the activation feature energy to locate the common objects in the images [3].

With the inspiration of the co-segmentation and motivation described above, our method has been proposed as the category-shared and category-specific feature extraction module (CSS-FEM) in Fig. 1 (b), which contains two parts: a category-shared feature extraction module (category-shared FEM) and a category-specific feature extraction module (category-specific FEM). When given an image pair from the same category, the category-shared FEM establishes the semantic reflection of the input [13]. By reducing the distance between them, the module extracts the category-shared features that contain high intra-class semantic similarity. In order to perform a better classification, the module needs to find the discriminative parts in them. The other part: category-specific FEM first focuses on the category-shared features by establishing the intra-class relationship mask by normalized correlation, then it performs the spatial-wise attention mechanism in category-shared features to obtain the discriminative information as category-specific features.

We assemble the two parts as the proposed CSS-FEM method. With comparison, we achieve an effective, novel attention mechanism on FGIC tasks. The whole method can be trained end-to-end in one stage and it's convenient to apply our method in other models.

The primary contributions of the paper are:

– We proposed a novel attention mechanism named category-shared and category-specific feature extraction module in the FGIC task. It extracts category-relevant information within image pairs as an intra-class relationship to achieve a better attention performance. The proposed CSS-FEM can extract category-shared features that contain high intra-class semantic similarity. Moreover, it excavates the discriminative parts of category-shared features as category-specific features to enhance the ability of classification.
– The experimental results show the effectiveness of the proposed method. In addition, ablation studies and visualizations are made to further evaluate the proposed CSS-FEM.

The rest of the paper is organized as follows. Section 2 introduces the details of the proposed method. Section 3 provides the experiments and discussions, along with the conclusion in Sect. 4.

2   Proposed Method

In this section, we propose the CSS-FEM as illustrated in the Fig. 2. The entire process is shown as: the model takes a pair of image from the same category as the input. It obtains the high-level feature map pair $F \in \mathbb{R}^{2 \times C \times H \times W}$. Then, F is sent to the CSS-FEM. The entire framework of the CSS-FEM is assembled by two parts: category-shared feature extraction module (category-shared FEM) and category-specific feature extraction module (category-specific FEM).

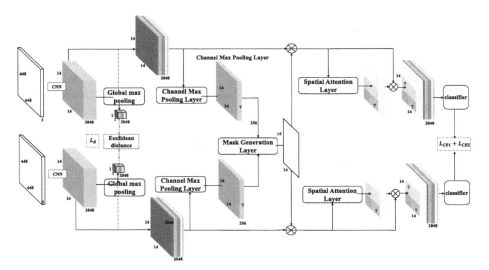

Fig. 2. The structure of the category-shared and specific feature extraction module (CSS-FEM). We take ResNet50 [12] as feature extraction base model. The proposed CSS-FEM acts as a block that is plugged between the last convolutional layer and classifier layer. The category-shared feature extraction module establishes the semantic reflection of the image pair. With the supervision of Euclidean distance, it extracts category-shared features that contain intra-class semantic similarity. Finally, category-shared FEM establishes the normalized correlation through a matrix accumulative algorithm, to model the intra-class relationship. As for the category-specific feature extraction module, it performs the spatial attention mechanism in the category-shared features. With spatial-mask applied, the discriminative information is obtained as category-specific features. In the last, CSS-FEM is supervised under the Euclidean distance and cross-entropy loss to get a better classification performance.

The first module: the category-shared feature extraction module (category-shared FEM) establishes the semantic reflection of the image pair. It measures the Euclidean distance within the reflections as semantic similarity. By reducing the distance, it constrains the model to focus on the category-shared features that contain high intra-class semantic similarity.

In the second module: the category-specific feature extraction module (category-specific FEM), we first focus on the category-shared features by the intra-class relationship mask to build the intra-class relationship within the image pair. Then the category-specific FEM realizes a spatial attention mechanism in the category-shared features, to find the inter-class discriminative parts as category-specific features.

The entire module can be trained end-to-end and easily plugged into other convolutional neural networks. In this part, we mainly use the ResNet50 [12] as the backbone.

2.1   Category-Shared Feature Extraction Module

Category-shared FEM contains a supervision block. After being given the high-level feature map pairs $F \in \mathbb{R}^{2 \times C \times H \times W}$, the category-shared FEM uses F to establish the semantic reflection of the image pair. The supervision block contains a global max-pooling layer and other operation layers. F is sent to the global max-pooling (**GMP**), to choose the max value of each channel-wise layer as the representation. It processes the feature map pair from F into feature vector pair as $F' \in \mathbb{R}^{2 \times C \times 1 \times 1}$. The information loss is inevitable, while the most representative value is selected. By doing this, we establish the semantic reflection of the feature map pairs approximately. The entire establishment is computed as

$$F' = \mathbf{GMP}(F), \tag{1}$$

where $F' \in \mathbb{R}^{2 \times C \times 1 \times 1}$ is the semantic reflection of the feature map pair. The semantic reflection of the image has the dimension of 2048. Then, F' is split into two vectors as $V_1 \in \mathbb{R}^C$ and $V_2 \in \mathbb{R}^C$ to calculate the Euclidean distance between them as the semantic similarity. The Euclidean distance $D_{Euclidean}$ represents the semantic difference of the extracted features. Each extracted feature delegates the semantic embedding of the pictures. In order to reduce the distance within images. We use the Euclidean distance as the variance of semantic difference D_E and define our loss function L_d to minimize the variance of semantic reflection and supervise the model to focus on more similar the channel-wise areas, as:

$$L_d(V_1, V_2) = min(D_E). \tag{2}$$

By reducing the Euclidean distance, we supervise the semantic reflection to approach each other and ensure the semantic similarity of F as extracted category-shared features. The supervision of the semantic reflection makes the whole category-shared features extraction module focus on the similar semantic parts of the image pair.

2.2   Category-Specific Feature Extraction Module

As the category-shared FEM has ensured the semantic similarity of the category-shared features F in the channel-wise, the goal of the category-specific feature extraction module (category-specific FEM) is to find the discriminative parts of the extracted category-shared features, to decrease the inter-class similarity. The category-specific FEM contains a mask generation block and a spatial-wise attention block.

The mask generation block aims to establish the intra-class relationship mask. It contains three layers: channel max-pooling layer, feature normalization layer, and mask generation layer. Category-shared features F first comes through the channel max-pooling layer. The channel max-pooling layer (CMP) divides each of the N channels into one channel block. Every channel block chooses the max feature value as the representation to reduce the channel dimension.

$F_{out} \in \mathbb{R}^{2 \times \frac{C}{N} \times H \times W}$ is the output of this layer. As the channel max-pooling layer has no learnable parameter, the training is accelerated without any parameter added. The channel max-pooling layer realizes a reduction in the channel dimension, which aims to reduce the calculation amount but maintain the information completeness of the category-shared features.

After the process of the channel max-pooling layer, F_{out} is put into the feature normalization layer, which is used to generate a normalized correlation between the feature map pair. It aims to establish intra-class relationship mask R_{mask}. As for the different channel-wise feature maps of any image, its distribution histogram has a certain difference. The feature normalization layer first realizes the normalization of the image distribution. We first reshape and split the $F_{out} \in \mathbb{R}^{2 \times \frac{C}{N} \times H \times W}$ into F_{out1} and $F_{out2} \in \mathbb{R}^{\frac{C}{N} \times HW}$. In the feature normalization layer, the operation formula for each channel of the feature map is shown as:

$$R_{n,c}^i = \frac{F_{outn,c}^i - \mu_{n,c}}{\sigma_{n,c}}, \tag{3}$$

$$R_n = \sum_{c=1}^{\frac{C}{N}} \sum_{i=1}^{HW} R_{n,c}^i, \tag{4}$$

where $n = 1$ or $n = 2$ represents F_{out1} and F_{out2}. c represents the cth channel dimension of the feature maps $(1 \le c \le \frac{C}{N})$. $\sigma_{n,c}$, $\mu_{n,c}$ represents the variance value and the mean value of $F_{outn,c}$. i represent the feature value position of $F_{outn,c}$ $(1 \le i \le HW)$. The final output is $R_n \in \mathbb{R}^{\frac{C}{N} \times HW}$.

The normalization operation converts the distribution of feature maps into the standard Gaussian distribution. After the normalization operation, the feature values of the feature location map are all converted to the $(-1, 1)$. Then, we can establish the normalized correlation within the image pair. Normalized correlation is usually used in measuring the similarity between matrixes. The closer the value is to 1, the more similar the matrix pair is. Therefore, we use the normalized correlation as the input to establish the intra-class relationship mask. The normalized correlation of the F_{out1} and F_{out2}: C is established as:

$$C = (R_1)^T \times R_2, \tag{5}$$

where $\times$ refers to matrix accumulative algorithm. The normalized correlation $C \in \mathbb{R}^{HW \times HW}$. Then mask generation layer uses normalized correlation C to establish an intra-class relationship mask. The mask generation layer contains a convolutional layer, a batch normalization layer, and an activation layer. The convolutional layer has an input channel of HW and output channel of 1 and the kernel size of it is 1×1. After the process of the mask generation layer and reshape, we get the intra-class relationship mask $F_{mask} \in \mathbb{R}^{1 \times H \times W}$. It is finally applied to the F as:

$$F_{mask} = F \otimes R_{mask}, \tag{6}$$

where $\otimes$ denotes the element-wise multiplication.

After the intra-class relationship mask is applied, the module has been constrained to focus on the category-shared features. Then the spatial-wise attention block aims to find the discriminative parts in them.

The spatial attention layer contains two convolutional layers, a batch normalization layer and a ReLU layer. The first layer's channel size of input and output is $H \times W$ and 64, and the second one's channel size of input and output are 64 and 1. The spatial attention layer aims to achieve the spatial-wise attention mechanism in the category-shared features and locate the discriminative parts as category-specific features to achieve a better classification performance. We finally use the fully connected layer and cross-entropy loss function to perform the classification task.

The differentiation loss works with the cross-entropy loss(L_{CE}) in the proposed CSS-FEM, which contains two cross-entropy losses L_{CE1} and L_{CE2} due to the two images of input. As there is L_d loss applied in the category-shared feature extraction module, we finally define our loss function $Loss$ as:

$$L = L_{CE1} + L_{CE2} + \alpha \times L_d, \tag{7}$$

where α is an hyper-parameter to balance the two part of the loss. In the experiment, we take $\alpha = 0.6$.

Table 1. Experiment result of CFEM applied ON CUB-200-2011 [22], FGVC-Aircraft [22], and Stanford Cars [15]. The red font represents the best results of the compared method and the blue font represents the second best result.

Method	Base model	CUB-200-2011	Stanford Cars	Aircrafts
KA (ICCV17 [4])	VGG16	85.3	88.3	91.7
KP (CVPR17 [9])	VGG16	86.2	86.9	92.4
DFL-CNN (CVPR18 [23])	VGG16	86.7	93.8	92.0
CIN (AAAI2020)	ResNet101	88.1	92.8	94.5
DBTNet (NeurIPS2019)	ResNet101	88.1	94.5	91.6
DFL-CNN (CVPR18 [23])	ResNet50	87.4	93.1	91.7
Cross-X (ICCV19 [16])	ResNet50	77.8	84.8	84.9
DCL (CVPR19 [7])	ResNet50	87.8	94.5	93.0
TASN (CVPR19 [25])	ResNet50	87.9	93.8	–
Cross-X (CVPR20 [16])	ResNet50	87.7	94.6	92.6
ACNet (CVPR20 [14])	ResNet50	88.1	94.6	92.4
CSS-FEM (Ours)	VGG16	87.2	93.8	92.8
CSS-FEM (Ours)	ResNet50	88.2	94.8	92.8
CSS-FEM (Ours)	ResNet101	88.4	95.1	93.8

3   Experiment

In this section, we evaluate the performance of our proposed method on three popular FGIC datasets, i.e., CUB-200-2011 [22], Stanford Cars [15], and FGVC-Aircraft [19]. This section will demonstrate the experiment in detail and analyze the experiment results. We also make the ablation study to prove the function of the proposed CSS-FEM.

3.1   Implementation Details

As the unique idea of our module, an image pair input is required. The input image size of our experiments is resized to 448×448 which is the same as the baseline methods. We build our module on ResNet50 [12] which is pre-trained on ImageNet by removing the last pooling layer and the fully-connected layer and the output of the backbone pair has the shape of $(2, 2048, 14, 14)$. We take $N = 8$ in the channel max pooling. As for the input image pair, we first select one image and then randomly select another image from the same category. The batch size in the dataloader is set to 64 in the CUB-200-2011 dataset and no annotations except for the label information is used. The pre progress of the input is limited to random cropping and horizontal flipping during training and center cropping during inference. We use Stochastic Gradient Descent optimizer and batch normalization as the regularizer. The learning rate of the pre-trained feature extraction layers is kept as 1×10^{-4}, while the learning rate of the fully connected layer is initially set at 0.01 and multiplied by 0.1 at 50th and 75th epoch, successively. We train our model for 100 epochs and the weight decay value is kept as 5×10^{-4}. Furthermore, the α in the loss function is empirically set to 0.2.

3.2   Experimental Results

The experimental results on CUB-200-2011, Stanford Cars, and FGVC-Aircraft are displayed in Table 1. In order to keep the comparisons fair enough, we compare the proposed common feature extractor module with recent methods with no extra annotations on these three popular used FGIC datasets. The proposed approach performs competitive results compared with the state-of-the-art methods with the $\alpha = 0.2$. Although our common feature extractor module has a certain parameter added to the model, there is no increase in the time-consuming. In addition, the CSS-FEM has no requirements for sophisticated initialization methods and works in a single stage. It helps in the training stage and takes no time-consuming in the prediction stage. The CSS-FEM can be easily added to CNN as a simple component, achieves competitive performance compared with the state-of-the-art methods. Taking these advantages, the CSS-FEM can enhance the learning ability of the convolutional neural network especially in the FGIC tasks.

188   H. Wang et al.

3.3   Ablation Studies

As for the ablation study, we take ResNet50 as the backbone and remove the
last max-pooling layer and classification layer and add the proposed module:
category-shared FEM part and category-specific FEM part. In order to prove the
contribution of category-shared FEM and category-specific FEM. The Table 2
indicates the significant meaning of the proposed module. We make the ablation
studies on the CUB-200-2011 [22], Stanford Cars [15], and FGVC-Aircraft [19].

Table 2. Contribution of the two parts in CSS-FEM (i.e., the category-shared FEM
and the category-specific FEM)

Method	CUB	Cars	Aircraft
Category-shared FEM	87.5	94.4	92.4
Category-specific FEM	87.8	94.6	92.6
CSS-FEM	88.2	94.8	92.8

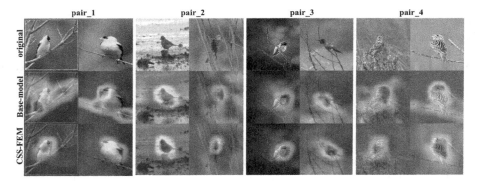

Fig. 3. Visualization results. There are four pairs of comparisons. As we take image
pairs as input in our proposed method, the results are also shown in pairs. The third
row is from our module CSS-FEM and the second row is from baseline. We achieve a
better attention performance on the category-shared features as the bird species. To
perform the classification task, we focus on the category-specific features in attention
areas.

3.4   Visualizations

The proposed category-shared and specific feature extraction module makes a
good performance in extracting the category-shared features of the image pairs
and fine-grained image classification. For a better understand, we apply the
Grad-CAM [8] to the proposed common feature extractor module using the

images from the FGIC dataset CUB-200-2011 for feature visualization. We compare the output visualization results of CSS-CFEM with the base model (ResNet-50) as is shown in the Fig. 3. From the visualization results, it can be observed that the proposed CSS-FEM can easily capture the category-shared features in the entire picture that have high intra-class similarity. It also can be shown that CSS-FEM locates the more discriminative parts in the category-shared feature areas to find the discriminative parts as category-specific features. The base model usually finds only one discriminative region for classification, leading to poor performance. In this case, the proposed approach will encourage the model to focus on more parts that contain discriminative information as much as possible.

4   Conclusions

We propose a useful and easy-to-apply method which is named as category-shared and specific feature extraction module (CSS-FEM). The proposed CSS-FEM is a novel attention mechanism. It first constrains the model to extract category-shared features that contain high intra-class semantic similarity. We use the category-shared feature to model the intra-class relationship, then in the category-shared features, we perform a spatial-wise attention mechanism to focus on the discriminative parts of the image pair as category-specific features. The process enables the model to focus on the object and find the discriminative parts to achieve a fine-grained image classification task.

Acknowledgement. This work was supported in part by the National Key R&D Program of China under Grant 2019YFF0303300 and under Subject II No. 2019YFF0303302, in part by National Natural Science Foundation of China (NSF) No. 62076031, 61922015, 61773071, U19B2036, in part by Beijing Natural Science Foundation Project No. Z200002, in part by the Beijing Academy of Artificial Intelligence (BAAI) under Grant BAAI2020ZJ0204, in part by the Beijing Nova Programme Interdisciplinary Cooperation Project under Grant Z191100001119140, and in part by BUPT Excellent Ph.D. Students Foundation No. CX2020105.

References

1. Akata, Z., Reed, S., Walter, D., Lee, H., Schiele, B.: Evaluation of output embeddings for fine-grained image classification. In: CVPR (2015)
2. Angelova, A., Zhu, S.: Efficient object detection and segmentation for fine-grained recognition. In: CVPR (2013)
3. Batra, D., Kowdle, A., Parikh, D., Luo, J., Chen, T.: icoseg: Interactive cosegmentation with intelligent scribble guidance. In: CVPR (2010)
4. Cai, S., Zuo, W., Zhang, L.: Higher-order integration of hierarchical convolutional activations for fine-grained visual categorization. In: ICCV (2017)
5. Chang, D., et al.: The devil is in the channels: Mutual-channel loss for fine-grained image classification. IEEE Trans. Image Process. (2020)
6. Chen, L., et al.: Sca-cnn: spatial and channel-wise attention in convolutional networks for image captioning. In: CVPR (2017)

7. Chen, Y., Bai, Y., Zhang, W., Mei, T.: Destruction and construction learning for fine-grained image recognition. In: CVPR (2019)
8. Cong, R., et al.: An iterative co-saliency framework for rgbd images. IEEE Transactions on Cybernetics (2017)
9. Cui, Y., Zhou, F., Wang, J., Liu, X., Lin, Y., Belongie, S.: Kernel pooling for convolutional neural networks. In: CVPR (2017)
10. Dubey, A., Gupta, O., Guo, P., Raskar, R., Farrell, R., Naik, N.: Pairwise confusion for fine-grained visual classification. In: ECCV (2018)
11. Gao, Y., Han, X., Wang, X., Huang, W., Scott, M.: Channel interaction networks for fine-grained image categorization. In: AAAI (2020)
12. He, K., Zhang, X., Ren, S., Sun, J.: Deep residual learning for image recognition. In: CVPR (2016)
13. Hu, J., Shen, L., Sun, G.: Squeeze-and-excitation networks. In: CVPR (2018)
14. Ji, R., Wen, L., Zhang, L., Du, D., Wu, Y., Zhao, C., Liu, X., Huang, F.: Attention convolutional binary neural tree for fine-grained visual categorization. In: CVPR (2020)
15. Krause, J., Stark, M., Deng, J., Fei-Fei, L.: 3d object representations for fine-grained categorization. In: ICCV workshops (2013)
16. Luo, W., et al.: Cross-x learning for fine-grained visual categorization. In: ICCV (2019)
17. Ma, Z., Lai, Y., Kleijn, W.B., Song, Y.Z., Wang, L., Guo, J.: Variational bayesian learning for dirichlet process mixture of inverted dirichlet distributions in non-gaussian image feature modeling. IEEE Trans. Neural Networks Learn. Syst. (2018)
18. Ma, Z., Xie, J., Lai, Y., Taghia, J., Xue, J.H., Guo, J.: Insights into multiple/single lower bound approximation for extended variational inference in non-gaussian structured data modeling. zIEEE Trans. Neural Networks Learn. Syst. (2019)
19. Maji, S., Rahtu, E., Kannala, J., Blaschko, M., Vedaldi, A.: Fine-grained visual classification of aircraft. HAL-INRIA (2013)
20. Pang, Y., Sun, M., Jiang, X., Li, X.: Convolution in convolution for network in network. IEEE Trans. Neural Networks Learn. Syst. (2017)
21. Rother, C., Minka, T., Blake, A., Kolmogorov, V.: Cosegmentation of image pairs by histogram matching-incorporating a global constraint into mrfs. In: CVPR (2006)
22. Wah, C., Branson, S., Welinder, P., Perona, P., Belongie, S.: The caltech-ucsd birds-200-2011 dataset. California Institute of Technology (2011)
23. Wang, Y., Morariu, V.I., Davis, L.S.: Learning a discriminative filter bank within a cnn for fine-grained recognition. In: CVPR (2018)
24. Zhao, B., Feng, J., Wu, X., Yan, S.: A survey on deep learning-based fine-grained object classification and semantic segmentation. Int. J. Autom. Comput. 14(2), 119–135 (2017). https://doi.org/10.1007/s11633-017-1053-3
25. Zheng, H., Fu, J., Zha, Z.J., Luo, J.: Looking for the devil in the details: Learning trilinear attention sampling network for fine-grained image recognition. In: CVPR (2019)
26. Zhuang, P., Wang, Y., Qiao, Y.: Learning attentive pairwise interaction for fine-grained classification. In: AAAI (2020)

Deep Mixture of Adversarial Autoencoders Clustering Network

Aofu Liu and Zexuan Ji[✉]

School of Computer Science and Engineering, Nanjing University of Science and Technology,
Nanjing 210094, China
jizexuan@njust.edu.cn

Abstract. As one of the most fundamental technologies in machine learning, unsupervised clustering methods have been vastly promoted due to the rapid development of deep learning. A widely utilized assumption is that the data is a set of low-dimensional manifolds, thus the key challenge for clustering methods is to ensure the potential features be completely separated. In this paper, we propose a mixture of adversarial autoencoders clustering (MAAE) network to solve the above problem. The data of each cluster is represented by one adversarial autoencoder. By introducing the adversarial information, the aggregated posterior of the hidden code vector of the autoencoder can better match with the prior distribution which is constructed based on the generalized Gaussian mixture model. So the latent features embedded in data can be well separated. The clustering network jointly learns the nonlinear data representation and the collection of adversarial autoencoders. The clustering results are obtained by minimizing the reconstruction loss of the mixture of adversarial autoencoders network. We perform experiments on popular standard datasets, and the results show that the proposed algorithm is effective in both clustering and feature separation.

Keywords: Clustering · Mixture autoencoders · Adversarial autoencoder

1 Introduction

Cluster analysis is a classic task in the fields of machine learning and computer vision. Many traditional clustering methods such as K-means [1] and GMM [2] are still widely used because of the simplicity and efficiency. However, due to the limitation of distance measurement, these methods can hardly obtain satisfactory results for high-dimensional data. For example, the conventional Euclidean distance is usually not suitable for distinguishing high-dimensional data.

Recently, many studies have been devoted to mapping high-dimensional data to low-dimensional space, which can obtain the latent representations for clustering algorithms. With the development of deep learning, many deep clustering methods based on unsupervised learning have been proposed [3]. Autoencoder [4] is a classic method of unsupervised learning, which can obtain low-dimensional features by reconstructing samples. Generative Adversarial Network (GAN) [5] is also an important unsupervised

© Springer Nature Switzerland AG 2021
H. Ma et al. (Eds.): PRCV 2021, LNCS 13019, pp. 191–202, 2021.
https://doi.org/10.1007/978-3-030-88004-0_16

method by generating new samples through adversarial training. Similar with GAN, adversarial autoencoder [6] makes the encoder distribution match the target distribution through adversarial learning. Variational autoencoder [7] introduces the constraints to force the encoder distribution to obey Gaussian distribution. Most deep clustering methods focus on clustering using the latent representation learned by various autoencoder models [8]. Deep embedding clustering (DEC) [9] first pre-trains an autoencoder with reconstruction loss and then optimize the clustering center in the embedded space by KL divergence. Deep clustering network (DCN) [10] jointly trains the reconstruction loss and k-means clustering loss. Although these methods have achieved satisfactory clustering results, they only map the data to a single feature space. For the complex data, the latent representations may not be well separated under only one feature space.

Therefore, it is intuitional to represent each cluster of data as a separate manifold. The whole data set can be represented as a mixture of low-dimensional nonlinear manifolds. MIXAE [11] utilizes multiple autoencoders to map different clusters to different low-dimensional spaces. The network is optimized by comparing the reconstruction loss between the input image and the output image of each decoder, and sample entropy is used to prevent trivial solutions. Besides, a clustering network is designed to calculate image weights to guide network optimization and clustering. DAMIC [12] designs a SoftMax-like loss function and a clustering layer for clustering. K-DAE [13] removes the clustering layer and adopts a simple mixture structure. The minimum reconstruction loss between input data and multiple reconstructed data is used to optimize the network. These methods try to treat each cluster as a separate manifold, but they cannot guarantee that the latent features belonging to different clusters are completely separated.

In order to better control the separation manifold of each cluster, the adversarial autoencoder is applicable by introducing the adversarial information, in which the latent representation obtained by the encoder can be trained as a specified target distribution. Therefore, each cluster of data will be embedded in different target distributions such as Gaussian distribution. However, the assumption, i.e., the latent representation follows Gaussian distribution, is not always satisfied for practical data. Therefore, it is critical to choose a more effective distribution to better fit the latent representations. Compared with Gaussian distribution, the generalized Gaussian distribution is a target distribution with more powerful representation ability. The generalized Gaussian mixture model (GGMM) [14] is more persuasive than the Gaussian mixture model (GMM) in modeling data because it is more extensive and easier to extend.

In this paper, we propose a mixture of adversarial autoencoder clustering (MAAE) network. The mixture of autoencoder network maps different clusters to different feature spaces to obtain the reconstructed samples. Cluster allocation is carried out according to the minimum reconstruction loss. Meanwhile, the adversarial training forces the original data to be embedded in the target distribution to separate the latent features of each cluster. The aggregated posterior of the hidden code vector of the autoencoder can better match with the prior distribution which is constructed based on the generalized Gaussian mixture model. Moreover, because the mean square error only considers the difference under pixel level, which is limiting in discriminative ability. To improve the accuracy of clustering, we introduce the image Euclidean distance (IMED) [15] to construct the clustering loss which can take the spatial relationship into account. The main contributions of our approach are summarized as follows:

(1) We propose a mixture of adversarial autoencoder clustering network, which introduces adversarial information into the original mixture of autoencoder network to embed the input data into different target distributions.
(2) We improve the model fitting ability for the latent features by utilizing the generalized Gaussian distribution as the target distribution to produce the prior information for each adversarial autoencoder.
(3) The image Euclidean distance is utilized to construct the clustering loss, which would further improve the clustering accuracy based on minimum reconstruction loss by considering the spatial information embedded in data.

2 Mixture of Adversarial Autoencoders

To solve the problem of dividing a dataset $\{x_1, x_2 \ldots x_n\} \in R^d$ into k clusters, we design a mixture of adversarial autoencoders clustering network. Figure 1 shows the network structure of MAAE. We assume that each cluster can be represented by one adversarial block. Therefore, MAAE consists of k adversarial blocks. Each adversarial block is an adversarial autoencoder, including encoder E, decoder G, and discriminator D. For each adversarial block, encoder E maps the input sample to the latent representation z, the decoder G uses z to generate a reconstructed sample. The discriminator D is used to distinguish the encoder distribution from the target distribution. Each adversarial block corresponds to a target distribution, which is sampled from a generalized Gaussian mixture model. Each input image is fed into k adversarial blocks to get k reconstructed images, and these images jointly calculate the clustering loss.

2.1 Adversarial Block

Autoencoder is a widely used network structure in unsupervised learning, which is composed of encoder and decoder. Given the input data $x \in R^n$, encoder parameters θ_E, and decoder parameters θ_G. the encoder E first maps x to its latent representation $z = f(x, \theta_E) \in R^d$, where $d < n$. The decoder G reconstructs z to $\tilde{x} = f(z, \theta_G)$. By calculating the reconstruction loss between x and $\tilde{x}$, the autoencoder optimizes its encoder and decoder parameters until the reconstruction loss converges. An appropriate low-dimensional representation of the data can be obtained.

Adversarial autoencoder introduces adversarial information based on the autoencoder and supplements a discriminator to the encoder and decoder. Suppose that $q(z|x)$ is the encoder distribution, $p(z|x)$ is the decoder distribution, and $p(z)$ is the target distribution. When the input data x is fed into the encoder, the latent layer features obey $q(z|x)$. The purpose of the adversarial autoencoder is to make the $q(z|x)$ approach the target distribution $p(z)$ while minimizing the reconstruction loss. The latent representation generated by the encoder can trick the discriminator into thinking that $q(z|x)$ comes

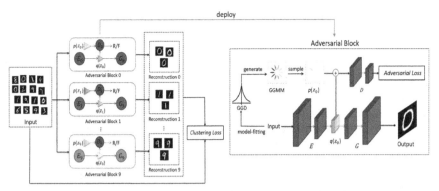

Fig. 1. The architecture of the mixture of adversarial autoencoders clustering (MAAE) network.

from the target distribution $p(z)$. The loss function of adversarial autoencoder consists of reconstruction loss and adversarial loss, where the reconstruction loss is:

$$L_{reconstruct} = \sum_{i=1}^{n} \|x_i - \breve{x}_i\|^2 \tag{1}$$

refer to GAN, the adversarial loss is:

$$L_{GAN} = \min_{\theta_E} \max_{\theta_D} E_{x \sim p_{data}}\big[\log D(x)\big] + E_{z \sim p_z}\big[\log(1 - D(G(z)))\big] \tag{2}$$

where θ_E and θ_D are the parameters of the encoder and discriminator, respectively. $D(\cdot)$ is the output of discriminator. By jointly optimizing the reconstruction loss and adversarial loss, the distribution generated by the encoder will approach the target distribution. In the case that only the decoder is retained, the decoder can still reconstruct the input target distribution into the original image like a generator.

2.2 Target Distribution

The target distribution $p(z)$ in the adversarial autoencoder can be manually chosen, where Gaussian distribution is widely used because of its simplicity and efficiency. However, in the case of complex data, the Gaussian distribution can hardly represent the characteristics of the underlying layer of the data.

In this paper, we replace the Gaussian distribution with the generalized Gaussian distribution to further improve the data fitting ability of the model. The generalized Gaussian distribution is defined as:

$$GGD\Big(x; \alpha, \mu, \sigma^2\Big) = \frac{\alpha}{2\beta\Gamma(1/\alpha)} exp\left(-\left(\frac{|x - \mu|}{\beta}\right)^\alpha\right)$$

$$\beta = \sigma\sqrt{\frac{\Gamma(1/\alpha)}{\Gamma(3/\alpha)}} and \ \Gamma(a) = \int_0^\infty t^{a-1}e^{-t}dt(a > 0) \tag{3}$$

where α is the shape parameter, which controls the speed of attenuation μ and σ^2 are the mean and variance of the distribution, respectively. Gaussian distribution is a special case when $\alpha = 2$, $\Gamma(1/2) = \sqrt{\pi}$ and $\Gamma(3/2) = \sqrt{\pi}/2$.

As shown in Fig. 2, by fitting the latent representations of the data extracted with autoencoder [16], the generalized Gaussian distribution has a higher degree of fitting to the data features than Gaussian distribution. Therefore, it is more reasonable and effective to choose generalized Gaussian distribution as the target distribution.

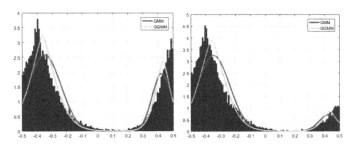

Fig. 2. Distribution fitting for the number 0 and 9 in MNIST with GMM and GGMM.

2.3 Loss Function

The total loss function of the proposed network consists of clustering loss and adversarial loss, which is defined as follows:

$$L_{MAAE} = L_{clustering} + \lambda L_{adversarial} \qquad (4)$$

where λ is the adversarial training startup variable, $\lambda = 1$ is set at a certain interval, otherwise $\lambda = 0$.

Given an input sample $x \in \{x_1, x_2 \ldots x_n\}$, it will enter into all adversarial blocks at the same time. The samples are firstly fed into each encoder to get the latent representations. Then a set of reconstructions $\tilde{x} = \{\tilde{x}_1, \tilde{x}_2, \ldots, \tilde{x}_k\}$ will be obtained from each decoder. By calculating the minimum loss between x and $\tilde{x}$, the conventional clustering loss is:

$$L_{clustering} = \sum_{t=1}^{n} \min_{i=1}^{k} d(x_t, \tilde{x}_t(i)) \qquad (5)$$

where $d(x, y)$ is the distance between x and y. Generally, the distance is calculated with the mean square error. However, since only the error of each pixel between images is considered, the mean square error is not entirely suitable for image distinguishing. The mean square error is sensitive to subtle changes in the image, and would misestimate the distance between images because it does not take the spatial structure of the samples into account.

To solve the above problem, we utilize the IMED as the distance measurement in Eq. (5). Specifically, given two M by N images $x = \{x^1, x^2, \ldots, x^{MN}\}$, $y = \{y^1, y^2, \ldots y^{MN}\}$, IMED is defined as:

$$d_{IME}^2(x, y) = \frac{1}{2\pi} \sum_{i,j=1}^{MN} exp\left\{-\frac{|P_i - P_j|^2}{2}\right\} (x^i - y^i)(x^j - y^j) \qquad (6)$$

where P is the pixel of the image, and $|P_i - P_j|$ indicates the distance between two pixels on the image.

In our method, IMED can reduce the clustering errors caused by distance measurement. Combined with Eq. (6), the clustering loss based on MAAE is defined as:

$$
\begin{aligned}
L_{clustering} &= \sum_{t=1}^{n} min \; d_{IME}^2\left(x_t, \hat{x}_t(i)\right) \\
&= \sum_{t=1}^{n} min \sum_{i=1}^{k} \sum_{j,l=1}^{MN} exp\left\{-\frac{|P_j - P_l|^2}{2}\right\} \left(x_t^j - \tilde{x}_t(i)^j\right)\left(x_t^l - \tilde{x}_t(i)^l\right)
\end{aligned}
\qquad (7)
$$

We minimize the clustering loss to optimize encoders and decoders in adversarial blocks. The cluster results can be obtained with:

$$c = \arg\min_{i=1}^{k} \sum_{j,l=1}^{MN} exp\left\{-\frac{|P_j - P_l|^2}{2}\right\}\left(x_t^j - \hat{x}_t(i)^j\right)\left(x_t^l - \hat{x}_t(i)^l\right) \qquad (8)$$

To further separate the latent representation of different cluster of data, the adversarial strategy is introduced in this paper. Assuming that the observed data is generated by a mixture model, where each cluster of data belongs to a separate distribution. We choose different target distributions for each adversarial autoencoder, and the adversarial process will make each encoder map the data to a different feature space to promote the separation of latent representations. Adversarial training is independent, each adversarial block does not interfere with each other when the discriminator discriminates the distribution generated by the encoder and the target distribution. The adversarial loss of MAAE is:

$$L_{adversarial} = \sum_{i=1}^{k} \min_{\theta_E} \max_{\theta_D} E_{x_i \sim p_{data_i}}\left[\log D_i(x_i) + E_{z_i \sim p_{z_i}}\left[\log(1 - D_i(G(z_i)))\right]\right] \qquad (9)$$

According to Eq. (6), the target distribution is chosen as k different generalized Gaussian distributions. The whole data set will be represented with GGMM.

2.4 Training Procedure

To initialize the parameters involved in the network, we firstly trained an independent autoencoder on the whole dataset. Then k-means was utilized to cluster the latent features extracted by the autoencoder to get an initial clustering results. Then, the collection of adversarial autoencoders were trained with different clusters of data. With the above pre-training, all the parameters of encoders and decoders were initialized. The network was jointly trained by minimizing the total loss function. Consequently, the training procedure of MAAE can be summarized as follows:

Step 1: Initialization by training a single autoencoder on the whole data set. The corresponding output of the encoder bottleneck layer was taken as the latent representation of the data. K-means was used to get the initial clustering for latent representation, and each cluster of data was used to train the corresponding cluster of adversarial autoencoders.

Step 2: Training the encoders and the decoders. The latent representation of the input sample was obtained by the encoder, and the reconstructed sample was obtained by the decoder. The minimum distance between the input image and each reconstructed image was calculated according to Eq. (7). For each input sample, the gradient caused by clustering loss will only affect the selected encoder and decoder during backpropagation.

Step 3: Training the encoders and discriminators. The latent representation of the input data obtained by the encoder and the target distribution generated by the sampling was used as the input of the discriminator. According to Eq. (9), the adversarial loss was calculated by the output of the discriminator. Each adversarial block corresponded to a different target distribution, and adversarial training was independent with each other.

3   Experiment

We tested the proposed method on three datasets including MNIST [17], Fashion [18], and USPS [19]. MNIST is a standard handwritten digital dataset, including 60,000 training samples and 10,000 test samples. Each sample is a 28 by 28 grayscale image with a label in 10 classes. Fashion is a dataset of Zalando's article images, which contains a training set of 60,000 samples and a test set of 10,000 samples. Each sample is a 28 by 28 grayscale images with a label in 10 classes. USPS is a digital dataset from the US Postal Service, including 7,291 training samples and 2,007 test samples. Each sample is a 16 by 16 grayscale image with a label in 10 classes.

To evaluate the proposed method, we compared the proposed method MAAE with seven methods, including k-means [1], AE + k-means, DEC [8], DCN [10], DCEC [20], DAMIC [12], K-DAE [13]. Furthermore, the ablation experiment was carried out and three versions of the proposed method MAAE were compared, i.e., $MAAE^1$ uses mean square error as the image distance without considering adversarial information, $MAAE^2$ uses IMED as the image distance on the basis of $MAAE^1$, and $MAAE^3$ uses Gaussian distribution as the target distribution based on $MAAE^2$. MAAE uses generalized Gaussian distribution as the target distribution on the basis of $MAAE^3$.

For the clustering quality evaluation, we adopt classification accuracy (ACC), normalized mutual information (NMI), and adjusted rand index (ARI) to evaluate all the comparison methods. For all the metrics, higher value indicates better performances.

All the experiments are implemented using Pytorch on a standard Windows 10 OS with an NVIDIA 1080Ti GPU and 16G memory.

3.1   Clustering Results

The evaluation results obtained by all the comparison methods are listed in Table 1. From this table, we can find that k-means can hardly obtain satisfactory results for all the testing datasets. Most deep learning based clustering models can obvious improve the classification results by mapping high-dimensional data to low-dimensional space. By mapping the data into one feature space, the results obtained by AE + k-means, DEC, DCN, and DCEC are still limited, because the latent representations may not be well separated under only one feature space. Both DAMIC and K-DAE can further improve the classification accuracies by representing each cluster as a separate manifold. However, both methods cannot guarantee that the latent features belonging to different clusters are completely separated. Comparatively, the proposed method outperforms all the comparison methods on all the testing datasets by introducing the adversarial information to embed the input data into different target distributions for each cluster.

For the ablation experiment, $MAAE^1$ can produce similar results with DAMIC and K-DAE due to the representation of each cluster as a separate manifold. With the utilization of IMED, $MAAE^2$ can obtain obvious improvements because IMED takes the spatial information into account. By introducing the adversarial information under the prior of Gaussian distribution, $MAAE^3$ can further improve the classification accuracy. Consequently, the proposed model can achieve the best results, which further indicates the effectiveness of each module involved in our network. The adversarial information can force the latent representations to match the target distribution and guarantee the separation among feature spaces. The GGMM can further improve the fitting ability of the model, and the utilization of IMED can further allocate clusters.

Table 1. Clustering performances on MNIST, Fashion, and USPS for all the comparison methods.

Methods	MNIST			Fashion			USPS		
	ACC	NMI	ARI	ACC	NMI	ARI	ACC	NMI	ARI
k-means	53.2	49.9	37.1	47.1	51.0	36.9	67.5	63.1	55.6
AE + km	81.3	74.6	69.4	59.4	62.2	46.0	71.1	67.8	61.3
DEC	84.3	80.2	75.2	51.5	53.9	40.5	76.0	77.1	-
DCN	83.0	81.0	74.7	50.6	55.1	42.2	69.0	68.0	-
DCEC	88.9	86.4	82.4	56.3	60.6	43.9	79.0	82.5	72.1
DAMIC	89.0	87.0	82.0	60.0	65.0	48.0	75.0	78.0	70.0
K-DAE	88.0	86.0	82.0	60.0	65.0	48.0	77.0	80.0	71.0
$MAAE^1$	87.9	85.5	81.1	57.6	61.7	44.5	77.1	80.0	71.0
$MAAE^2$	92.3	87.1	82.9	60.5	65.1	48.4	82.1	81.9	72.5
$MAAE^3$	93.5	88.4	84.5	61.2	65.3	48.9	83.9	82.7	73.3
MAAE	**94.8**	**89.9**	**85.8**	**61.8**	**65.9**	**49.3**	**84.5**	**87.4**	**74.2**

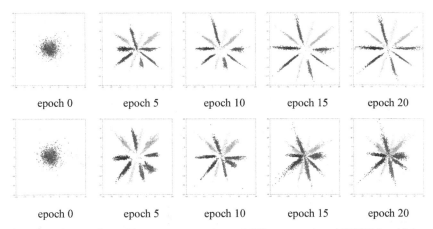

| epoch 0 | epoch 5 | epoch 10 | epoch 15 | epoch 20 |

| epoch 0 | epoch 5 | epoch 10 | epoch 15 | epoch 20 |

Fig. 3. Clustering results and latent representations of different epoch on MNIST, in which epoch 0 represents the initializations. Top row shows the results by setting the target distribution set as the generalized Gaussian distribution. Second row shows the results obtained by using Gaussian distributions as the target distribution.

Figure 3 shows the spatial scatter diagram of the latent representations of the clustering results. As can be seen, each cluster of data is embedded in the expected target distribution and the latent representations of different clusters can be well separated. To distinguish two different distributions, we set $\alpha = 0.9$ in generalized Gaussian distribution. Obviously, generalized Gaussian distribution shows more powerful representation ability than Gaussian distribution. The results verify the effectiveness of the proposed method in clustering and feature separation.

3.2 Reconstruct and Generate

Figure 4 shows the examples of reconstruction results on MNIST, Fashion, and USPS obtained by the proposed model. Figure 4(a) shows ten examples of different clusters. Figure 4(b) shows the reconstructions of decoders in different adversarial blocks. We can find that each decoder will only reconstruct the corresponding cluster of data, no matter which cluster the input sample come from. This is because all decoders only accept the encoder distribution which is similar to the corresponding target distribution and reconstruct it into their own cluster sample. Although the reconstruction of a single decoder is similar, it has the best reconstruction effect for sample that belongs to its own cluster.

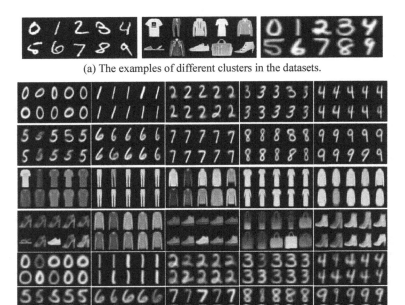

(a) The examples of different clusters in the datasets.

(b) Reconstructions of different adversarial blocks.

Fig. 4. Reconstruction results on MNIST, Fashion, and USPS.

Removing the encoder and discriminator, we only retain the decoder and use the respective target distribution of the adversarial block as its input. Figure 5(a) shows a good generation effect. The inputs of each decoder are all sampled from the corresponding target distribution and the style of generated sample varies with different sampling point. To ensure that each decoder only accepts the corresponding target distribution, that is, the latent representations of different sample are separated, Fig. 5(b) shows the generation results of different decoders with the target distribution of the first cluster in each dataset. As can be seen, only the first cluster in each dataset can obtain satisfactory generations and the others decoders crashed. We are convinced that the latent representations are well separated in the proposed model, because the decoder does not accept other target distributions.

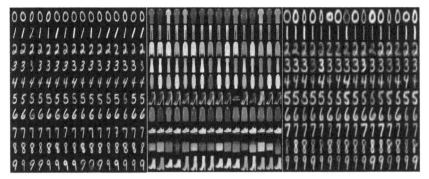

(a) Generation results of different decoders with the corresponding target distribution.

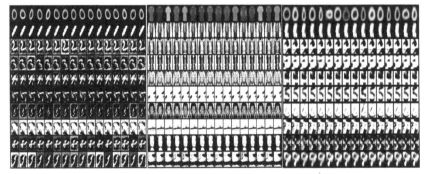

(b) Generation results of different decoders with the target distribution of the first cluster in each dataset.

Fig. 5. Generated results on MNIST, Fashion and USPS.

4   Conclusion

In this paper, we proposed a mixture of adversarial autoencoders clustering (MAAE) network. To obtain a set of separable latent representations, each cluster of data was represented with an independent adversarial autoencoder. By introducing the adversarial information, the aggregated posterior of the hidden code vector of the autoencoder could better match with the prior distribution which was constructed based on the generalized Gaussian mixture model. The clustering network jointly learned the nonlinear data representation and the collection of adversarial autoencoders. The clustering results were obtained by minimizing the improved reconstruction loss. The proposed model could produce more separable latent codes and outperformed other methods. How to improve the robustness to initializations will subject to our future research.

Acknowledgments. This work was supported by National Natural Science Foundation of China under Grant No. 62072241, and in part by Natural Science Foundation of Jiangsu Province under Grants No. BK20180069, and in part by Six talent peaks project in Jiangsu Province under Grant No. SWYY-056.

References

1. Ding, C., He, X.: K-means clustering via principal component analysis. In: Proceedings of the Twenty-First International Conference on Machine Learning (2004)
2. Reynold, D.: Gaussian mixture models. Encycl. Biometrics **741** (2009)
3. Aljalbout, E., Golkov, V., Siddiqui, Y., Strobel, M., Cremers, D..: Clustering with deep learning: taxonomy and new methods. arXiv preprint arXiv:1801.07648 (2018)
4. Ng, A.: Sparse autoencoder. CS294A Lecture notes, 72.2011, 1–19 (2011)
5. Goodfellow, I., et al.: Generative adversarial nets. In: NIPS, pp. 2672–2680 (2014)
6. Makhzani, A., Shlens, J., Jaitly, N., Goodfellow, I.: Adversarial autoencoders. arXiv preprint arXiv:1511.05644 (2015)
7. Kingma, D.P., Welling, M.: Auto-encoding variational bayes. arXiv preprint arXiv:1312.6114 (2013)
8. Min, E., Guo, X., Liu, Q., Zhang, G., Cui, J., Long, J.: A survey of clustering with deep learning: from the perspective of network architecture. IEEE Access **6**, 39501–39514 (2018)
9. Xie, J., Girshick, R., Farhad, A.: Unsupervised deep embedding for clustering analysis. In: ICML (2016)
10. Yang, B., Fu, X., Sidiropoulos, N.D., Hong, M.: Towards Kmeans-friendly spaces: simultaneous deep learning and clustering. In: ICML (2017)
11. Zhang, D., Sun, Y., Eriksson, B., Balzano, L.: Deep unsupervised clustering using mixture of autoencoders. arXiv preprint arXiv:1712.07788 (2017)
12. Chazan, S.E., Gannot, S., Goldberge, J..: Deep Clustering based on a mixture of Autoencoders. In: 2019 IEEE 29th International Workshop on Machine Learning for Signal Processing, pp. 1–6 (2019)
13. Opochinsky, Y., Chazan, S.E., Gannot, S., Goldberger, J.: K-autoencoders deep clustering. In: IEEE International Conference on Acoustics, Speech and Signal Processing, pp. 4037–4041 (2020)
14. Mohamed, O.M.M., Jaïdane-Saïdane, M.: Generalized gaussian mixture model. In: 17th European Signal Processing Conference, pp. 2273–2277 (2009)
15. Wang, L., Zhang, Y., Feng, J.: On the euclidean distance of images. IEEE Trans. Pattern Anal. Mach. Intell. **27**(8), 1334–1339 (2005)
16. Thanh Minh Nguyen, Q.M., Wu, J., Zhang, H.: Bounded generalized Gaussian mixture model. Pattern Recogn. **47**(9), 3132–3142 (2014)
17. Deng, L.: The mnist database of handwritten digit images for machine learning research. IEEE Signal Process. Mag. **29**(6), 141–142 (2012)
18. Xiao, H., Vollgraf, R.: Fashion-mnist: a novel image dataset for benchmarking machine learning algorithms: arXiv preprint arXiv:1708.07747 (2017)
19. Hull, O.J.: A database for handwritten text recognition. IEEE Trans. Pattern Anal. Machine Intell. **16**(5), 550–554 (1994)
20. Guo, X., Liu, X., Zhu, E., Yin, J.: Deep clustering with convolutional autoencoders. In: International Conference on Neural Information Processing, pp. 373–382 (2017)

SA-InterNet: Scale-Aware Interaction Network for Joint Crowd Counting and Localization

Xiuqi Chen, Xiao Yu, Huijun Di$^{(\boxtimes)}$, and Shunzhou Wang

Beijing Laboratory of Intelligent Information Technology, School of Computer Science and Technology, Beijing Institute of Technology, Beijing 100081, China
`ajon@bit.edu.cn`

Abstract. Crowd counting and crowd localization are essential and challenging tasks due to uneven distribution and scale variation. Recent studies have shown that crowd counting and localization can complement and guide each other from two different perspectives of crowd distribution. How to learn the complementary information is still a challenging problem. To this end, we propose a Scale-aware Interaction Network (SA-InterNet) for joint crowd counting and localization. We design a dual-branch network to regress the density map and the localization map, respectively. The dual-branch network is mainly constructed with scale-aware feature extractors, which can obtain multi-scale features. To achieve mutual guidance and assistance of the two tasks, we design a density-localization interaction module by learning the complementary information. Our SA-InterNet can obtain accurate density map and localization map of an input image. We conduct extensive experiments on three challenging crowd counting datasets, including ShanghaiTech Part_A, ShanghaiTech Part_B and UCF-QNRF. Our SA-InterNet achieves superior performance to state-of-the-art methods.

Keywords: Crowd counting · Crowd localization · Multi-scale feature learning · Mutual learning

1 Introduction

Crowd counting is an active research topic in computer vision, aiming to obtain an accurate number of people in a crowd scene. It has attracted a lot of research interests due to its wide application in public safety [4], video surveillance [24], urban planning [25], and so on. Recently, crowd localization has become popular as an essential auxiliary task to get head position. The main problems of crowd counting and crowd localization are scale variations and uneven distribution, especially in highly congested scenes.

Many crowd counting methods [1,10,16,18,20,26] have shown that multi-scale features are helpful to solve this problem and obtain more accurate counting

---

H. Di—This is a student paper.

© Springer Nature Switzerland AG 2021
H. Ma et al. (Eds.): PRCV 2021, LNCS 13019, pp. 203–215, 2021.
https://doi.org/10.1007/978-3-030-88004-0_17

results. However, the predicted density map is sometimes inconsistent with the actual crowd distribution and cannot obtain an accurate head position, which has been verified in [11]. Getting precise head position of each person in the crowd also has essential meaning and value. Moreover, density map and localization map reflect two different perspectives of crowd distribution, their features can complement and guide each other. Some methods [11,12,15] have attempted to use the complementary information. However, the model in [11] is a redundant multi-branch network without mutual learning and guidance of density information and localization information. The model proposed by [12] requires data labeled with rectangular boxes, and the graph neural network proposed in [15] is challenging to be optimized. So it's necessary to design an easy-to-learn network for crowd counting and localization.

In this paper, we propose a novel Scale-aware Interaction Network (SA-InterNet) for joint crowd counting and localization, which adopts encoder-decoder architecture. Our encoder is based on VGG-16 [19] to extract the semantic features of the image, and our decoder is a dual-branch network, which can regress the density map and localization map, respectively. To extract multi-scale features, we design a scale-aware feature extractor (SAFE) via dilated convolution, which can enlarge receptive fields without increasing computation. To achieve mutual help of crowd counting and localization, we develop a density-localization interaction module (DLIM) to obtain the complementary information. We evaluate our method on three crowding counting datasets, and experiment results show the superiority of our SA-InterNet.

In summary, the main contributions of this paper are as follows:

1) We develop a scale-aware feature extractor (SAFE) to extract the multi-scale features, which can expand receptive fields without increasing computation.
2) We design a density-localization interaction module (DLIM) to achieve mutual assistance and guidance between crowd counting and crowd localization by learning their complementary information.
3) Based on SAFE and DLIM, we propose a Scale-aware Interaction Network (SA-InterNet) for joint crowd counting and localization. We conduct extensive experiments on three challenging datasets, including ShanghaiTec Part_A [26], ShanghaiTec Part_B [26] and UCF-QNRF [7]. Compared with other advanced methods, our SA-InterNet achieves outstanding performance in both crowd counting and crowd localization.

2   Related Work

Traditional crowd counting methods can be generally divided into two categories: detection-based methods [5,17] and regression-based methods [2,3,9]. The former category adopts detectors to get each person's position and the number of detection results as the number of people, which performs well in sparse crowd scenes. The latter category learns a mapping relationship between hand crafted features and the number of people, which can solve the occlusion problem to a certain extent but ignores spatial information. Although the above methods have achieved promising results, they are not applicable to real-world applications due to limited generalization ability. Benefited from the feature representation ability of convolutional neural network (CNN), many computer vision tasks (*e.g.,* video object segmentation [27,29], human parsing [23,30], human object interaction detection [28], etc.) have been prompted, and many CNN-based crowd counting methods [1,10–12,14–16,18,26] have been proposed and make significant progresses. Instead of directly predicting the number of people, most of them formulate the counting task into a density map regression problem and obtain the crowd number by summing up all the pixel values of the density map.

To solve the problem of scale variations, many CNN-based methods try to extract multi-scale features. Zhang *et al.* [26] proposed a Multi-Column Convolutional Neural Network (MCNN), which obtained multi-scale information by different size convolution kernels in three branches. Sam *et al.* [18] proposed a multi-column Switching-CNN and used a density level classifier to select the most optimal regressor for input patches. However, multi-column or multi-branch networks required heavy computation. To this end, Li *et al.* [10] proposed CSRNet, a single column network that used dilated convolution to enlarge the receptive field. Cao *et al.* [1] and Miao *et al.* [16] proposed the scale aggregation module that used inception module [21] to extract scale-aware feature.

To get both the number of people and the head position in a crowd scene, some methods attempt to combine crowd counting and crowd localization and take advantage of their complementary information. Idrees *et al.* [7] proposed a composition loss for counting, density map estimation, and localization in dense crowds, they integrated the density map to obtain crowd count and sharpened the density map to obtain the localization map. Liu *et al.* [12] obtained density map and head's position by combining detection-based method and regression-based method. Liu *et al.* [11] proposed RAZ-Net containing a multi-branch network to obtain the density map and localization map, respectively. Luo *et al.* [15] proposed a graph neural network to learn the complementary features of crowd counting and crowd localization, and got the density map and localization map simultaneously.

3   Proposed Method

Figure 1 shows our proposed Scale-aware Interaction Network (SA-InterNet), which adopts an encoder-decoder structure. We apply the first ten layers of the pre-trained VGG-16 [19] as our encoder to extract low-level features and design a dual-branch network as the decoder to regress the crowd density map and localization map, respectively. Specifically, the dual-branch network consists of density estimation branch and localization branch, which are constructed by scale-aware feature extractor (SAFE) to extract multi-scale information. To achieve mutual assistance and guidance between crowd counting and localization, we design the density-localization interaction module (DLIM) to obtain the complementary information. We will introduce the detailed components of our proposed method in the following sections.

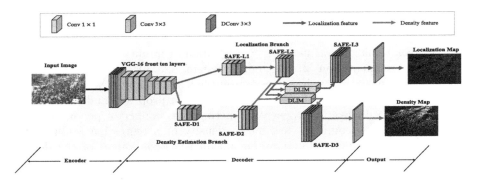

Fig. 1. The architecture of our proposed SA-InterNet. Image is send to the encoder to extract the low-level features, and a dual-branch network as the decoder to output the corresponding density map and localization map, respectively. SAFE can extract multi-scale features, and DLIM is used to learn the complementary information between the two branches.

3.1   Scale-Aware Feature Extractor

To handle scale variations, we propose a scale-aware feature extractor (SAFE) to obtain multi-scale features, which contains four columns with the same input and the same number of output channels. Give the input feature x, the output feature F of SAFE is as follow:

$$F = concat \left[Conv_{1 \times 1}(x), Conv_{3 \times 3}(x), DConv_{3 \times 3}(x), DConv_{3 \times 3}(x) \right], \qquad (1)$$

where $Conv_{r \times r}(\cdot)$ represents the standard $r \times r$ convolution, $DConv_{r \times r}(\cdot)$ represents the dilated convolution with filter size of $r \times r$, which can enlarge receptive fields without increasing computation. To get more scale information, the dilated rates of the two $DConv_{3 \times 3}(\cdot)$ are 2 and 3, respectively. Compared with the modules [1,16] mentioned above, SAFE only contains two kinds of small convolution kernels and obtains multi-scale information with a simpler structure and less computation.

As shown in Fig. 1, in the decoder network, both the density estimation branch and the localization branch are constructed by three SAFE modules in a cascade way, and the number of their output feature channels is 128, 64, and 16, respectively. Since our encoder is based on VGG-16 [19], the feature map input to the decoder is 1/8 of the original image, we upsample the feature map with a scale factor of 2 after each SAFE to restore the feature map to it's original size.

3.2   Density-Localization Interaction Module

The density map and the localization map reflect crowd distribution from two perspectives. The density map represents crowd distribution in a coarse-grained manner, and only the number of people can be obtained from it, while the localization map represents the crowd distribution in a fine-grained way with accurate head positions. Furthermore, crowd localization information can alleviate local inconsistencies in density map, while the crowd density map can offer guidance information for crowd localization. Thus, we can use the information of the density map and the localization map to improve crowd counting and localization results, respectively. Based on this motivation and inspired by self-calibration module [13], we design a density-localization interaction module (DLIM). To achieve mutual guidance of crowd counting and localization, we apply two symmetrical DLIMs, one of which is shown in Fig. 2. It is the process in which density feature map x_1 guides localization feature map x_2 to learn the complementary information, and for simplicity, we only describe this module. Given the input density feature map x_1 and localization feature map x_2, we first downsample x_2 to enlarge the receptive field and then upsample to restore the feature map to its original size for subsequent operations, as follows:

$$F_1 = Up\left(Conv\left(Avg\left(x_2\right)\right)\right), \tag{2}$$

where $Avg\left(\cdot\right)$ is the average pooling function, $Up\left(\cdot\right)$ represents the bilinear interpolation operation and $Conv(\cdot)$ represents the standard 3×3 convolution. To realize density information supplement and guide localization information, we first use x_1 to supplement the localization feature map F_1 obtained in Eq. 2, and then generate a guiding weight for localization feature map, as follows:

$$F_2 = Conv\left(x_2\right) \otimes \sigma(x_1 \oplus F_1), \tag{3}$$

where σ is the sigmoid function, $\otimes$ denotes element-wise multiplication and $\oplus$ denotes element-wise summation. We use x_2 as the residual to prevent the loss of localization information, and the final output localization feature map $x_2{}'$ is as follows:

$$x_2{}' = Conv(F_2) + x_2. \tag{4}$$

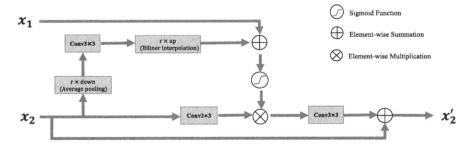

Fig. 2. The architecture of density-localization interaction module (DLIM). It's the process of supplementing and guiding localization feature map x_2 by density feature map x_1. Another symmetrical DLIM swaps the input position to realize localization information to guide and supplement density information.

3.3   Loss Function

The loss function consists of crowd counting loss (L_{cou}) and crowd localization loss (L_{loc}). For the crowd counting task, we use $L2$ loss to calculate the difference between the estimated density map and the ground truth map, which can be defined as follows:

$$L_{cou} = \frac{1}{2N} \sum_{i=1}^{N} \| F(X_i; \theta) - F_i \|_2^2, \tag{5}$$

where $F(X_i; \theta)$ and F_i are estimated density map and ground truth map of the i-th image X_i, and N is the number of training images, θ is the parameter learned by our network.

For the crowd localization task, human head pixels in the image are classified into positive category, while the other pixels are classified into negative category, so binary cross entropy (BCE) loss is adopted. Considering the imbalance between the number of positive pixels and the number of negative pixels, we use a constant μ (set to 100 as in [11]) to adjust localization loss, defined as follows:

$$L_{loc} = \sum_{j=1}^{M} -(\mu \cdot y_j \log(\hat{y}_j) + (1 - y_j) \log(1 - \hat{y}_j)), \tag{6}$$

where M is the number of pixels in this image, y_j is the label of pixel j, which is 1 if the pixel is of positive category and 0 otherwise, $\hat{y}_j$ represents the probability of pixel j is classified into positive category. The final loss function L is defined as follows:

$$L = l_{cou} + \lambda L_{loc}, \tag{7}$$

where λ is a hyperparameter to balance L_{cou} and L_{loc}, we empirically set λ to 0.1 in our experiments.

4   Experiments

4.1   Datasets

We conduct experiments on three public datasets (*i.e.,* ShanghaiTech Part_A, ShanghaiTech Part_B, and UCF-QNRF). The details are as follows:

ShanghaiTech Part_A. ShanghaiTech Part_A dataset [26] contains 482 images with different resolutions and a total of 24.1K head annotations. There are 300 images for training and 182 images for testing. ShanghaiTech Part_A is a challenging crowd counting dataset due to highly congested scenes and significant crowd density variations.

ShanghaiTech Part_B. ShanghaiTech Part_B dataset [26] contains 716 annotated images with relatively sparse crowd scenes taken from the streets in Shanghai. There are 400 images for training and 316 images for testing, and all the images have the same resolution of 768 × 1024.

UCF-QNRF. UCF-QNRF dataset [7] is a large-scale crowd counting dataset with 1535 high-resolution images and a total of 1.25 million head annotations. There are 1201 images for training and 334 images for testing. The most diverse viewing perspectives, densities, and lighting variations make this dataset realistic and difficult to count.

4.2 Evaluation Metrics

For the crowd counting task, we use the widely-adopted mean absolute error (MAE) and mean squared error (MSE) to evaluate the performance of our SA-InterNet. The definitions are as follows:

$$MAE = \frac{1}{n} \sum_{i=1}^{n} \left| y_i - y_i^{gt} \right|, \tag{8}$$

$$MSE = \sqrt{\frac{1}{n} \sum_{i=1}^{n} \left| y_i - y_i^{gt} \right|^2}, \tag{9}$$

where n is the number of test images, y_i and y_i^{gt} are the predicted count and the ground truth count of the i-th image, respectively.

Following [7,22], we use the precision (P), recall (R), F1-measure (F_1) to evaluate crowd localization performance. The head localization is successful when the distance between the predicted head point and the actual point is less than the distance threshold. We use the average accuracy at various distance thresholds (*i.e.*, 1, 2, 3, ..., 100 pixels) to quantify the localization error. The evaluation metrics are defined as follows:

$$P = \frac{TP}{TP + FP}, \tag{10}$$

$$R = \frac{TP}{TP + FN}, \tag{11}$$

$$F_1 = \frac{2 * P * R}{P + R}, \tag{12}$$

where TP denotes that the head point is truly predicted when the distance between the predicted head location point and the ground truth point is less than the pixel threshold; FP means that the head point is falsely predicted when the distance between them is larger than the pixel threshold; FN indicates that the predicted point which should match the ground truth point does not exist.

4.3 Implementation Details

Our network is end-to-end trainable. We choose the first ten layers of the pre-trained VGG-16 [19] as our encoder, and the other layers are initialized by Gaussian distribution with mean of zero and standard deviation of 0.01. The Adam optimizer is adopted to optimize our network with a fixed learning rate of 1×10^{-5} and weight decay of 1×10^{-4}. We perform all experiments on a NVIDIA RTX 2080 Ti GPU card with the Pytorch framework.

4.4   Comparison with State-of-the-Arts

Crowd Counting. To evaluate our proposed model, we conduct experiments on three crowd counting datasets and compare our results with other recent methods. The comparison results are shown in Table 1 and Table 2. From Table 1, we can see that our method achieves the best MAE of 62.8 on ShanghaiTech Part_A and the best MSE of 12.4 on ShanghaiTech Part_B. In Table 2, compared with the state-of-the-art method HyGNN [15], we get the lowest MSE of 178.6 and comparable MAE of 102.5 on UCF-QNRF. Although our MAE is a little higher, our model is simpler and easier to train. These results demonstrate that our SA-InterNet can perform well in both congested and sparse crowd scenes.

Table 1. Comparison results for crowd counting on ShanghaiTech Part_A and Shang-haiTech Part_B datasets.

Method	PartA_MAE	PartA_MSE	PartB_MAE	PartB_MSE
MCNN [26]	110.2	173.2	26.4	41.3
Switch-CNN [18]	90.4	135	21.6	33.4
CSRNet [10]	68.2	115	10.6	16
SANet [1]	67.0	104.5	8.4	13.5
RA2-Net [11]	65.1	106.7	8.4	14.1
TEDNet [8]	64.2	109.1	8.2	12.8
ADCrowNet [14]	63.2	**98.9**	**7.7**	12.9
SA-InterNet (**ours**)	**62.8**	102.5	8	**12.4**

Table 2. Comparison results for crowd counting on UCF-QNRF dataset.

Method	MAE	MSE
MCNN [26]	277	426
Switch-CNN [18]	228	445
RA2-Net [11]	116	195
TEDNet [8]	113	188
HyGNN [15]	**100.8**	185.3
SA-InterNet (**Ours**)	102.5	**178.6**

Crowd Localization. We conduct crowd localization experiments on UCF-QNRF dataset and the results are shown in Table 3. Compared with other recent methods, our SA-InterNet achieves the best localization results. Although our precision is lower than SCLNet [22], we get higher recall and F1-measure, indicating that the generated localization map is more accurate. Figure 3 presents the visualization results of crowd localization. Red dots are ground truth head locations, and green dots are predicted head locations. We can see that our method can predict the precise position of most heads.

Table 3. Comparison results for crowd localization on UCF-QNRF dataset.

Method	Av. Precision	Av. Recall	F1-measure
MCNN [26]	59.93%	63.50%	61.66%
DenseNet [6]	70.19%	58.10%	63.87%
CL [7]	75.80%	59.75%	66.82%
SCLNet [22]	**83.99%**	57.62%	67.36%
SA-InterNet (**ours**)	72.23%	**64.73%**	**68.27%**

Fig. 3. The visualization results of crowd localization on the UCF-QNRF dataset. Red points denote ground truths and green points denote predict head points. (Color figure online)

4.5   Ablation Study

Efficacy of Scale-Aware Feature Extractor. We first explore the settings of filter size and dilated rate for SAFE and the results are shown in Table 4, the networks of all these experiments are dual-branched without DLIM. We find that the counting results increase as the number of columns increase, and a better result might exist when the number of columns is 8 or more. To avoid excessive parameters or computations, we choose the current best combination with MAE of 62.2 and MSE of 107.1, and the corresponding SAFE has been introduced in Sect. 3.1.

To verify the effectiveness of SAFE, we conduct ablation experiments on ShanghaiTech Part_A dataset, and the results are shown in Table 5. Compared with the results in the first row, we find that SAFE can significantly improve

Table 4. Ablation studies of different settings of SAFE on ShanghaiTech Part_A dataset.

Setting			ShanghaiTech Part_A	
Column	Dilated rate	Filter size	MAE	MSE
1	1	3	66.4	113.9
2	1, 1	3, 3	65.9	110.3
2	1, 1	1, 3	65.2	117.1
2	1, 2	3, 3	64.8	114.5
4	1, 1, 1, 2	1, 3, 5, 3	65.2	117.1
4	1, 1, 2, 2	1, 3, 3, 3	63.6	108.8
4	1, 1, 2, 3	1, 3, 3, 3	**62.2**	**107.1**
8	1, 1, 1, 1, 2, 2, 3, 3	1, 1, 3, 3, 3, 3, 3, 3	62.3	107.5

Table 5. Ablation studies of SAFE on ShanghaiTech Part_A dataset.

Method	Counting	Localization	MAE	MSE
Single counting	×	–	72.7	123.3
Single counting	√	–	67.7	107.7
Dual-branch	×	×	66	114.2
Dual-branch	×	√	66.3	116.3
Dual-branch	√	×	63.5	112.4
Dual-branch	√	√	**62.2**	**107.1**

the counting performance in the second row, and the localization branch provides valuable complementary information for crowd counting to help improve counting results in the third row. We get the best results in the last row when both the density estimation branch and localization branch are configured with SAFE.

Efficacy of Density-Localization Interaction Module. To explore the effectiveness of the density-localization interaction module (DLIM), we conduct ablation experiments on ShanghaiTech Part_A dataset. As shown in Table 6, the best results are obtained when DLIM is applied to the second layer of the dual-branch network. At the cost of a slight decrease in MAE, MSE increases significantly, indicating that the localization information is helpful for density map generation. When DLIM is applied to the first or third layer, the results are not improved. It may be that the first layer features are too messy to obtain helpful information, and the third layer features have rich semantic information, which will be suppressed by DLIM. Figure 4 presents the visualization results of estimated density maps generated by different network. Compared with single counting network and dual-branch network without DLIM, our SA-InterNet obtains more accurate results, especially in dense crowd areas.

Table 6. Ablation studies of DLIM on ShanghaiTech Part_A dataset.

Location			ShanghaiTech Part_A	
1	2	3	MAE	MSE
			62.2	107.1
√			65.2	113.8
	√		**62.8**	**102.5**
		√	63.5	109.5

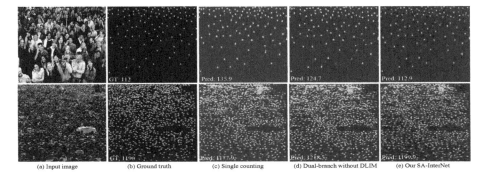

| (a) Input image | (b) Ground truth | (c) Single counting | (d) Dual-branch without DLIM | (e) Our SA-InterNet |

Fig. 4. Visualization comparisons. From left to right: input images, ground truth density maps, density maps generated by single counting network, dual-branch network without DLIM and SA-InterNet. Our proposed SA-InterNet can generate high-quality density maps and predict accurate counting results both in sparse and dense crowd scenes.

5 Conclusion

In this paper, we propose a SA-InterNet for joint crowd counting and crowd localization, which adopts a dual-branch network to regress the density map and the localization map, respectively. The dual-branch network is composed of multiple SAFEs and DLIMs. SAFE extracts multi-scale features, and DLIM learns the complementary information to achieve mutual assistance of crowd counting and localization. We conduct extensive experiments on three challenging datasets and the experiment results show that the effectiveness of our method.

References

1. Cao, X., Wang, Z., Zhao, Y., Su, F.: Scale aggregation network for accurate and efficient crowd counting. In: Ferrari, V., Hebert, M., Sminchisescu, C., Weiss, Y. (eds.) ECCV 2018. LNCS, vol. 11209, pp. 757–773. Springer, Cham (2018). https://doi.org/10.1007/978-3-030-01228-1_45
2. Chan, A.B., Liang, Z.J., Vasconcelos, N.: Privacy preserving crowd monitoring: counting people without people models or tracking. In: CVPR, pp. 1–7 (2008)
3. Chan, A.B., Vasconcelos, N.: Bayesian Poisson regression for crowd counting. In: ICCV (2009)
4. Chen, J., Liang, J., Lu, H., Yu, S.I.: Videos from the 2013 Boston Marathon: An Event Reconstruction Dataset for Synchronization and Localization (2016)
5. Dalal, N., Triggs, B.: Histograms of oriented gradients for human detection. In: CVPR, pp. 886–893 (2005)
6. Huang, G., Liu, Z., Van Der Maaten, L., Weinberger, K.Q.: Densely connected convolutional networks. In: CVPR, pp. 2261–2269 (2017)
7. Idrees, H., et al.: Composition loss for counting, density map estimation and localization in dense crowds. In: Ferrari, V., Hebert, M., Sminchisescu, C., Weiss, Y. (eds.) ECCV 2018. LNCS, vol. 11206, pp. 544–559. Springer, Cham (2018). https://doi.org/10.1007/978-3-030-01216-8_33
8. Jiang, X., et al.: Crowd counting and density estimation by trellis encoder-decoder networks. In: CVPR, pp. 6133–6142 (2019)
9. Lempitsky, V., Zisserman, A.: Learning to count objects in images. In: NIPS, pp. 1324–1332 (2013)
10. Li, Y., Zhang, X., Chen, D.: CSRNet: dilated convolutional neural networks for understanding the highly congested scenes. In: CVPR, pp. 1091–1100 (2018)
11. Liu, C., Weng, X., Mu, Y.: Recurrent attentive zooming for joint crowd counting and precise localization. In: CVPR, pp. 1217–1226 (2019)
12. Liu, J., Gao, C., Meng, D., Hauptmann, A.G.: DecideNet: counting varying density crowds through attention guided detection and density estimation. In: CVPR, pp. 5197–5206 (2018)
13. Liu, J., Hou, Q., Cheng, M., Wang, C.: Improving convolutional networks with self-calibrated convolutions. In: CVPR, pp. 10093–10102 (2020)
14. Liu, N., Long, Y., Zou, C., Niu, Q., Pan, L., Wu, H.: ADCrowdNet: an attention-injective deformable convolutional network for crowd understanding. In: CVPR, pp. 3225–3234 (2019)
15. Luo, A., Yang, F., Li, X., Nie, D., Jiao, Z.: Hybrid graph neural networks for crowd counting. In: AAAI, pp. 11693–11700 (2020)
16. Miao, Y., Lin, Z., Ding, G., Han, J.: Shallow feature based dense attention network for crowd counting. In: AAAI, pp. 11765–11772 (2020)
17. Sabzmeydani, P., Mori, G.: Detecting pedestrians by learning shapelet features. In: CVPR, pp. 1–8 (2007)
18. Sam, D.B., Surya, S., Babu, R.V.: Switching convolutional neural network for crowd counting. In: CVPR, pp. 4031–4039 (2017)
19. Simonyan, K., Zisserman, A.: Very deep convolutional networks for large-scale image recognition. In: ICLR (2015)
20. Sindagi, V.A., Patel, V.M.: Generating high-quality crowd density maps using contextual pyramid CNNs. In: ICCV, pp. 1879–1888 (2017)
21. Szegedy, C., et al.: Going deeper with convolutions. In: CVPR, pp. 1–9 (2015)

22. Wang, S., Lu, Y., Zhou, T., Di, H., Lu, L.: SCLNet: spatial context learning network for congested crowd counting. Neurocomputing **404**, 227–239 (2020)

23. Wang, W., Zhou, T., Qi, S., Shen, J., Zhu, S.C.: Hierarchical human semantic parsing with comprehensive part-relation modeling. IEEE TPAMI (2021)

24. Xiong, F., Shi, X., Yeung, D.Y.: Spatiotemporal modeling for crowd counting in videos. In: ICCV, pp. 5161–5169 (2017)

25. Zhan, B., Monekosso, D.N., Remagnino, P.: Crowd analysis: a survey. Mach. Vis. Appl. **19**, 345–357 (2008)

26. Zhang, Y., Zhou, D., Chen, S., Gao, S.: Single-image crowd counting via multi-column convolutional neural network. In: CVPR, pp. 589–597 (2016)

27. Zhou, T., Li, J., Wang, S., Tao, R., Shen, J.: MATNet: motion-attentive transition network for zero-shot video object segmentation. IEEE TIP **29**, 8326–8338 (2020)

28. Zhou, T., Qi, S., Wang, W., Shen, J., Zhu, S.C.: Cascaded parsing of human-object interaction recognition. IEEE TPAMI (2021)

29. Zhou, T., Wang, S., Zhou, Y., Yao, Y., Li, J., Shao, L.: Motion-attentive transition for zero-shot video object segmentation. In: AAAI, pp. 13066–13073 (2020)

30. Zhou, T., Wang, W., Liu, S., Yang, Y., Van Gool, L.: Differentiable multi-granularity human representation learning for instance-aware human semantic parsing. In: CVPR, pp. 1622–1631 (2021)

Conditioners for Adaptive Regression Tracking

Ding Ma and Xiangqian Wu[✉]

School of Computer Science and Technology, Harbin Institute of Technology,
Harbin, China
{madingcs,xqwu}@hit.edu.cn

Abstract. Given the input of the whole search area, the goal of regression tracking is to estimate the location of the target object by a calculated response map which usually generates by a Gaussian function. Due to object appearance variation caused by factors such as lighting change and deformation, the update is essential during tracking process. To this end, recent deep regression trackers usually fine-tune the model using hundreds of iterations of gradient descent. Although such trackers achieve pretty good performance, the fine-tuning process is inefficient and limits the tracker's speed which is a relatively important evaluation metric. To solve this problem, we propose an adaptive regression model that utilizes a single forward pass to adapt the model to the specific target object. During forward pass process, two additional networks are designed to adjust the parameters of intermediate layer of the regression model. Such two networks are referred to as conditioners which pay attention to the context and spatial information simultaneously. Extensive experiments illustrate that our approach is much faster than the fine-tuned trackers while achieving competitive performance on four public tracking benchmarks including OTB2015, TC128, UAV123 and VOT2018.

Keywords: Conditioner · Regression tracking · Visual tracking

1 Introduction

With the powerful representation of deep features, much progress has been made in the tracking literature. depending on the sampling strategy, recent deep trackers can be roughly classified as two-stage [13, 20, 25] and one-stage tracker [7, 16, 18, 24, 27].

This work was supported in part by the National Key R&D Program of China under Grant 2018YFC0832304 and 2020AAA0106502, by the Natural Science Foundation of China under Grant 62073105, by the Distinguished Youth Science Foundation of Heilongjiang Province of China under Grant JC2018021, by the State Key Laboratory of Robotics and System (HIT) under Grant SKLRS-2019-KF-14 and SKLRS-202003D, and by the Heilongjiang Touyan Innovation Team Program.

H. Ma et al. (Eds.): PRCV 2021, LNCS 13019, pp. 216–228, 2021.
https://doi.org/10.1007/978-3-030-88004-0_18

By definition, the process of the two-stage tracker is consist of the sampling stage and classification stage. In the sampling stage, a set number of candidates are extracted depending on the location of the target object in the previous frame. The task of the classification stage is to estimate the position of the target object in the current frame depending on the score ranking of a pre-trained classifier. Although some two-stage trackers can achieve high accuracy, the number of candidates and heavy computational cost in classifier would also impact the performance of two-stage trackers.

Different from the two-stage trackers, one-stage trackers formulate visual tracking as a specific object searching problem. One-stage trackers can be broadly divided into two categories: discriminative correlation filters (DCFs) based trackers [7,16,27] and deep regression networks (DRNs) based trackers [18,24]. Compared with the DRNs, DCFs trackers got top performance on several popular tracking benchmarks [17,26]. However, the main drawback among DCFs trackers is that the DCFs trackers take few advantages of end-to-end learning. On the contrary, DRNs trackers have pay more attention due to it has the potential to take full advantage of end-to-end learning. Most of DRNs trackers share a similar paradigm: Firstly, offline fine-tune a network based on the annotation in the first frame for several hundred forward-backward iterations to make the model more responsive to the specific target object; Secondly, online fine-tune the network to handle the appearance variations of the target object. Especially, some DRNs trackers are updated frame-by-frame to adapt the appearance variations.

We found that the real-time capacity largely fades away as these DRNs trackers performing hundreds of iterations for fine-tuning. That is, the DRNs should minimize the number of fine-tuning to accelerate the tracking speed which is an important indicator to measure the quality of the tracker. And this paper attempts to solve this problem.

In this paper, we propose a novel approach to adapt the generic regression network to the target object in a single feed-forward pass, which is referred to as CARM. Another two conditioners are proposed to adjust the parameters of intermediate layers in the generic regression network during the tracking phase. We implement the conditioners with neural networks that pay attention to the information of relevant important spatial-temporal context and trajectory, respectively. The reasons for choosing such information are listed as follows.

– In practice, the context is the surrounding background of the target object within a determined region. Most of the context is invariably between two consecutive frames (e.g., Even if the appearance of the target object changes significantly, the context containing the object does not change much as the overall scene remains similar and only a small portion of the context is changed). Hence, there exist some spatial-temporal relationships between consecutive frames. Meanwhile, to extract more relevant spatial-temporal context which contributes to locating the object precisely, the redundant spatial-temporal context should be purified.
– During the tracking process, most of the time the object moves smoothly and its appearance changes little or slowly. Therefore, the trajectory is an important cue to locate the object in the current frame.

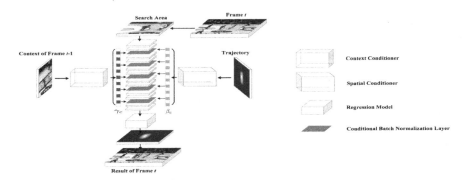

Fig. 1. An overview of our model with three subnetwork: a generic regression network, a visual context conditioner, and a trajectory conditioner. The two conditioners generates a set of parameters which adjust the feature maps of the generic regression network. The figure is best viewed in color. (Color figure online)

With the above analysis, we illustrate the basic flow of our architecture in Fig. 1. By extracting the relevant important spatial-temporal context and trajectory information, an attentive context prediction network and a trajectory network produce a set of parameters, which are injected into the regression network for layer-wise feature manipulation. Without the fine-tuning process, our model is able to adapt the generic regression network to the specific object. Our CARM model has several interesting properties. Firstly, there is no need to fine-tune the regression network during the test phase. That is, the parameters produced by the visual context network will guide the generic regression network to focus on the specific object instance. Secondly, the visual prediction network encodes the relevant important spatial-temporal context from past frames, which helps to discriminate the target from the background. Thirdly, the trajectory provides a piece of essential prior knowledge to locate the target object in the current frame. The key contributions are summarized as follows:

(1) We propose an adaptive regression tracking model with two conditioners, which facilitates the tracking speed by removing the fine-tuning process.
(2) We devise two conditioners, i.e., visual context network and trajectory network. Especially, visual context network guides the generic regression network to focus on the specific object. Trajectory network provides a piece of essential prior knowledge to locate the target object in the current frame.
(3) We verify the effectiveness of our tracker on four published benchmarks OTB2015 [26], TC128 [17], UAV123 [19] and VOT2018 [1].

The rest of the paper is organized as follows. We briefly review related work in Sect. 2. The detailed configuration of the proposed algorithm is described in Sect. 3. Section 4 illustrates experimental results on four large tracking benchmarks. Finally, conclusions are drawn in Sect. 5.

2   Related Work

2.1   One-Stage Visual Tracking

The one-stage trackers formulate visual tracking as a specific object searching problem, which directly calculates a response map through a regression model. One-stage trackers can be broadly classified into two categories: discriminative correlation filters (DCFs) based trackers [7,16,27] and deep regression networks (DRNs) based trackers [18,24]. Although the DCFs based trackers achieve top performance, the features and correlation filters are optimized independently. As opposed to DCFs based trackers, DRNs have the potential opportunity to take full advantage of end-to-end learning. CREST [24] fused the outputs of baseline and another two residuals to estimate the location of the target object. DSLT [18] proposed a shrinkage loss and an ensemble strategy to improve the performance of the DRNs based trackers. We note that current DRNs based trackers are fine-tuned online frequently to adapt the appearance variations of the target object, and this process is time-consuming to drag speed down. To solve this problem, we propose a unified architecture that requires only one forward pass from the conditioners to generate all parameters needed for the regression model to adapt to the specific object.

2.2   Conditional Instance Learning

Conditional instance learning is a recently introduced technique for normalizing the activities of neurons in deep neural networks for adaptive neural modeling in some vision and audio tasks. For image stylization, Golnaz Ghiasi *et al.* [11] present an artistic stylization network with the CBN to allow real-time stylization using any content/style image pair. For speech recognition, a Dynamic Layer Normalisation (DLN) [14] is proposed for adaptive neural acoustic modeling in speech recognition. And our approach is also motivated by conditional Instance learning, where the regression model is adaptive to the target by batch normalization parameters conditioned on a guidance input.

3   The Proposed Conditional Regression Tracking

In the proposed framework, we utilize two conditioners to instantly adapt the generic regression network to a specific object instance. The one conditioner is a visual context, and the other is a continuous trajectory. We employ two networks to inject these two cues to the main regression network, based on the search region and bounding box in the previous frame. Before introducing the designed architecture, we briefly review the Conditional Batch Normalization (CBN) which is widely used in conditional instance learning.

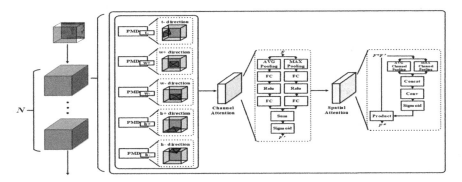

Fig. 2. The architecture of visual context network: Each Multi-Dimensional LSTM module contains 5 direction Multi-Dimensional LSTM units followed by context blending block. The blending block contains a channel attention and a spatial attention module. N corresponds to the number of Multi-Dimensional LSTM layers in the generic regression network.

3.1   Conditional Batch Normalization

Batch normalization (BN) is introduced by [12], which is efficient for training the feed-forward networks by normalizing feature statistics. Mathematically, given an input batch $x \in \mathbb{R}^{N \times C \times H \times W}$, the mean and standard deviation for each feature channel is normalized by BN:

$$y_c = \gamma_c \left(\frac{x_c - \mu(x_c)}{\sigma(x_c)} \right) + \beta_c \tag{1}$$

where γ_c and β_c are affine parameters learned from data; $\mu(x_c)$ and $\sigma(x_c)$ are the mean and standard deviation, which are computed across batch for each individual feature channel. BN uses mini-batch statistics during training and replace them with discrepancy during inference. Recently, alternative normalization schemes e.g. Conditional Batch Normalization (CBN) [9–11], which can alleviate domain shifts by recomputing statistics in the target domain. And γ_c and β_c could be replaced by other parameter generators. Inspired by this, we utilize two networks to produce the value of γ_c and β_c depending on the cues of visual context and trajectory.

3.2   Visual Context and Trajectory Formulating

In Eq. (1), the input x is normalized across the affine parameters γ_c and β_c in scale and shift domains. γ_c is a 1D vector for channel-wise weighting, and β_c is a 2D matrix for generating point-wise biases. So, we design the two networks such that the visual context produces channel-wise scale parameters to adjust the weights of different channels, while the trajectory generates element-wise bias parameters to inject spatial prior to the regression model.

3.3 Visual Context Network

In visual tracking, a context consists of a target object and its immediate surrounding background within a determined region. Most of the local contexts remain unchanged as changes between two consecutive frames can be reasonably assumed to be smooth. Therefore, there exists a strong spatial-temporal relationship between the local scenes containing the object in consecutive frames. Therefore, the visual context network provides the consecutive spatial-temporal information channel-wise weights, which guides the generic regression network to locate the object.

The architecture of the proposed visual context network is shown in Fig. 2. Each Multi-Dimensional LSTM units includes 5 direction Multi-Dimensional LSTM, e.g. time-direction, right-direction, left-direction, up-direction and down-direction. After each Multi-Dimensional LSTM unit, there is an attentive context blending unit that combines the output of Multi-Dimensional LSTM units and re-weights the outputs for selecting the important relevant context. And the visual context network is constructed by stacking N blocks. Mathematically, the process of each block is shown as follows:

$$
\begin{aligned}
\mathbf{i}_k &= \sigma(\mathbf{W}_i * \mathbf{x}_k + \mathbf{H}_i * \mathbf{s}_{k-1} + \mathbf{b}_i), \\
\mathbf{f}_k &= \sigma(\mathbf{W}_f * \mathbf{x}_k + \mathbf{H}_f * \mathbf{s}_{k-1} + \mathbf{b}_f), \\
\mathbf{o}_k &= \sigma(\mathbf{W}_o * \mathbf{x}_k + \mathbf{H}_o * \mathbf{s}_{k-1} + \mathbf{b}_o), \\
\tilde{\mathbf{c}}_k &= \tanh(\mathbf{W}_{\tilde{c}} * \mathbf{x}_k + \mathbf{H}_{\tilde{c}} * \mathbf{s}_{k-1} + \mathbf{b}_{\tilde{c}}), \\
\mathbf{c}_k &= \mathbf{f}_k \odot \mathbf{c}_{k-1} + \mathbf{i}_k \odot \tilde{\mathbf{c}}_k, \\
\mathbf{s}_k &= \mathbf{o}_k \odot \tanh(\mathbf{c}_k).
\end{aligned}
\tag{2}
$$

where $\mathbf{x}_k$ is the input. $\mathbf{i}_k$, $\mathbf{f}_k$ and $\mathbf{o}_k$ is the input, forget and output gate, respectively. The parallel LSTM unit computes the current and hidden state $\mathbf{c}_k$, $\mathbf{s}_k$. $\tilde{\mathbf{c}}_k$ is the input to the $\mathbf{c}_k$. $*$ is the convolution operation, and $\odot$ is the element-wise multiplication. The non-linear function σ and tanh are applied element-wise, where $\sigma(x) = \frac{1}{1+e^{-x}}$. W and H are the weights for the input-output state. More local context can be contained with lager kernel size in weight matrices.

To directly sum outputs of Multi-Dimensional LSTM units assumes that the information from each direction is equally important. We propose an attentive blending to learn the relative importance of each direction for precisely tracking. At first, the information from all directions are gathered in $\mathbf{S}$:

$$
\mathbf{S} = \begin{bmatrix} \mathbf{s}^{t-} & \mathbf{s}^{h-} & \mathbf{s}^{h+} & \mathbf{s}^{w-} & \mathbf{s}^{w+} \end{bmatrix}^T
\tag{3}
$$

the vector $\mathbf{S}$ is processed by an attention module:

$$
\begin{aligned}
\mathbf{S}' &= M_c(\mathbf{S}) \otimes \mathbf{S}, \\
\mathbf{S}'' &= M_s(\mathbf{S}') \otimes \mathbf{S}',
\end{aligned}
\tag{4}
$$

where $M_c \in \mathbb{R}^{C \times 1 \times 1}$ is a 1D channel attention map and $M_s \in \mathbb{R}^{1 \times H \times W}$ is a 2D spatial attention map. $\otimes$ is element-wise multiplication. This process guarantees the channel attention values broadcasted along the spatial dimension and vice versa. For channel attention, a channel attention map is utilized to exploit the inter-channel relationship of features. Two different spatial context descriptors $\mathbf{S}_{avg}^c$ and $\mathbf{S}_{max}^c$, which represent average-pooled features and max-pooled features respectively. The channel attention map $M_c \in \mathbb{R}^{C \times 1 \times 1}$ is generated by a shared FC layer with the above descriptors. After that, each descriptor is fed into the shared FC layer, the final representations are produced by fusing the output feature vectors with element-wise summation. The channel attention is expressed as follows:

$$M_c(\mathbf{S}) = \sigma(W_1(W_0(\mathbf{S}_{avg}^c)) + W_1(W_0(\mathbf{S}_{max}^c))) \tag{5}$$

where σ denotes the sigmoid function, W_0 and W_1 are the weights of FC layers.

After obtained the channel attention map, we generate a spatial attention map by utilizing the inter-spatial relationship of features. To compute the spatial attention, average-pooling and max-pooling operation along the channel axis and concatenate them to get efficient feature descriptors. Then, a convolution layer is applied to generate a spatial map $M_s(S) \in \mathbb{R}^{H \times W}$ which encodes where to emphasize. The process can be illustrated as:

$$M_s(\mathbf{S}) = \sigma(f([\mathbf{S}_{avg}^s; \mathbf{S}_{max}^s])) \tag{6}$$

where σ denotes the sigmoid function, and $[;]$, $f(\cdot)$ represents concatenation and convolution operation. $\mathbf{S}_{avg}^s \in \mathbb{R}^{1 \times H \times W}$ and $\mathbf{S}_{max}^s \in \mathbb{R}^{1 \times H \times W}$ denote average-pooled and max-pooled features across the channel. In the end, γ_c is obtained by performing 1×1 convolution on $\mathbf{S}''$.

3.4   Trajectory Network

The trajectory network plays a role of motion prior, which is depended on the fact that the object moves continuously between frames. Specifically, we set the prior to be the prediction location of the object in the previous frame. As the final regression map is a Gaussian-like map, we encode the trajectory in previous frames with a two-dimensional Gaussian distribution. To match the resolution of different feature maps in the regression network, we down-sample the trajectory map into different scales. and then different scaled trajectory map is used to generate the β_c of the corresponding condition layer:

$$\beta_c = \tilde{\gamma}_c M_t + \tilde{\beta}_c \tag{7}$$

where M_t is the down-sampled trajectory map, $\tilde{\gamma}_c$ and $\tilde{\beta}_c$ are the scale-and-shift parameters learned by a 1×1 convolution layer.

3.5   Implementation, Training and Inference

Network Structure
1. Generic Regression Network. Our generic regression network follows the one used in [18], which utilizes the VGG-16 [23] model as the backbone feature part. The output response of the *conv4_3* and *conv5_3* are fused via residual

connection to generate the final response by the regression part. So, We add CBN layer in-between the convolution layers (from $conv4\_1$ to $conv5\_3$) in VGG-16, which contains 6 CBN layers totally.

2. Context Prediction Network. The context prediction network consists of a stack of multi-Dimensional LSTM module with kernels of size 3×3. The number of layers is set as 3 to cover a larger context. The number of hidden units is 64. The states of the LSTMs are initialized to zero.

3. Trajectory Network. The baseline consists 3 convolutional layers with kernels 3×3. Each layers have 128, 128 and 64 units, respectively. We add ReLU layer in-between the convolution layers. Then, 6 convolution layers with kernels 1×1 are designed to down-sample the β_c for matching different scaled feature maps in the generic regression network. The parameters are initialized randomly.

Network Training. We train the whole network through two stages. In the first stage, we train the generic regression network with a multi-domain strategy [20]. In each iteration, the network is updated based on a mini-batch that consists of the training samples from one sequence, where a single branch (i.e. the regression part) is enabled. Sequently, the generic regression network is optimized with ILSVRC 2015 [22] video object detection dataset by the shrinkage loss $\mathcal{L}_{shrinkage}$ [18]. At training time, a search area is sampled from a random frame, which is 5 times the maximum value of object width and height. Then, soft labels are generated by a two-dimensional Gaussian function with a kernel width proportional (0.1) to the target size. The loss function is optimized with Adam optimizer with the learning rate 10^{-5} using mini-batch of 8, and run for 80000 iterations. Then, the generic regression network is trained with two conditioners (i.e. visual context network and trajectory network). Different from the random strategy in the first stage, we train the whole network sequentially. Given a pair of frames (x_i, x_j), we generate two search areas (sa_i, sa_j) and a soft labels sl_j. We feed the search area sa_i into the visual context network for generating the γ_c and the predicted search area $\hat{sa}_i$ (by adding a sigmoid layer at the end of the visual context network). Here, we minimize the loss between the predicted search area $\hat{sa}_i$ and search area sa_j by:

$$\begin{aligned} \mathcal{L}(sa_j, \hat{sa}_i) &= \mathcal{L}_p(sa_j, \hat{sa}_i) + \mathcal{L}_{gdl}(sa_j, \hat{sa}_i), \\ \mathcal{L}_p(sa_j, \hat{sa}_i) &= \|sa_j - \hat{sa}_i\|_p, \end{aligned} \tag{8}$$

where $\mathcal{L}_{gdl}$ the Image Gradient Difference Loss (GDL). $p = 1$ in all experiments. Sequently, the soft label sl_i is fed into the trajectory network to generate β_c. Then, the generic regression network with the generated parameters γ_c, β_c is optimized by:

$$\mathcal{L} = \mathcal{L}_{shrinkage}(sl_j, \hat{sl}_j) + \alpha \mathcal{L}(sa_j, \hat{sa}_i) \tag{9}$$

where $\hat{sl}_j$ is the predicted output. In our experiments, α is set to 1. ILSVRC video dataset are also used and Adam optimizer with the learning rate of 10^{-6} with batches of 2 are used. Training is performed for 11000 iterations.

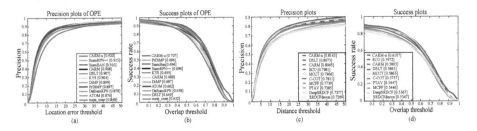

Fig. 3. (a) and (b) are the precision and success plots on OTB2015, respectively. (c) and (d) are the precision and success plots on TC128, respectively.

Table 1. Quantitative comparisons of DRNs based trackers on OTB2015.

Trackers	CREST	meta_crest	DSLT	CARM-u	CARM
Succ (%)	62.3	63.2	65.8	70.7	68.8
Prec (%)	83.8	84.8	90.7	92.0	90.8
FPS	2.4	32	5.7	8	33

Inference. Given a new frame, a search patch is cropped by the estimated position in the last frame. The whole network takes this search patch, context area and response map of the last frame as inputs. Then, it outputs a response map, where the location of the maximum value indicates the position of target objects. Once obtaining the estimated position, we carry out scale estimation using the scale pyramid strategy as in [8]. We set the ratio of scale changes to 1.03 and the levels of scale pyramid to 3.

4   Experiments

We also propose another version of CARM, named CARM-u. Especially, in CARM-u, we update the model every 10 frames. Then, we present the comparisons with other state-of-the-art trackers on four benchmark datasets including OTB2015 [26], TC128 [17], UAV123 [19] and VOT2018 [1]. For the OTB2015 dataset, the tracking performance is measured by conducting a one-pass evaluation (OPE) based on two metrics: center location error and overlap ratio. These two evaluation metrics are also utilized in TC128 [17] and UAV123 [19] datasets. In the VOT2018 dataset [1], each tracker is measured by the metrics of Accuracy Ranks (AR), Robustness Ranks (RR) and Expected Average Overlap (EAO). Finally, we verify the effectiveness of components in CARM by ablation studies.

OTB2015 Dataset [26]. We first analyze our method on OTB2015 [26]. We evaluate our CARM-u and CARM with several state-of-the-art trackers including: SiamRPN++ [15], SiamBAN [4], DSLT [18], KYS [2], DiMP [3], PrDiMP [5], DaSiamRPN [28], ATOM [6] and meta_crest [21]. As shown in Fig. 3 and

Table 2. Quantitative comparisons on UAV123. (The **first** and *second* best results are shown in bold and italic.)

	CARM	CARM-u	SiamRPN++	DSLT	DSST	ECO	SRDCF
Succ (%)	53.2	**62.5**	*61.3*	53.0	35.6	52.5	46.4
Prec (%)	*77.1*	**82.2**	-	74.6	58.6	74.1	67.6

Table 3. Quantitative comparisons on VOT2018. (The **first** and *second* best results are shown in bold and italic.)

	CARM	CARM-u	SiamRPN++	DSLT	CREST	SiamFC	SRDCF
EAO	0.374	**0.417**	*0.414*	0.341	0.326	0.235	0.308
AR	0.55	**0.60**	*0.60*	0.58	0.56	0.53	0.42
RR	*0.23*	**0.15**	0.23	0.31	0.33	0.24	0.74

Table 1, our CARM achieves competitive results among all the compared trackers. Surprisingly, the updated version (CARM-u) achieves the best precision and success score (92.0% and 70.7%) compared to the state-of-the-art regression trackers. We attribute the favorable performance by two reasons. Firstly, the proposed visual context network captures both the variations of the target and its surrounding background. Secondly, the trajectory network provides the motion cue for locating the target precisely.

TC128 Dataset [17]. TC128 dataset [17] consists of 128 colorful video sequences. We fairly compare our CARM and CARM-u with several state-of-the-art trackers. Figure 3 shows that the proposed method achieves the best success score by a large margin compared to the aforementioned trackers. Since TC128 contains a large number of small target objects, which demonstrates that our model performs well for tracking small-scaled objects.

UAV123 Dataset [19]. UAV123 dataset [19] consists of more than 110K frames captured from a UAV. Results in terms of precision and success plots are shown in Table 2. Among previous methods, CARM-u significantly outperforms DSLT, achieving precision and success scores of 82.2 and 62.5, respectively. In particular, our CARM beats DSLT, on the precision and success scores with an approximately 6 times speed-up. It indicates that our CARM is more robust with better generalization capability.

VOT2018 Dataset [1]. VOT2018 [1] is the recently released challenging benchmark. We compare our tracker with the representative approaches in Table 3. According to Table 3, compared with the DSLT [18] and CREST [24] from the regression category, our tracker has the highest EAO and ranks first in terms of AR and RR. This exhibits that meaningful features extracted from both conditioners can not only accurately locate the target but also show strong robustness.

Table 4. Ablation study of our method on OTB2015.

	CARM	w/o visual context	w/o trajectory
Succ (%)	68.8	63.7	64.8
Prec (%)	90.8	83.8	84.7

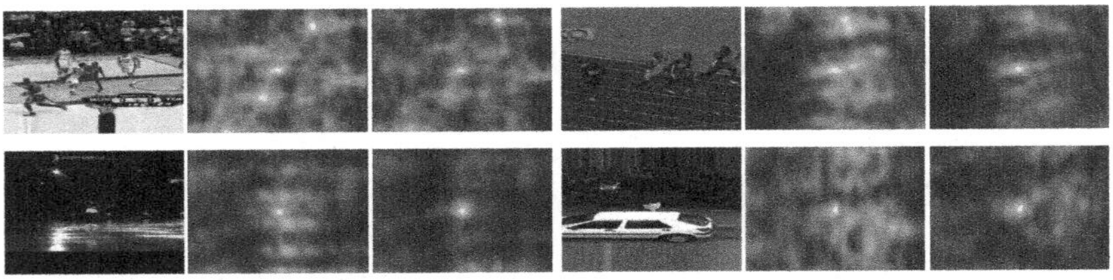

Fig. 4. Visualization for the effect of the conditioners. Some search area (1st and 4th columns) with the changes in response maps before (2nd and 5th columns) and after (3rd and 6th columns) applying the conditioned parameters.

4.1 Ablation Study

To investigate the effectiveness of individual components in our method, we perform several ablation studies on OTB2015. We first perform the impact of the two conditioners on the quality of the generic regression network. Table 4 summarizes the results of this comparison. According to this Table 4, all results support that the two conditioners make a meaningful contribution to locating the target precisely. Besides, Fig. 4 shows some examples of how the conditioners modify the response maps, thus verifying the proposed conditioners can be beneficial in the task of visual tracking. Figure 4 shows that the conditioners help the tracker to adapt to various target appearance changes, and are also effective suppressing distractors in the background.

5 Conclusions

We propose a novel framework to process regression-based tracker efficiently. To alleviate the slow speed of fine-tuning developed by previous regression-based methods, we proposed to use two conditioners to modulate the parameters in the regression network with one forward pass. We achieved outstanding performance in four large public tracking benchmarks with real-time speed 33 FPS.

References

1. Kristan, M., et al.: The sixth visual object tracking VOT2018 challenge results. In: Leal-Taixé, L., Roth, S. (eds.) ECCV 2018. LNCS, vol. 11129, pp. 3–53. Springer, Cham (2019). https://doi.org/10.1007/978-3-030-11009-3_1

2. Bhat, G., Danelljan, M., Van Gool, L., Timofte, R.: Know your surroundings: exploiting scene information for object tracking. In: Vedaldi, A., Bischof, H., Brox, T., Frahm, J.-M. (eds.) ECCV 2020. LNCS, vol. 12368, pp. 205–221. Springer, Cham (2020). https://doi.org/10.1007/978-3-030-58592-1_13
3. Bhat, G., Danelljan, M., Van Gool, L., Timofte, R.: Learning discriminative model prediction for tracking. arXiv (2019)
4. Chen, Z., Zhong, B., Li, G., Zhang, S., Ji, R.: Siamese box adaptive network for visual tracking (2020)
5. Danelljan, M., Gool, L.V., Timofte, R.: Probabilistic regression for visual tracking (2020)
6. Danelljan, M., Bhat, G., Khan, F.S., Felsberg, M.: Atom: accurate tracking by overlap maximization. In: CVPR (2019)
7. Danelljan, M., Bhat, G., Shahbaz Khan, F., Felsberg, M.: ECO: efficient convolution operators for tracking. In: CVPR (2017)
8. Danelljan, M., Häger, G., Khan, F., Felsberg, M.: Accurate scale estimation for robust visual tracking (2015)
9. De Vries, H., Strub, F., Mary, J., Larochelle, H., Pietquin, O., Courville, A.C.: Modulating early visual processing by language. In: NIPS (2017)
10. Dumoulin, V., Shlens, J., Kudlur, M.: A learned representation for artistic style. In: ICLR (2017)
11. Ghiasi, G., Lee, H., Kudlur, M., Dumoulin, V., Shlens, J.: Exploring the structure of a real-time, arbitrary neural artistic stylization network. arXiv (2017)
12. Ioffe, S., Szegedy, C.: Batch normalization: Accelerating deep network training by reducing internal covariate shift. arXiv (2015)
13. Jung, I., Son, J., Baek, M., Han, B.: Real-time MDNet. In: Ferrari, V., Hebert, M., Sminchisescu, C., Weiss, Y. (eds.) ECCV 2018. LNCS, vol. 11208, pp. 89–104. Springer, Cham (2018). https://doi.org/10.1007/978-3-030-01225-0_6
14. Kim, T., Song, I., Bengio, Y.: Dynamic layer normalization for adaptive neural acoustic modeling in speech recognition. arXiv (2017)
15. Li, B., Wu, W., Wang, Q., Zhang, F., Xing, J., Yan, J.: SiamRPN++: evolution of Siamese visual tracking with very deep networks. In: CVPR (2019)
16. Li, P., Chen, B., Ouyang, W., Wang, D., Yang, X., Lu, H.: GradNet: gradient-guided network for visual object tracking. In: ICCV (2019)
17. Liang, P., Blasch, E., Ling, H.: Encoding color information for visual tracking: algorithms and benchmark. TIP **24**, 5630–5644 (2015)
18. Lu, X., Ma, C., Ni, B., Yang, X., Reid, I., Yang, M.-H.: Deep regression tracking with shrinkage loss. In: Ferrari, V., Hebert, M., Sminchisescu, C., Weiss, Y. (eds.) Computer Vision – ECCV 2018. LNCS, vol. 11218, pp. 369–386. Springer, Cham (2018). https://doi.org/10.1007/978-3-030-01264-9_22
19. Mueller, M., Smith, N., Ghanem, B.: A benchmark and simulator for UAV tracking. In: Leibe, B., Matas, J., Sebe, N., Welling, M. (eds.) ECCV 2016. LNCS, vol. 9905, pp. 445–461. Springer, Cham (2016). https://doi.org/10.1007/978-3-319-46448-0_27
20. Nam, H., Han, B.: Learning multi-domain convolutional neural networks for visual tracking. In: CVPR (2016)
21. Park, E., Berg, A.C.: Meta-tracker: fast and robust online adaptation for visual object trackers. In: Ferrari, V., Hebert, M., Sminchisescu, C., Weiss, Y. (eds.) ECCV 2018. LNCS, vol. 11207, pp. 587–604. Springer, Cham (2018). https://doi.org/10.1007/978-3-030-01219-9_35
22. Russakovsky, O., et al.: ImageNet large scale visual recognition challenge. Int. J. Comput. Vis. **115**(3), 211–252 (2015). https://doi.org/10.1007/s11263-015-0816-y

23. Simonyan, K., Zisserman, A.: Very deep convolutional networks for large-scale image recognition. Computer Science (2014)
24. Song, Y., Ma, C., Gong, L., Zhang, J., Lau, R.W.H., Yang, M.H.: CREST: convolutional residual learning for visual tracking. In: ICCV (2017)
25. Song, Y., et al.: Vital: Visual tracking via adversarial learning. In: CVPR (2018)
26. Wu, Y., Lim, J., Yang, M.H.: Object tracking benchmark. TPAMI **37**, 1834–1848 (2015)
27. Xu, T., Feng, Z.H., Wu, X.J., Kittler, J.: Joint group feature selection and discriminative filter learning for robust visual object tracking. In: ICCV (2019)
28. Zhu, Z., Wang, Q., Li, B., Wu, W., Yan, J., Hu, W.: Distractor-aware Siamese networks for visual object tracking. In: Ferrari, V., Hebert, M., Sminchisescu, C., Weiss, Y. (eds.) ECCV 2018. LNCS, vol. 11213, pp. 103–119. Springer, Cham (2018). https://doi.org/10.1007/978-3-030-01240-3_7

Attention Template Update Model for Siamese Tracker

Fengshou Jia, Zhao Tang, and Yun Gao$^{(\boxtimes)}$

School of Information Science and Engineering, Yunnan University, Kunming, China

Abstract. Visual tracking is defined as a template-matching task in current Siamese approaches. The tracker needs to locate the target by matching the template with the search area in each frame. Most current Siamese methods either do not use an update strategy or use a linear update method with a fixed learning rate. Neither of the above two strategies allows the target template to dynamically adapt to frequent and dramatic changes in appearance. To solve this problem, we propose a template update model based on the attention mechanism. Our model updates the template in a nonlinear manner. It can fully explore the weight relationship of various features in the template, so that the template can pay more attention to features that are more beneficial to determine the target in different situations. In addition, by adding an adjustment block, the error and invalid information in the old template can be removed before updating. Extensive experiments on several datasets demonstrated the effectiveness of our update model. We used SiamFC++ as our basic tracker and achieved state-of-the-art performance by adding our model. Moreover, our model has lightweight structure, and thus it can be easily applied to most Siamese trackers with minimal computational cost.

Keywords: Object tracking · Siamese networks · Template update

1 Introduction

Visual tracking is an essential task in computer vision. In this task, a given initial frame is used as a template to predict the target location of all subsequent video frames and mark the target by a rectangular box. Object tracking remains challenging because of frequent interference factors, such as occlusion and changes in shape, illumination, or scale. Current mainstream object tracking algorithms can be roughly divided into two categories: correlation filtering methods [1–6] and deep Siamese network methods [7–12]. The former uses a Gaussian kernel to perform linear binary classification in a high-dimensional space to distinguish the background and target. It achieves a high-speed by transferring the calculation

---

F. Jia—Graduate student.

This work is supported by National Natural Science Foundation of China (No. 61802337).

H. Ma et al. (Eds.): PRCV 2021, LNCS 13019, pp. 229–241, 2021.
https://doi.org/10.1007/978-3-030-88004-0_19

process to the Fourier domain. The latter uses convolutional neural networks (CNNs) as backbone networks to extract the features of the template and the search area, and obtains the location response map using cross-correlation operations. The Siamese method is more advantageous in terms of accuracy because it uses more advanced learnable deep CNN features, which have strong feature representation capabilities.

The correlation filtering method [2,3] generally adopts a linear method to update the template. The predicted target information is added to the accumulated template with a fixed learning rate as the matching template for the next frame. The advantage of this method is that it can record the overall information of the target in the past. However, once there is a tracking error, incorrect feature information is added, resulting in target drift, and it is difficult to recover from this error. The deep Siamese method usually has no update strategy; it always uses the ground truth of the first frame as a template. There are three main reasons for using this approach. First, adding the accumulated template mechanism would force the training to be recurrent, making the procedure cumbersome and inefficient. Second, template updates affect the real-time performance of the algorithm, and it goes against the high-speed requirement of online tracking. Third, the initial template contains the most important and accurate feature information of the target. In contrast, the template obtained by the linear update method cannot be more reliable than the initial template because of the inevitable introduction of noise.

With the strategy of not updating the template [7,8], the tracker ignores the changes in the appearance of the target; thus, it cannot refer to the target's latest appearance information. This may cause tracking errors when the target undergoes drastic changes in appearance. The linear update method updates the target appearance information at a fixed rate, which makes it difficult to adapt to frequent interference. Therefore, a more intelligent and nonlinear update strategy is necessary. Recently, some scholars proposed a new template update model called UpdateNet [13]. The model uses the CNN for offline learning, allowing the model to automatically learn a nonlinear method to update the template. However, this method has the following problem: each feature extracted from the backbone network is regarded as equally important; thus the model cannot dynamically adapt to frequent and dramatic changes in appearance during tracking process. We believe that the importance of each feature should change with the appearance of the target.

The attention model (AM) was originally used in machine translation [14] and has become a significant concept in the field of neural networks. In computer vision, some parts of the input may be more helpful for decision-making than others. Hu et al. proposed a channel attention module SE [15], which effectively mines the relationship between various features. We believe that it is necessary to assign different weights to template features by using an attention mechanism.

In this study, we propose an effective attention update model. The model can fully mine the weight relationship between the various features of the template. Inspired by the gating mechanism of the LSTM network [16], before each template update, an adjustment module is used to selectively remove part of the invalid information in the previous accumulated template, which reduces the influence of error information on the template. The effectiveness of our model was evaluated on several benchmarks. Our model achieved state-of-the-art performance for several datasets and significantly outperformed the traditional template update method. In addition, the model can be easily applied to most Siamese trackers owing to its simple structure, and low computational cost.

2   Related Work

Tracking Frameworks. At present, object tracking algorithms are mainly divided into two categories:detection-based correlation filtering methods [1–3] and Siamese methods [7,8,10] based on correlation matching [1-6]. MOSSE [1] first introduced the concept of cross-correlation in object tracking and used the least squares method to find the optimal solution. KCF [2] uses kernelized ridge regression as a classifier to perform linear binary classification in a high-dimensional space to distinguish the foreground and background. KCF uses the nature of the circulant matrix to transfer the calculation process to the Fourier domain and obtains extremely high computational efficiency.Methods such as SRDCF [6], C-COT [4], ECO [3], CSR-DCF [5], etc. reduce the response of non-target areas and effectively suppress background interference by spatial regularization and adding a mask matrix to the loss function.

The other branch is based on the Siamese structure. SiamFC [7] first used the Siamese architecture for template matching for object tracking. Inspired by Faster-RCNN [17], follow-up models, including CFNet [9], RASNet [12], SiamRPN [8], DasiamRPN [10], and SiamRPN++ [11], introduced the regional proposal network (RPN) into the field of object tracking, which significantly improved the accuracy but has a high computational cost owing to a large number of anchors. The score between the anchor and the target has a large ambiguity, and thus, the robustness is affected. This structure is no longer mainstream. FCOS [18] first proposed the concept of anchor-free structures in object detection, advocating the abandonment anchors. Methods such as SiamFC++ [19], SiamCAR [20], SiamBan [21], and FCAF [22] used the idea of FCOS to perform regression prediction on the corresponding points between the response graph and the original graph instead of a series of predesigned anchor points.

In addition, some new object tracking algorithms [23–26] based on the transformer [27] have reached the state-of-the-art. The transformer is a promising structure; for example, the best algorithm in natural language processing, Bert [28] is based on transformers. We expect breakthrough achievements of transformers in the field of object tracking in the near future.

Methods of Template Updates. Most trackers either use a simple linear strategy to update the template or avoid updating the initial template. The accumulated error of the online template update may cause template drift, which eventually leads to tracking failure. If the method uses no template updates, it cannot track the latest appearance information of the template. To further improve the robustness of the tracking algorithm, the problem of template updating has gradually attracted attention. The MOSSE method [1] is based on the peak-to-sidelobe ratio(PSR) for confidence judgment to determine whether the template should be updated. Memtrack[29] used a dynamic memory network, the gated residual template, and the initial template to generate the final matching template. MemTrack can effectively use the first few frames to learn target appearance variation and background suppression online. Choiet et al. [30] proposed a template update method based on reinforcement learning; however, it did not consider accumulated information.

3   Proposed Method

In this section, we introduce the proposed template update model. We apply the attention and gating mechanism so that the model can fully explore the relationship between the template features. The model can better adapt to the appearance changes of the target owing to nonlinear template updates. Moreover, our model has a simple structure, thus it can be easily applied to most Siamese trackers with low computational complexity.

3.1   Traditional Update

Several traditional trackers use a simple average strategy to update the template:

$$TA_i = (1 - \gamma)TA_{i-1} + \gamma T_i, \tag{1}$$

where, TA_i is the current template, TA_{i-1} is the accumulated template obtained in the last frame, and T_i is the template of the predicted target in the current frame. The learning rate γ is set to a fixed small value (e.g. $\gamma = 0.0102$) following the assumption that the appearance of the object changes smoothly and consistently in consecutive frames. However, such assumptions are not realistic. Severe scale changes, deformations, and occlusions cause drastic changes in appearance. Another disadvantage is that it lacks a deletion mechanism because it keeps adding new template information. We believe that a module with a removal mechanism is necessary for object tracking, because it can mitigate the influence of erroneous features by filtering the old template before updating it.

3.2   Network Architecture

The framework of the proposed template update model is presented in Fig. 1. We first used the accumulated template TA_{i-1} to predict the target location and extract features T_i from this region. The accumulated template TA_{i-1} is then processed through an adjustment block to eliminate invalid information. We concatenate the initial features T_0, accumulated features TA_{i-1}, and current template T_i to form the input of the update block TC. A channel attention block is necessary before this step because the three templates above have unequal importance. The update block references the weighted concatenated features to predict the accumulated template TA_i. Finally, the new template output from the update block is processed through the last attention block to redistribute the channel weight. Note that the above-mentioned initial template T_0 and current frame template T_i are features extracted by the backbone in the basic tracker. Considering the case of template update error, we set up a skip connection for T_0 so that the updater can access the initial template T_0 when the accumulated template TA_{i-1} is unreliable. Similarly, we also set a skip connection for the accumulated template TA_{i-1}, so that this stage can be skipped when there is no need to adjust the template. There is also a skip connection in TC for the case of incorrectness of weight distribution. The entire template update process can be defined as follows:

$$TA_i = \psi(T_0, TA_{i-1}, T_i), \tag{2}$$

where, ψ donates our entire update model; it receives the following input: the initial template T_0, accumulated template TA_{i-1}, and template T_i extracted from the predicted target location in the current frame.

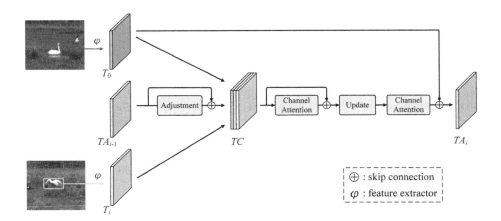

Fig. 1. Pipeline of template update model.

3.3 Adjustment and Update Blocks

Both adjustment and update blocks are composed of two simple convolutional layers. For the accumulated template TA_i, it is inevitable to add incorrect template information during the update process. In addition, the outdated target appearance information becomes invalid. To reduce the influence of invalid and incorrect information on the template update process, the adjustment block performs a nonlinear transformation on the old template TA_{i-1} before the template update, extracts useful information and removes useless information. The update block receives the weighted concatenated template TC as input and predicts the accumulated template TA_i of the next frame. However, the importance of various features in the newly accumulated template TA_i cannot be adaptively changed with changes in the target appearance. Therefore, the channel weight of the new template must be redistributed.

3.4   Channel Attention Block

Certain features are helpful to determine the target appearance in different situations. We use a channel attention module to dynamically assign different weights to features so that the template can adapt to changes in the appearance of the target. We used this block twice, before and after the template update. We believe that assigning channel weights before the template update can help the subsequent update modules to extract useful feature information and predict more accurate new templates. In addition, assigning the channel weights again after the template is updated helps the tracker to perform subsequent template matching. The architecture of the channel attention is shown in Fig. 2. To integrate the information of each channel, the average value z_c of c-th channel is calculated as follows:

$$z_c = \frac{1}{H \times W} \sum_{i=1}^{H} \sum_{j=1}^{W} x_c(i,j), \tag{3}$$

where x_c is the feature map of c-th channel of x that obtained from the CNN layer, with a size of $H \times W$. c is a channel index in subject to $1 \leq c \leq C$, and (i,j) represent the index of pixel values in x_c. The feature map x is compressed into a feature vector z with a size $1 \times 1 \times C$, by operating the global average pooling in each channel. The first fully connected layer is used to reduce the dimensionality. Next the second fully connected layer is used to perform nonlinear transformation and restore the number of channels. Subsequently, the vector is mapped to interval [0,1] by the sigmoid function, and the weight coefficients of each channel are obtained as follows:

$$s = \sigma(W_2 Relu(W_1 z)). \tag{4}$$

Finally, the weight coefficients are correspondingly multiplied to each channel:

$$\tilde{x}_c = s_c \times x_c. \tag{5}$$

After the above operations, the weighted features $\tilde{x}$ are obtained. Features with weights can better adapt to changes in the appearance of the target.

3.5   Training Model

We train our update model to output the same features as the ground truth. Therefore, our loss function is defined as the L2 loss of TA_i and TG_i:

$$Loss = ||\psi(T_0, TA_{i-1}, T_i) - TG_i||^2, \tag{6}$$

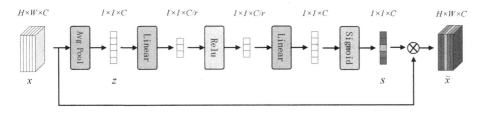

Fig. 2. Structure of the channel attention block, where $\otimes$ denotes channel-wise multiplication, r is reduction ratio.

where ψ represents the output of the entire updated model, TG_i is the feature extracted from the location of the ground truth, T_0 is the feature that originates from the initial frame, and TA_{i-1} is the output obtained by updating the model in the last frame. The template features T_0 from the initial frame and the target features TG_i can be easily obtained by extracting features from the ground-truth locations in the corresponding frames. The predicted target Ti of each frame can be obtained by running the basic tracker on the training set. We cannot obtain the accumulated template TA_{i-1} directly from the training set because we do not know what to expect. One of the methods is to randomly initialize the parameters of the updated model, and run the basic tracker with the updated model on the training set to obtain the accumulated template in the corresponding frames. However, this method suffers from serious problems. First, the accumulated template obtained by the randomly initialized updater is of low quality, which would considerably impair the performance of the original tracker. In addition, this would force the training to be recurrent, making the procedure cumbersome and inefficient.

Two-Stage Training Method. Inspired by the training method in UpdateNet, we propose a simpler and more effective training method. In the first stage, the tracker using the linear update method is run on the training set to generate an accumulated template for each frame. Although the quality of this linear accumulated template is low, it is better than the template obtained by randomly initializing the updater. In this manner, we can obtain the three training sample pairs (T_0, TA_{i-1}, T_i) and the training label TG_i simultaneously. The training samples obtained in the first stage are used to train the updating model, and the model with the best performance is selected to continue the second stage of training. After repeated training, the performance of the tracker in the original video was greatly improved. If the original video set is used again, many positive samples are obtained because of the high success rate but negative samples are insufficient. Therefore, we cannot use the previous video set again in the second stage; otherwise, it causes a serious overfitting problem. To avoid this situation, we use the updated model obtained in the previous stage of training and run it on a new batch of videos to obtain a new batch of training data.

4 Experiment

4.1 Implementation Details

In this study, we used a PC with an Ubuntu 16.04 operating system, with an I7-9800X CPU (64 GB RAM) and two NVIDIA RTX 5000 GPUs (16 GB memory) for training and testing. We used PyTorch 1.3.1 as our neural network framework. The optimizer is Adam, and the learning rate is set to $1e-6$ in the first stage and $1e-7$ or $1e-8$ in the second stage. We selected one video from each of the first 20 categories of Lasot and one video from each of the last 20 categories to create training sets for the first and second stages, respectively. In our work, we chose SiamFC++ of AlexNet version as our basic tracker, whose backbone output size $H \times W \times C$ is $4 \times 4 \times 256$. The first convolutional layer parameters of the adjustment block in our updated model are (256, 96, 1), which represent the input channel, output channel, and convolution kernel size, respectively. The second convolutional layer parameter are (96, 256, 1), and there is a ReLU activation function in between. The update block also contains two convolutional layers, whose parameters are (768, 96, 1) and (96, 256, 1). The reduction ratio r in channel attention block is 16.

4.2 Checkpoint Selection

The model parameters need to be saved for each training epoch with different network structures and hyperparameters. It takes a significant amount of time to perform a benchmark test with a checkpoint. Therefore, we propose a simple index to improve the efficiency of screening of best parameters. For example, in OTB-100, we first use the basic tracker without any update strategy to select the 10 worst-performing videos and 10 best-performing videos, and we define their average success rates as $\tilde{R}_{worst}$ and $\tilde{R}_{best}$. We calculate the corresponding R_{worst} and R_{best} values for each epoch. We believe that if the algorithm gains overall performance improvement, it must have as much improvement as possible in the worst video, and simultaneously, there must be as little reduction as possible in the best video. We can estimate the final overall performance by observing the values of R_{worst} and R_{best}. We only selected the checkpoints with higher scores in the OTB test and discarded the checkpoints with lower scores. Meanwhile, the training process can be terminated when the performance of this index is poor. In this manner, our entire working time is reduced to less than one-fifth of the original. In this study, the values of $\tilde{R}_{worst}$ and $\tilde{R}_{best}$ were 0.25 and 0.84, respectively. We only tested the checkpoint with $R_{worst} > 0.33$ and $R_{best} > 0.83$.

4.3 Performance in Several Benchmarks

We selected SiamFC++ (AlexNet) [19] as our basic tracker, used the pretrained model provided by the author, and added the trained update model to form our tracker. We evaluated the proposed model using OTB-100 [31] and UAV123 [32], and confirmed that it achieved excellent performance, especially when compared with the traditional linear update model.

OTB-100. is the most commonly used benchmark in object tracking, and contains 100 video sequences. The test set is challenging because of frequent interferences such as scale transformation, lighting changes, occlusion, and motion blur. In Table 1, our tracker is compared with state-of-the-art trackers. Compared with the basic tracker SiamFC++ with no update strategy, our model achieved an improvement of approximately 1.5%. Additional comparison results are presented in Fig. 3.

UAV123. contains 123 videos taken by drones. Owing to long video sequences, it is necessary to maintain a long-term template update, which is a significant challenge to the stability of the updater. Results in Fig. 4 demonstrates that our model achieved a competitive performance.

Table 1. Comparison with several state-of-the-art trackers on OTB-100. Bold indicates the basic tracker, and italic indicates the method after adding our update model.

Tracker	ECO	SiamRPN++	PrDiMP	SiamFC++	TransT	Ours
Source	CVPR17	CVPR19	CVPR20	AAAI20	CVPR21	
AUC	0.691	0.696	0.684	**0.685**	0.694	*0.698*

4.4 Ablation Studies

In this section, we conduct ablation experiments to verify the necessity of each part of the network. We also compare different template update strategies.

Update Strategy. As shown in Table 2, we evaluated the impact of different update strategies on the final results for OTB-100 and UAV123. (1) The first line is our basic tracker, SiamFC++, which does not use any update strategy. (2) The second line is the traditional linear update method added to the basic tracker. (3) The third line is the basic tracker with our proposed update model.

According to the results, the traditional linear update model cannot cope with the frequent appearance changes of the target owing to the limitation of its fixed learning rate. The results in the Table 2. show that the linear update strategy generates a negative effect, and our new updated model significantly outperforms the traditional linear model.

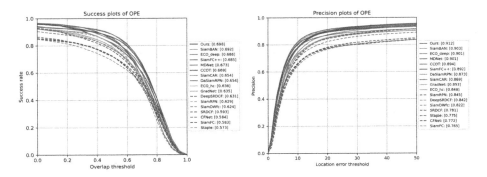

Fig. 3. Comparison with more trackers on OTB-100 in terms of success and precision plots of one-pass evaluation (OPE).

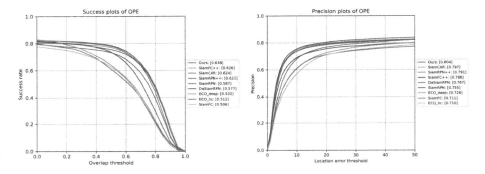

Fig. 4. Comparison with state-of-the-art trackers on UAV123 in terms of success and precision plots of OPE.

Table 2. Comparison of our update method with the traditional linear update.

Update strategy	OTB100(AUC)	UAV123(AUC)
Basic	0.685	0.626
Basic + Linear	0.675	0.615
Basic + Ours	0.698	0.638

Necessity and Sequence of Each Components. In this section, we conduct ablation experiments to explore the influence of different components in the update model and the order of components. Table 3 compares the results of our experiment. As shown in Fig. 1, our entire update model has three main types of components: (1) "Up" is the update block, (2) "CA" is the channel attention block, and (3) "AD" is the adjustment block. "None" means that the template update strategy is not used. The stage is the number of stages for which our model is trained.

Table 3. Comparison results of different model structures on OTB-100.

Stage	Model architecture	AUC
1	None	0.685
1	Up	0.640
1	CA + Up	0.634
1	Up + CA	0.688
1	CA + Up + CA	0.691
1	AD + CA + Up + CA	0.693
2	AD + CA + Up + CA	0.698

According to Table 3, a single update module does not produce a good update effect. As in our previous analysis, we speculate that the features of different channels are of varying importance to the tracking task, and the new template must be given a new weight. However, although a simple weight redistribution can improve the ability to locate the target, it does not achieve a significant improvement compared to the original algorithm with no update strategy. After adding the adjustment block, the old template is filtered in each frame. Thus the most useful information in the accumulated template is extracted, and invalid or erroneous information is eliminated. The performance was significantly improved compared with the original algorithm when two channel attention blocks were added, and the success rate exceeded 0.69 for the first time. Further improvements can be achieved based on our proposed two-stage training method. The accumulated template used in the first training set is obtained by linear updating, and training with such a training set may lead to model learning through linear updating. If the original video set is used again in the second stage, it will cause serious overfitting problems. It can be proven that our newly proposed training method is simple but very effective.

5   Conclusion

In this paper, we proposed a new template update model for object tracking. This model can remove invalid or erroneous information in the old template as much as possible before updating. After the update is completed, the channel attention block is used to assign different weights to each feature of the template, adapting it to different appearance changes of the target. We also simplify and improve the previous training methods that have the problem of overfitting, and obtain better performance. Moreover, our model can be easily applied to most Siamese trackers because of its high portability. Extensive evaluation proves that our update model has improved the performance of the basic tracker on multiple benchmarks, and its performance far exceeds that of the traditional linear update model.

References

1. Bolme, D.S., Beveridge, J.R., Draper, B.A., Lui, Y.M.: Visual object tracking using adaptive correlation filters. In: 2010 IEEE Computer Society Conference on Computer Vision and Pattern Recognition, pp. 2544–2550. IEEE (2010)
2. Henriques, J.F., Caseiro, R., Martins, P., Batista, J.: High-speed tracking with kernelized correlation filters. IEEE Trans. Pattern Anal. Mach. Intell. **37**(3), 583–596 (2014)
3. Danelljan, M., Bhat, G., Shahbaz Khan, F., Felsberg, M.: ECO: efficient convolution operators for tracking. In: Proceedings of the IEEE Conference on Computer Vision and Pattern Recognition, pp. 6638–6646 (2017)
4. Danelljan, M., Robinson, A., Shahbaz Khan, F., Felsberg, M.: Beyond correlation filters: learning continuous convolution operators for visual tracking. In: Leibe, B., Matas, J., Sebe, N., Welling, M. (eds.) ECCV 2016. LNCS, vol. 9909, pp. 472–488. Springer, Cham (2016). https://doi.org/10.1007/978-3-319-46454-1_29
5. Lukezic, A., Vojir, T., Cehovin Zajc, L., Matas, J., Kristan, M.: Discriminative correlation filter with channel and spatial reliability. In: Proceedings of the IEEE Conference on Computer Vision and Pattern Recognition, pp. 6309–6318 (2017)
6. Danelljan, M., Hager, G., Shahbaz Khan, F., Felsberg, M.: Learning spatially regularized correlation filters for visual tracking. In: Proceedings of the IEEE International Conference on Computer Vision, pp. 4310–4318 (2015)
7. Bertinetto, L., Valmadre, J., Henriques, J.F., Vedaldi, A., Torr, P.H.S.: Fully-convolutional Siamese networks for object tracking. In: Hua, G., Jégou, H. (eds.) ECCV 2016. LNCS, vol. 9914, pp. 850–865. Springer, Cham (2016). https://doi.org/10.1007/978-3-319-48881-3_56
8. Li, B., Yan, J., Wu, W., Zhu, Z., Hu, X.: High performance visual tracking with Siamese region proposal network. In: Proceedings of the IEEE Conference on Computer Vision and Pattern Recognition, pp. 8971–8980 (2018)
9. Valmadre, J., Bertinetto, L., Henriques, J., Vedaldi, A., Torr, P.H.: End-to-end representation learning for correlation filter based tracking. In: Proceedings of the IEEE Conference on Computer Vision and Pattern Recognition, pp. 2805–2813 (2017)
10. Zhu, Z., Wang, Q., Li, B., Wu, W., Yan, J., Hu, W.: Distractor-aware Siamese networks for visual object tracking. In: Ferrari, V., Hebert, M., Sminchisescu, C., Weiss, Y. (eds.) ECCV 2018. LNCS, vol. 11213, pp. 103–119. Springer, Cham (2018). https://doi.org/10.1007/978-3-030-01240-3_7
11. Li, B., Wu, W., Wang, Q., Zhang, F., Xing, J., Yan, J.: SiamRPN++: evolution of Siamese visual tracking with very deep networks. In: Proceedings of the IEEE/CVF Conference on Computer Vision and Pattern Recognition, pp. 4282–4291 (2019)
12. Wang, Q., Teng, Z., Xing, J., Gao, J., Hu, W., Maybank, S.: Learning attentions: residual attentional Siamese network for high performance online visual tracking. In: Proceedings of the IEEE Conference on Computer Vision and Pattern Recognition, pp. 4854–4863 (2018)
13. Zhang, L., Gonzalez-Garcia, A., Weijer, J., Danelljan, M., Khan, F.S.: Learning the model update for Siamese trackers. In: Proceedings of the IEEE/CVF International Conference on Computer Vision, pp. 4010–4019 (2019)
14. Bahdanau, D., Cho, K., Bengio, Y.: Neural machine translation by jointly learning to align and translate. arXiv preprint arXiv:1409.0473 (2014)
15. Hu, J., Shen, L., Sun, G.: Squeeze-and-excitation networks. In: Proceedings of the IEEE Conference on Computer Vision and Pattern Recognition, pp. 7132–7141 (2018)

16. Schmidhuber, J., Hochreiter, S.: Long short-term memory. Neural Comput. **9**(8), 1735–1780 (1997)
17. Ren, S., He, K., Girshick, R., Sun, J.: Faster R-CNN: towards real-time object detection with region proposal networks. arXiv preprint arXiv:1506.01497 (2015)
18. Tian, Z., Shen, C., Chen, H., He, T.: FCOS: fully convolutional one-stage object detection. In: Proceedings of the IEEE/CVF International Conference on Computer Vision, pp. 9627–9636 (2019)
19. Xu, Y., Wang, Z., Li, Z., Yuan, Y., Yu, G.: SiamFC++: towards robust and accurate visual tracking with target estimation guidelines. In: Proceedings of the AAAI Conference on Artificial Intelligence, vol. 34, pp. 12549–12556 (2020)
20. Guo, D., Wang, J., Cui, Y., Wang, Z., Chen, S.: SiamCAR: Siamese fully convolutional classification and regression for visual tracking. In: Proceedings of the IEEE/CVF Conference on Computer Vision and Pattern Recognition, pp. 6269–6277 (2020)
21. Chen, Z., Zhong, B., Li, G., Zhang, S., Ji, R.: Siamese box adaptive network for visual tracking. In: Proceedings of the IEEE/CVF Conference on Computer Vision and Pattern Recognition, pp. 6668–6677 (2020)
22. Han, G., Du, H., Liu, J., Sun, N., Li, X.: Fully conventional anchor-free Siamese networks for object tracking. IEEE Access **7**, 123934–123943 (2019)
23. Chen, X., Yan, B., Zhu, J., Wang, D., Yang, X., Lu, H.: Transformer tracking. arXiv preprint arXiv:2103.15436 (2021)
24. Wang, N., Zhou, W., Wang, J., Li, H.: Transformer meets tracker: exploiting temporal context for robust visual tracking. arXiv preprint arXiv:2103.11681 (2021)
25. Chu, P., Wang, J., You, Q., Ling, H., Liu, Z.: Spatial-temporal graph transformer for multiple object tracking. arXiv preprint arXiv:2104.00194 (2021)
26. Yan, B., Peng, H., Fu, J., Wang, D., Lu, H.: Learning spatio-temporal transformer for visual tracking. arXiv preprint arXiv:2103.17154 (2021)
27. Vaswani, A., et al.: Attention is all you need. arXiv preprint arXiv:1706.03762 (2017)
28. Devlin, J., Chang, M.W., Lee, K., Toutanova, K.: BERT: pre-training of deep bidirectional transformers for language understanding. arXiv preprint arXiv:1810.04805 (2018)
29. Yang, T., Chan, A.B.: Learning dynamic memory networks for object tracking. In: Ferrari, V., Hebert, M., Sminchisescu, C., Weiss, Y. (eds.) ECCV 2018. LNCS, vol. 11213, pp. 153–169. Springer, Cham (2018). https://doi.org/10.1007/978-3-030-01240-3_10
30. Choi, J., Kwon, J., Lee, K.M.: Real-time visual tracking by deep reinforced decision making. Comput. Vis. Image Underst. **171**, 10–19 (2018)
31. Wu, Y., Lim, J., Yang, M.H.: Object tracking benchmark. IEEE Trans. Pattern Anal. Mach. Intell. **37**(9), 1834–1848 (2015). https://doi.org/10.1109/TPAMI.2014.2388226
32. Mueller, M., Smith, N., Ghanem, B.: A benchmark and simulator for UAV tracking. In: Leibe, B., Matas, J., Sebe, N., Welling, M. (eds.) ECCV 2016. LNCS, vol. 9905, pp. 445–461. Springer, Cham (2016). https://doi.org/10.1007/978-3-319-46448-0_27

Insight on Attention Modules for Skeleton-Based Action Recognition

Quanyan Jiang[1], Xiaojun Wu[1(✉)], and Josef Kittler[2]

[1] School of Artificial Intelligence and Computer Science,
Jiangnan University, Wuxi, China
quanyan_jiang@stu.jiangnan.edu.cn, wu_xiaojun@jiangnan.edu.cn
[2] Centre for Vision, Speech and Signal Processing,
University of Surrey, Guildford, UK

Abstract. Spatiotemporal modeling is crucial for capturing motion information in videos for action recognition task. Despite of the promising progress in skeleton-based action recognition by graph convolutional networks (GCNs), the relative improvement of applying the classical attention mechanism has been limited. In this paper, we underline the importance of spatio-temporal interactions by proposing different categories of attention modules. Initially, we focus on providing an insight into different attention modules, the Spatial-wise Attention Module (SAM) and the Temporal-wise Attention Module (TAM), which model the contexts interdependencies in spatial and temporal dimensions respectively. Then, the Spatiotemporal Attention Module (STAM) explicitly leverages comprehensive dependency information by the feature fusion structure embedded in the framework, which is different from other action recognition models with additional information flow or complicated superposition of multiple existing attention modules. Given intermediate feature maps, STAM simultaneously infers the feature descriptors along the spatial and temporal dimensions. The fusion of the feature descriptors filters the input feature maps for adaptive feature refinement. Experimental results on NTU RGB+D and Kinetics-Skeleton datasets show consistent improvements in classification performance, demonstrating the merit and a wide applicability of STAM.

Keywords: Action recognition · Attention mechanism · Spatiotemporal modeling

1 Introduction

The task of action recognition is a fundamental and challenging problem. Its goal is to recognize and classify video into different categories, including motion (e.g., running, walking, climbing) and interaction (e.g., handshake, embrace). In spite of the recent progress in deep neural networks, the capability of video analysis advances slowly, mostly due to the diversity of human-object interactions and complexity of spatio-temporal inferences. The core technical problem in action

© Springer Nature Switzerland AG 2021
H. Ma et al. (Eds.): PRCV 2021, LNCS 13019, pp. 242–255, 2021.
https://doi.org/10.1007/978-3-030-88004-0_20

recognition is to design an effective spatiotemporal module, that is expected to be able to capture complex motion information with high flexibility, while yet to be of low computational consumption for processing high dimensional video data efficiently.

Recently, state-of-the-art methods based on Graph Convolutional Networks (GCNs) [12,17,18,25] have been proposed to generate higher-level feature maps on the graph. GCNs have significantly pushed the performance of skeleton-based action recognition methods thanks to the representation power of the graph structure. The critical prerequisite of this approach is to determine an appropriate graph structure for the skeleton data at the respective convolutional layers, so that the GCN is empowered to extract correlated features. However, how to devise a spatiotemporal module with high efficiency and strong flexibility still remains to be an unsolved problem in action recognition. Consequently, we aim at advancing the current attention modules along this direction.

Apart from the graph structures, attention mechanisms have been widely used in the literature [2,3,13,19,21–24]. In this paper, we focus on devising a principled adaptive module to capture spatiotemporal information in a more flexible way. Attention can be interpreted as a method of unveiling long-range temporal dependencies. It also suggests where the focus of information extraction should be directed to improve the feature representation. We aim to increase the representation power by using novel attention modules, so as to improve the accuracy of state-of-the-art methods for skeleton-based action recognition. Instead of capturing channel-wise dependencies as in traditional models [5,20, 22], we focus on the two principal dimensions of feature maps in the temporal and spatial axes. Since convolution operations cannot highlight the frame and the joint containing important motion information, we adopt our modules to learn 'when' and 'where' to enhance the information along the temporal and spatial axes, respectively. As a result, our modules improve the recognition performance by learning which information to emphasize or suppress.

To demonstrate the effectiveness of our attention modules, we evaluate them on two distinct large-scale datasets for skeleton-based action recognition tasks, i.e., NTU RGB+D [16] and Kinetics-Skeleton [25]. We obtain accuracy improvements by plugging attention modules into different baseline GCNs, revealing the efficacy and wide applicability of proposed methods. STAM is able to yield a very competitive accuracy on both datasets.

The main contributions of this work can be summarized as follows: (**1**) We propose effective attention modules (SAM, TAM) that can be widely applied to boost the representation power of GCNs in skeleton-based action recognition. (**2**) An attention module(STAM) consisting of SAM and TAM is proposed to extract motion information adaptively among context features. (**3**) We demonstrate the consistent merits of STAM on benchmarking datasets with different characteristics.

2   Related Work

2.1   Skeleton-Based Action Recognition

Action recognition is a popular vision research topic. Different from direct video data analysis, action recognition from skeleton data has been studied widely in recent years. Different architectures have been proposed since the successful introduction of GCNs for skeleton-based action recognition [12,18,25]. Since skeleton data represents the human body with 2D or 3D coordinates of joints, learning the spatio-temporal dependence of each joint in the clips is crucial. An intuitive and simple way of constructing a spatio-temporal graph is to connect the vertices corresponding to the same joint in consecutive frames [25]. AS-GCN [12] proposes an encoder-decoder structure to capture action-specific latent dependencies. Shi *et al.* [18] introduce a two-stream adaptive GCNs using a multi-branch architecture, which considers joint and bone information simultaneously. The bone information represents the difference of coordinates between adjacent joints. While most of the recent methods mainly focus on two factors, graph structure and information representation, we concentrate on the spatio-temporal dependence in modelling information rather than improving the accuracy by increasing the computational burden.

2.2   Attention Mechanisms

According to the research on the human brain and the vision system, attention is essential in human perception [1,6,15]. The human vision system expresses the visual structure of the entire scene by selectively focusing on salient areas [9]. Hu *et al.* [5] introduce a channel-wise attention module to exploit the spatial correlations of the feature maps. In their SE (squeeze - excitation) block, they use a global average pooling operation to squeeze spatial information into a channel descriptor. The excitation operation contains a dimensionality-reduction layer and a dimensionality-expansion layer to reduce the parameter overhead. The procedure aims to capture channel-wise dependencies by analyzing internal relationships of the channel descriptors. However, ECA-Net [20] demonstrates that the dimensionality reduction in SE-Net introduces side effects on the reliability of the channel attention prediction. Besides the problems of existing attention modules, no temporal-wise attention mechanism has been studied thoroughly in the field of skeleton-based action recognition. Given an input skeleton sequence, our attention modules compute complementary attention map, focusing on 'when' and 'where' attention is warranted. In our attention modules (SAM, TAM, STAM), we exploit both spatial and temporal attention thanks to an efficient architecture without resorting to redundant convolution and pooling operations.

3   Multi-category Attention Modules

The success of models that utilize spatio-temporal information and motion features [7,22] suggests that it is crucial to incorporate attention processing for action recognition in both the spatial and temporal dimensions.

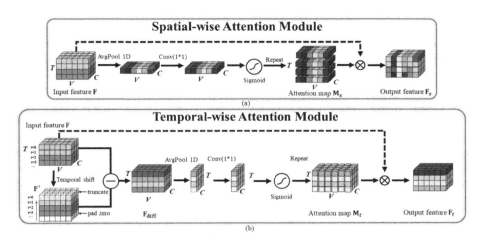

Fig. 1. Diagram of spatial and temporal attention modules. (a) The spatial-wise attention module. (b) The temporal-wise attention module. $\otimes$ represents the element-wise multiplication. $\ominus$ represents the element-wise subtraction. $\mathbf{F}$ denotes the original feature map. $\mathbf{F}'$ represents the novel feature map after moving. The difference between $\mathbf{F}$ and $\mathbf{F}'$ is the spatial difference feature $\mathbf{F}_{\mathbf{diff}}$.

We start by describing the basic operation of our proposed attention modules. These modules can effectively enhance original GCNs arithmetics in skeleton-based action recognition tasks. For a clearer illustration, we first introduce the spatial-wise attention module (SAM) and temporal-wise attention module (TAM) in terms of their structure and motivation. We then explain the spatiotemporal attention module (STAM) and discuss specific integration options in the context of datasets with different characteristics.

Unlike video data, the original skeleton information is represented by $\mathbf{x} \in \mathbb{R}^{N \times C \times T \times V \times M}$, where N is the number of clips, C initially consists of the joint positions and a confidence score (similar to RGB channels information in image), T denotes the number of frames in a sequence, V and M represent the number of joints and people. The network usually reshapes the original dimension of data to $(N \times M, C, T, V)$. In the following content, we omit the dimension of batch size. The feature $\mathbf{F} \in \mathbb{R}^{C \times T \times V}$ is used as the input feature map of attention mechanisms.

3.1 Spatial-Wise Attention Module

With the aim of setting the response intensity in the spatial dimension, we propose a spatial attention module, as shown in Fig. 1(a). It produces a spatial attention map by exploiting the inter-joint relationship of feature maps. Given an intermediate feature map $\mathbf{F} \in \mathbb{R}^{C \times T \times V}$ as input, SAM sequentially infers a spatial-wise attention map $\mathbf{M_s} \in \mathbb{R}^{C \times 1 \times V}$. The overall attention process of SAM can be summarized as:

$$\mathbf{F_s} = \mathbf{M_s} \otimes \mathbf{F},\tag{1}$$

where C is the number of channels for feature mapping generated by GCN filters. T and V represent the number of frames and joints. The number of V in different datasets is not the same because of the design of graph structures. $\otimes$ represents the element-wise multiplication. $\mathbf{F_s}$ is a novel feature map obtained from SAM. The following describes the details of calculating the attention map $\mathbf{M_s}$.

As each joint of a feature map is considered as a spatial detector, spatial attention focuses on 'where' lies the meaningful information, given a skeleton sequence. Generally, the global information of each joint compressed in the temporal dimension describes the relative importance of the joints. We first aggregate global temporal information of each joint by using average-pooling operation in the temporal dimension, generating a spatial context descriptor: $\mathbf{F_{avg}^s} \in \mathbb{R}^{C \times 1 \times V}$, which denotes average-pooled features. We explain in detail the process of calculating the elements of $\mathbf{F_{avg}^s}$ as follows:

$$\mathbf{F_{avg}^s} = f_{avg}^t(\mathbf{F}) = \frac{1}{T}\sum_{i=1}^{T} F_{c,v}(i),\tag{2}$$

where $F_{c,v}(i)$ represents a single frame of the input feature. f_{avg}^t refers to the average pooling operation in the temporal dimension.

To combine the global information efficiently, we squeeze the temporal dimension of the input feature and extract the spatial descriptor by compressing the global motion information in the temporal dimension for each joint in each channel. The spatial descriptor is then forwarded to a nonlinear unit to produce our spatial attention map $\mathbf{M_s} \in \mathbb{R}^{C \times 1 \times V}$. The nonlinear unit is composed of one convolutional layer and a sigmoid activation function, as described below:

$$\mathbf{M_s} = F_{ac}(\mathbf{F_{avg}^s}, \mathbf{W_1}) = \sigma(\mathbf{W_1}\mathbf{F_{avg}^s} + \mathbf{b_1}),\tag{3}$$

where F_{ac} denotes the activation operation, σ denotes the sigmoid function. The elements of $\mathbf{W_1}$ and $\mathbf{b_1}$ are convolutional parameters.

To capture the interdependencies between joints and to minimize the computational overhead, we use a convolution operation with the filter size of 1×1 in the spatial and temporal dimensions. Our ultimate objective is to make the joint importance, $\mathbf{M_s}$, properly reflected by weighting. The sigmoid function successfully helps us scale the weights of joint importance. After the nonlinear unit is applied to the spatial descriptor $\mathbf{F_{avg}^s}$, we merge the output feature maps using element-wise multiplication as described in Eq. 1.

The overall architecture of SAM is shown in Fig. 1(a). With the spatial attention module, we capture the importance of each joint based on its global information in the temporal dimension.

3.2 Temporal-Wise Attention Module

Despite focusing on capturing dependencies between the channels of feature maps, the existing algorithms ignore dependencies between the frames. We generate a temporal attention map by utilizing the inter-frame relationship of feature maps. Different from SAM, temporal-wise attention focuses on motion information, which is complementary to SAM. Given a feature map $\mathbf{F} \in \mathbb{R}^{C \times T \times V}$ as input, TAM infers a temporal attention map $\mathbf{M_t} \in \mathbb{R}^{C \times T \times 1}$, as shown in Fig. 1. The attention process is similar to SAM, as described below:

$$\mathbf{F_t} = \mathbf{M_t} \otimes \mathbf{F}, \tag{4}$$

where $\otimes$ represents the element-wise multiplication. $\mathbf{F_t}$ is the output feature map obtained from TAM. The following describes the details of generating the attention map $\mathbf{M_t}$.

The changed information of each frame compressed in the spatial dimension describes the different importance of frames. To build the dependencies between frames, we first obtain the temporal descriptor, which is generated by the temporal shift and average-pooling operations. From the spatial dimension of the feature map, the spatial difference for the adjacent frames can be calculated by the temporal shift operation. We can decouple the operation of capturing the spatial difference feature into two steps: shift and subtraction. We shift the input feature $\mathbf{F}$ by -1 along the temporal dimension and calculate the difference between $\mathbf{F}$ and $\mathbf{F}'$ to be $\mathbf{F_{diff}}$. As shown in Fig. 1(b), the spatial difference $\mathbf{F_{diff}}$ is defined as follows:

$$\mathbf{F_{diff}} = \mathbf{F} - \mathbf{F}'. \tag{5}$$

Each frame of the feature map $\mathbf{F_{diff}}$ conveys the temporal difference information for the adjacent frames. We then apply the average-pooling operation along the spatial axis to generate a temporal descriptor. The pooling operation captures the global dynamic spatial information of each frame. In short, the temporal context descriptor $\mathbf{F_{avg}^t} \in \mathbb{R}^{C \times T \times 1}$ is computed as:

$$\mathbf{F_{avg}^t} = f_{avg}^s(\mathbf{F_{diff}}) = \frac{1}{V} \sum_{j=1}^{V} F_{c,t}(j), \tag{6}$$

where $F_{c,t}(j)$ represents a single joint of the spatial difference feature $\mathbf{F_{diff}} \in \mathbb{R}^{C \times T \times V}$ conveying the channel and temporal information. f_{avg}^s refers to the average pooling operation in the spatial dimension. To capture the interdependencies between the frames and scaling the weights of frame relevance, we feed the temporal descriptor to a convolution layer and an activation function. The above operations generate a temporal attention map $\mathbf{M_t} \in \mathbb{R}^{C \times T \times 1}$ which encodes 'what' and 'when' to emphasize or suppress. We describe the detailed operation below:

$$\mathbf{M_t} = F_{ac}(\mathbf{F_{avg}^t}, \mathbf{W_2}) = \sigma(\mathbf{W_2}\mathbf{F_{avg}^t} + \mathbf{b_2}), \tag{7}$$

where F_{ac} denotes the activation operation, σ denotes the sigmoid function. The elements of $\mathbf{W_2}$ and $\mathbf{b_2}$ are convolutional parameters.

After the element-wise multiplication in Eq. 4, we have a novel feature map $\mathbf{F_t}$, which is generated by the temporal-wise attention module. Figure 1(b) depicts the computation process of the temporal-wise attention module. We use

TAM to respond to the importance of each frame based on the global information in the spatial dimension.

3.3 Spatiotemporal Attention Module

Given an input skeleton clip, the spatial-wise attention and temporal-wise attention compute complementary attention maps, focusing on 'where' and 'when' to apply it. Considering this, two attention modules can be placed in a parallel or sequential manner to constitute a novel spatiotemporal attention module.

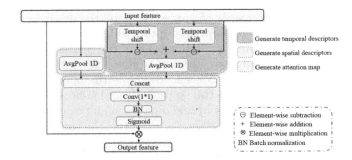

Fig. 2. Diagram of spatiotemporal attention module with parallel combination. $+$ represents the element-wise addition. $\ominus$ represents the element-wise subtraction. $\otimes$ represents the element-wise multiplication.

We experimented with both of combinations and considered the parallel manner as the final choice. Because we found that the parallel manner has stable performance. Figure 2 shows a diagram of STAM with the parallel structure. Spatial and temporal feature descriptors are combined together by concatenation in the channel dimension. STAM uses a 1×1 convolution layer to reduce the number of channels, then applies a normalization layer to normalize the synthesized feature descriptor. Unlike TAM in Fig. 1(b), which generates the difference feature between the current frame and one adjacent frame, STAM captures the sum of difference features between the current frame and two adjacent frames. STAM is able to capture the stable temporal dependencies because of its ability to recognize dynamic information. The attention process of STAM can be summarized as:

$$\mathbf{F_{st}} = \mathbf{M_{st}} \otimes \mathbf{F} = \sigma(Bn(\mathbf{W_3}[\mathbf{F_{avg}^s}; \mathbf{F_{avg}^t}] + \mathbf{b_3})) \otimes \mathbf{F}, \qquad (8)$$

where $\otimes$ represents the element-wise multiplication, σ denotes the sigmoid function. $\mathbf{F_{st}}$ is the output feature map obtained from STAM. $\mathbf{M_{st}}$ represents a feature map that is produced by STAM with a parallel combination. The elements of $\mathbf{W_3}$ and $\mathbf{b_3}$ are convolutional parameters. The parameters represent a convolution operation for dimensionality reduction. $\mathbf{F_{avg}^s}$ represents the spatial context descriptor calculated by Eq. 2. $\mathbf{F_{avg}^t}$ represents the temporal context descriptor calculated by Eq. 6. The descriptors of spatial and temporal are combined together by concatenation in the channel dimension. We find that STAM have significant advantages for datasets with different characteristics. This situation is also due to we use a parallel approach to avoid multiple activation functions.

4   Experiments

In this section, we validate our attention modules on two large-scale datasets: NTU RGB+D [16] and Kinetics-Skeleton [25] for skeleton-based action recognition. In the following experiments, we report the top-1 classification accuracy on these datasets. In order to demonstrate the general applicability of our attention modules across different architectures, we reproduce the networks [18,25] in the PyTorch framework and report the results by adding our attention modules in the experiments. Figure 3 shows a diagram of our attention modules integrated with a GCN block in AGCN [18] as an example.

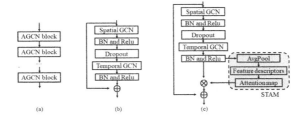

Fig. 3. Our attention module integrated with a AGCN block in 2s-AGCN [18]. (a) AGCN framework; (b) Details of AGCN block; (c) STAM integrated. This figure shows the exact position of our attention module when integrated within an AGCN block. We apply attention module on the GCN unit outputs in each block.

4.1   Datasets

NTU RGB+D. NTU RGB+D is the most popular and extensive dataset of 3D joint coordinates. It contains 56880 human action video sequences in 60 classes. There are two types of activities: single person action, e.g., picking up, and interaction by two persons, e.g., hugging. These videos are captured by three cameras of equal height but different horizontal angles. Kinect depth sensors detect 25 3D joint locations of each person in the camera coordinate system. This dataset consists of two benchmarks: 1) Cross-subject (X-Sub): All volunteers are divided into a subset and its complement. The actors of each subset perform 40320 actions to form a training set, and the complement subset containing 16560 clips to form an evaluation set. 2) Cross-view (X-View): This benchmark includes 37920 and 18960 clips for training and evaluation, respectively.

Kinetics-Skeleton. The original kinetics is a large dataset of recordings of human action recognition tasks collected from YouTube, containing 300k human action video clips in 400 classes. Yan et al. [25] proposed capturing the skeleton information in each frame of the clip by using the OpenPose toolbox. Unlike the joint label of the NTU RGB+D dataset, the skeleton graph in Kinetics-Skeleton consists of 18 joints. Each joint is represented as (X, Y, C), which includes 2D coordinates (X, Y), and confidence score C.

4.2 Ablation Studies

In this subsection, we analyze each attention module by experiments on NTU RGB+D and Kinetics-Skeleton datasets. The accuracy of classification is used as the evaluation criterion. For this ablation study, we adopt AGCN [18] as the base architectures. All the experiments were conducted using two 2080ti GPUs. We train AGCN on NTU RGB+D for 50 epochs with an initial learning rate of 0.1. The learning rate is divided by 10 at the 30th epoch and 40th epoch respectively. For Kinetics-Skeleton, we train the networks for 65 epochs and adopt the learning rate of 0.01 at 45th epoch and 0.001 at 55th epoch.

Table 1. Comparison of different attention methods on the Kinetics-Skeleton test set. **F** represents TAM without temporal shift operation. $\mathbf{F_{diff}}$ represents TAM incorporating a temporal shift operation. 's' represents the sequential spatial-temporal order. 'p' represents the parallel combination.

Methods	Top-1 (%)	Top-5 (%)
2s-AGCN	35.1	57.1
2s-AGCN+block	35.0	57.6
2s-AGCN+SAM	35.5	58.2
2s-AGCN+TAM (**F**)	35.2	57.5
2s-AGCN+TAM ($\mathbf{F_{diff}}$)	35.4	57.8
2s-AGCN+STAM (s)	35.2	58.1
2s-AGCN+STAM (p)	36.1	58.4

Spatial-Wise Attention. Here we focus on demonstrating the benefits of applying the spatial attention module. We experimentally verify the impact of the network depth by adding a AGCN block after the original AGCN framework. The operation of simply increasing the depth of the network does not capture the global joint dependencies. Table 1 shows that the network effect after increasing a AGCN block is not satisfactory (2s-AGCN+block). But the spatial attention module steadily improves the accuracy.

Temporal-Wise Attention. To enhance the important frames in the sequence, we apply an effective method to compute temporal attention. The design philosophy is symmetric with the spatial attention branch. We use the shift operation

along temporal dimension and the subtraction operation to generate the spatial difference information for the adjacent frames. We compare two methods of
generating the temporal descriptor: encoding spatial difference information and
using the original feature map. The original spatial information is static spatial
information of each frame without temporal shift operation. Table 1 shows the
experimental results. We can observe that the temporal shift operation produces
better accuracy with zero parameters, indicating that encoding spatial difference
information leads to finer attention inference, compared with using just static
information.

Spatiotemporal Attention. In this experiment, we compare two different ways
of arranging the spatial and temporal attention modules: sequential spatial-
temporal and parallel use of both attention modules. As the spatial and temporal
attentions focus on enhancing the information in different dimensions, the unique
combination manner may affect the overall performance. Table 1 summarizes the
experimental results for different attention arranging methods. We analyze the
results of the sequential configuration and the parallel configuration, and we
find that the parallel configuration has advantage on Kinetics-Skeleton. This situation is similar on NTU RGB+D dataset. The skeleton information in NTU
RGB+D is more fine-grained and accurate, which can effectively be accentuated
by applying sophisticated attention model. In the following experiments, we use
the parallel STAM on datasets.

Two-Stream Framework. The two-stream network architecture is widely used
in the field of video analysis. We rigorously evaluate our attention modules by
using two-stream information in both datasets. The information provided by
skeleton datasets, such as NTU RGB+D and Kinetics-Skeleton, is generally the
coordinate information of the joint points. To generate the bone information, Shi
et al. [18] calculate the vector difference for the adjacent joints. To confirm the
effectiveness of our attention modules, we show the performance achieved using
bone information and fusing two-stream AGCNs on two datasets in Table 2. As
demonstrated in experiments, it implies our STAM is an efficient yet effective
spatiotemporal adaptive scheme.

Stacking on Different Networks. We evaluate our attention modules in various network architectures, including ST-GCN [25] and 2s-AGCN [18]. Table 2
and Table 3 summarize the experimental results on the two datasets. The networks with our attention modules outperform all the baselines, demonstrating
that they can generalize well on large-scale dataset. STAM also achieves the
highest performance among these baselines, which strongly prove the combination of spatial and temporal information is more beneficial for our adaptive
scheme.

Table 2. The effect of using different information flows on NTU RGB+D and Kinetics-Skeleton datasets. We adopt the 2s-AGCN [18] framework and apply SAM, TAM, STAM to the AGCN blocks respectively.

Datasets	Methods	Joint top-1 (%)	Bone top-1 (%)	Combine top-1 (%)
Kinetics-Skeleton	2s-AGCN	35.1	33.3	36.1
	2s-AGCN+SAM	35.5	35.1	37.5
	2s-AGCN+TAM	35.4	34.3	36.6
	2s-AGCN+STAM	**36.1**	**35.3**	**37.7**
NTU RGB+D (cross-subject)	2s-AGCN	86.5	86.1	88.5
	2s-AGCN+SAM	87.4	87.8	89.3
	2s-AGCN+TAM	87.5	87.9	89.7
	2s-AGCN+STAM	**87.6**	**88.0**	**89.9**
NTU RGB+D (cross-view)	2s-AGCN	93.7	93.2	95.1
	2s-AGCN+SAM	94.7	94.5	95.8
	2s-AGCN+TAM	94.3	94.3	95.6
	2s-AGCN+STAM	**94.8**	**94.6**	**95.9**

Table 3. The impact of our proposed attention modules on NTU RGB+D and Kinetics-Skeleton datasets. We adopt the ST-GCN [25] framework and apply SAM, TAM, STAM to the GCN blocks respectively.

Datasets	Methods	Top-1 (%)
Kinetics-Skeleton	ST-GCN	30.7
	ST-GCN+SAM	32.1
	ST-GCN+TAM	31.0
	ST-GCN+STAM	**32.6**
NTU RGB+D (cross-subject)	ST-GCN	81.5
	ST-GCN+SAM	81.6
	ST-GCN+TAM	81.6
	ST-GCN+STAM	**82.2**
NTU RGB+D (cross-view)	ST-GCN	88.3
	ST-GCN+SAM	89.2
	ST-GCN+TAM	89.5
	ST-GCN+STAM	**90.8**

4.3 Comparison with the State-of-the-Art

To demonstrate the advantage and validity of our attention module, we compare our method with state-of-the-art algorithms on Kinetics-Skeleton and NTU RGB+D, respectively. As shown in Table 4, we report the performance of these methods on both cross-subject and cross-view benchmarks of the NTU RGB+D dataset. And we also show the top-1 recognition accuracy on the Kinetics-Skeleton dataset. Note, 2s-AGCN [18] with STAM gains in accuracy significantly. This implies that our proposed attention module is powerful, showing its efficacy in capturing rich dependence and enhancing the features in different dimensions respectively.

Table 4. Performance comparison on NTU RGB+D and Kinetics-Skeleton datasets with current state-of-the-art methods.

Datasets	Methods	Top-1 (%)
Kinetics-Skeleton	TCN [8]	20.3
	ST-GCN [25]	30.7
	AS-GCN [12]	34.8
	2s-AGCN [18]	36.1
	GT-LSTM [11]	36.6
	NAS [14]	37.1
	2s-AGCN+STAM (ours)	**37.7**
NTU RGB+D (cross-subject)	ST-GCN [25]	81.5
	STGR-GCN [10]	86.9
	GR-GCN [4]	87.5
	AS-GCN [12]	86.8
	2s-AGCN [18]	88.5
	GT-LSTM [11]	89.2
	NAS [14]	89.4
	2s-AGCN+STAM (ours)	**89.9**
NTU RGB+D (cross-view)	ST-GCN [25]	88.3
	STGR-GCN [10]	92.3
	GR-GCN [4]	94.3
	AS-GCN [12]	94.2
	2s-AGCN [18]	95.1
	GT-LSTM [11]	95.2
	NAS [14]	95.7
	2s-AGCN+STAM (ours)	**95.9**

5   Conclusions

We have proposed the spatial attention module (SAM), the temporal attention module (TAM), and the spatiotemporal attention module (STAM) to capture the interdependence in different dimensions. These attention modules achieve a considerable performance improvement while keeping the computational overhead small. We enhance the performance by exploiting the spatial dependences with SAM. For the temporal attention module, we use the temporal shift operation to capture the spatial difference information without any additional parameters. Our spatiotemporal attention module simultaneously captures the spatial and temporal dependence. We conducted extensive experiments with various models and confirmed that all the attention modules enhance the performance of baselines on two different benchmark datasets: Kinetics-Skeleton, NTU RGB+D. In the future research, we will investigate the merit of enhancing features capturing long-range temporal information.

References

1. Corbetta, M., Shulman, G.L.: Control of goal-directed and stimulus-driven attention in the brain. Nat. Rev. Neurosci. **3**(3), 201–215 (2002)

2. Du, W., Wang, Y., Qiao, Y.: Recurrent spatial-temporal attention network for action recognition in videos. TIP **27**(3), 1347–1360 (2017)
3. Fu, J., et al.: Dual attention network for scene segmentation. In: CVPR, pp. 3146–3154 (2019)
4. Gao, X., Hu, W., Tang, J., Liu, J., Guo, Z.: Optimized skeleton-based action recognition via sparsified graph regression. In: Proceedings of the 27th ACM International Conference on Multimedia, pp. 601–610 (2019)
5. Hu, J., Shen, L., Sun, G.: Squeeze-and-excitation networks. In: CVPR, pp. 7132–7141 (2018)
6. Itti, L., Koch, C., Niebur, E.: A model of saliency-based visual attention for rapid scene analysis. IEEE Trans. Pattern Anal. Mach. Intell. **20**(11), 1254–1259 (1998)
7. Jiang, B., Wang, M., Gan, W., Wu, W., Yan, J.: STM: spatiotemporal and motion encoding for action recognition. In: ICCV, pp. 2000–2009 (2019)
8. Kim, T.S., Reiter, A.: Interpretable 3D human action analysis with temporal convolutional networks. In: CVPRW, pp. 1623–1631. IEEE (2017)
9. Larochelle, H., Hinton, G.E.: Learning to combine foveal glimpses with a third-order Boltzmann machine. In: Advances in Neural Information Processing Systems, pp. 1243–1251 (2010)
10. Li, B., Li, X., Zhang, Z., Wu, F.: Spatio-temporal graph routing for skeleton-based action recognition. In: AAAI, vol. 33, pp. 8561–8568 (2019)
11. Li, H., Zhu, G., Zhang, L., Song, J., Shen, P.: Graph-temporal LSTM networks for skeleton-based action recognition. In: Peng, Y., et al. (eds.) PRCV 2020. LNCS, vol. 12306, pp. 480–491. Springer, Cham (2020). https://doi.org/10.1007/978-3-030-60639-8_40
12. Li, M., Chen, S., Chen, X., Zhang, Y., Wang, Y., Tian, Q.: Actional-structural graph convolutional networks for skeleton-based action recognition. In: CVPR, pp. 3595–3603 (2019)
13. Miech, A., Laptev, I., Sivic, J.: Learnable pooling with context gating for video classification. arXiv preprint arXiv:1706.06905 (2017)
14. Peng, W., Hong, X., Chen, H., Zhao, G.: Learning graph convolutional network for skeleton-based human action recognition by neural searching. arXiv preprint arXiv:1911.04131 (2019)
15. Rensink, R.A.: The dynamic representation of scenes. Vis. Cogn. **7**(1–3), 17–42 (2000)
16. Shahroudy, A., Liu, J., Ng, T.T., Wang, G.: NTU RGB+D: a large scale dataset for 3D human activity analysis. In: CVPR, pp. 1010–1019 (2016)
17. Shi, L., Zhang, Y., Cheng, J., Lu, H.: Skeleton-based action recognition with directed graph neural networks. In: CVPR, pp. 7912–7921 (2019)
18. Shi, L., Zhang, Y., Cheng, J., Lu, H.: Two-stream adaptive graph convolutional networks for skeleton-based action recognition. In: CVPR, pp. 12026–12035 (2019)
19. Vaswani, A., et al: Attention is all you need. In: Advances in Neural Information Processing Systems, pp. 5998–6008 (2017)
20. Wang, Q., Wu, B., Zhu, P., Li, P., Zuo, W., Hu, Q.: ECA-Net: efficient channel attention for deep convolutional neural networks. arXiv preprint arXiv:1910.03151 (2019)
21. Wang, X., Girshick, R., Gupta, A., He, K.: Non-local neural networks. In: CVPR, pp. 7794–7803 (2018)
22. Woo, S., Park, J., Lee, J.-Y., Kweon, I.S.: CBAM: convolutional block attention module. In: Ferrari, V., Hebert, M., Sminchisescu, C., Weiss, Y. (eds.) ECCV 2018. LNCS, vol. 11211, pp. 3–19. Springer, Cham (2018). https://doi.org/10.1007/978-3-030-01234-2_1

23. Xiao, T., Fan, Q., Gutfreund, D., Monfort, M., Oliva, A., Zhou, B.: Reasoning about human-object interactions through dual attention networks. In: ICCV, pp. 3919–3928 (2019)
24. Xie, S., Sun, C., Huang, J., Tu, Z., Murphy, K.: Rethinking spatiotemporal feature learning: speed-accuracy trade-offs in video classification. In: Ferrari, V., Hebert, M., Sminchisescu, C., Weiss, Y. (eds.) ECCV 2018. LNCS, vol. 11219, pp. 318–335. Springer, Cham (2018). https://doi.org/10.1007/978-3-030-01267-0_19
25. Yan, S., Xiong, Y., Lin, D.: Spatial temporal graph convolutional networks for skeleton-based action recognition. In: AAAI (2018)

AO-AutoTrack: Anti-occlusion Real-Time UAV Tracking Based on Spatio-temporal Context

Hongyu Chu[1], Kuisheng Liao[1], Yanhua Shao[1(✉)], Xiaoqiang Zhang[1],
Yanying Mei[1], and Yadong Wu[2]

[1] Southwest University of Science and Technology, Mianyang 621010, China
`syh@cqu.edu.cn`, {`chuhongyu,xqzhang,myy930`}`@swust.edu.cn`
[2] School of Computer, Sichuan University of Science and Engineering,
Zigong 643000, China

Abstract. Correlation filter (CF)-based methods have shown extraordinary performance in visual object tracking for unmanned aerial vehicle (UAV) applications, but target occlusion and loss is still an urgent problem to be solved. Aiming at this problem, most of the existing discriminant correlation filter (DCF)-based trackers try to combine with object detection, thereby significantly improving the anti-occlusion performance of the visual object tracking algorithm. However, the significantly increases the computational complexity and limits the speed of the algorithm. This work proposes a novel anti-occlusion target tracking algorithm using the target's historical information, *i.e.*, the AO-AutoTrack tracker. Specifically, after the target is occluded or lost, the target information contained in the spatio-temporal context is used to predict the subsequent position of the target, and the historical filter is employed to re-detect the target. Extensive experiments on two UAV benchmarks have proven the superiority of our method and its excellent anti-occlusion performance.

Keywords: UAV object tracking · Spatio-temporal context · Anti-occlusion · Correlation filters

1 Introduction

Visual object tracking is one of the hot topics in the field of computer vision, aiming to estimate the state in subsequent frames of the target only with the information (position, size, *etc.*) obtained from the first frame. Empowering UAV (unmanned aerial vehicle) with visual object tracking brings many applications, such as aerial cinematography [1], person following [2], aircraft tracking [3], and traffic patrolling [4].

Supported by National Natural Science Foundation of China (61601382), Project of Sichuan Provincial Department of Education (17ZB0454), The Doctoral Fund of Southwest University of Science and Technology (19zx7123), Longshan talent of Southwest University of Science and Technology (18LZX632).

At present, there are two main research focuses in this field: discriminative correlation filter (DCF)-based approaches and deep learning-based methods [5]. Although deep feather-based and convolutional neural network (CNN)-based methods greatly improve the accuracy and robustness of the tracker. However, the convolution operations bring a huge computational burden, leading to a decrease in speed and hindering practical application. Therefore, take into consideration UAV hardware limitations and DCF-based methods' high computational efficiency, most UAV visual object tracking methods are based on DCF.

Recently there have been great advances in DCF-based approaches, but those performances are still constrained by part occlusion, full occlusion, and out-of-view, *etc*. To address these issues, on the one hand, regularization terms are used to improve the learning of the target to prevent the model from being polluted. For example, STRCF [6] and AutoTrack [5] introduced predefined regularization terms or adaptive learning regularization terms, respectively. These methods have made some progress in short-term occlusion, but they are not enough for long-term occlusion or out-of-view.

On the other hand, the combination of tracker and object detection method has become a choice, such as LCT [7] and JSAR [8], and gain some effect in re-detection. However, due to the computational burden of the object detection method, the speed has been greatly reduced.

In this work, an efficient re-detection strategy is proposed to deal with the occlusion problem. Specifically, we designed a simple and effective tracking failure monitoring mechanism by using the target's information hidden in the spatio-temporal context; further, and a re-detection method for multi-search region prediction is proposed. It can improve the anti-occlusion ability of the tracker with almost no decrease in speed. Our main contributions are summarized as follows:

- A novel robust anti-occlusion real-time UAV tracking method based on spatio-temporal context is proposed, namely AO-AotoTrack. It fully exploits the rich target information obtained from spatio-temporal context to boost the accuracy and robustness of the UAV tracking process.
- A simple and efficient re-detection mechanism is proposed to solve the problem of long-term occlusion and out-of-view in the process of target tracking.
- Large-scale experiments are conducted on one short-term UAV benchmark and one long-term benchmark to validate the outstanding performance of the proposed method.

2   Related Work

2.1   Discriminative Correlation Filter Tracking Algorithm

Since D. S. Bolme *et al.* first introduced correlation filter into the visual object tracking field. Due to its amazing speed and efficiency, it attracted many researchers to made in-depth research and produced a lot of achievements: scale estimate [9,10], boundary effect [5,6,11,12], long-term tracking [7,13,14], and multi-feature fusion [15,16]. However, the existing methods seem to be unable to solve the long-term occlusion problem perfectly, and the occlusion problem is common in practical application. Hence, it is still an open problem to design a tracker with both excellent anti-occlusion performance and satisfactory running speed.

2.2   Anti-occlusion Object Tracking

Aiming at filter degradation caused by occlusion, SRDCF [17], STRCF [6], ASRCF [11], AutoTrack [5] have obtained a certain effect by introducing regularization terms. Although the performance is not satisfactory with long-time occlusion. To solve this problem, literature [7,8,14] combined with object detection algorithms for re-detection. However, by using additional detectors, although the performance is improved, there could be a large loss of speed in the process of re-detection. First, this work utilizes spatio-temporal context information to quickly generate multiple proposals, and then we utilize a historical filter to find and select the most likely bounding box to re-initialize the tracker. The re-detection strategy proposed in this work is more efficient.

2.3   DCF Onboard UAV

Generally, there are three reasons why the DCF-based tracking method is more suitable for UAV than most other trackers: adaptability, robustness, and efficiency. In latest years, the researchers have investigated it deeply and improved the performance of the CF-based method in UAV to an even higher level [5,8,12,18]. But it is still a little weak to deal with occlusion problems in real scenes. By contrast, our AO-AutoTrack has excellent anti-occlusion performance and only occupies a small amount of computing resources to obtain satisfactory performance.

3   Proposed Tracking Approach

In this section, we first give a brief introduction to our baseline AutoTrack [5], and then give a detailed explanation of our work.

3.1   Review AutoTrack

Our baseline AutoTrack [5], published in the CVPR2020, is revisited. The optimal filter H_t and temporal regularization parameter θ_t in frame t are learned by minimizing the objective function as shown in the Eq. (1):

$$
\begin{aligned}
\mathcal{E}(\mathbf{F}_t, \theta_t) = &\frac{1}{2}\|\sum_{k=1}^{K} \mathbf{x}_t^k \circledast \mathbf{f}_t^k - \mathbf{y}\|_2^2 + \frac{1}{2}\sum_{k=1}^{K} \|\tilde{\mathbf{w}} \odot \mathbf{f}_t^k\|_2^2 \\
&+ \frac{\theta_t}{2}\sum_{k=1}^{K} \|\mathbf{f}_t^k - \mathbf{f}_{t-1}^k\|_2^2 + \frac{1}{2}\|\theta_t - \tilde{\theta}\|_2^2
\end{aligned}
\tag{1}
$$

where $\mathbf{x}_t^k \in \mathbb{R}^{T \times 1} (k = 1, 2, 3, ..., K)$ is the extracted feature with length T in frame t, and K denotes the number of channels, $\mathbf{y} \in \mathbb{R}^{T \times 1}$ is the desired Gaussian-shaped response. $\mathbf{f}_t^k$, $\mathbf{f}_{t-1}^k \in \mathbb{R}^{T \times 1}$ respectively denote the filter of the k-th channel trained in the t-th and $(t-1)$-th frame, $\odot$ indicates the convolution operator. Noted that $\mathbf{F}_t = [\mathbf{f}_t^1, \mathbf{f}_t^2, \mathbf{f}_t^3, ..., \mathbf{f}_t^K]$. $\tilde{\mathbf{w}} \in \mathbb{R}^{T \times 1}$ indicates the spatial regularization parameter, it's used to restrain the boundary effect. Temporal regularization is the third and fourth term in Eq. (1), where $\tilde{\theta}$ is θ_t's reference value used to restrict filter's variation by penalizing the difference between the current and previous filters.

However, online automatically and adaptively learn spatio-temporal regularization term could be adapted to complex scenarios, to a certain extent, but there are some deficiencies as follows:

Exception Tracking Monitoring Mechanism: AutoTrack [5] define local response variation vector Π to learn spatio-temporal regularization term automatically and adaptively. The Π was compared with the predefined parameter to determine the trace state, which prone to misjudgment without using target response information.

Cannot Deal with Long-Term Occlusion: To a certain extent, AutoTrack [5] solved the problem of short-term occlusion. Due to the lack of search target strategy, it seems that the AutoTrack cannot deal with the long-term occlusion problem.

3.2   Temporal Regularization Analysis and Improvement

AutoTrack [5] has achieved a great improvement in accuracy and robustness because of automatic regularization. It also brings a certain computational burden, especially in temporal regularization. Therefore, this work abandons automatic temporal regularization, so the optimal filter H_t in frame t is learned by minimizing the following objective function:

$$\mathcal{E}(\mathbf{F}_t) = \frac{1}{2} \| \sum_{k=1}^{K} \mathbf{x}_t^k \circledast \mathbf{f}_t^k - \mathbf{y} \|_2^2 + \frac{1}{2} \sum_{k=1}^{K} \| \tilde{\mathbf{w}} \odot \mathbf{f}_t^k \|_2^2$$
$$+ \frac{\theta_t}{2} \sum_{k=1}^{K} \| \mathbf{f}_t^k - \mathbf{f}_{t-1}^k \|_2^2 \qquad (2)$$

3.3   Re-detection Mechanism

As shown in Fig. 1, an example of video sequence *bake2* of UAV123@10 fps [19], is the center idea of our re-detection mechanism, when observed tracking failure, the re-detection will be implemented immediately. There are three parts of the proposed re-detection mechanism: storing spatio-temporal context information, object proposals generation, and candidates scoring. Specifically, the first part is the preparation of the re-detection mechanism. And spatio-temporal context information includes the target's historical location, bounding box's size, and filter. Further, object proposals generation and candidates scoring are the first and second stages of the re-detection process, respectively. Specifically, in the first stage, n candidate regions are generated in the re-detection state. And then, in the second stage, candidate regions are scored through historical filters for re-initializing filters. The detailed illustration is as follows. As also shown in step 18 to step 25 of Algorithm 1.

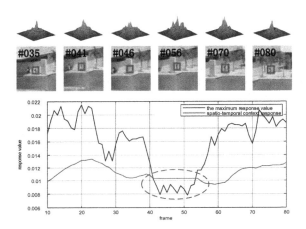

Fig. 1. Central idea of re-detection mechanism. When the maximum response value in frame t (ζ_t) is larger than the threshold value (spatio-temporal context response value ζ_e), tracking procedure is normally implemented; if not, tracking procedure is in a abnormal state, activate the re-detection mechanism and stop updating filter and ζ_e. When the peak value of the response map (ζ_t equals to ζ_{max}) generated by selected proposals and historical filter $\mathbf{F}_{t-k}$ exceeds ζ_e, the bounding box is re-initialized and re-starts to be tracked normally.

Tracking Failure Monitoring Mechanism: First, we define a spatio-temporal context response value ζ_e, the form is shown as follows:

$$\zeta_e = \frac{\eta * \sum_{n=1}^{k} \zeta_n}{k} \tag{3}$$

where η is a hyperparameter, ζ_n is the historical maximum response value of frame n, k represents the number of the historical maximum response value. When the current maximum response value $\zeta_t < \zeta_e$, it is in an abnormal state, we stop updating the filter and activate the re-detection mechanism. Otherwise, filter updates will continue.

Object Proposal Generation: When the re-detection starts, the first stage generates n target candidate regions with the same size as before, its center point position P_{t+1}^n is determined by Eq. (4),

$$P_{t+1}^n = \begin{cases} P_t & n = 1 \\ P_t + [V_x^t, V_y^t] * R_{\theta_n} & n > 1 \end{cases}, \tag{4}$$

where, P_t is target position when tracking failure, $[V_x^t, V_y^t]$ denotes spatio-temporal context target local pixel speed, R_{θ_n} represents the velocity offset rotation matrix, the size of proposals is same as before. The $[V_x^t, V_y^t]$ and R_{θ_n} can be denoted by,

$$[V_x^t, V_y^t] = [[x_{t-k}, y_{t-k}] - [x_t, y_t]]/k , \tag{5}$$

$$R_{\theta_n} = [cos\theta_n, -sin\theta_n; sin\theta_n, cos\theta_n], \tag{6}$$

Object Proposals Scoring: In the second stage, for the n candidate regions. We first extract features as normal. Further, the filter of the $(t-k)$-th frame was used to calculate the response map $\mathbf{R}_t$, and finally, according to the maximum response value ζ_{max} to locate the target location. However, when the target is occluded, there is usually no target in the selected search area, and the maximum response value will be reduced greatly. Therefore, we also use the maximum response value as the confidence level. And we utilize the response value of spatio-temporal context ζ_e as the threshold to determine the state of tracking. If the maximum response value of proposals $\zeta_{max} > \zeta_e$, it means the re-detection is successful, and then re-initialize the filter; or else, filter updates are continued in the next frame. Where the formulation of response map $\mathbf{R}_t$ is shown as follows:

$$\mathbf{R}_t = \mathcal{F}^{-1}\Big(\sum_{d=1}^D \widetilde{\mathbf{f}}_{t-1}^{d*} \odot \widetilde{\mathbf{m}}_t^d \Big) . \tag{7}$$

Where $\mathcal{F}^{-1}$ represents inverse discrete Fourier transform, and $\mathbf{m}_t^d$ is the d-th feature representation of the search region. The overall of the proposed method is presented in Algorithm 1.

4   Experiment

In this section, the proposed method is evaluated on one challenging short-term UAV benchmark UAV123@10 fps [19], and one long-term benchmark UAV20L [19], which are especially captured by UAV from the aerial view, including 143 challenging image sequences. The results are compared with 8 state-of-the-art trackers, *i.e.*, AutoTrack [5], STRCF [6], JSAR_Re [8], LCT [7], ARCF-H [12], KCF [16], ECO-HC [20], and fDSST [21].

Algorithm 1: AO-AutoTrack

Input: Location and size of Initial bounding box

Output: Location and size of object in frame t

1 if $t = 1$ then
2 | Extract training samples $\mathbf{X}_t$
3 | Use Eq. (2) to initialize $\mathbf{F}_t$
4 else
5 | Extract search region feature maps $\mathbf{X}_t$
6 | Generate $\mathbf{R}_t$ by Eq. (7)
7 | find ζ_t, estimate location of target P_t
8 | if $t < k$ then
9 | | store $\mathbf{F}_t$, ζ_t and P_t
10 | else
11 | | Use Eq. (3) to generate ζ_e
12 | | if $\zeta_t > \zeta_e$ then
13 | | | Estimate P_t and extract $\mathbf{M}_i$
14 | | | Update $\mathbf{F}_t$ by Eq. (2)
15 | | | Delete $\mathbf{F}_{t-k}$, ζ_{t-k}, and P_{t-k}
16 | | | store $\mathbf{F}_t$, ζ_t, and P_t
17 | | else
18 | | | Generate proposals with search area by Eq. (4)
19 | | | Generate response map of proposals by Eq. (7) and $\mathbf{F}_{t-k}$
20 | | | Find the largest peak value ζ_{max}
21 | | | if $\zeta_{max} > \zeta_e$ then
22 | | | | Enable re-detection, initialize the object
23 | | | else
24 | | | | Continue to re-detect next frame
25 | | | end
26 | | end
27 | end
28 end

4.1 Implementation Details

Our AO-AutoTrack is based on AutoTrack [5]. ζ_e and k in Eq. (3) is set to 2/3 and 10. The number of search region n is set to 4. The angle of rotation θ is set to 0.25π and -0.25π. The other hyperparameter is the same as our baseline AutoTrack [5]. All experiments are implemented with MATLAB R2018b and performed on a machine running Windows 10 Operating System with AMD Ryzen 7 4800H (2.90 GHz) CPU, 16 GB RAM, and an NVIDIA RTX 1660Ti (6 GB RAM) GPU for fair comparisons.

4.2   Comparison with Hand-Crafted Based Trackers

Overall Evaluation: As displayed in Table 1, it shows the overall performance of AO-AutoTrack and other top eight CPU-based trackers on two benchmarks. AO-AutoTrack has outperformed all other CPU-based realtime trackers in precision. AO-AutoTrack has outstanding performance in precision with the third fast speed of 56.7 fps (frames per second), only slower than fDSST [21] (135.8 fps) and KCF [16] (475.9 fps). AO-AutoTrack obtained the third place in AUC(Area Under The Curve), however, AO-AutoTrack has a tiny gap compared with the best tracker JSAR_Re [8] (0.005 in AUC) and STRCF [6](0.015 in AUC), and it has remarkably improved the speed by 183.5% and 98.9% compared with JSAR_Re [8] and STRCF [6]. Averagely, AO-AutoTrack gains an improvement of 2% in AUC and 2.4% in precision compared with the baseline AutoTrack.

Table 1. Average precision, success rate and speed comparison between top 8 hand-crafted trackers on UAV123@10 fps and UAV20L [19]. Bold, italic and bold italic indicate the first, second and third place, respectively.

Tracker	Precision	Success rate	Speed	Venue
AO-AutoTrack	**0.607**	*0.514*	*56.7*	Ours
JSAR_Re [8]	*0.601*	*0.519*	20.0	IROS'20
AutoTrack [5]	0.593	0.504	46.2	CVPR'20
STRCF [6]	*0.600*	**0.529**	28.5	CVPR'18
ECO-HC [20]	0.572	0.505	54.4	CVPR'17
ARCF-H [12]	0.585	0.492	51.2	ICCV'19
fDSST [21]	0.447	0.392	*135.8*	PAMI'17
LCT [7]	0.405	0.290	44.0	CVPR'15
KCF [16]	0.359	0.247	**475.9**	PAMI'15

Attribute-Based Comparison: Figure 2 exhibits the precision and success plots of real-time trackers on five challenging attributes from UAV123@10 fps and UAV20L [19]. It can be seen that AO-AutoTrack has outstanding performance in the attributes of background clutter and full occlusion. In UAV123@10 fps [19], AO-AutoTrack has respectively improved the precision by 4.5% and 9.6% compared with the second-best trackers in the attributes of background clutter and full occlusion; In out-of-view, partial occlusion, and similar object, AO-AutoTrack have respectively obtained an advantage of 3.0%, 3.3%, and 3.6% compared with AutoTrack. As for UAV20L [19], AO-AutoTrack improves Auto-Track by 17.5%, 20.5%, and 10.1% in the attributes of background clutter, full occlusion, and out-of-view. The experimental results show that our algorithm outperforms the state-of-the-art methods.

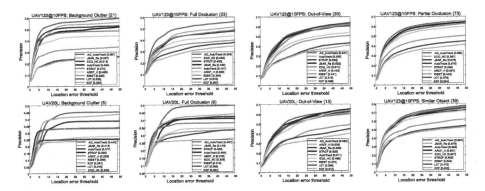

Fig. 2. Attribute-based comparison with hand-crafted real-time trackers. Five attributes, *i.e.*, background clutter, full occlusion, partial occlusion, out-of-view and similar object from 2 UAV datasets are displayed.

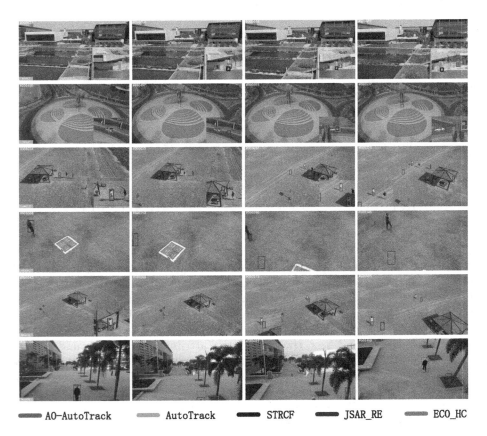

Fig. 3. Tracking results demonstration of 5 hand-crafted trackers on five video sequences, *i.e.*, *bake2*, *car2_1*, *group1_2*, *person8_1*, and *person14_1* respectively.

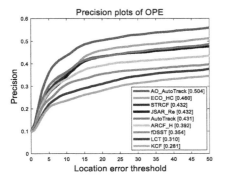
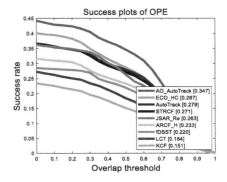

Fig. 4. Precision plots and Success plots of AO-AutoTrack and eight state-of-the-art trackers on 33 Full Occlusion sequences of UAV123@10 fps dataset [19].

4.3   Re-detection Evaluation

To verify the effectiveness of the re-detection mechanism of AO-AutoTrack, we conduct experiments on AO-AutoTrack with eight state-of-the-art trackers on subsets (include 33 sequences) of full occlusion of UAV123@10 fps [19], the precision plot and AUC plot are reported in Fig. 4. AO-AutoTrack ranks No.1 and improves the tracking precision and AUC by 16.9% and 24.3% compared with AutoTrack. Some typical results are exhibited in Fig. 3, only AO-AutoTrack successfully re-detect the target in the attributes of long-term occlusion and out-of-view.

5   Conclusions

In this work, a novel robust anti-occlusion real-time UAV tracking method based on spatio-temporal context is proposed. Introduced rich object location information to generate object proposals, and then utilize historical filters to re-detect target. Our method improved the performance in the respect of the anti-occlusion and robustness in complex scenes. Comprehensive experiments have validated that AO-AutoTrack is the state-of-the-art CPU-based tracker with a speed of 56.7 fps.

References

1. Rogerio, B., Cherie, H., Wenshan, W., Sanjiban, C., Sebastian, S.: Towards a robust aerial cinematography platform: Localizing and tracking moving targets in unstructured environments. In: IROS, pp. 229–236 (2019)
2. Rui, L., Minjian, P., Cong, Z., Guyue, Z., Fang, L.: Monocular long-term target following on UAVs. In: CVPR Workshops, pp. 29–37 (2016)

3. Fu, C., Carrio, A., Olivares-Mendez, M.A., Suarez-Fernandez, R., Campoy, P.: Robust real-time vision-based aircraft tracking from unmanned aerial vehicles. In: ICRA, pp. 5441–5446 (2014)
4. Karaduman, M., Çınar, A., Eren, H.: UAV traffic patrolling via road detection and tracking in anonymous aerial video frames. J. Intell. Robot. Syst. **95**(2), 675–690 (2019)
5. Li, Y., Fu, C., Ding, F., Huang, Z., Lu, G.: AutoTrack: towards high-performance visual tracking for UAV with automatic spatio-temporal regularization. In: Proceedings of the IEEE/CVF Conference on Computer Vision and Pattern Recognition, pp. 11 923–11 932 (2020)
6. Li, F., Tian, C., Zuo, W., Zhang, L., Yang, M.-H.: Learning spatial-temporal regularized correlation filters for visual tracking. In: CVPR Conference Proceedings, pp. 4904–4913 (2018)
7. Ma, C., Yang, X., Zhang, C., Yang, M.-H.: Long-term correlation tracking. In: Proceedings of the IEEE Conference on Computer Vision and Pattern Recognition, pp. 5388–5396 (2015)
8. Ding, F., Fu, C., Li, Y., Jin, J., Feng, C.: Automatic failure recovery and re-initialization for online UAV tracking with joint scale and aspect ratio optimization. arXiv preprint arXiv:2008.03915 (2020)
9. Li, F., Yao, Y., Li, P., Zhang, D., Zuo, W., Yang, M.-H.: Integrating boundary and center correlation filters for visual tracking with aspect ratio variation. In: ICCV Workshops, pp. 2001–2009 (2017)
10. Li, Y., Zhu, J., Hoi, S.C., Song, W., Wang, Z., Liu, H.: Robust estimation of similarity transformation for visual object tracking. In: Proceedings of the AAAI Conference on Artificial Intelligence, vol. 33, pp. 8666–8673 (2019)
11. Dai, K., Wang, D., Lu, H., Sun, C., Li, J.: Visual tracking via adaptive spatially-regularized correlation filters. In: Proceedings of the IEEE/CVF Conference on Computer Vision and Pattern Recognition, pp. 4670–4679 (2019)
12. Huang, Z., Fu, C., Li, Y., Lin, F., Lu, P.: Learning aberrance repressed correlation filters for real-time UAV tracking. In: Proceedings of the IEEE International Conference on Computer Vision, pp. 2891–2900 (2019)
13. Fan, H., Ling, H.: Parallel tracking and verifying: a framework for real-time and high accuracy visual tracking. In: Proceedings of the IEEE International Conference on Computer Vision, pp. 5486–5494 (2017)
14. Zhang, Y., Wang, D., Wang, L., Qi, J., Lu, H.: Learning regression and verification networks for long-term visual tracking. arXiv preprint arXiv:1809.04320 (2018)
15. Bertinetto, L., Valmadre, J., Golodetz, S., Miksik, O., Torr, P.H.: Staple: complementary learners for real-time tracking. In: Proceedings of IEEE Conference on Computer Vision and Pattern Recognition (CVPR), pp. 1401–1409 (2016)
16. Henriques, J.F., Caseiro, R., Martins, P., Batista, J.: High-speed tracking with kernelized correlation filters. IEEE Trans. Pattern Anal. Mach. Intell. **37**, 583–596 (2015)
17. Danelljan, M., Hager, G., Shahbaz Khan, F., Felsberg, M.: Learning spatially regularized correlation filters for visual tracking. In: ICCV Conference Proceedings, pp. 4310–4318 (2015)
18. Fu, C., Li, B., Ding, F., Lin, F., Lu, G.: Correlation filter for UAV-based aerial tracking: a review and experimental evaluation. arXiv preprint arXiv:2010.06255 (2020)

19. Mueller, M., Smith, N., Ghanem, B.: A benchmark and simulator for UAV tracking. In: Leibe, B., Matas, J., Sebe, N., Welling, M. (eds.) ECCV 2016. LNCS, vol. 9905, pp. 445–461. Springer, Cham (2016). https://doi.org/10.1007/978-3-319-46448-0_27
20. Danelljan, M., Bhat, G., Shahbaz Khan, F., Felsberg, F.: ECO: efficient convolution operators for tracking. In: CVPR Conference Proceedings, pp. 6931–6939 (2017)
21. Danelljan, M., Häger, G., Khan, F.S., Felsberg, M.: Discriminative scale space tracking. IEEE Trans. Pattern Anal. Mach. Intell. **39**(8), 1561–1575 (2016)

Two-Stage Recognition Algorithm for Untrimmed Converter Steelmaking Flame Video

Yi Chen, Jiyuan Liu, and Huilin Xiong$^{(\boxtimes)}$

Shanghai Key Laboratory of Intelligent Sensing and Recongnition,
Shanghai Jiao Tong University, Shanghai, China
yi_chen@sjtu.edu.cn

Abstract. In the process of converter steelmaking, identification of converter status is the basis of subsequent steelmaking control, directly affecting the cost and quality of steelmaking. Usually, the status of converter can be identified according to the flame of furnace port. In this paper, we propose a two-stage recognition algorithm to identify converter status using furnace flame video. In the first stage, we design a 2D feature extractor based on the shift module, aiming to capture temporal information in real time. In the second stage, an attention-based network is designed to get a more discriminative temporal attention, in which we model the distribution of temporal attention using conditional Variational Auto-Encoder (VAE) and a generative attention loss. We collect the video data and construct the corresponding data set from real steelmaking scene. Experimental results show that our algorithm can meet the accuracy and speed requirements.

Keywords: Converter steelmaking · Feature extraction · Shift module · Attention-based · VAE

1 Introduction

Steel industry is one of the most important industries in the world. It is the material basis for a nation's economy and defense. Converter steelmaking is the principal way of steel production, and about 70% of steel in the world is produced via converters nowadays. In the process of converter steelmaking, the identification of converter status is of great importance and can directly affect the steelmaking cost and quality [12,23]. In practice, four status in converter steelmaking need to be identified, namely drying, normal, slagging, and splashing. With different converter status, flame in the furnace will present different forms [9]. As shown in Fig. 1, the flame in the normal state is usually dark red. In the slagging state, as the name implies, there are some slag particles thrown out from the furnace; In the splashing state, a large amount of slag particles

---

Y. Chen—The first author is a student.

© Springer Nature Switzerland AG 2021
H. Ma et al. (Eds.): PRCV 2021, LNCS 13019, pp. 268–279, 2021.
https://doi.org/10.1007/978-3-030-88004-0_22

are thrown out from the furnace, or in even worse case, more molten steel can be splashed out; In the drying state, the flame is bright and thin. Traditional methods for status identification mostly rely on human experience. Experienced workers can recognize the status of converter with high accuracy, according to their own experience. However, this is also a very timing job. The goal of this paper is to investigate the possibility of using machine to recognize the four status of converter from furnace flame video. Essentially, this is a video recognition task. However, compared to the general video recognition works, the video here is what we called untrimmed video [15].

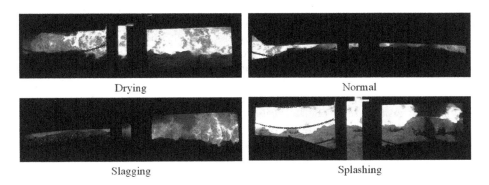

Fig. 1. Four typical status of the converter

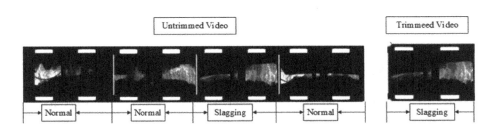

Fig. 2. Difference between trimmed and untrimmed videos with label slagging

Existing video recognition methods are generally based on trimmed video. But for untrimmed video, their performance is not satisfied. As shown in Fig. 2, a trimmed video has only one status, while an untrimmed video contains additional context. For example, in a slagging video, the slagging part may only account for a small part. If we directly sample untrimmed videos for training, the keyframes of slagging are likely to be evaded. The recognition results obtained in this way is difficult to meet the requirements. Fortunately, we can draw inspiration from research on action location [10,11,15,19], and complete the recognition task for untrimmed flame video in two stages. The general idea of action location is: train a feature extractor based on the trimmed data set, and then apply the extractor to extract the segmented features of the untrimmed video. At last, an appropriate location algorithm is designed to complete the location task.

We can apply the idea of action location to the recognition of untrimmed video. But for practical applications, there are still two problems to solve, one is that it's hard to realize real-time performance. We have to rely on trimmed video recognition methods to extract video features in advance, and then perform recognition for untrimmed video. However, most of the current video recognition methods rely on 3D convolution, which is extremely time-consuming and almost impossible to achieve real-time performance. Another problem is that the action location method is still not effective in the direct application and we need to design the corresponding targeted recognition algorithm to optimize.

To address the above two issues, we proposed a two-stage attention-based recognition algorithm for untrimmed flame video. In the first stage, we design a 2D feature extractor, which can dramatically increase speed without losing temporal information. In the second stage, we designed a model based on attention and generation network. The attention part is used to aggregate all features for classification while the generation part is designed for enhancing the discrimination of attention. Extensive experiments are conducted on FireDataSet collected from the real steelmaking scene. Experimental results show that our method can meet the speed and precision requirements.

Main contributions of this paper are listed as follows:

1. For the first time, video recognition method is applied to the converter steelmaking.
2. A faster feature extractor based on 2D convolution is proposed.
3. A method for untrimmed video recognition is proposed, and an attention optimization method based on generated network is presented.

2   Related Works

At present, video based recognition methods have not been applied to converter steelmaking. But some advanced scientific research has provided us with ideas. Follow the pipeline of our algorithm, related works can be divided into two parts: feature extraction and action location.

Feature Extraction. We have to rely on video recognition methods to extract features. It can be divided into two categories. Methods in the first category often adopt 3D CNNs that stack 3D convolutions to jointly model temporal and spatial semantics [2]. I3D [1] networks use two stream CNNs with inflated 3D convolution to achieve the best performance at the time on the Kinetics data set. Non-local Network [17] introduces a special non-local operation to make better use of the long-range temporal dependencies between video frames. Other modifications for the 3D CNNs, including irregular convolution/pooling [14] and the decomposing 3D convolution kernels [8], can also boost the performances of 3D

CNNs. The second type of approach typically adopts the 2D + 1D paradigm, where 2D convolution is applied to each input frame, followed by a 1D module that summarizes the features of each frame. Resnet(2 + 1)D networks [13] are a typical example of this principle. TSN [16] realizes the classification and recognition of long video by learning and aggregating the representation frames sampled from the evenly distributed video. TRN [22] and TSM [4] replace average pool operations with interpretable relational modules and make use of shift modules to better capture information along the temporal dimensions. However, these two methods are generally based on trimmed videos. For the untrimmed flame videos, missing the KeyFrames may lead to unsatisfactory results of recognition.

Action Location. Researches on action location have provided us with inspiration for untrimmed video recognition. Early action location was inspired by target detection methods. A typical example is the SCNN network [11] proposed by Zheng Shou in 2016. He generated candidate regions of video clips through sliding window, and then positioned and selected sliding window using Overlap Loss. Subsequently, fully supervised positioning algorithms such as R-C3D [18], SSN [21], P-GCN [20], etc. are proposed. However, the temporal boundary annotations are expensive to obtain for fully-supervised setting. This motivates efforts in developing models that can be trained with weaker forms of supervision such as video-level labels. Untrimmednets [15] use a classification module to perform action classification and a selection module to detect important temporal segments. Inspired by the class activation mapping, STPN [5] introduces an attention module to learn the weight of segmented feature representation. The method generates the detection by thresholding temporal class activation mapping(T-CAM). Nguyen et al. propose to penalize the discriminative capacity of background and self-guide loss [6], which is also utilized in our classification module. Baifeng Shi et al. analyzed the defects of previous models from the perspective of probability theory, and introduced the generative model [10] to action location, which is also the main inspiration for our model. However, most of these methods require feature extraction in advance. In order to apply it to untrimmed flame video, we also need to solve the problem of real-time feature extraction.

3   Method

In this section, we will introduce how to recognize the untrimmed flame videos. It is mainly divided into two parts. In the first part, we will introduce the feature extraction algorithm. In the second part, an attention-based recognition algorithm is designed based on video-level labels and features. And at the end of the second part, parameter training method of the entire network is presented.

3.1   Feature Extraction

For untrimmed videos, we can't directly sample the videos for training, which would cause missing keyframes. Therefore, we consider to extract features for untrimmed video in advance, and then design the corresponding feature-based recognition module. In this section, we present our method of training the feature extractor.

Inspired by the TSM network [4], we found that video recognition can be achieved by making use of shift modules, speeding up computation while preserving temporal information. As shown on the right of Fig. 3, by moving part of the channel up or down at random, the temporal information between frames is exchanged and can be learned in back propagation. We choose ResNet as the backbone of the network and add temporal shift module to the convolution branch. To train a 2D feature extractor, we constructed a trimmed flame video data set, and obtained 2D feature extractor through sampling training. Figure 3 show a schematic diagram of the network structure and a flowchart of feature extraction.

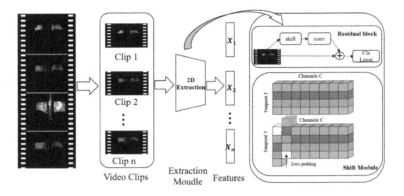

Fig. 3. Feature extraction flow chart and 2D Extraction structure

The number of floating-point operations for a 2D convolution is: $2 \times C_{in} \times K^2 \times C_{out} \times W \times H$. The 3D convolution has one more dimension, and the output data has one more time dimension, so the total number of operations is: $2 \times C_{in} \times K^3 \times C_{out} \times W \times H \times T$. Every time a convolution is performed, the 3D convolution time is K * T times that of the 2D convolution. When extracting feature expressions for T frames, 3D convolution takes K times as long as 2D convolution. As a result, extraction time is greatly reduced in 2D convolution way.

Finally, we apply the 2D feature extractor to extract a feature expression every eight(or other fixed number) frames for the given untrimmed flame video $V = \{v_t\}$.

3.2 Recognition for Untrimmed Flame Videos

Now we are provided with a training set of videos and video-level labels $y \in \{0, \cdots C\}$, where C denotes the number of possible flame type and 0 indicates the background. Through Sect. 3.1, We can write each training video as a tuple of feature vectors and video-level label:

$$(\mathbf{x}_t, y), \quad \mathbf{x}_t \in R^d, \quad y \in \{0, \ldots, C\} \tag{1}$$

Inspired by the weakly-supervised action location methods, we learn the attention $\lambda = \{\lambda_t\}$ directly from pre-extracted frame-level features and make video classification subsequently, where $\lambda_t \in [0, 1]$ is the attention of frame t.

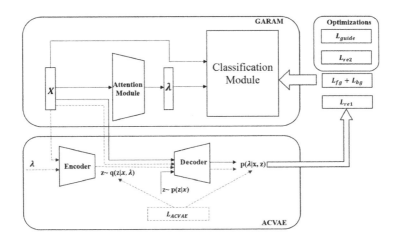

Fig. 4. Recognition for untrimmed flame video. The solid line is the data stream of GARAM, and the dashed line is the data stream of ACVAE

Attention-Based Framework. In attention-based video recognition problem, the target is to learn the frame attention $\boldsymbol{\lambda}$ for a better long untrimmed video recognition. The whole schematic diagram of our method called **GARAM** is shown in Fig. 4, and We will elaborate it in the following sections.

we equate attention-based action location with solving the maximum a posteriori problem:

$$\max_{\lambda \in [0,1]} \log p(y \mid \mathbf{X}, \boldsymbol{\lambda}) \tag{2}$$

This part $\log p(y \mid \mathbf{X}, \boldsymbol{\lambda})$ emphasizes learning discriminative $\boldsymbol{\lambda}$ to complete video classification, which is similar to previous action location works, such as TCAM [5] and Background Modeling [6]. Specifically, we utilize attention $\boldsymbol{\lambda}$ and $1 - \boldsymbol{\lambda}$ to perform temporal average pooling over all frame-level features. A video-level foreground $x_{fg} \in Rd$ and background feature $x_{ig} \in Rd$ is generated:

$$\mathbf{x}_{fg} = \frac{\sum_{t=1}^{T} \lambda_t \mathbf{x}_t}{\sum_{t=1}^{T} \lambda_t} \tag{3}$$

$$\mathbf{x}_{bg} = \frac{\sum_{t=1}^{T}(1-\lambda_t)\mathbf{x}_t}{\sum_{t=1}^{T}(1-\lambda_t)} \tag{4}$$

Encouraging the discriminative capability of the foreground features $\mathbf{x}_{fg}$ while punishing the capability of the background features $\mathbf{x}_{bg}$, we feed the pooled feature to fully-connected softmax layer, parameterized by $\mathbf{w}_c \in \mathrm{R}_d$ for class c:

$$p(c \mid \mathbf{x}_{fg}) = \frac{e^{w_c \cdot \mathbf{x}_{\mathrm{fg}}}}{\sum_{i=0}^{C} e^{w_i \cdot \mathbf{x}_{\mathrm{fg}}}} \tag{5}$$

As a result, the discriminative model part $\log p(y \mid \mathbf{X}, \boldsymbol{\lambda})$ can be solved by minimizing the following discriminative loss:

$$\mathcal{L}_d = \mathcal{L}_{fg} + \alpha \cdot \mathcal{L}_{bg} = -\log p\left(y \mid \mathbf{x}_{fg}\right) - \alpha \cdot \log p\left(0 \mid \mathbf{x}_{bg}\right) \tag{6}$$

Self-guided Attention Loss. In the attention framework, the value of attention is particularly important to distinguish between foreground and background. We added a self-guided loss to fine-tune attention. The attention λ_t measures the generic action of a frame from the bottom up while the temporal class activation maps (TCAM) [5] act as the top-down, class-aware information. Specifically, the top-down attention maps from class y computed from TCAM is given by:

$$\hat{\lambda}_t^{fg} = G\left(\sigma_s\right) * \frac{\exp^{\mathbf{w}_y^T \mathbf{x}_t}}{\sum_{c=0}^{C} \exp^{\mathbf{w}_c^T \mathbf{x}_t}} \tag{7}$$

$$\hat{\lambda}_t^{bg} = G\left(\sigma_s\right) * \frac{\exp^{\mathbf{w}_y^T \mathbf{x}_t}}{\sum_{c=0}^{C} \exp^{\mathbf{w}_c^T \mathbf{x}_t}} \tag{8}$$

where $G\left(\sigma_s\right)$ refers to a Gaussian filter used to smooth the class-specific, top-down attention signals. $\hat{\lambda}_t^{fg}$ and $\hat{\lambda}_t^{bg}$ are foreground and background top-down attention maps, respectively. Hence the $\mathcal{L}_{\mathrm{guide}}$ can be formulated as:

$$\mathcal{L}_{\mathrm{guide}} = \frac{1}{T} \sum_{t=1}^{T} \left|\lambda_t - \hat{\lambda}_t^{fg}\right| + \left|\lambda_t - \hat{\lambda}_t^{bg}\right| \tag{9}$$

Generative Attention Loss. After self-guide attention loss, the attention $\boldsymbol{\lambda}$ has already been optimized once. However, a more discriminative and robust attention is expected. We train on fixed length videos, but we expect that it work for any length of video for testing. Here we propose to utilize generative networks to enhance the quality of attention. As shown in Fig. 4, $\boldsymbol{\lambda}$ is generated from $\mathbf{X}$. The network optimizes $\boldsymbol{\lambda}$ through $\hat{\boldsymbol{\lambda}}^{fg}$ while back propagation. Here $\boldsymbol{\lambda}$ is specific to each sample $\mathbf{X}$, that is, one sample in the distribution $p(\boldsymbol{\lambda} \mid \mathbf{X})$. We expect to model the distribution $p(\boldsymbol{\lambda} \mid \mathbf{X})$ before back propagation and select a more robust and Less noisy $\boldsymbol{\lambda}_{recon}$ for optimization.

We utilize a Conditional Variational AutoEncoder (CVAE) [3] to model the distribution $p(\boldsymbol{\lambda} \mid \mathbf{X})$ called **ACVAE**. By introducing the continuous variable z, the original distribution can be written as:

$$p(\boldsymbol{\lambda} \mid \mathbf{x}) = \int_z p(z \mid x)p(\boldsymbol{\lambda} \mid z, x)dz \tag{10}$$

where the distribution of z is known, e.g. $z \sim N(0, 1), \lambda \mid z \sim N(\mu(z), \sigma(z))$. We hope to solve the $(\mu(z), \sigma(z))$ and generate $p(\lambda \mid z, x)$ by designing the decoder. Then through the above formula, the generation of the distribution of $\lambda \mid x$ is completed. The core idea behind is to sample values of z that are likely to produce $\lambda \mid x$, which means that we need an approximation $q(\lambda \mid z, x)$ to the intractable posterior $p(\lambda \mid z, x)$. ACVAE integrates two parts of encoder and decoder, and optimized by maximizing the following formula:

$$
\begin{aligned}
\mathcal{L}_{ACVAE} &= - \int_z q(z \mid x, \lambda) \log p(\lambda \mid z, x)dz \\
&\quad + \beta \cdot KL(q(\mathbf{z} \mid \mathbf{x}, \boldsymbol{\lambda}) \| p(\mathbf{z} \mid \mathbf{x})) \\
&\simeq - \frac{1}{L} \sum_{l=1}^{L} \log p(x_t \mid \lambda_t, z_t^{(l)}) \\
&\quad + \beta \cdot KL(q(\mathbf{z} \mid \mathbf{x}, \boldsymbol{\lambda}) \| p(\mathbf{z} \mid \mathbf{x}))
\end{aligned} \tag{11}
$$

In the GARAM model, we fix the ACVAE model and minimize the generative attention loss:

$$
\begin{aligned}
\mathcal{L}_{re1} &= - \sum_{t=1}^{T} \log \left\{ \int_z p(z \mid x)p(\hat{\lambda}^{fg} \mid z, x)dz \right\} \\
&\simeq \sum_{t=1}^{T} \log \left\{ -\frac{1}{L} \sum_{l=1}^{L} \log p(\hat{\lambda}_t^{fg} \mid x_t, \mathbf{z}_t^{(l)}) \right\}
\end{aligned} \tag{12}
$$

where $\mathbf{z}$ is sampled from the prior $p(\mathbf{z} \mid \mathbf{x})$. In our experiments, with L set to 1 and the output of decoder defined by $f(\mathbf{x}, \mathbf{z})$, (11) can be written as:

$$\mathcal{L}_{re1} \simeq \sum_{t=1}^{T} \log \left\{ \log p(\hat{\lambda}_t^{fg} \mid \mathbf{x}, \mathbf{z}^{(l)}) \right\} \propto \sum_{t=1}^{T} \left\| \hat{\boldsymbol{\lambda}}^{fg} - f(\mathbf{x}, \mathbf{z}) \right\|^2 \tag{13}$$

In our model, due to the lack of a true $\boldsymbol{\lambda}$ value, we cannot directly and independently train the ACVAE model. Therefore, we train GARAM and ACVAE alternately. We hope to use the real $\mathbf{X}$ and the $\boldsymbol{\lambda}$ given by GARAM to train a more robust and less noisy λ_{recon} generator ACVAE. And by adding Eq. (13) to loss, GARAM is optimized in turn.

Generative Representation Loss. DGAM [10] proposes that the representation of frames have to be accurately predicted from the attention $\boldsymbol{\lambda}$. It is called Generative Attention Modeling in DGAM. In order to distinguish it from our work, we call it Generative Representation Model according to the problem it solves.

Similar to the previous part, we obtain $p(\mathbf{x} \mid \boldsymbol{\lambda})$ by optimizing RCVAE:

$$
\begin{aligned}
\mathcal{L}_{RCVAE} \simeq -\frac{1}{L} \sum_{l=1}^{L} \log p(x_t \mid \lambda_t, \mathbf{z}_t^{(l)}) \\
+ \beta \cdot KL(q(\mathbf{z} \mid \mathbf{x}, \boldsymbol{\lambda}) \| p(\mathbf{z} \mid \boldsymbol{\lambda}))
\end{aligned}
\tag{14}
$$

Fix RVCAE when training GARAM and mimimize:

$$
\mathcal{L}_{re2} = -\sum_{t=1}^{T} \log p(x_t \mid \lambda_t, z_t) \propto \sum_{t=1}^{T} \|x_t - f(x_t, z_t)\|^2
\tag{15}
$$

At last, we optimize the whole framework by performing the following three steps alternately.

1. Traing GARAM model with loss:

$$
\mathcal{L} = \mathcal{L}_d + \gamma_1 \mathcal{L}_{re1} + \gamma_2 \mathcal{L}_{re2} + \gamma_3 \mathcal{L}_{\text{guide}}
\tag{16}
$$

where γ_1, γ_2 denote the hyper-parameters.
2. Update ACVAE with loss $\mathcal{L}_{ACVAE}$.
3. Update RCVAE with loss $\mathcal{L}_{RCVAE}$.

4   Experiments

4.1   Datasets

We conduct experiments on the video data set collected from real steelmaking scene. It has two version. Version 1(v1) is a trimmed data set made for training feature extractor. Version 2(v2) is an untrimmed data set made for recognition task. Table 1 shows the details of our FireDataSet.

Table 1. Details of our FireDataSet

	Version 1(v1)	Version 2(v2)
Categories	4	4
Type	Trimmed	Untrimmed
Total number	11435	2917
Video length (s)	1.5	10
fps	25	25
Establish for	Feature extractor	Recognition

4.2   Implemented Details

In the first stage, based on V1, We trained 3D feature extractor ResNet3d and 2D feature extractor based on ResNet2d and shift module respectively. And the two feature extractors are used to extract a feature expression every 8 frames from the video of V2. In the second stage, we set $\alpha = 1$ in Eq. 6 while $\gamma_1, \gamma_2, \gamma_3$ in Eq. 16 is set differently for best accuracy. The exact values are given in the Table 2. We train the ACVAE and GARAM alternately and use Adam optimizer with learning rate 10^{-3}. All the experiments is conducted on NVIDIA GEFORCE RTX 3090 and implemented with PyTorch [7]. The specific experimental data are summarized in Table 2.

4.3   Data Analysis

Table 2. Feature extraction speed comparison

Extraction module	Time (s)	Loss	Accuracy (%)
ResNet18(3D)	22.70	$\mathcal{L}_d + 0.3\mathcal{L}_{re2} + 0.3\mathcal{L}_{\text{guide}}$	85.9
ResNet18(3D)	22.70	$\mathcal{L}_d + 5\mathcal{L}_{re1} + 0.3\mathcal{L}_{re2} + 0.3\mathcal{L}_{\text{guide}}$	**87.3**
R182D+shift	7.17	$\mathcal{L}_d + 0.1\mathcal{L}_{re2} + 0.1\mathcal{L}_{\text{guide}}$	86.6
R182D+shift	7.17	$\mathcal{L}_d + \mathcal{L}_{re1} + 0.1\mathcal{L}_{re2} + 0.1\mathcal{L}_{\text{guide}}$	**88.1**

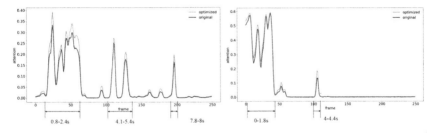

Fig. 5. The discrimination of attention under different loss. The optimized line is the result with $\mathcal{L}_{re1}$, while the original is the result without $\mathcal{L}_{re1}$. Ground truth is marked below.

Accuracy and Speed Comparison. The length of sample videos is 10 s. The experimental results show that in the case of using the same loss optimization, our feature extraction method can reduce the test time to 7.17 s while maintaining the accuracy, which is of great importance to practical application. In the case of using the same feature extraction method, our optimization for the attention part can increase accuracy by 1.5%. In all, compared with the two-stage recognition algorithm based on 3D convolution features, our method can not only improve the accuracy, but also greatly shorten the test time.

Impact of $\mathcal{L}_{rel}$ to Attention. We choose two untrimmed slagging videos for display. Figure 5 shows the attention values under different loss of the two videos. With $\mathcal{L}_{rel}$ added to the model, the key frame is encouraged with a larger value while the background is suppressed with a small value. It means that $\mathcal{L}_{rel}$ plays a positive role, and it indeed enhances the discrimination of attention.

5 Conclusion

For the first time, we successfully applied a video recognition algorithm to converter status identification. Inspired by the pipeline line of action location, we have to extract video feature in advance. It is 3D convolution feature, which is extremely time-consuming in practical applications. By using temporal shift module, we design a fast 2D feature extraction method without losing temporal information. In this way, we can achieve not only real-time speed but also promote subsequent recognition.

In the part of the recognition algorithm, to enhance the robustness and discrimination of λ, we propose to utilize a generative model to model the distribution of $\log p(\lambda \mid \mathbf{X})$, and a generative attention loss is designed. We train and test on the flame video data sets collected from the real steelmaking scene. It is worth mentioning that our work can fully meet accuracy and speed requirements well.

References

1. Carreira, J., Zisserman, A.: Quo Vadis, action recognition? A new model and the kinetics dataset. In: Proceedings of the IEEE Conference on Computer Vision and Pattern Recognition, pp. 6299–6308 (2017)
2. Ji, S., Xu, W., Yang, M., Yu, K.: 3D convolutional neural networks for human action recognition. IEEE Trans. Pattern Anal. Mach. Intell. **35**(1), 221–231 (2012)
3. Kingma, D.P., Welling, M.: Auto-encoding variational Bayes. arXiv preprint arXiv:1312.6114 (2013)
4. Lin, J., Gan, C., Han, S.: TSM: temporal shift module for efficient video understanding. In: Proceedings of the IEEE/CVF International Conference on Computer Vision, pp. 7083–7093 (2019)
5. Nguyen, P., Liu, T., Prasad, G., Han, B.: Weakly supervised action localization by sparse temporal pooling network. In: Proceedings of the IEEE Conference on Computer Vision and Pattern Recognition, pp. 6752–6761 (2018)
6. Nguyen, P.X., Ramanan, D., Fowlkes, C.C.: Weakly-supervised action localization with background modeling. In: Proceedings of the IEEE/CVF International Conference on Computer Vision, pp. 5502–5511 (2019)
7. Paszke, A., et al.: PyTorch: an imperative style, high-performance deep learning library. arXiv preprint arXiv:1912.01703 (2019)

8. Qiu, Z., Yao, T., Mei, T.: Learning spatio-temporal representation with pseudo-3D residual networks. In: Proceedings of the IEEE International Conference on Computer Vision, pp. 5533–5541 (2017)
9. Seetharaman, S., Mclean, A., Guthrie, R., Sridhar, S.: Treatise on process metallurgy (2013)
10. Shi, B., Dai, Q., Mu, Y., Wang, J.: Weakly-supervised action localization by generative attention modeling. In: Proceedings of the IEEE/CVF Conference on Computer Vision and Pattern Recognition, pp. 1009–1019 (2020)
11. Shou, Z., Wang, D., Chang, S.F.: Temporal action localization in untrimmed videos via multi-stage CNNs. In: Proceedings of the IEEE Conference on Computer Vision and Pattern Recognition, pp. 1049–1058 (2016)
12. Terpak, J., Laciak, M., Kacur, J., Durdan, M., Trefa, G.: Endpoint prediction of basic oxygen furnace steelmaking based on gradient of relative decarburization rate. In: 2020 21th International Carpathian Control Conference (ICCC) (2020)
13. Tran, D., Wang, H., Torresani, L., Ray, J., LeCun, Y., Paluri, M.: A closer look at spatiotemporal convolutions for action recognition. In: Proceedings of the IEEE conference on Computer Vision and Pattern Recognition, pp. 6450–6459 (2018)
14. Wang, L., Qiao, Y., Tang, X.: Action recognition with trajectory-pooled deep-convolutional descriptors. In: Proceedings of the IEEE Conference on Computer Vision and Pattern Recognition, pp. 4305–4314 (2015)
15. Wang, L., Xiong, Y., Lin, D., Van Gool, L.: UntrimmedNets for weakly supervised action recognition and detection. In: Proceedings of the IEEE Conference on Computer Vision and Pattern Recognition, pp. 4325–4334 (2017)
16. Wang, L., et al.: Temporal segment networks: towards good practices for deep action recognition. In: Leibe, B., Matas, J., Sebe, N., Welling, M. (eds.) ECCV 2016. LNCS, vol. 9912, pp. 20–36. Springer, Cham (2016). https://doi.org/10.1007/978-3-319-46484-8_2
17. Wang, X., Girshick, R., Gupta, A., He, K.: Non-local neural networks. In: Proceedings of the IEEE Conference on Computer Vision and Pattern Recognition, pp. 7794–7803 (2018)
18. Xu, H., Das, A., Saenko, K.: R-C3D: region convolutional 3D network for temporal activity detection. In: Proceedings of the IEEE International Conference on Computer Vision, pp. 5783–5792 (2017)
19. Zeng, R., Huang, W., Gan, C., Tan, M., Huang, J.: Graph convolutional networks for temporal action localization. In: 2019 IEEE/CVF International Conference on Computer Vision (ICCV) (2019)
20. Zeng, R., et al.: Graph convolutional networks for temporal action localization. In: Proceedings of the IEEE/CVF International Conference on Computer Vision, pp. 7094–7103 (2019)
21. Zhao, Y., Xiong, Y., Wang, L., Wu, Z., Tang, X., Lin, D.: Temporal action detection with structured segment networks. In: Proceedings of the IEEE International Conference on Computer Vision, pp. 2914–2923 (2017)
22. Zhou, B., Andonian, A., Oliva, A., Torralba, A.: Temporal relational reasoning in videos. In: Proceedings of the European Conference on Computer Vision (ECCV), pp. 803–818 (2018)
23. Zhou, M., Zhao, Q., Chen, Y., Shao, Y.: Carbon content measurement of BOF by radiation spectrum based on support vector machine regression. Spectrosc. Spectr. Anal. **038**(006), 1804–1808 (2018)

Scale-Aware Multi-branch Decoder for Salient Object Detection

Yang Lin[1], Huajun Zhou[1], Xiaohua Xie[1,2,3]($\boxtimes$), and Jianhuang Lai[1,2,3]

[1] School of Computer Science and Engineering, Sun Yat-sen University,
Guangzhou, China
{liny239,zhouhj26}@mail2.sysu.edu.cn,
{xiexiaoh6,stsljh}@mail.sysu.edu.cn
[2] Guangdong Key Laboratory of Information Security Technology,
Guangzhou, China
[3] Key Laboratory of Machine Intelligence and Advanced Computing,
Ministry of Education, Guangzhou, China

Abstract. Recently, Salient Object Detection (SOD) has been witnessed remarkable advancements owing to the introduction of Convolution Neural Networks (CNNs). However, when faced with complex situations, such as salient objects of different sizes appearing in the same image, the detection results are far away from satisfaction. In this paper, we propose a scale-aware multi-branch decoder to detect salient objects at different scales. Specifically, different branches of the decoder are assigned to extract distinctive features and predict maps for objects of different sizes. Meanwhile, we design two flows between branches for feature aggregation. Global Distinction Flow (GDF) allows each branch to focus on the semantic information of specific sizes object, while Hierarchical Integration Flow (HIF) combines different information on each branch and forms the final detection result. In addition, we generate new masks by decomposing the ground true saliency map and use them in different branches for supervision. Comparing with existed state-of-the-art methods, experiments conducted on five datasets demonstrate the effectiveness of our method.

Keywords: Salient object detection · Multi-branch · Feature aggregation

1 Introduction

The human visual system can distinguish the most important part from visual scene then focus on it. In order to simulate this mechanism, salient object detection was developed to find the critical part of an image. The main challenge of this task is to find the salient objects and maintain their contour information. As a basic step of visual processing, it is often applied in other fields of computer vision, such as visual tracking [17], instance segmentation [22], and so on.

© Springer Nature Switzerland AG 2021
H. Ma et al. (Eds.): PRCV 2021, LNCS 13019, pp. 280–292, 2021.
https://doi.org/10.1007/978-3-030-88004-0_23

In recent years, Convolutional Neural Networks (CNNs) have shown impressive capability in extracting semantic information from images, and they have been widely used in various computer vision tasks. As one of the most representative networks, U-net [8] has achieved remarkable results in solving the SOD problem due to the construction of enriched feature maps. Most of the current methods [9,16,20,26] are improved on the basis of U-net. Some researches [16,18] use the local information to refine high-level features in a bidirectional or recursive way. Besides, recent works [20,26] pay attention to object boundary by proposing boundary-aware objective functions or using contour as a supervised signal. However, when there are multiple objects with large scale differences in the image, small scale objects are easy to be ignored. This is due to the difficulty of handling objects of different scales simultaneously in a single U-shaped network. On the one hand, in order to extract the semantic information of large-scale objects, deep network is needed to obtain a large receptive field. On the other hand, multiple pooling operations in deep network, which are used to reduce computational complexity, can easily lose the information of small-scale objects. Therefore, it is necessary to find a method that can detect objects at different scales simultaneously.

To address the above problems, we propose a novel scale-aware multi-branch decoder, which includes multiple branches for detecting objects of different scales in the image. Specifically, the number of blocks in each branch is different. The deeper branch is used to detect objects of larger size to prevent the problem of information loss. In the training stage, we decompose the ground true saliency map into multiple partial masks, based on the scales of the salient objects. By using different generated masks for supervision, each branch can focus on objects of specific scales. Between different branches, we designed two flows for feature fusion. In the detection process of each branch, we propose a Global Distinction Flow (GDF) to reduce the interference from the semantic information of its non-target objects. After preliminary processing of the features from the encoder, each branch subtracts its output from the features of the corresponding block at the previous branch, aiming at paying attention to other smaller objects. To combine the information on different branches and form the final detection result, we propose a Hierarchical Integration Flow (HIF). Each branch receives the final feature map from the next branch and adds it to its own feature map. By gradually accumulating the semantic information of each branch, salient objects of different sizes can be detected finally.

The main contributions of our work are summarized as follows:

1. We propose a novel scale-aware multi-branch decoder for saliency detection. Using different masks generated by decomposing the saliency map for supervision, multiple branches in the decoder tend to detect objects of different scales.
2. We design two flows between branches for feature aggregation. Global Distinction Flow (GDF) is used for each branch to focus on different objects by ignoring the information from other branches, while Hierarchical Integration Flow (HIF) is used to combine all features in branches to form the final detection result.

3. We conduct extensive experiments on five datasets to demonstrate the effectiveness of the proposed model.

2   Related Work

Traditional Methods: In the past years, a large number of salient object detection methods have been proposed. Early methods are mainly based on hand-craft features, such as image contrast [1], color [4] and so on. More details about traditional methods are introduced in [13]. Here we mainly discuss the saliency detection methods based on deep learning.

Patch-Wise Deep Methods: Recently, as Convolutional neural networks (CNNs) have achieved competitive performance on recognition tasks, early works utilized them to determine whether an image region is salient. Li et al. [12] extracted features from three image regions of different scales in generic CNNs, and obtained final results through fully connected networks. In [6], Zhao et al. took both global context and local context integrated by multi-context deep CNNs for saliency detection. Although these models surpass the performance of traditional methods, taking image patches as the inputs tends to lose spatial information and often produces coarse results. Besides, the use of multiple fully connected layers makes detection time-consuming.

FCN-based Methods and Feature Aggregation: Compared with the patch-wise deep method, researchers proposed more effective models based on FCN [7]. Li et al. [10] proposed a unified framework that combined a pixel-level fully convolutional stream and a segment-level spatial pooling stream. Hariharan et al. [11] attempted to integrate the features from multiple middle layers. U-net [8] applies skip-connections between every two corresponding layers of encoder and decoder, to construct a contracting path to capture context and a symmetric expanding path that enables precise localization. Many methods follow this structure due to its remarkable performance. Zhang et al. [16] integrated multilevel features into multiple resolutions, and fused them to predict the saliency map in a recursive manner. In [18], Zhang et al. extracted multi-scale context-aware features and passed messages between them through a gated bi-directional module. Different from the above methods, this paper attempts to fuse the features from different branches in the decoder to detect objects of different sizes.

Use of Contour Information: Since capturing precise boundary is one of the main challenges for saliency detection, numerous methods begin to utilize contour information. Liu et al. [9] proposed a deep hierarchical saliency network, which first generated coarse map from global view and integrated local contexts to hierarchically improve image details. In [19], Chen et al. developed a Contour loss to pay more attention to boundary pixels and presented the hierarchical global attention module to attend to global contexts. Qin et al. [20] used a residual refine module after generating coarse saliency map, and proposed a hybrid loss including BCE, structural similarity index (SSIM) and Intersection over Union (IOU). Zhou et al. [31] proposed an Interactive Two-Stream Decoder consists of a saliency branch and a contour branch, which interactively transmitting the features to explore their correlation.

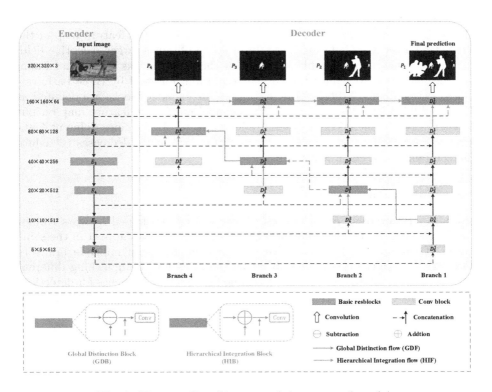

Fig. 1. The overall architecture of the proposed model.

3   Method

As shown in Fig. 1, the proposed model follows the encoder-decoder structure. The encoder is used to extract multi-level features of the input image. The decoder includes multiple branches with different depths to detect objects of different sizes. Each branch of the decoder and part of the encoder form a U-shaped network. Meanwhile, there are two flows between branches: Global Distinction Flow (GDF) and Hierarchical Integration Flow (HIF). The GDF allows different branches to ignore the semantic information on other branches then focus on objects of specific sizes. The HIF gradually combines the semantic information from each branch, and finally forms a complete detection result. Using the mask generated by decomposing the saliency map for supervision, each branch is assigned to detect objects in a specific size range.

3.1 Encoder

Similar to other popular saliency detection methods, we use ResNet-50 pre-trained on ImageNet as our backbone network. Our encoder has a total of six blocks, of which the first four blocks are obtained from the feature map of the ResNet-50 after a convolution layer. In order to obtain a larger receptive field, the encoder has two more blocks. Each of these two blocks uses a maximum pooling operation to reduce the spatial resolution, followed by three basic res-blocks containing 512 filters. The outputs of six encoder blocks are denoted as E_i (i = 1,2,3,4,5,6). As we can see, the channel number of feature map output by each block in the encoder gradually increases to obtain more different high-level semantic information, while the resolution gradually decreases to reduce the computational complexity.

3.2 Multi-branch Decoder

Multiple Branches. As mentioned above, a single U-shaped network cannot handle the situation where multiple objects of different sizes appear in the same image. Therefore, we hope to solve this problem by using multiple branches decoder. We use the size of the object as the basis for distinguishing different objects. Objects of different sizes are assigned to different branches for detection, which means that each branch only focuses on objects with a specific range of sizes. Besides, deep branches are used to detect large objects and shallow ones are used to detect small objects to prevent the problem of information loss.

In order to detect objects of different sizes, we have designed a total of four branches in the decoder. The depth of each branch varies from 3 to 6, which means that each branch contains a different number of blocks. Each convolution block of the branch includes a 3×3 convolution, a batch normalization, and a ReLU activation function. For convenience, we define the function α as:

$$\alpha(D) = upsample(relu(bn(conv(D)))) \tag{1}$$

where bn and $conv$ are the operators of batch normalization and convolution, while D denotes the input feature map. So the process of each convolution block can be formulated as:

$$D_i^j = \alpha(concat(D_{i+1}^j, E_{i+1})) \tag{2}$$

where $concat$ is the operator of concatenation and D_i^j denotes the output of the block i in Branch j. In particular, D_i^j is none for $i + j > 6$. In other words, except for the first block in each branch whose input comes from the feature of the corresponding encoder block, the input of the other convolution block is the concatenation of two parts: the output from the previous block in the same branch and the feature of corresponding encoder block. Except for the last block of each branch whose upsample factor is set as 1, the other blocks will increase the spatial resolution by an upsampling operation after the ReLU function. Finally, the output of each branch is fed to a 3×3 convolution, followed by linear upsampling and a sigmoid function to generate a saliency map, which can be formulated as:

$$P_j = sigmoid(upsample(conv(D_0^j))) \qquad (3)$$

where P_j denotes the predicted map of Branch j for $j = 1, 2, 3, 4$.

Global Distinction Flow (GDF). The GDF starts from Branch 1 and passes information to the shallow one. As shown in Fig. 1, GDF contains three blocks, named as Global Distinction Block (GDB). Each GDB subtracts the features of corresponding block in the previous branch from the output of the previous block. Then perform the same operation as the convolution block. The process of each GDB can be formulated as:

$$D_i^j = \alpha(concat((D_{i+1}^j - D_{i+1}^{j-1}), E_{i+1})) \qquad (4)$$

for $(i, j) = (1, 4), (2, 3), (3, 2)$. In this way, each branch can ignore the semantic information of the previous branch, which means ignoring the relatively large salient objects and focus on detecting other smaller objects.

Hierarchical Integration Flow (HIF). Contrary to the direction of the GDF, the HIF starts from Branch 4 and passes information to the deep one. Note that the output of the last block in each branch has not been upsampled, so its spatial resolution is the same as the output of the previous block. As shown in Fig. 1, HIF contains three blocks, named as Hierarchical Integration Block (HIB). Each HIB adds the features of corresponding block in the previous branch to the output of the previous block. Then perform the same operation as the convolution block. The process of each HIB can be formulated as:

$$D_0^j = \alpha(concat((D_1^j + D_0^{j+1}), E_{i+1})) \qquad (5)$$

for $j = 1, 2, 3$. In this way, the detection information of each branch is gradually accumulated. The information of objects with different sizes is compared, and the salient objects of each size can be detected, thus forming a complete detection result.

3.3   Supervision

Masks Generation. To make each branch focus on detecting objects of different sizes, the supervision information needs to be divided. Given the ground truth saliency map of an image, we generate 4 different masks whose initial value is all zero through the following steps. First, we binarize the saliency map to 0

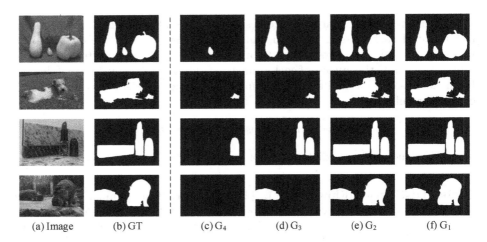

| (a) Image | (b) GT | (c) G_4 | (d) G_3 | (e) G_2 | (f) G_1 |

Fig. 2. Different examples of our generated masks. The first two columns are input images and ground true, while last four columns represent generated multi-scale masks.

and 1 according to its pixel value. The threshold in our experiment is set to 128. Then, we find out all the connected components in the map and calculate the scale of each connected component. The connected components with a scale less than 50 are considered as noise, so they are ignored. Next, we set the maximum scale for the 4 masks as 8000, 15000, 30000, and ∞ (infinity) respectively. All connected components smaller than the maximum scale value are added to the corresponding mask. The maximum scale of the fourth mask is set to ∞, which means that the fourth mask is exactly the same as the ground truth map. The four generated masks are denoted as G_4, G_3, G_2, and G_1 respectively. Finally, we use these 4 masks to supervise the saliency maps generated by the 4 branches. Examples of generated masks are shown in Fig. 2.

Loss Function. We adopt an effective hybrid loss function proposed in BASNet [20], which is a combination of BCE, SSIM, and IoU loss:

$$L(P_j, G_j) = L_{BCE}(P_j, G_j) + L_{SSIM}(P_j, G_j) + L_{IoU}(P_j, G_j) \tag{6}$$

where P_j and G_j denote the predicted map and generated mask respectively. The three loss are defined as:

$$L_{BCE} = -(G_j log P_j + (1 - G_j) log(1 - P_j)) \tag{7}$$

$$L_{SSIM} = 1 - \frac{(2\mu_{P_j}\mu_{G_j} + C_1)(2\sigma_{P_jG_j} + C_2)}{(\mu_{P_j}^2 + \mu_{G_j}^2 + C_1)(\sigma_{P_j}^2 + \sigma_{G_j}^2 + C_2)} \tag{8}$$

$$L_{IoU} = 1 - \frac{TP}{TN + TP + FP} \tag{9}$$

where μ_{P_j}, μ_{G_j} and σ_{P_j}, σ_{G_j} are the mean and standard deviations of P_j and G_j respectively, while $\sigma_{P_j G_j}$ is their covariance. TP, TN and FP represent true-positive, true-negative and false-positive, respectively.

Table 1. Comparison of the proposed method and other 9 state-of-the-art methods on 5 benchmark datasets. Evaluation Metrics include F-measure, S-measure, E-measure (larger is better) and MAE (smaller is better). Bold, Italic and Bolditalic indicate the best, second best and third best results respectively. We also compare the number of Parameter (# Param.) and Frame Per Second (FPS) for each method.

Datasets & Metrics		Ours	SCRN [25]	EGNet [26]	CPD [21]	BASNet [20]	BANet [30]	ICNet [27]	DGRL [28]	PiCANet [29]	SRM [15]
DUTS-TE [14]	F_{max}	***0.885***	*0.888*	**0.889**	0.865	0.859	0.872	0.855	0.828	0.860	0.826
	F_{avg}	**0.827**	***0.809***	*0.815*	0.805	0.791	*0.815*	0.767	0.794	0.759	0.753
	F_β^w	**0.836**	0.803	*0.816*	0.795	0.803	***0.811***	0.762	0.774	0.755	0.722
	E_m	**0.911**	0.901	*0.907*	***0.904***	0.884	*0.907*	0.880	0.899	0.873	0.867
	S_m	***0.880***	*0.885*	**0.887**	0.869	0.866	0.879	0.865	0.842	0.869	0.836
	MAE	**0.036**	*0.040*	*0.039*	0.043	0.048	***0.040***	0.048	0.050	0.051	0.059
DUT-O [3]	F_{max}	*0.813*	*0.811*	**0.815**	0.797	0.805	0.803	*0.813*	0.779	0.803	0.769
	F_{avg}	**0.770**	0.746	*0.756*	***0.747***	*0.756*	0.746	0.739	0.709	0.717	0.707
	F_β^w	*0.749*	0.720	**0.738**	0.719	**0.751**	0.736	0.730	0.697	0.695	0.658
	E_m	**0.881**	0.869	*0.874*	***0.873***	0.869	0.865	0.859	0.850	0.848	0.843
	S_m	0.833	*0.837*	**0.841**	0.825	***0.836***	0.832	*0.837*	0.810	0.832	0.798
	MAE	**0.051**	*0.056*	*0.053*	***0.056***	*0.056*	0.059	0.061	0.063	0.065	0.069
HKU-IS [12]	F_{max}	*0.932*	**0.935**	**0.935**	0.925	***0.930***	***0.930***	0.925	0.914	0.919	0.906
	F_{avg}	**0.908**	0.897	*0.901*	0.891	0.898	***0.899***	0.880	0.882	0.870	0.873
	F_β^w	**0.897**	0.878	*0.887*	0.876	*0.890*	***0.887***	0.858	0.865	0.842	0.835
	E_m	0.953	*0.954*	**0.956**	0.952	0.947	***0.955***	0.943	0.947	0.941	0.939
	S_m	0.911	*0.917*	**0.918**	0.906	0.908	***0.913***	0.908	0.896	0.905	0.887
	MAE	**0.029**	0.033	*0.031*	0.034	0.033	***0.032***	0.037	0.038	0.044	0.046
ECSSD [2]	F_{max}	0.941	**0.950**	*0.947*	0.939	0.942	***0.945***	0.938	0.925	0.935	0.917
	F_{avg}	0.890	*0.918*	*0.920*	0.917	0.879	**0.923**	0.880	0.903	0.886	0.892
	F_β^w	*0.905*	0.899	0.903	0.898	***0.904***	**0.908**	0.881	0.883	0.867	0.853
	E_m	0.926	0.942	*0.947*	***0.949***	0.921	**0.953**	0.923	0.943	0.927	0.928
	S_m	0.916	**0.927**	*0.925*	0.918	0.916	***0.924***	0.918	0.906	0.917	0.895
	MAE	*0.037*	*0.037*	*0.037*	*0.037*	*0.037*	**0.035**	***0.041***	0.043	0.046	0.054
PASCAL-S [4]	F_{max}	*0.880*	**0.890**	0.878	0.872	0.863	***0.879***	0.866	0.860	0.870	0.850
	F_{avg}	0.794	**0.839**	*0.831*	*0.831*	0.781	***0.838***	0.786	0.814	0.804	0.804
	F_β^w	**0.819**	*0.816*	0.807	0.803	0.800	***0.817***	0.790	0.792	0.782	0.762
	E_m	0.862	*0.888*	0.879	***0.887***	0.853	**0.889**	0.860	0.881	0.862	0.861
	S_m	0.851	**0.867**	*0.853*	0.847	0.837	***0.853***	0.850	0.839	*0.854*	0.833
	MAE	*0.067*	**0.065**	0.075	0.072	0.077	***0.070***	0.071	0.075	0.076	0.085
# Param. (M)		72.8	25.2	111.7	47.9	95.5	–	–	–	106.1	61.2
FPS		30.1	19.3	10.2	22.7	32.8	–	–	–	14.8	34.3

4   Experiments

4.1   Experimental Setup

Datasets: We evaluate the proposed method on 5 datasets: PASCAL-S [4], ECSSD [2], HKU-IS [12], DUTS [14] and DUT-O [3], including 850, 1000, 4447,

15572 and 5168 images respectively. DUTS [14] contains 10553 images for training (DUTS-TR) and 5019 images for testing (DUTS-TE). Following previous works [20,21,26,29], we train the proposed model on DUTS-TR dataset.

Evaluation Metrics: In this paper, 6 standard metrics are used to evaluate the proposed model. Mean Absolute Error (MAE) calculates the average pixel-wise difference between the predicted map P_j and generated mask G_j. The calculation formula is

$$MAE = \frac{1}{w \times h} \sum_{i=1}^{w \times h} |P_j(i) - G_j(i)| \tag{10}$$

where w and h are the spatial sizes of the map P_j. F-measure is the weighted harmonic mean of precision and recall, denoted as F_β:

$$F_\beta = \frac{(1 + \beta^2) \times Precision \times Recall}{\beta^2 \times Precision + Recall} \tag{11}$$

where β^2 is set to 0.3 to weight precision more than recall. By adjusting the threshold from 0 to 255, we can get different F_β values, where F_{max} and F_{avg} represent the maximum and average F_β values respectively. The weighted F-measure F_β^w proposed in [5] defines a weighted precision and a weighted recall according to the spatial positions of pixels. In addition, we also use Similarity Measure (S-Measure) [23] and Enhanced-alignment Measure (E-Measure) [24]. For the specific meaning, please refer to the corresponding paper.

Implementation Details: For training, we apply multi-scale for data augmentation. In general, there are three sizes of our training images: 256×256, 320×320, 384×384. The default size of the test image is 320×320. We implement our network based on the publicly available framework Pytorch 1.0.0. A GTX1080 Ti GPU is used to train our model with batch size of 8. The parameters of the first four blocks in the encoder are initialized by the weight of ResNet-50 pre-trained on ImageNet. Adam is utilized to train our model with a total of 25 epochs. The learning rate is initially set to 1e–5 and decayed with a factor of 0.1 after 20 epochs.

4.2   Comparison with State-of-the-Arts

We compare the proposed method with 9 state-of-the-art salient object detection methods, including SCRN [25], EGNet [26], CPD [21], ICNet [27], BANet [30], BASNet [20], DGRL [28], PiCANet [29] and SRM [15]. We use the results published in [32], which are all provided by authors or computed by their released codes. All methods use ResNet-50 as the backbone for training, which is the same setting in this paper.

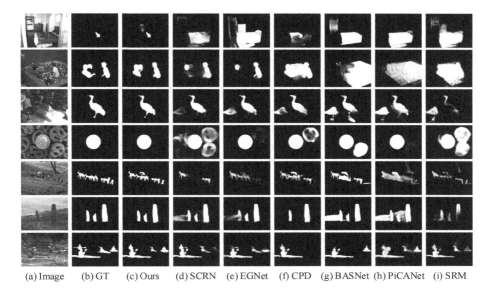

(a) Image   (b) GT   (c) Ours   (d) SCRN   (e) EGNet   (f) CPD   (g) BASNet (h) PiCANet   (i) SRM

Fig. 3. Example results of different methods.

Quantitative Evaluation. The evaluation results of 6 metrics on the 5 benchmark datasets are shown in Table 1. It can be seen that the method proposed in this paper has achieved competitive performance. Compared with other methods, the method proposed has obtained the largest number of optimal metrics, with a total of 11. In particular, the proposed method has outstanding performance on large datasets such as HKU-IS [12], DUTS-TR [14], and DUT-O [3], where all metrics can basically reach the best or close to the best. While on the datasets ECSSD [2] and PASCAL-S [4], our method has achieved less competitive performance. Here we speculate that it may be caused by the fact that ECSSD [2] contains fewer images of objects with different scales and the saliency maps in PASCAL-S [4] are non-binary. The performance of multiple metrics on multiple datasets proves the effectiveness of our method.

Qualitative Evaluation. In Fig. 3, we show the detection results of the proposed method and the other six methods as a comparison. It can be seen that our method is able to accurately segment salient objects under various scenarios, including small objects in the complex backgrounds (1st and 2nd rows) and the recognition of relationships between different objects (3rd and 4th rows). In particular, with the help of multiple branches, our method achieves more robust results in multiple objects scenarios (5th, 6th, and 7th rows). It can not only find the large salient objects in the image, but also find out the small ones.

Table 2. Ablation analysis with different experimental settings on the DUTS-TE dataset.

Ablation	F_{max}	F_{avg}	F_β^w	E_m	S_m	MAE
Single branch	0.879	0.807	0.827	0.899	0.878	0.040
Same supervision	0.881	0.811	0.826	0.902	0.875	0.037
w/o HIF	0.879	0.820	0.826	0.905	0.874	0.037
w/o GDF	0.876	0.823	0.819	0.905	0.868	0.037
Ours	**0.886**	**0.825**	**0.837**	**0.910**	**0.879**	**0.036**

4.3 Ablation Study

In this section, we illustrate the effectiveness of each component used in our model. We conduct ablation study on the DUTS-TE dataset. Different from the previous experimental settings, there is no operation for data augmentation. In addition to the verification of the complete architecture, we also conduct 4 other experiments. First, in order to verify the benefits of multiple branches in decoder, we set up a single branch structure for comparison. Second, to compare with the settings of different branches detecting different objects, supervision information is directly used so that each branch detects all salient objects. Finally, we verify the effectiveness of the GDF and HIF by removing them respectively. From Table 2, it can be seen that the various modules and settings of the proposed model contribute to better detection results.

5 Conclusion

In this paper, we propose a scale-aware multi-branch decoder framework for salient object detection. By dividing the ground true saliency map, different branches use the generated masks for supervision and focus on detecting objects of different sizes. In feature fusion, we design Global Distinction Flow (GDF) and Hierarchical Integration Flow (HIF). GDF helps each branch ignore the semantic information of other branches, so as to focus on detecting objects of other sizes. HIF gradually combines the information of each branch to form the final detection result. Finally, extensive experiments demonstrate the effectiveness of the proposed model.

References

1. Perazzi, F., Krähenbühl, P., Pritch, Y., et al.: Saliency filters: contrast based filtering for salient region detection. In: 2012 IEEE Conference on Computer Vision and Pattern Recognition, pp. 733–740. IEEE (2012)
2. Yan, Q., Xu, L., Shi, J., et al.: Hierarchical saliency detection. In: Proceedings of the IEEE Conference on Computer Vision and Pattern Recognition, pp. 1155–1162 (2013)

3. Yang, C., Zhang, L., Lu, H., et al.: Saliency detection via graph-based manifold ranking. In: Proceedings of the IEEE Conference on Computer Vision and Pattern Recognition, pp. 3166–3173 (2013)
4. Li, Y., Hou, X., Koch, C., et al.: The secrets of salient object segmentation. In: Proceedings of the IEEE Conference on Computer Vision and Pattern Recognition, pp. 280–287 (2014)
5. Margolin, R., Zelnik-Manor, L., Tal, A.: How to evaluate foreground maps? In: Proceedings of the IEEE Conference on Computer Vision and Pattern Recognition, pp. 248–255 (2014)
6. Zhao, R., Ouyang, W., Li, H., et al.: Saliency detection by multi-context deep learning. In: Proceedings of the IEEE Conference on Computer Vision and Pattern Recognition, pp. 1265–1274 (2015)
7. Long, J., Shelhamer, E., Darrell, T.: Fully convolutional networks for semantic segmentation. In: Proceedings of the IEEE Conference on Computer Vision and Pattern Recognition, pp. 3431–3440 (2015)
8. Ronneberger, O., Fischer, P., Brox, T.: U-net: convolutional networks for biomedical image segmentation. In: Navab, N., Hornegger, J., Wells, W.M., Frangi, A.F. (eds.) MICCAI 2015. LNCS, vol. 9351, pp. 234–241. Springer, Cham (2015). https://doi.org/10.1007/978-3-319-24574-4_28
9. Liu, N., Han, J.: Dhsnet: deep hierarchical saliency network for salient object detection. In: Proceedings of the IEEE Conference on Computer Vision and Pattern Recognition, pp. 678–686 (2016)
10. Li, G., Yu, Y.: Deep contrast learning for salient object detection. In: Proceedings of the IEEE Conference on Computer Vision and Pattern Recognition, pp. 478–487 (2016)
11. Hariharan, B., Arbeláez, P., Girshick, R., et al.: Hypercolumns for object segmentation and fine-grained localization. In: Proceedings of the IEEE Conference on Computer Vision and Pattern Recognition, pp. 447–456 (2015)
12. Li, G., Yu, Y.: Visual saliency based on multiscale deep features. In: Proceedings of the IEEE Conference on Computer Vision and Pattern Recognition, pp. 5455–5463 (2015)
13. Borji, A., Cheng, M.M., Hou, Q., et al.: Salient object detection: a survey. Comput. Vis. Media 5(2), 117–150 (2019)
14. Wang, L., Lu, H., Wang, Y., et al.: Learning to detect salient objects with image-level supervision. In: Proceedings of the IEEE Conference on Computer Vision and Pattern Recognition, pp. 136–145 (2017)
15. Wang, T., Borji, A., Zhang, L., et al.: A stagewise refinement model for detecting salient objects in images. In: Proceedings of the IEEE International Conference on Computer Vision, pp. 4019–4028 (2017)
16. Zhang, P., Wang, D., Lu, H., et al.: Amulet: aggregating multi-level convolutional features for salient object detection. In: Proceedings of the IEEE International Conference on Computer Vision, pp. 202–211 (2017)
17. Lee, H., Kim, D.: Salient region-based online object tracking. In: 2018 IEEE Winter Conference on Applications of Computer Vision (WACV), pp. 1170–1177. IEEE (2018)
18. Zhang, L., Dai, J., Lu, H., et al.: A bi-directional message passing model for salient object detection. In: Proceedings of the IEEE Conference on Computer Vision and Pattern Recognition, pp. 1741–1750 (2018)
19. Chen, Z., Zhou, H., Xie, X., et al. Contour loss: Boundary-aware learning for salient object segmentation. arXiv preprint arXiv:1908.01975 (2019)

20. Qin, X., Zhang, Z., Huang, C., et al.: Basnet: boundary-aware salient object detection. In: Proceedings of the IEEE/CVF Conference on Computer Vision and Pattern Recognition, pp. 7479–7489 (2019)
21. Wu, Z., Su, L., Huang, Q.: Cascaded partial decoder for fast and accurate salient object detection. In: Proceedings of the IEEE/CVF Conference on Computer Vision and Pattern Recognition, pp. 3907–3916 (2019)
22. Wang, Y., Xu, Z., Shen, H., et al.: Centermask: single shot instance segmentation with point representation. In: Proceedings of the IEEE/CVF Conference on Computer Vision and Pattern Recognition, pp. 9313–9321 (2020)
23. Fan, D.P., Cheng, M.M., Liu, Y., et al.: Structure-measure: a new way to evaluate foreground maps. In: Proceedings of the IEEE International Conference on Computer Vision, pp. 4548–4557 (2017)
24. Fan, D.P., Gong, C., Cao, Y., et al.: Enhanced-alignment measure for binary foreground map evaluation. arXiv preprint arXiv:1805.10421 (2018)
25. Wu, Z., Su, L., Huang, Q.: Stacked cross refinement network for edge-aware salient object detection. In: Proceedings of the IEEE/CVF International Conference on Computer Vision, pp. 7264–7273 (2019)
26. Zhao, J.X., Liu, J.J., Fan, D.P., et al.: EGNet: edge guidance network for salient object detection. In: Proceedings of the IEEE/CVF International Conference on Computer Vision, pp. 8779–8788 (2019)
27. Wang, W., Shen, J., Cheng, M.M., et al.: An iterative and cooperative top-down and bottom-up inference network for salient object detection. In: Proceedings of the IEEE/CVF Conference on Computer Vision and Pattern Recognition, pp. 5968–5977 (2019)
28. Wang, T., Zhang, L., Wang, S., et al.: Detect globally, refine locally: a novel approach to saliency detection. In: Proceedings of the IEEE Conference on Computer Vision and Pattern Recognition, pp. 3127–3135 (2018)
29. Liu, N., Han, J., Yang, M.H.: Picanet: learning pixel-wise contextual attention for saliency detection. In: Proceedings of the IEEE Conference on Computer Vision and Pattern Recognition, pp. 3089–3098 (2018)
30. Su, J., Li, J., Zhang, Y., et al.: Selectivity or invariance: Boundary-aware salient object detection. In: Proceedings of the IEEE/CVF International Conference on Computer Vision, pp. 3799–3808 (2019)
31. Zhou, H., Xie, X., Lai, J.H., et al.: Interactive two-stream decoder for accurate and fast saliency detection. In: Proceedings of the IEEE/CVF Conference on Computer Vision and Pattern Recognition, pp. 9141–9150 (2020)
32. Pang, Y., Zhao, X., Zhang, L., et al.: Multi-scale interactive network for salient object detection. In: Proceedings of the IEEE/CVF Conference on Computer Vision and Pattern Recognition, pp. 9413–9422 (2020)

Densely End Face Detection Network for Counting Bundled Steel Bars Based on YoloV5

Huajie Liu and Ke Xu[✉]

Collaborative Innovation Center of Steel Technology, University of Science and Technology Beijing, Beijing 100083, China
xuke@ustb.edu.cn

Abstract. Steel bars are important materials in infrastructure construction. It is necessary to know the number of steel bars that are bundled in production, storing, and trading, while manually counting bundled steel bars is time-consuming and error-prone. A densely end face detection network for counting bundled steel bars is developed. Cross stage partial connection is introduced to reduce the duplicate gradient information of PeleeNet. An attention module is added to enhance the feature extraction ability of the network. Features at different levels of the backbone network are adaptively spatial fused and used as inputs of the network for end face detection of bundled steel bars. The F1 score of the densely end face detection network for counting bundled steel bars is 98.84%, while the model size is 0.52M, and the inference time is 0.022 s.

Keywords: Steel bar · Densely object · Object detection · Lightweight network

1 Introduction

Recently, many fields have produced the demand for countings, such as fruit counting in agriculture [1], population counting in security [2], and cell counting in medicine [3]. Steel bar counting exists in the process of production, storing, and trading. Construction companies also need to count steel bars when they check and accept the purchased steel bars. Count of bundled steel bars is very important for steel bar manufacturers and construction companies. Xiaohu et al. [4] used adaptive Otsu binarization to segment the end face of the steel bars and detect the profile. Monsura et al. [5] manually processed the background of the steel bar image and used k-means clustering to segment the steel bars and background. Due to the diverse colors of the steel bar end faces, it is difficult to select the cluster center. Hao et al. [6] used the concave point matching method to segment the end face of the bundled steel bars, but it cannot deal with the severe adhesion. Zhang et al. [7] used morphological calculations to solve the adhesion of the end face of the steel bars, and detected it through the Hough transform. Because the radius of the steel bar needs to be known, it is difficult to apply. Su et al. [8] used gradient Hough circle transformation to detect the end face of steel bars with different radius. The template matching method [8–11] was used to detect the end face of the steel bar. Due to the

H. Ma et al. (Eds.): PRCV 2021, LNCS 13019, pp. 293–303, 2021.
https://doi.org/10.1007/978-3-030-88004-0_24

irregular shape of the end face of the steel bar, it is difficult to determine the center of the end face, which affects the detection effect. Yan et al. [12] used the area method to detect the end face of the steel bar, which has a large amount of calculation and vulnerable to environmental interference. Liu et al. [13] extracted the hog features of the image, and used SVM (Support Vector Machine) classifier to detect the end face of the steel bar. However, the above works only use the shape, color, and other low-level features of the steel bar's end face, the detection effect is easily affected by the environment. The convolutional neural network can extract high-level semantic information of the image, it is more robust to changes of the environment. Faster R-CNN [14] performs better in the detection of steel bar's end face than the traditional methods, while the inference speed is slow. Therefore, there are many single-stage works. Improved RetinaNet [15] for steel bar's end face detection, the F1 score is only 93%. YONGJIAN et al. [16] proposed Inception-RFB-FPN model reaches 98.20%. Zeng et al. [17] proposed a lightweight network for steel bar's end face detection, with 4.7M parameters, while the F1 score is only 83.4%. The improved Nested Unet [18] performs better in detecting the end face of steel bars, and the F1 score is 98.8%. However, the existing convolutional neural networks for steel bar's end face detection can only achieve good performance in one aspect of accuracy or speed, but they are difficult to balance high precision and high speed, and also they are difficult to be deployed on low-power devices. Therefore, it is necessary to develop a lightweight high-precision steel bar's end face detection network.

We designed a lightweight network for densely end face detection of bundled steel bars based on YoloV5 [19]. The original backbone parameters and computational cost of YoloV5 are relatively large, and need to be replaced with lightweight backbones. Two dense layers in PeleeNet are partially connected across stages, and SE (Squeeze-and-Excitation) attention module is added to build SEPeleeNetCSP module for feature extraction. The original "neck" part of YoloV5 is computationally intensive, and the feature fusion method does not consider that the steel bar's end face is a small object, which leads to a poor detection effect. Therefore, ASFF (Adaptive Spatial Feature Fusion) [20] is used to fuse features of different stages and using P3 level feature map for detection. As traditional NMS (Non-Maximum Suppression) does not perform well in the detection of dense objects, applying Cluster-weighted NMS [21] in the post-processing stage greatly improves the detection effect of the densely end face detection network.

2   Densely End Face Detection Network

2.1   Cross Stage Partial Backbone with Attention Module

Although the smallest model size of YoloV5 is 4.3M, the computational cost of the input image with 640 × 832 resolution is 18.9 GFLOPs. The parameters and calculation of the backbone network are relatively large, and the effect of feature extraction needs to be improved. We choose PeleeNet as the baseline, as PeleeNet [22] is one of the lightweight backbones with better performance in recent years. The 2-way dense layer is the basic module for feature extraction in PeleeNet and has fully inherited the excellent structure of DensNet, while the problem of gradient duplication also exists in PeleeNet. CSPNet [23] uses a cross stage partial connection operation to solve the problem of DensNet gradient duplication, the cross stage partial connection operation allows one part of the feature map to pass through the stage, and the other part to concat with the feature map who passing through the stage, which is still valid for PeleeNet. SENet [24] used the SE attention module to succeed in the ImageNet competition. The Squeeze operation in the SE attention module takes the global spatial characteristics of each channel as the representation of the channel and uses global average pooling to generate statistics for each channel. Excitation operation is to learn the degree of dependence of each channel and adjust different feature maps to obtain the final output according to the degree of dependence. SE attention module is inserted into the PeleeNet connected by the Cross Stage Partial, and information of unimportant channels is suppressed, while information of important channels is enhanced and a stronger feature extraction network is generated. The SE attention module is inserted behind the output of the 2-way dense layer, the growth rate is 16, and the final SEPeleeNetCSP3 module is shown in Fig. 1, which means there are three SEmodules.

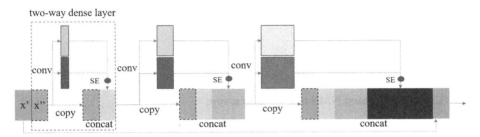

Fig. 1. SEPeleeNetCSP3 module

2.2 Lightweight Module of Adaptively Spatial Feature Fusion

The original "neck" of YoloV5 is the PAN (Path Aggregation Network) module, the shallow feature map contains more location information, and the PAN module transfers the shallow features into the deep layers. PAN is conducive to the positioning of large objects, The deep feature map contains more semantic information, which transfers deep features to the shallow layer, which is conducive to the classification of small objects. Therefore, the PAN module transfers shallow features to deep features, which is not helpful for small object detection, and PAN feature fusion does not consider the importance of different levels of features for detection. To make features more effective for detection tasks, the PAN module is abandoned, and ASFF is directly fusion the features of different levels in the backbone. The fused features have rich high-level semantic information for classification and have location information for positioning. The P3 level feature map of ASFF is used for detection, and obtained a lighter weight and better performance model than using the PAN module. The overall structure of the network is shown in Fig. 2. Stem Layer and Transition Layer are used for down-sampling.

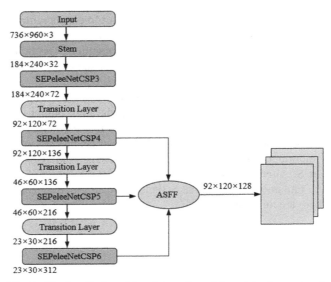

Fig. 2. The bundled steel bars' densely end face detection network

2.3 Cluster-Weighted NMS in Post-processing

The detected boxes generated by the detector have a lot of redundancy, and NMS is needed for post-processing. There are N detected boxes $\boldsymbol{B} = [\mathcal{B}_1, \mathcal{B}_2, \cdots, \mathcal{B}_N]^{\mathrm{T}}$, and they have been sorted by classification scores. Cluster-weighted NMS compute the IoU (Intersection over Union) matrix $\boldsymbol{X} = \{x_{ij}\}_{N \times N}$ in parallel, where $x_{ij} = \mathrm{IoU}(\mathcal{B}_i, \mathcal{B}_j)$, and triangulate X up. Then take the maximum value by column, and obtain the one-dimensional tensor $\boldsymbol{b} = \{b_i\}_{1 \times N}$, $b_i \in \{0, 1\}$, for iteration t, the NMS result b^{t-1}, then use formula (1) and (2) to comput$\boldsymbol{C}^t$, when b^t is equal to b^{t-1}, stop the iteration and get C.

$$A^t = \mathrm{diag}\left(\boldsymbol{b}^{t-1}\right) \tag{1}$$

$$C^t = A^t \times X \tag{2}$$

Use classification scores $s = [s_1, s_2, \cdots, s_N]^{\mathrm{T}}$ to multiply C by every column to get C', and then use formula (3) to perform weighted average of coordinates.

$$B = \frac{C' \times B}{\sum_j C'} \tag{3}$$

Cluster-weighted NMS is Cluster NMS combined with weighted average methods, and it performs better than traditional NMS in dense scenarios.

3   Experiments and Results

3.1   Experiment Configuration

Data used in the experiments are from a public data set [25], which has 200 images for training, 50 images for the validation set, and 200 images for testing. The operating system is Ubuntu 18.04, and the hardware environment is AMD 3600 and NVIDIA GeForce GTX 2060 (6G).

3.2   Backbone Training and Evaluation

The evaluation indicators are the size of parameters, GFLOPs (Giga floating point of operations), F1 score. The calculation formulas for precision, recall, and F1 score is as formula (4)–(6).

$$Precision = \frac{TP}{TP + FP} \tag{4}$$

$$Recall = \frac{TP}{TP + FN} \tag{5}$$

$$F1 = 2 \times \frac{Recall \times Precision}{Recall + Precision} \tag{6}$$

TP is true positive, FP is false positive, FN is false negative, and TN is true negative.

The backbone of YoloV5 is replaced with different improved versions of PeleeNet for experimentation. The input image resolution is 640 × 832, the batch size is 8, the learning rate is 0.02, SGD optimization is used, and epochs is 600.

The results of ablation experiments are shown in Table 1. The original model is PeleeNet which uses Torchvision NMS for post-processing. Torchvision NMS has the same effect as traditional NMS, while it is faster. PeleeNet-A uses Cluster-weighted NMS for post-processing, and the detection effect is clearly shown according to the F1 Score, and Cluster-weighted NMS is better than Torchvision NMS when the object is dense.

Therefore, the post-processing of subsequent experiments uses Cluster-Weighted NMS. The growth rate of PeleeNet-B is half as much as PeleeNet-A, which reduces the number of network channels and narrows the network, and significantly reduces GFLOPs used to measure the computational cost. The size of parameters is also reduced. PeleeNetCSP performs CSP operations in PeleeNet-B, and further reduces the size of parameters and computational cost by removing the repetition of gradient information. SEPeleeNetCSP adds the SE module in PeleeNetCSP, and makes full use of the effective information in the channels through the SE module, and the F1 Score is the same as the original PeleeNet. SE module does not cause too many parameters and computational cost is almost not increased. Results of the experiments show that the improvement on peleenet can greatly reduce the size of parameters and computational costs, while remains the same performance as the original model.

Table 1. Ablation experiment

Backbone	Growth Rate	Parameters	F1 Score	GFLOPs
PeleeNet	32	2.1M	82.24%	14.2
PeleeNet-A	32	2.1M	97.12%	14.2
PeleeNet-B	16	0.42	96.44%	3.7
PeleeNetCSP	16	0.27M	96.10%	3.4
SEPeleeNetCSP	16	0.27M	97.05%	3.4

3.3 Performance Comparison Experiment for Different Backbones

In this experiment, YoloV5's backbone is replaced with a different backbone, and the same training method as SEPeleeNetCSP is used. The evaluation indicators are the size of parameters, GFLOPs, F1 score, and inference time.

Figure 3(a) shows F1 scores of different backbones. From Fig. 3(a), it is shown that PeleeNet has the highest F1 score, and SEPeleeNetCSP is closely followed by PeleeNet in second place. F1 scores of Yolov5(small), VovNet, SEPeleeNetCSP, and PeleeNet are almost the same. Other backbones are significantly lower than SEPeleeNetCSP, which shows that SEPeleeNetCSP's performance is competitive.

Figure 3(b) shows GFLOPs of different backbones. From Fig. 3(b), it is shown that GhostNet has an extremely low computational cost, as GFLOPs is only 0.5. GFLOPs of MobileNet V3-small is 1.8 higher than GhonstNet, and SEPeleeNetCSP ranked third with 3.4 GFLOPs. Other backbones' GFLOPs are higher than SEPeleeNetCSP, which shows that SEPeleeNetCSP has the potential to be deployed in low-power devices.

Figure 3(c) shows the sizes of parameters of different backbones. From Fig. 3(c), it is shown that the size of parameters of GhostNet is 0.06M, while SEPeleeNetCSP is 0.27M, which is slightly more than GhostNet. A small number of parameters will make the size of the model very small, which is particularly beneficial when deploying to low-power devices.

From Fig. 3, it is shown that only SEPeleeNetCSP has high performance, small size of parameters, and low computational costs. Other backbones don't have high performance, low parameter amount, and low calculation at the same time.

3.4 Training and Evaluation of the Network

In this experiment, PAN and ASFF are used as the neck for SEPeleeNetCSP backbone respectively. ASFF only uses P3 level feature map for detection, and uses 736×960 resolution images for training. The learning rate is 0.04, the epochs are 300, and the rest of the settings are the same as SEPeleeNetCSP. Figure 4(a) and (b) show the recall and precision of the network with PAN and ASFF as the neck on the verification set respectively. After 300 epochs of training, the accuracy and recall of the model on the verification set tend to be stable, the precision and recall are close to 100%.

As shown in Table 2, the F1 score of the model with ASFF as the neck is 1.72% higher than the model with PAN as neck, the parameters are reduced by 2.65m, and the

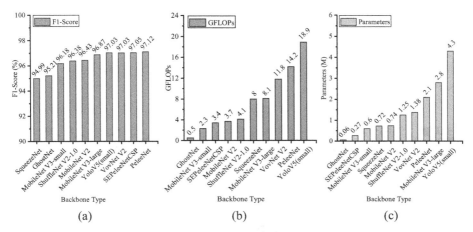

Fig. 3. F1 scores, GFLOPs, and Parameters of different backbones

calculation is reduced by 4.7 GFLOPs, which shows the superiority of the model with ASFF as neck and using P3 level features for detection.

Figure 5 shows the detection results of bundled steel bars. From Fig. 5, it is shown that although the model has high precision and recall, the actual detection results still have false detection and miss detection. Figure 5(a) shows false detection of the ground as a steel bar. Figure 5(b) shows miss detection because of strong illumination. Figure 5(c) shows miss detection due to occlusion. In Fig. 5(d) and Fig. 5(e), there is no miss detection and false detection, as the end of the steel bars in the image is relatively regular. Figure 5(f) shows that the network has a certain degree of robustness to occlusion.

Figure 6 shows the detection results on stacked wood. Figure 6(a) is the result of using the network directly, and there are some false detection and miss detection. 16 images of stacked woods are used for training the network, the final detection result is shown in Fig. 6(b), and all the end faces of the wood can be identified.

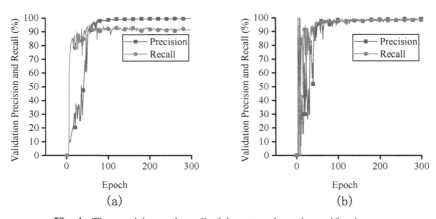

Fig. 4. The precision and recall of the network on the verification set

Table 2 Performance comparison experiment between PAN and ASFF

Neck	Parameters	F1 Score	GFLOPs
PAN	3.17M	97.19%	13.2
ASFF	0.52M	98.84%	8.5

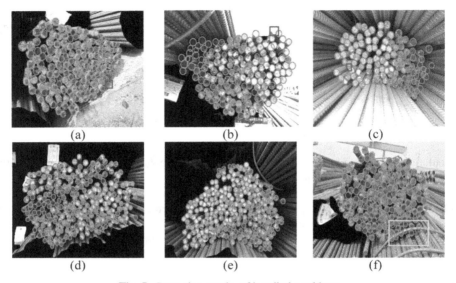

(a)   (b)   (c)
(d)   (e)   (f)

Fig. 5. Detection results of bundled steel bars

(a)   (b)

Fig. 6. Detection results of stacked wood

3.5 Model Comparison

The densely end face detection network is compared with the results from other references. The comparison standards are the size of parameters, F1-Score, inference time, and resolution. As shown in Table 3, the densely end face detection network has a competitive F1-Score with small image resolution while maintaining excellent performance.

The network's parameters are much smaller than other networks, and the inference speed is also very fast.

Table 3. Compared the results with other references.

Method	Resolution	Parameters	F1 Score	Inference Time
Shi et al.[14]	1333 × 1000	-	99.15%	-
MING et al.[15]	-	-	93%	-
Zhu et al.[16]	-	-	98.17%	0.0306s
Zeng et al. [17]	-	4.7M	83.4%	-
Xie et al.[18]	1024 × 1024	9.2M	98.8%	-
Ours	736 × 960	0.52M	98.84%	0.022s

4   Discussion

Traditional NMS has poor performance in dealing with dense object detection, and cluster-weight NMS is used to improve it. Our method can identify the end faces of dense steel bars that are not severely blocked, but it is still difficult to identify the end faces with a blocked area greater than 1/2. This situation can be manually supplemented in practical applications to make the detection more accurate. As the dataset has fewer images, it is difficult to cover complex changes in the scene, which will lead to false detection and miss detection, if more images can be obtained for training, the effect will be improved.

5   Conclusions

The detection of steel bar's end faces has important applications in the industry. It can free the workers from the heavy work of steel bar counting, and it is also beneficial to the intelligent warehouse management of steel bar production. The dense detection network has a good effect not only on the bundled steel bars, but also on the stacked wood. the F1 Score of the dense detection network for end faces of bundled steel bars is 98.84%, the size of parameters is 0.52M and the computational cost is 8.5 GFLOPs, which is very suitable for deployment in low-power devices.

References

1. Bellocchio, E., Ciarfuglia, T.A., Costante, G., Valigi, P.: Weakly supervised fruit counting for yield estimation using spatial consistency. IEEE Robot. Autom. Lett. **4**, 2348–2355 (2019)
2. Zhang, Y., Zhou, D., Chen, S., Gao, S., Ma, Y.: Single-image crowd counting via multi-column convolutional neural network. In: 2016 IEEE Conference on Computer Vision and Pattern Recognition (CVPR), pp. 589–597 (2016)

3. Zhang, D., Zhang, P., Wang, L.: Cell counting algorithm based on YOLOv3 and image density estimation. In: 2019 IEEE 4th International Conference on Signal and Image Processing (ICSIP), pp. 920–924 (2019)
4. Xiaohu, L., Jineng, O.: Research on steel bar detection and counting method based on contours. In: 2018 International Conference on Electronics Technology (ICET), pp. 294–297 (2018)
5. Ablidas, M.R.M., Monsura, A.S., Ablidas, L.A., dela Cruz, J.: An application of image processing technology in counting rebars as an alternative to manual counting process. Int. J. Simul.: Syst. Sci. Technol. **20**, 1–9 (2019)
6. Jingzhong, W., Hao, C., Xiaoqing, X.: Pattern recognition for counting of bounded bar steel. In: Fourth International Conference on the Applications of Digital Information and Web Technologies (ICADIWT 2011), pp. 173–176 (2011)
7. Zhang, Y., Jiang, M., Wu, Y., Zhou, X.: An automatic steel bar splitting system based on two-level of the chain transmission. In: 2015 IEEE International Conference on Cyber Technology in Automation, Control, and Intelligent Systems (CYBER), pp. 587–590 (2015)
8. Ying, X., Wei, X., Pei-xin, Y., Qing-da, H., Chang-hai, C.: Research on an automatic counting method for steel bars' image. In: 2010 International Conference on Electrical and Control Engineering, pp. 1644–1647 (2010)
9. Zhang, D., Xie, Z., Wang, C.: Bar section image enhancement and positioning method in on-line steel bar counting and automatic separating system. In: 2008 Congress on Image and Signal Processing, pp. 319–323 (2008)
10. Lei, Y.L., Yang, J.Z., Zhang, Y.H.: Efficient steel bar count scheme based on image recognition in mechanical engineering. AMM **340**, 805–808 (2013)
11. Hou, W., Duan, Z., Liu, X.: A template-covering based algorithm to count the bundled steel bars. In: 2011 4th International Congress on Image and Signal Processing, pp. 1813–1816 (2011)
12. Yan, X., Chen, X.: Research on the counting algorithm of bundled steel bars based on the features matching of connected regions. In: 2018 IEEE 3rd International Conference on Image, Vision and Computing (ICIVC), pp. 11–15 (2018)
13. Liu, C., Zhu, L., Zhang, X.: Bundled round bars counting based on iteratively trained SVM. In: Huang, D.-S., Bevilacqua, V., Premaratne, P. (eds.) ICIC 2019. LNCS, vol. 11643, pp. 156–165. Springer, Cham (2019). https://doi.org/10.1007/978-3-030-26763-6_15
14. Jinglie, S.: Steel bar counting algorithm research based on convolutional neural network. Mater dissertation. Huazhong University of Science & Technology, Wuhan (2019)
15. Ming, H., Chen, C., Liu, G., Deng, H.: Improved counting algorithm for dense steel bars based on RetinaNet. Transducer Microsyst. Technol. **39**, 115–118 (2020)
16. Zhu, Y., Tang, C., Liu, H., Huang, P.: End face localization and segmentation of steel bar based on convolution neural network. IEEE Access. **8**, 74679–74690 (2020)
17. Qu, F., Li, C., Peng, K., Qu, C., Lin, C.: Research on detection and identification of dense rebar based on lightweight network. In: Zeng, J., Jing, W., Song, X., Lu, Z. (eds.) ICPCSEE 2020. CCIS, vol. 1257, pp. 440–446. Springer, Singapore (2020). https://doi.org/10.1007/978-981-15-7981-3_32
18. Haizhen, X.: Research and application of steel bar counting algorithm in complex scene. Mater dissertation. University of Electronic Science and Technology of China, Chengdu (2020)
19. ultralytics/yolov5. https://github.com/ultralytics/yolov5. Accessed 4 June 2021
20. Liu, S., Huang, D., Wang, Y.: Learning spatial fusion for single-shot object detection. arXiv: 1911.09516 [cs]. (2019)
21. Zheng, Z., et al.: Enhancing geometric factors in model learning and inference for object detection and instance segmentation. arXiv:2005.03572 [cs]. (2020)

22. Wang, R.J., Li, X., Ling, C.X.: Pelee: a real-time object detection system on mobile devices. In: Bengio, S., Wallach, H., Larochelle, H., Grauman, K., Cesa-Bianchi, N., Garnett, R. (eds.) Advances in Neural Information Processing Systems, pp. 1963–1972. Curran Associates Inc, Montréal, Canada (2018)
23. Wang, C., Liao, H.M., Wu, Y., Chen, P., Hsieh, J., Yeh, I.: CSPNet: a new backbone that can enhance learning capability of CNN. In: 2020 IEEE/CVF Conference on Computer Vision and Pattern Recognition Workshops (CVPRW), pp. 1571–1580 (2020)
24. Hu, J., Shen, L., Sun, G.: Squeeze-and-excitation networks. In: 2018 IEEE/CVF Conference on Computer Vision and Pattern Recognition, pp. 7132–7141 (2018)
25. Datafountain.Cn: Digital China innovation contest (2019). https://www.datafountain.cn/com petitions/332

POT: A Dataset of Panoramic Object Tracking

Shunda Pei, Zhihao Chen, and Liang Wan$^{(\boxtimes)}$

College of Intelligence and Computing, Tianjin University, Tianjin, China
{peishunda,zh_chen,lwan}@tju.edu.cn

Abstract. Object tracking in spherical panoramic videos is an important yet less studied problem. Besides the common challenges in planar video tracking, panoramic video tracking should address the issues of image distortion and split caused by the sphere-to-plane projection. In this paper, we build a panoramic object tracking dataset, namely POT, containing 40 fully annotated spherical panoramic videos, which are annotated on the pixel level. Moreover, we propose a simple yet effective panoramic tracking framework based on local field-of-view projection, which considers the spherical geometry explicitly and performs object tracking on the tangent plane. Our framework can be easily integrated with well-studied planar video trackers. Besides, we boost five mainstream trackers with our framework. Comprehensive experimental results on the POT dataset demonstrate that the proposed framework is well behaved in panoramic object tracking.

Keywords: Panoramic tracking dataset · Single-object tracking · Tangent plane projection

1 Introduction

The past decade has witnessed the increasing trend of spherical panoramic images and videos being more popular due to their 360-degree field-of-view. Many studies on panoramic images and videos have been explored, including spherical convolution [5,19], spherical feature description [7], movement prediction [23], scene viewpoint recognition [21], etc.

However, as an important task in computer vision, object tracking in spherical panoramic videos is still less studied. With the increasing use of panoramic cameras in scenarios such as intelligent monitor, street view maps, etc., it is a common task to track a target in panoramic videos. For example, we may need to track a suspect in panoramic monitor. In this case, a tracking algorithm will lighten our burden. There are some differences on object tracking between planar videos and panoramic videos. Figure 1 (a) shows two equirectangular images, which are the most common sphere-to-plane panoramic representation. It can be seen that when the target goes near the image top or bottom, longitude-latitude spherical expansion will introduce object distortion and when the target goes

© Springer Nature Switzerland AG 2021
H. Ma et al. (Eds.): PRCV 2021, LNCS 13019, pp. 304–317, 2021.
https://doi.org/10.1007/978-3-030-88004-0_25

(a)   (b)

Fig. 1. Examples of panoramic images. (a) shows two panoramic images of equirectangular format, (b) shows the interested regions of (a), which is obtained by local FOV projection.

near left or right boundaries, this expansion will introduce object split. These two situations will disturb the tracking progress.

The lack of public dataset of object tracking in panoramic videos is one of the main reasons why panoramic object tracking is less studied. In recent years, the emergence of tracking datasets like OTB [20], VOT2015 [11], TrackingNet [18] and GOT-10k [8], along with the rapid development of deep convolutional neural network approaches, have brought great advancements to object tracking. However, these datasets are all composed of 2D planar videos, which are shot by traditional single lens cameras. It is in urgent need to develop a dataset of panoramic videos for studing panoramic object tracking.

In this paper, we build a panoramic object tracking dataset, namely POT, by collecting 40 panoramic videos to facilitate the research of panoramic object tracking. Moreover, to represent the target status accurately, we annotate the sequences with a pixel level method, which is similar to that of the image segmentation task. Besides, inspired by previous works [11,20], we tag every sequence with several attributes for better evaluation and analysis of tracking approaches.

Based on our proposed dataset, we attempt to seek for a resolution for panoramic object tracking. In this regard, considering the tracking process and characteristic of panoramic images, we develop a general framework based on local FOV projection. This framework can be combined with mainstream trackers and enable them to track in spherical panoramic videos without object distortion and split. Specifically, during the tracking process in panoramic videos, we transform the equirectangular frame to spherical form and then project the local region of interested target to tangent plane. As shown in Fig. 1 (b), the tangent plane of local region is not subjected to distortion and split. Our contributions are summarized as:

- We propose a panoramic object tracking dataset POT with 40 fully annotated sequences. To the best of our knowledge, we are the first to annotate object tracking dataset on pixel level. We will release the dataset to facilitate the research in community.
- We propose a general framework based on local FOV projection for object tracking in panoramic videos, which can circumvent the influences of target distortion and split led by latitude-longitude spherical expansion.
- We enhance five classical trackers SAMF [14], Staple [1], ECO [3], STRCF [13] and Siamfc [2] with our framework, and evaluation results on POT dataset verify the superiority of our method for panoramic object tracking.

2   Related Work

Object Tracking Dataset. Since 2013 many datasets of object tracking have been proposed to evaluate and compare trackers. The OTB [20] and VOT [11]

Fig. 2. Annotated panoramic videos for performance evaluation of panoramic tracking. The first annotated frame of each video is shown with its name at top left and visual attributes at bottom. Furthermore, we make local amplification of the object area at top right.

datasets represent the initial attempts to unify the test data and performance measurements for tracking algorithms. Later on, several other datasets have been proposed for some specific issues, including the large-scale people and rigid object tracking dataset NUS PRO [12], color tracking dataset TColor128 [15], high frame-rate tracking dataset NfS [10] and so on. These datasets also play an important role in facilitating the research of tracking methods. Recently, more large-scale datasets have been proposed, such as TrackingNet [18], LaSOT [6] and GOT-10k [8], and they are large enough for training and testing trackers and have greatly promoted the development of object tracking. However, these datasets all consist of 2D planar videos.

Panoramic Object Tracking. Object tracking in spherical panoramic videos is an important yet less studied problem. Among several existing works, it has been a general consensus to eliminate the influences of image distortion and split lead by sphere-to-plane expansions. These works focus on different forms of panoramic videos. Zhou et al. [22] tracked object in six sides of cubic panorama images and used padding to fetch up the image boundary when tracking in a single side, but splicing several sides is time consuming and may introduce extra distortion. Ahmad et al. [4] applied Training-Learning-Detection (TLD) [9] in 360-degree videos of horizontal direction, which lacks the view of two poles on the spherical image. Similarly, Liu et al. [16] claimed that the top and bottom regions contain fewer interested objects and removed them in equirectangular images. Obviously, the above two methods are not suitable for many situations where objects exist near top or bottom boundaries. Besides, Liu et al. [17] tracked single object in fisheye panoramic videos and marked object with trapezoid box rather than traditional rectangle box. However, fisheye panoramic videos only include hemisphere FOV thus missing the integral description for whole scene. In general, these methods have restrictions for object tracking in spherical panoramic videos.

3   POT Dataset

In this section, we introduce the details of POT, which is short for panoramic object tracking dataset, including its characteristics, attributes, annotation issues and evaluation methodology. It should be noted that we preserve the sequences as equirectangular images because they are the most common sphere-to-plane panoramic representation which is utilized by many spherical panoramic cameras (Insta360, Samsung Gear 360, Ricoh Theta S).

308   S. Pei et al.

Table 1. Attributes of annotated panoramic sequence

Attr	Description
DIST	Distorted-When objects move near the poles, they will be distorted
SP	Split-When objects move to left or right sides, they will be split to two parts
CM	Camera Moving-The moving of the camera causes the two poles and both sides of the panoramic frame change continuously
SV	Scale Variation—The scale change of the target size between the first frame and current frame is large
MB	Motion Blur—The target region is blurred due to the motion of the target or the camera
FM	Fast Motion-The motion of the target is fast
BC	Background Clutters—The background near the target has similar color or texture as the target
PO	Partial Occlusion-The target is partially sheltered

3.1 Database Characteristics

Existing tracking datasets such as OTB [20] and VOT [11] mainly consist of planar videos. There is a lack of public dataset of panoramic videos. We build up a panoramic object tracking dataset POT. The dataset is composed of 40 fully annotated sequences. The frame number of a sequence ranges from 61 to 421, and the average number of all sequence is 201. Figure 3 (b) shows the length distribution of POT dataset, we can see most sequences are between 100 and 300 frames. It contains targets of different classes including cars, person, face, doll and so on. Most videos are shot by ourselves with panoramic cameras in various scenes, such as road, classroom, library, etc. except for sequence 'dancer', which is obtained from the internet. The resolution of each frame is 1152ˆ576, where the width is twice the height in equirectangular representation of spherical image. The first annotated frame of each video is shown in Fig. 2.

3.2 Attributes

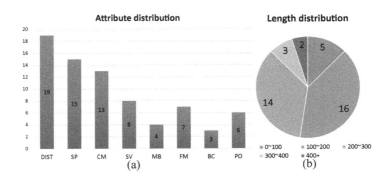

Fig. 3. (a) shows the attribute distribution of POT dataset, (b) shows the length distribution.

As we know, many factors may influence the strength and weakness of different trackers. Inspired by former works [11,20], we category the sequences according to 8 attributes defined in Table 1 for better evaluation and analysis of the trackers in panoramic videos. Each attribute represents a common challenge in panoramic tracking. As shown in Fig. 2, we annotate the attributes of each sequence at left bottom of the first frame and we find that one sequence may be tagged with more than one attribute. The attribute distribution of our POT dataset is shown in Fig. 3 (a). We can see that some attributes occur more frequently, e.g., DIST and SP, which are unique to panoramic vision. Here we describe the DIST and SP attributes in detail.

DIST attribute. As discussed in Sect. 1, when a target comes close to the both poles of a spherical image, it will distort sharply in the equirectangular representation due to sphere-to-plane expansions. Note that the distortion is caused by spherical geometry instead of deformation of the target itself.

SP attribute. As we know, the equirectangular representation is spread from the 360° spherical image, so its left and right sides are connected in the real scene. When a target comes close to left or right side, it is going to be split to two parts and then cross the side to the other one. It is essential for an excellent tracker in panoramic videos to deal with this situation.

3.3   Annotation Methodology

In former works of single object tracking dataset [6,8,10–12,15,18,20], they annotated the object with a rectangular bounding box. While this annotation method does not apply to panoramic videos. It can be seen from top of Fig. 4 (a) that when the target comes close to top or bottom, there may exist large non-target areas in the bounding box owing to the characteristic of spherical geometry. It can be calculated that in this image, the intersection-over-union of the rectangular area and the accurate target area is only 43.52%. Figure 4 (b) shows that target will be split apart when it comes close to left or right sides of equirectangular iamge, one bounding box can not cover it completely. In this regard, we annotate the target on pixel level similar to the image segmentation task. We reserve the regions in the form of binary masks for eval- uation as shown in bottom of Fig. 4 (a) and Fig. 4 (b). In this way, we can get the accurate position and size of target no matter where it moves and it is fair for evaluation of different tracking algorithms. It should be mentioned that for the first frame, we annotate it with a rectangular bounding box. Then trackers can get the initial position of target and carry out the tracking.

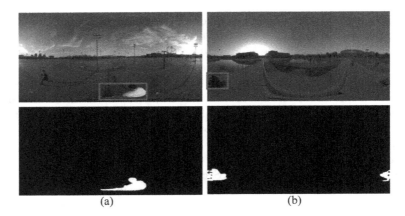

Fig. 4. The comparison of two annotation methods. Red rectangular represents the traditional method and the binary image denotes the result of our proposed method. (Color figure online)

3.4 Evaluation Methodology

Inspired by OTB [20], we exploit the success plot to represent the bounding region overlap. Specifically, we define r_t and r_g as the tracked bounding box and the pixel level ground truth respectively, and they are both reserved in the form of binary mask. Then the overlap score is denoted as $S = \frac{|r_t \bigcap r_g|}{|r_t \bigcup r_g|}$, where $\bigcap$ and $\bigcup$ denote the intersection and union of two regions, respectively, and $|\cdot|$ represents the pixel numbers in the region. One more thing, the precision plot is generally also used in tracking evaluation, which represents the center location error precision. However, we can not record the center position of the target with our annotation method, so we will not adopt this evaluation criterion.

4   The Local FOV Based Framework

In this section, we mainly introduce our proposed local FOV projection based framework. We utilize the equirectangular videos as inputs because they are the most common sphere-to-plane panoramic representation which is utilized by many spherical panoramic cameras (Insta360, Samsung Gear 360, Ricoh Theta S). Also note that in our work, the equirectangular image is expanded from the internal surface of sphere, which is equivalent to shooting the spherical image with a virtual camera at the sphere center.

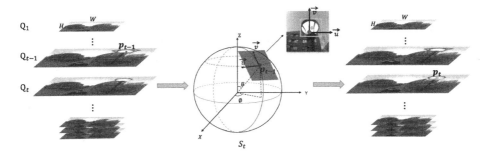

Fig. 5. The overview our framework. Red point and bounding box represent the target position and size at frame $t-1$ and green ones correspond to those at frame t. (Color figure online)

4.1 Framework Overview

Before explaining our tracking framework, we first briefly summarize the single object tracking problem. We denote the N-frame video as $\mathbf{I} = \{\mathbf{I}_t\}_{t=1,...,N}$. Given an initial position of an interested target which is usually marked by a bounding box, a tracker aims to estimate tight bounding boxes wrapping the target at subsequent frames. In general, the estimation can be formulated as an energy function, given by

$$E(\mathbf{box}_t) = \mathrm{dif}(\mathbf{T}_{t-1}, \mathrm{crop}(\mathbf{I}_t; \mathbf{box}_t)), \tag{1}$$

with the candidate bounding box as independent variable. The templete $\mathbf{T}_{t-1}$ represents the embedding of temporal knowledge generated from tracking results of the last $t-1$ frames. $\mathrm{crop}(\mathbf{I}_t; \mathbf{box}_t)$ means cropping a candidate region from

---

Algorithm 1: Local FOV projection based Tracking

Input: Initial bounding box $\mathbf{box}_1 \in \Re^4$ of interested object, equirectangular
    input is set to $\mathbf{Q}$
Output: $\{\mathbf{box}_t | t = 2, ..., \mathrm{nFrames}\}$
Initialize position p_1, target size s_1 and object templete $\mathbf{T}_1$
while $t <= \mathrm{nFrames}$ **do**
    Local FOV projection:
    1: Transform $\mathbf{Q}_t$ to spherical image $\mathbf{S}_t$ via Eq. (3);
    2: Project $\mathbf{S}_t$ to tangent plane with tangent point p_{t-1} via Eq. (5);
    Target tracking:
    1: Crop candidates from tangent plane projection;
    2: Obtain $\mathbf{box}_t$ via optimizing Eq. (2);
    Model update:
    1: Use $\mathbf{T}_{t-1}$ and $\mathbf{box}_t$ to update $\mathbf{T}_t$ with proper method;
    $t = t + 1$;
end

$\mathbf{I}_t$ with bounding box $\mathbf{box}_t$. dif($\cdot$) measures the similarity of templete $\mathbf{T}_{t-1}$ and candidate region which is extracted from current frame t.

We denote the equirectangular inputs as $\mathbf{Q} = \{\mathbf{Q}_t\}_{t=1,...,N}$ with width of W and height of H, where $W = 2H$. When we straightforward track in $\mathbf{Q}$ with planar video trackers, $\mathbf{I}_t = \mathbf{Q}_t$ in Eq. (1). However, this strategy may be subjected to object distortion and split, which makes trackers easily fail. To solve this problem, in our framework shown as Fig. 5, we firstly transform the equirectangular input to the spherical domain and project the local FOV of target to the tangent plane, then we can track in the tangent plane. Specifically, we denote $\mathbf{S}_t$ as the spherical representation of $\mathbf{Q}_t$ and it can be transformed from $\mathbf{Q}_t$ with Eq. (3). Then we project the local region of $\mathbf{S}_t$ to the tangent plane with the tangent point of p_{t-1}, which is the detection of target position at frame $t-1$. Since the sequence has temporal continuity, we can safely assume that target position does not change too much between adjacent frames. Based on the assumption, we get a search region from the tangent plane whose size varies with different trackers. Then we maintain the tracking process in the search region at each frame. In this case, $\mathbf{I}_t$ in Eq. (1) can be denoted as this search region. We will have $\mathbf{I}_t = \text{tang}(\mathbf{S}_t; p_{t-1})$, where $\text{tang}(\cdot; p)$ is the tangent projection operation with tangent point p. Thus, we rewrite Eq. (1) as

$$E(\mathbf{box}_t) = \text{dif}(\mathbf{T}_{t-1}, \text{crop}(\text{tang}(\mathbf{S}_t; p_{t-1}); \mathbf{box}_t)). \tag{2}$$

After that, we return tracking results back to equirectangular frame and get the target center p_t for tracking at next frame.

4.2   Tangent Plane Projection

In this subsection, we first introduce the basic process of equirectangular-to-sphere and then present the details of obtaining the search region with certain size at frame t. Firstly, we transform the equirectangular input to spherical form via interior plane expansion of spherical image:

$$\theta = \pi \frac{x}{H}, \phi = 2\pi(1 - \frac{y}{W}), \tag{3}$$

where (x, y) are plane coordinates on equirectangular input with width of W and height of H and (θ, ϕ) is the polar coordinates representation in spherical image. In three-dimensional world coordinate system, the point on the sphere with polar coordinate of (θ, ϕ) can be rewritten as $(r \sin\theta \cos\phi, r \sin\theta \sin\phi, r \cos\theta)$ with radius $r = \frac{W}{2\pi}$.

Now given $P_0 = (r \sin \theta_0 \cos \phi_0, r \sin \theta_0 \sin \phi_0, r \cos \theta_0)$ as the tangent point, which is the three dimensional representation of p_{t-1}, we try to get a search region from the tangent plane with tangent point of P_0. The size of the search region varied from different trackers. Now we assign pixel values to each point in the search region. As shown in Fig. 5, we denote $\vec{u}$, $\vec{v}$ as a pair of unit base vectors at point P_0 on the tangent plane, where $\vec{u}$ goes along the direction of the latitude and $\vec{v}$ the direction of longitude. We can get

$$\begin{cases} \vec{u} = (\sin \phi_0, -\cos \phi_0, 0) \\ \vec{v} = (-\cos \phi_0 \cos \theta_0, -\sin \phi_0 \cos \theta_0, \sin \theta_0) \end{cases} \tag{4}$$

via relationship of space geometry. Next, we can represent all points in tangent plane to three dimensional coordinate system with $\vec{u}$, $\vec{v}$ and p_{t-1}

$$P = P_0 + i\vec{u} + j\vec{v}, \tag{5}$$

where (i, j) are two-dimensional coordinates in tangent plane with origin of P_0 and their ranges depend on size of the search region. We then connect point P and the sphere origin to get the intersection P' on sphere, which shares a common polar coordinate with P. The pixel of P' is easily obtained via sphere-to-equirectangular transformation. Then we get pixel of P assigned with P'. We get the search region with all points assigned by pixels via the above method. Algorithm 1 describes our framework in details.

4.3   Target Description with Irregular Bounding Box

In our framework, we perform the panoramic tracking process in the search region obtained from tangent plane. As we know, when tracked in planar videos, interested target is described with a rectangle bounding box. As shown in Fig. 5, we describe the target with rectangle bounding box in tangent plane in the same way. We then transform the rectangle bounding box in tangent plane to irregular (trapezoid-like) bounding box at equirectangular frame. The intention is to present and evaluate the tracking result and get the target center of current frame for tracking at next frame. In detail, we go through all points on the boundary line of the bounding box in tangent plane, and get their corresponding points in equirectangular frame with the similar method as shown in Subsect. 4.2. Then we can obtain an irregular region with the same visual field as the rectangle bounding box in tangent plane.

5   Experiment

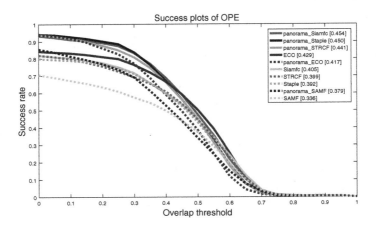

Fig. 6. Success plots of 10 trackers on POT dataset.

5.1   Setup

Our method is implemented on Matlab and runs on Intel i7 2.2GHz laptop. We boost five mainstream trackers: SAMF [14], Staple [1], ECO [3], STRCF [13] and Siamfc [2] with our framework and get five boosted trackers, i.e. panoSAMF, panoStaple, panoECO, panoSTRCF and panoSiamfc. It is noted that as mentioned in Sect. 1, existing studies of panoramic object tracking either have different panorama formats of input with ours or have certain defects, and we do not have their implementation details. Therefore, we mainly focus on the comparison between the trackers of original version and pano version on POT dataset. We use success plot to represent the performance of the trackers.

5.2   Overall Performance

In this subsection, we discuss the overall tracking results of all the 40 sequences in POT. Results of success plots are shown in Fig. 6. It clearly illustrates that except ECO, trackers combined with our framework outperform the original ones. See the AUC scores of success rate illustrated in the success plot, panoSAMF, panoStaple, panoSTRCF and panoSiamfc outperform SAMF, Staple, STRCF and Siamfc with 12.8%, 14.8%, 10.5%, 12.1% relative improvements respectively.

5.3   Performance Analysis by Attributes

For the detailed performance analysis, we perform experiments on various challenge attributes listed in Table 1. Table 2 illustrates the success rate of each tracker on different attributes. We note that for each attribute, the top three trackers are almost the pano versions. Specially, as for DIST and SP attributes which are unique in panoramic videos, the five pano trackers are far better than their corresponding original ones in both measures. As for the reasons, when there is a split in the process of crossing the left side to right, traditional trackers can not track the target anymore for losing view of target. While in pano ones, there is no such situation, because the search region produced by local FOV projection is complete. In another situation, when the target walks near bottom of the frame, traditional trackers may loses track of the target gradually for the serious target distortion. While with the search region of real shape, the performance of pano trackers is unaffected. The experiment result illustrates that trackers enhanced with our framework are well behaved in handling object distortion and split in panoramic object tracking.

Table 2. Each entry contains the AUC scores of success rate for each tracker on different attributes. The trackers are ordered by overall average overlap scores, and the top three methods for each attribute are denoted by different emphasis: bold, italic, *blue*.

Tracker	ALL	DIST	SP	CM	SV	MB	FM	BC	PO
panoSiamfc	**45.4**	**46.4**	42.6	**51.3**	**44.1**	*48.5*	**41.7**	*40.7*	*50.3*
panoStaple	*45.0*	*43.6*	**45.4**	48.4	38.1	*47.9*	38.1	**44.4**	**55.4**
panoSTRCF	*44.1*	*42.7*	*44.2*	48.2	*42.9*	**52.6**	34.7	*42.6*	*52.4*
ECO	42.9	40.8	38.2	*49.8*	38.3	45.0	*40.9*	34.9	43.5
panoECO	41.7	37.6	*43.8*	46.5	**39.5**	47.2	**39.1**	34.1	45.5
Siamfc	40.5	40.3	32.6	*48.8*	38.5	41.5	28.6	36.1	40.1
STRCF	39.9	39.2	32.9	47.2	37.7	38.2	25.8	37.8	41.8
Staple	39.2	38.7	31.7	47.7	34.9	35.8	28.6	40.6	42.6
panoSAMF	37.9	37.4	33.7	46.6	29.9	39.7	26.9	39.9	42.0
SAMF	33.6	30.0	28.4	46.7	31.9	33.0	21.4	36.1	34.5

6   Conclusion

In this paper, we have analyzed the challenges of object tracking in spherical panoramic videos and proposed a panoramic object tracking dataset POT with 40 fully annotated sequences, which are annotated on pixel level. Moreover, we have proposed a general framework based on local FOV projection for panoramic tracking, which can be combined with arbitrary trackers of planar videos. We have enhanced five mainstream trackers, which are well-studied in planar video object tracking, with our framework. Experimental results of the enhanced trackers and original versions on POT have validated the superiority of our proposed framework in panoramic object tracking.

References

1. Bertinetto, L., Valmadre, J., Golodetz, S., Miksik, O., Torr, P.H.: Staple: Complementary learners for real-time tracking. In: CVPR (2016)
2. Bertinetto, L., Valmadre, J., Henriques, J.F., Vedaldi, A., Torr, P.H.S.: Fully-convolutional Siamese networks for object tracking. In: Hua, G., Jégou, H. (eds.) ECCV 2016. LNCS, vol. 9914, pp. 850–865. Springer, Cham (2016). https://doi.org/10.1007/978-3-319-48881-3_56
3. Danelljan, M., Bhat, G., Shahbaz Khan, F., Felsberg, M.: ECO: efficient convolution operators for tracking. In: CVPR (2017)
4. Delforouzi, A., Tabatabaei, S.A.H., Shirahama, K., Grzegorzek, M.: Unknown object tracking in 360-degree camera images. In: ICPR (2016)
5. Esteves, C., Allen-Blanchette, C., Makadia, A., Daniilidis, K.: Learning SO(3) equivariant representations with spherical CNNs. In: ECCV (2018)
6. Fan, H., et al.: LaSOT: a high-quality benchmark for large-scale single object tracking. In: The IEEE Conference on Computer Vision and Pattern Recognition (CVPR), June 2019
7. Guan, H., Smith, W.A.: BRISKS: binary features for spherical images on a geodesic grid. In: CVPR (2017)
8. Huang, L., Zhao, X., Huang, K.: GOT-10k: a large high-diversity benchmark for generic object tracking in the wild. IEEE Trans. Pattern Anal. Mach. Intell. (2019)
9. Kalal, Z., Mikolajczyk, K., Matas, J.: Tracking-learning-detection. TPAMI **34**(7), 1409–1422 (2011)
10. Kiani Galoogahi, H., Fagg, A., Huang, C., Ramanan, D., Lucey, S.: Need for speed: A benchmark for higher frame rate object tracking. In: The IEEE International Conference on Computer Vision (ICCV), October 2017
11. Kristan, M., et al.: The visual object tracking VOT2015 challenge results. In: The IEEE International Conference on Computer Vision (ICCV) Workshops, December 2015
12. Li, A., Lin, M., Wu, Y., Yang, M., Yan, S.: NUS-PRO: a new visual tracking challenge. IEEE Trans. Pattern Anal. Mach. Intell. **38**(2), 335–349 (2016)
13. Li, F., Tian, C., Zuo, W., Zhang, L., Yang, M.H.: Learning spatial-temporal regularized correlation filters for visual tracking. In: CVPR (2018)
14. Li, Y., Zhu, J.: A scale adaptive kernel correlation filter tracker with feature integration. In: Agapito, L., Bronstein, M.M., Rother, C. (eds.) ECCV 2014. LNCS, vol. 8926, pp. 254–265. Springer, Cham (2015). https://doi.org/10.1007/978-3-319-16181-5_18
15. Liang, P., Blasch, E., Ling, H.: Encoding color information for visual tracking: Algorithms and benchmark. IEEE Trans. Image Process. **24**(12), 5630–5644 (2015)
16. Liu, K.C., Shen, Y.T., Chen, L.G.: Simple online and realtime tracking with spherical panoramic camera. In: ICCE (2018)
17. Liu, L., Yan, Z., Liu, Q.: Panoramic visual tracking based on adaptive mechanism. JVCIR **57**, 99–106 (2018)
18. Muller, M., Bibi, A., Giancola, S., Alsubaihi, S., Ghanem, B.: TrackingNet: a large-scale dataset and benchmark for object tracking in the wild. In: The European Conference on Computer Vision (ECCV) (2018)
19. Su, Y.C., Grauman, K.: Learning spherical convolution for fast features from 360 imagery. In: NIPS (2017)
20. Wu, Y., Lim, J., Yang, M.H.: Online object tracking: a benchmark. In: The IEEE Conference on Computer Vision and Pattern Recognition (CVPR), June 2013

21. Xiao, J., Ehinger, K.A., Oliva, A., Torralba, A.: Recognizing scene viewpoint using panoramic place representation. In: CVPR (2012)
22. Zhou, Z., Niu, B., Ke, C., Wu, W.: Static object tracking in road panoramic videos. In: ISM (2010)
23. Zhu, Y., Zhai, G., Min, X.: The prediction of head and eye movement for 360 degree images. SPIC **69**, 15–25 (2018)

DP-YOLOv5: Computer Vision-Based Risk Behavior Detection in Power Grids

Zhe Wang[✉], Yubo Zheng, Xinhang Li, Xikang Jiang, Zheng Yuan, Lei Li, and Lin Zhang

Beijing University of Posts and Telecommunications, Beijing 100876, China
tylorwang@bupt.edu.cn

Abstract. The level of safety in power grid construction has been improved by Computer Vision (CV) recently with deep Convolutional Neural Networks (CNN). However, due to environmental complexity and risk behaviors diversity, the current detection algorithms still have false and missing detection problems. This paper quantitatively analyses these practical problems and proposes a Double Precise YOLOv5 (DP-YOLOv5) method. Compared to other state-of-art detectors, DP-YOLOv5 highlights three points: integrating multi objects for each classification to avoid false detection in complex environments, adding standard operation samples for guidance to reduce the missing detection caused by risk behaviors diversity, and using Depthwise Separable convolutional networks to reduce model parameters. The proposed DP-YOLOv5 method is evaluated on a dataset with 2.5k images generated in real power grid operation environments provided by State Grid Jiangsu Electric Power Co., Ltd. Compared with the state-of-art YOLOv5s detector. Experimental results show that the precision of DP-YOLOv5 is 7.1% higher, while the model size is 20% less.

Keywords: Risk behavior detection · Power grids · DP-YOLOv5 · Real-time detection

1 Introduction

1.1 Computer Vision

As one of the applications in Artificial Intelligence (AI), CV plays an essential role in the intelligent healthcare system, automatic driving, face recognition, and behavior recognition. With powerful feature extraction capability, Convolutional Neural Networks (CNN) and Recurrent Neural Networks (RNN) have been continuously used in the CV area and achieved outstanding performance. For instance, [1] proposes NOise Tolerant Ensemble RCNN (NOTE-RCNN) consists of two classification heads and a distribution head to avoid overfitting caused by noisy and false-negative labels. To enhance the robustness and accuracy of

---

Z. Wang—Student.

© Springer Nature Switzerland AG 2021
H. Ma et al. (Eds.): PRCV 2021, LNCS 13019, pp. 318–328, 2021.
https://doi.org/10.1007/978-3-030-88004-0_26

the model detection in different environments, [2] proposes an approach named Improved Faster Regions with Convolutional Neural Network Features (IFaster R-CNN) to detect the presence of engineers and excavators at a great accuracy under different backgrounds. [3] uses multi-relation detection and contrastive training strategies to achieve good target recognition results when the training samples are insufficient. Also, for the Incremental Few-Shot Detection (IFSD) problem, [4] proposes an Open-ended Centre Net (ONCE) algorithm, a detector designed for incrementally learning to detect novel class objects with merely a few samples. For the problem of sample's unbalanced distribution, [5] uses a concurrent softmax layer to handle the multi-label issues in object detection and proposes a soft sampling method with a hybrid training scheduler to deal with the label imbalance problem.

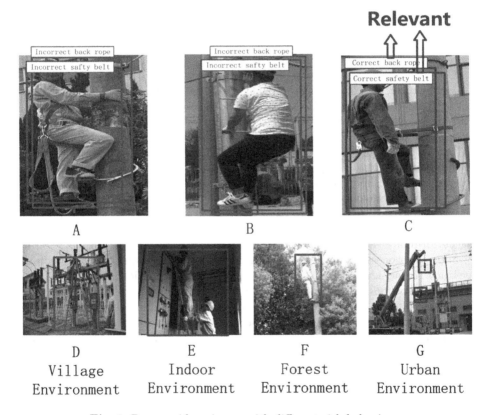

Fig. 1. Power grid engineers with different risk behaviors.

1.2 Automatic Risk Behavior Detection

Automatic risk behavior detection ensures workplace safety and productivity is one of CV areas' most critical challenges. In actual power grid operation worksites, power companies often make a series of strict operation regulations to guarantee a safe working environment for engineers. However, skilled engineers often ignore safety regulations. They use safety equipment according to their habits for convenience or comfort, which increases the difficulties of risk behavior detection. As shown in Fig. 1, skilled engineers with different working habits may tie the back rope anywhere near their bodies. In addition, due to the ubiquitous power grid facilities and quantities of operations, the risk behavior detection in power grid construction has the characteristics of diverse operating environments. The geographical environment and operation mode of urban cities, villages, and forest areas are quite different. Typical power grid operation scenes are shown in Fig. 1. It can be seen that the background of forest area grid construction has much more noise than that in urban and indoor environments. The quantities of background noise and diverse operation types of worksites will decrease the accuracy of risk behavior detection.

Generally, the risk behavior identification methods are based on image feature extraction. In [2], an Improved Faster Regions with Convolutional Neural Network Features (IFaster R-CNN) method is proposed by introducing deep-learning-based object detection to demonstrate the possibility of detecting workers without relying on assumptions about backgrounds used in previous research. [6] proposes an automated CV-based method that uses two CNN models to determine whether workers are wearing their harnesses while in operation. However, all these methods fail to detect different types of risk behaviors at one time, which is not practical in the real construction scene. Recently, the powerful image recognition algorithm You Only Look Once (YOLO) has been widely concerned by researchers. Scholars have proposed some improvements to achieve outstanding performance in object classification and risk behavior detection based on it. [7] adds preprocessing operation based on YOLOv5 to improve the detector's accuracy. To achieve a better trade-off among accuracy, efficiency, and memory storage, [8] proposes a modified YOLO-tiny method with an insulator (MTI-YOLO) network for insulator detection in complex aerial images. While YOLO-based detectors are very powerful in some industrial scenarios, they still face the challenges of localization difficulty in the complex environment [9], which is necessary for the construction of power grids.

In summary, the existing detection methods fail to meet the requirements of different risk behavior detection due to the considerable differences in background and various risk behaviors in the grid construction scene. To address these problems, a DP-YOLOv5 method based on YOLOv5s is proposed in this paper to integrate multiple objects of each risk behavior. With the correct operation samples guided, the DP-YOLOv5 can provide a more accurate and detailed location of risk behavior by comparing the object's spatial location and confidence level. Besides, a Depthwise Separable convolution network [10] is also being added to the backbone of DP-YOLOv5 to improve the detection speed. The innovations of this paper are summarized as follows:

1) A DP-YOLOv5 method is proposed to detect various unsafe behaviors in power grid construction, including ladder-climbing without support, incorrect back rope equipping, incorrect safety belt equipping, and hook unlocking.
2) The standard samples guidance and multi-object integration methods are used to improve the accuracy of different risk behaviors detection.
3) The Depthwise Separable Convolution networks are used in DP-YOLOv5 to achieve the roughly same performance in precision with merely 80% of the parameters, which is more suitable for the power grid scenario with insufficient computing power.

2 Method

This section proposes the baseline and enhancement of the proposed DP-YOLOv5 method. As shown in Fig. 2, the framework of the DP-YOLOv5 is mainly composed of three parts: (1) A Depthwise Separable convolution-based CNN feature extraction module. (2) A multi-objects integration algorithm to improve detection accuracy. (3) A standard samples guidance algorithm to decrease missing detection.

2.1 Revisit YOLOv5s

Baseline. The proposed DP-YOLOv5 method is based on the open-source object-detection project YOLOv5. Based on the scale of the parameters, the YOLOv5 network can be divided into four models named YOLOv5s, YOLOv5m, YOLOv5l, and YOLOv5x, whose parameters scale increases in sequence. The YOLOv5s is used as the DP-YOLOv5 base model for its high run-time speed. The structure of the YOLOv5s model is mentioned in [11]. In the YOLOv5s model, the Focus module performs slice and concatenate operations to the input image. The CBL module consists of the convolution layer, BatchNormal, and leakyRELU, which is the basic convolution module. The CSP (Cross Stage Partial) module mainly performs feature extraction from feature maps, contributing most of the model's parameters. The SPP module is spatial pyramid pooling to prevent image distortion. The Concate module is to concatenate layers together. YOLOv5s also adds a bottom-up feature pyramid structure based on the PAN (Path Aggregation Network) structure to convey strong semantic features from top to bottom and robust positioning features from the bottom to up. Combining feature aggregation from different feature layers to improve the network's ability to detect different scales' targets.

Training Schedule. On our custom dataset, the network is trained with stochastic gradient descent (SGD) for 100 iterations on an RTX3070 GPU. The IoU is set to 0.6, and the batch size is set from 16 to 32. The initial learning

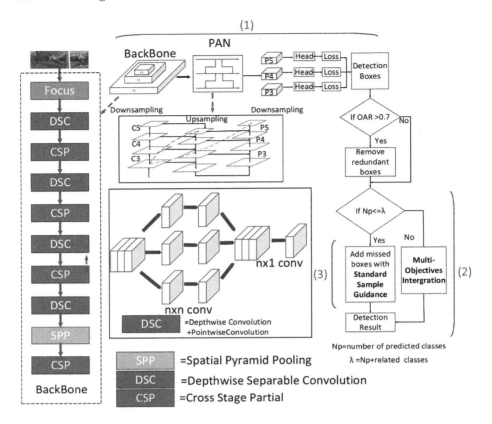

Fig. 2. Detecting Process of DP-YOLOv5

rate is set to 2 e–4. The experimental environment is a PyTorch deep learning framework based on the Ubuntu system. We use CUDA11 + cuDNN8 to accelerate GPU operations. The hardware configuration for training and testing is Intel i7-10700 CPU, Nvidia GeForce RTX3070 GPU, and 32G memory.

2.2 Selection of Enhancement

Depthwise Separable Convolution. Depthwise Separable convolution (DS) network, which consists of Depthwise convolution (DW) and Point-wise convolution (PW), is usually used to decrease the amount of calculation in feature extraction. By splitting the original image into different feature maps, the number of parameters decreases to 1/3 of the calculation amount of the convolution. Therefore, all convolution networks in the YOLOv5 backbone are replaced by DS convolution networks to ensure the speed of feature extraction. The detailed structure of DS convolution networks is shown in Fig. 2.

Smaller Input Size. Decrease the input size to minimize the area of objects. Thus, information of the objects will be calculated faster than before. As a result, Recognition efficiency will be increased. A Smaller input size occupies less memory, which will also in turn to speed up the training process by increasing batch size. To be more specific, the batch size is increased from 16 images per GPU to 32 images per GPU, and the input size is decreased from 640 to 480. Although this process will reduce the recall rate of risk behaviors, the proposed standard samples guidance and multi-objects integration algorithm in the risk evaluation mechanism can effectively improve the accuracy of risk behavior identification (Fig. 3).

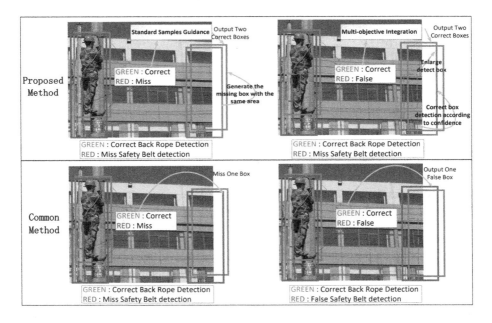

Fig. 3. Compared with common methods.

Standard Samples Guidance. Because of the enormous types of risk behaviors, CNN may fail to extract enough features from images. It is inevitable for detectors to have lower precision and even omit detection. However, the required standard behaviors are generally the same, and some have a strong correlation. When the expected behaviors are not detected, the behaviors of the engineers are generally risky. Besides, it is often necessary to wear a variety of safety equipment for grid operations. For example, in Fig. 1, engineers need to equip safety ropes and safety belts while climbing a power pole. Thus the safety rope equipping and safety belt equipping are behaviors with solid correlation. This prior knowledge can also be used to improve detection accuracy. Therefore, with the help of numerous standard operations and other associated safety behaviors, the proposed Standard Samples Guidance can reduce the probability of missing recognition. The specific steps are as follows:

Algorithm 1: Standard Samples Guidance Algorithm

---

1 **Input:** Risk behavior objects B={B_1,B_2,...,B_j}
2 **Output:** B with missed behavior objects M={M_1,M_2,...,M_j}

   for i=1:length(B) **do**
      Check whether the behavior B_i has associated behaviors
      if Behavior B_i is associated with Behavior M_i **then**
         if Operation M_i is not detected **then**
            Add operation M_i to objects set B
         end if
      end if
   end for
   while condition **do**
      ...
   end while

   return objects set B={B_1,B_2,...,B_j,...,M_i}

Multi-objects Integration. With the complexity and diversity of the construction environment, the irrelevant background noise will significantly affect the identification accuracy, resulting in false identification of risk behavior. To solve this problem, a Multi-objects integrate algorithm, which can output the target location of both detail and overall boxes of risk behaviors, is being used to improve the accuracy of risk behavior identification. The algorithm flow is as follows:

1) $R_r = \{R_r(1), R_r(2), ..., R_r(i)\}$ and $R_d = \{R_d(1), R_d(2), ..., R_d(j)\}$ represent the collection of risk behavior objects and specific parts of risk behavior objects from one image. Sort all candidate objects in R_r and R_d by center location from left to right.
2) The objects Overlapping Area Ratio (OAR) is calculated by traversing the candidate objects $R_d(i)$ in R_d and comparing it with elements in R_r: OAR = area of overlap area/area of $R_d(i)$.
3) Comparing the OAR with the set threshold of $\gamma = 0.7$. If the OAR is greater than γ, the object $R_d(i)$ is considered to belong to risk behavior object $R_r(j)$. Thus all the objects are sorted as:
   $R_{pairs}(k) = \{(R_r(1), R_d(1)), (R_r(2), R_d(2)), ..., (R_r(k), R_d(k))\}$
4) The class names of objects $R_r(k)$ and $R_d(k)$ will be compared, once they are the same, the $R_r(k)$ object will be deleted to avoid redundant. When the class names of a pair of objects are different, comparing their confidence and deleting the object with a lower value.

3   Experiments

3.1   Dataset

The training dataset consists of 2.5k images generated in a real power grid operation environment given by State Grid Jiangsu Electric Power Co., Ltd. It includes correct/incorrect safety belt equipping, correct/incorrect back rope equipping, climbing ladder with/without support, hook locking/unlocking. Each image represents one or several correct/incorrect behaviors in different working environments. 15% of images are separated as the testing set, while the rest 85% of images are used as the training set. Risk behaviors presented on the images were surrounded with the bounding boxes by qualified members of our team. The training dataset also includes negative samples without a complete panorama of risk behaviors.

3.2   Ablation Studies

To demonstrate the effectiveness of proposed DP-YOLOv5 method, the effectiveness of each module is presented incrementally. Experimental results are shown in Table 1, where the effectiveness of modules is evaluated by precision, recall, parameters(size of the model), inference time, and total processing time per image. With the same IoU set as 0.6, the detection results are shown in Fig. 4. For each pair, DP-YOLOv5 can detect more risk behaviors and output a more precise position. The recall and precision rate are calculated as fellow:

$$\text{Recall} = \frac{TP}{TP + FN} \tag{1}$$

$$\text{Precision} = \frac{TP}{TP + FP} \tag{2}$$

TP means that specific behavior in the detection image is correctly identified; FP represents that other behaviors are misidentified as the behavior to be detected; TN represents that other behavior is recognized as other behavior; FN means that the behavior to be detected is identified as another behavior.

Table 1. The ablation study of DP-YOLOv5 module.

	Methods	Precision	Recall	Parameters	Infer time	Total process time (3070)
A	YOLOv5s	0.858	0.752	13.7 MB	1.6	2.3
B	A + DS Convolution	0.849	0.773	11.9 MB	1.4	2.1
C	B + Smaller Input Size	0.838	0.726	11.9 MB	0.9	1.6
D	DP-YOLOv5	0.935	0.746	11.9 MB	0.9	1.9

Fig. 4. Qualitative detection examples on test dataset with YOLOv5s and DP-YOLOv5. The left is the result of YOLOv5s and right is the result of DP-YOLOv5.

A. First of all, the original design of YOLOv5s is chosen to work as the baseline. Training settings are described in Sect. 3.2 entirely. As shown in Table 1, the YOLOv5s achieves about 85% and 75.3% in Precision and Recall, respectively. With 13.7 MB parameters, the detection speed of YOLOv5s can reach 2.3 ms.

A → B. The first improvement is to change all the CNN in the YOLOv5s backbone to Depthwise Separable convolution networks. The detailed structure of the Depthwise Separable convolution network is shown in Fig. 2. The DW (Depthwise Convolution network) and PW(Point-wise Convolution network) are a group of convolution structures. By splitting the image into different channels, the Depthwise Separable convolution network decreases the processing time from 2.3 ms to 2.1 ms. Although model B is a bit inaccurate compared with model A, with an approximate 1% and 2% decrease in precision and recall, respectively, the efficiency of multi-channel feature extraction promotes Depthwise Separable convolution to be adopted in the final model.

B → C. Since the input size of YOLOv5 during evaluation is 640, the training and evaluation input size is decreased to 480 to increase efficiency. The processing time improves from 2.1 ms to 1.6 ms, and the precision and recall decrease about 1% and 5%, respectively.

C → D. The proposed risk evaluation mechanism consists of standard samples guidance and multi-objects integration at the end of output to decrease the false and omission of detection, increasing about 10% and 2% in precision and recall, respectively.

3.3 Comparison with Other State-of-art Detectors

The DP-YOLOv5 method is also compared with other state-of-art detect methods to verify the proposed method's effectiveness. The infer time, Precision, Recall, and total processing time are calculated to assess the quality and efficiency of objects detected by the trained network.

Table 2. Comparison of detection accuracy of DP-YOLOv5 and other detectors.

Methods	Image size	Precision	Recall	Parameters	Infer time	Total process time (3070)
YOLOv5s	640	0.858	0.752	13.7 MB	1.6	2.3
YOLOv5s	480	0.87475	0.68575	13.7 MB	0.9	1.6
YOLOv5s	320	0.88275	0.6385	13.7 MB	0.5	1.1
YOLOv5s+Mobilenetv2	640	0.653	0.804	7.2 MB	2.0	3.0
YOLOv5s+Mobilenetv2	480	0.705	0.738	7.2 MB	2.0	3.0
YOLOv5s+Mobilenetv2	320	0.750	0.733	7.2 MB	2.0	3.0
YOLOv5l	640	0.849	0.772	89.4 MB	5.4	6.0
YOLOv5l	480	0.878	0.662	89.4 MB	3.2	3.9
YOLOv5l	320	0.79	0.756	89.4 MB	1.7	2.3
YOLOv5l+Mobilenetv2	640	0.684	0.71	29.6 MB	3.1	4.1
YOLOv5l+Mobilenetv2	480	0.75	0.722	29.6 MB	1.8	2.8
YOLOv5l+Mobilenetv2	320	0.73	0.69	29.6 MB	1.0	1.9
YOLOv5m	640	0.887	0.778	40.5 MB	3.4	4.0
YOLOv5m	320	0.863	0.699	40.5 MB	1.1	1.7
YOLOv5x	640	0.889	0.80	167 MB	10.3	11.0
YOLO-Mobile	640	0.67	0.749	42.3 MB	2.7	3.7
YOLO-Mobile	480	0.702	0.733	42.3 MB	1.6	2.6
YOLO-Mobile	320	0.734	0.644	42.3 MB	0.9	1.8
DP-YOLOv5	480	0.935	0.746	11.9 MB	1.2	2.1

As shown in Table 2, the DP-YOLOv5 method is compared with the YOLOv5 series method (s, l, m, x) and a YOLO-Mobile method proposed in [12] because of their similar structure. With the same IoU set as 0.6, it clearly shows that the DP-YOLOv5 outperforms the YOLO-Mobile 36% in precision and about 30% decrease in processing time. Besides, the backbone of YOLOv5s and YOLOv5l is changed to MobileNetv2 for experiments to evaluate the efficiency of DP-YOLOv5. Experiment results show that, although the parameters of YOLOv5s+Mobilenetv2 are 30% smaller than DP-YOLOv5. The DP-YOLOv5 is 28% and 25% higher in precision compared with YOLOv5s+Mobilenetv2 and YOLOv5l+mobilenetv2, respectively.

Therefore, It can be concluded that compared with other state-of-art methods, the proposed DP-YOLOv5 has certain advantages in the balance of speed and accuracy in power grid risk behavior detection.

4   Conclusion

This paper presents a method called DP-YOLOv5 to detect risk behaviors in the power grid construction, which quantitatively analyzes practical problems in worksites and proposes a risk evaluation mechanism to handle these problems. Compared to other state-of-art detectors, it contains three highlights: integrate multi-objects from each classification to avoid false detection in complex environments, add standard operation samples to assist risk operation identification and reduce the risk behaviors omission, and the convolutional network in the backbone module is substituted with the Depthwise Separable Convolution networks, which obtains detection efficiency. Evaluations were done on a dataset provided by State Grid Jiangsu Electric Power Co., Ltd. The effectiveness of the proposed methods shows that it has achieved state-of-art performance. In the future, more risk behaviors will be added for further evaluation, and it's believed that the proposed DP-YOLOv5 can help to ensure engineers' safety in the actual power grid construction.

References

1. Gao, J., Wang, J., Dai, S., Li, L.-J., Nevatia, R.: Note-RCNN: noise tolerant ensemble RCNN for semi-supervised object detection. In: Proceedings of the IEEE/CVF International Conference on Computer Vision, pp. 9508–9517 (2019)
2. Fang, W., Ding, L., Zhong, B., Love, P.E.D., Luo, H.: Automated detection of workers and heavy equipment on construction sites: a convolutional neural network approach. Adv. Eng. Inform. **37**, 139–149 (2018)
3. Fan, Q., Zhuo, W., Tang, C.-K., Tai, Y.-W.: Few-shot object detection with attention-RPN and multi-relation detector. In: Proceedings of the IEEE/CVF Conference on Computer Vision and Pattern Recognition, pp. 4013–4022 (2020)
4. Perez-Rua, J.-M., Zhu, X., Hospedales, T.M., Xiang, T.: Incremental few-shot object detection. In: Proceedings of the IEEE/CVF Conference on Computer Vision and Pattern Recognition, pp. 13846–13855 (2020)
5. Peng, J., Bu, X., Sun, M., Zhang, Z., Tan, T., Yan, J.: Large-scale object detection in the wild from imbalanced multi-labels. In: Proceedings of the IEEE/CVF Conference on Computer Vision and Pattern Recognition, pp. 9709–9718 (2020)
6. Fang, W., Ding, L., Luo, H., Love, P.E.D.: Falls from heights: a computer vision-based approach for safety harness detection. Autom. Constr. **91**, 53–61 (2018)
7. Derakhshani, M.M., et al.: Assisted excitation of activations: a learning technique to improve object detectors. In: Proceedings of the IEEE/CVF Conference on Computer Vision and Pattern Recognition, pp. 9201–9210 (2019)
8. Liu, C., Yiquan, W., Liu, J., Han, J.: MTI-YOLO: a light-weight and real-time deep neural network for insulator detection in complex aerial images. Energies **14**(5), 1426 (2021)
9. Redmon, J., Farhadi, A.: YOLO9000: better, faster, stronger. In: Proceedings of the IEEE Conference on Computer Vision and Pattern Recognition, pp. 7263–7271 (2017)
10. Lin, Y., et al.: A high-speed low-cost CNN inference accelerator for depthwise separable convolution. In: 2020 IEEE International Conference on Integrated Circuits, Technologies and Applications (ICTA), pp. 63–64. IEEE (2020)
11. Liu, Y., BingHang, L., Peng, J., Zhang, Z.: Research on the use of YOLOv5 object detection algorithm in mask wearing recognition. World Sci. Res. J. **6**(11), 276–284 (2020)
12. Howard, A.G. et al.: Efficient convolutional neural networks for mobile vision applications

Distillation-Based Multi-exit Fully Convolutional Network for Visual Tracking

Ding Ma and Xiangqian Wu$^{(\boxtimes)}$

School of Computer Science and Technology, Harbin Institute of Technology,
Harbin, China
{madingcs,xqwu}@hit.edu.cn

Abstract. Obtaining a trade-off between accuracy and efficiency for a convolutional neural network is highly desired in the deep classification-based trackers. However, it is observed that existing methods make the predictions with the latest exits strategy for all the samples, making such strategy a time-consuming solution. Motivated by this, we propose a multi-exit architecture based on the principle of knowledge distillation to improve the speed of prediction by encouraging early exits to imitate later and more accurate exits. Specifically, we propose a distillation-based multi-exit fully convolutional network (FCN), named DMENet, for visual tracking. In DMENet, different types of attention mechanisms are embedded into different representation levels of FCN to capture more discriminative information. Then, three exits augment at different levels of FCN to handle the processing of a frame to stop early. The DMENet is trained offline with knowledge distillation to improve the accuracy of early exits. The confidence score of an exit is utilized to decide whether to locate the target with high confidence on this exit or continue processing the next exit. The extensive evaluation performed on OTB-100, UAV123, LaSOT and VOT2018 benchmarks demonstrate the proposed tracker outperforms state-of-the-art approaches with a high speed (36 FPS).

Keywords: Knowledge distillation · Multi-exit fully convolutional network · Visual tracking

This work was supported in part by the National Key R&D Program of China under Grant 2018YFC0832304 and 2020AAA0106502, by the Natural Science Foundation of China under Grant 62073105, by the Distinguished Youth Science Foundation of Heilongjiang Province of China under Grant JC2018021, by the State Key Laboratory of Robotics and System (HIT) under Grant SKLRS-2019-KF-14 and SKLRS-202003D, and by the Heilongjiang Touyan Innovation Team Program.

Electronic supplementary material The online version of this chapter (https://doi.org/10.1007/978-3-030-88004-0_27) contains supplementary material, which is available to authorized users.

© Springer Nature Switzerland AG 2021
H. Ma et al. (Eds.): PRCV 2021, LNCS 13019, pp. 329–341, 2021.
https://doi.org/10.1007/978-3-030-88004-0_27

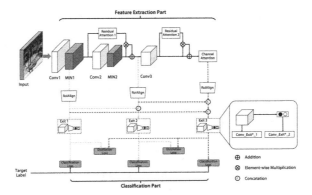

Fig. 1. An overview of the proposed multi-exit network by distillation training. The proposed framework consists of a feature extraction part and a classification part (including 3 exits). The green, yellow and red lines indicate the data stream for each exit. The detailed architecture is shown in the Supplementary Material. (Color figure online)

1 Introduction

With the delightful performance in extracting high-level feature representation, the Convolutional Neural Networks (CNNs) are effective in visual tracking tasks. However, in addition to the high accuracy, most CNN-based trackers are too slow. In this paper, we aim to develop a tracking-by-detection tracker which achieve both accuracy and speed.

State-of-the-art CNN-based trackers can be divided into two groups. The first group of CNN-based trackers follows the similarity tracking strategy [2,6,7,16,17,28]. Siamese network architecture is used in these trackers. The target is located by the similarity measure between the target template and search area. Benefit from its no update setting, similarity trackers achieve relatively high speed with low accuracy. To further improve the speed of similarity trackers, EAST [13] proposes to forward the image through a multi-exit architecture similar to the one used in our paper. However, our work has two distinct features: first, we optimize the multi-exit architecture in an end-to-end manner, while in [13], the mixed EAST independently learn the correlation filters with cheap pixel features and optimize the Siamese network with deep features. In the deep learning era, EAST can hardly benefit from end-to-end training. Secondly, we do not process the challenging frames until later exits. Take advantage of the knowledge distillation, we transfer the information from late to early exits by encouraging early exits to deal with challenging frames easy. Finally, we stack two residual attention modules with top and bottom layers of the FCN to boost the target-specific information from early to late exits.

The second group of approaches utilizes the classification strategy. An online learned classifier network is used in these trackers and the target is located by the tracking-by-detection technique. Even such trackers achieve state-of-the-art accuracy, their tracking speed is very low due to the computational cost of the latest exits strategy for all the samples. [15] is proposed to speed up the [19] with RoIAlign technique, but sacrificed accuracy. A robust classification-based tracker ideally should have state-of-the-art accuracy while maintaining real-time tracking speed. The accuracy of classification-based tracking approaches can be improved by more discriminative features, and the higher speed can be achieved by multi-exits architecture. To achieve these goals, we propose a distillation based multi-exits network (that we refer to as DMENet) for classification-based tracking (see Fig. 1). The proposed approach takes advantage of knowledge distillation for multi-exit classification architecture. This encourages early exits to mimic to probabilistic outputs of later exits and hence the discriminative power of early exits is improved. Besides, the proposed approach leverages multi-RoIAlign to exploit multi-level and multi-resolution information from multiple convolutional layers and then fuse them in each exit to improve the discrimination capacity. Furthermore, the proposed approach uses different kinds of attention modules to capture different target-specific information. The proposed approach achieves state-of-the-art accuracy in classification-based tracking without compromising the tracking speed. In summary, the main contributions of this work are summarized as follows:

I, We propose an end-to-end multi-exits learning framework based on knowledge distillation, which improves the discriminative ability of early exits to cope with challenging frames easy.

II, To improve the discriminative power of later exits, we use multi-RoIAlign to exploit multi-level information from multiple convolutional layers and then fuse them in later exits. Besides, different kinds of attention modules are introduced to improve the discriminative power further.

III, Quantitative and qualitative evaluations demonstrate the outstanding performance of our tracking algorithm compared to the state-of-the-art techniques in four public benchmarks with real-time speed of 36 FPS.

2   Related Work

2.1   Visual Tracking

Here, we will review the most relevant visual object trackers and techniques: similarity-based trackers and classification-based trackers. Similarity-based trackers [2,6,7,16,17,28] formulate visual tracking as a similarity tracking problem by searching the similarity of a target template in subsequent search area. In this group of methods, EAST [13] use a multi-exit architecture similar to the one used in our paper, their high speed relies on hand-crafted features and limited uses of deep representation. Contrary to EAST, the whole pipeline in our tracker within a pure deep neural network. Different from similarity-based trackers, the classification-based tracking framework is consists of a sampling stage and a classification stage. MDNet [19] is a popular classification-based tracking algorithm with state-of-the-art accuracy. MDNet passes the candidate regions

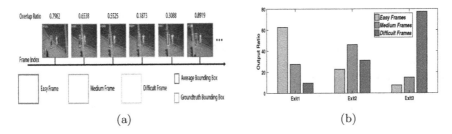

Fig. 2. (a) is the visual examples of easy, medium and difficult frames in the *Jogging-1* sequence. (b) is the ratio of each category coming out from each exit.

through a CNN pre-trained on a large-scale dataset. To solve the problem of appearance variations and class imbalance in MDNet, Song *et al.* [24] use adversarial learning to obtain the most robust features of the objects over a long period and proposed a high-order cost-sensitive loss to decrease the effect of class imbalance. However, the speed of such trackers is usually around 1 FPS which is too slow for practical use. Motivated by the RoI technique in Fast R-CNN [10], RT-MDNet [15] is proposed to speed up the MDNet by accelerating the process of candidate regions extraction procedure. Even though RT-MDNet achieves high speed, the tracking accuracy is decreased. Our objective is to improve accuracy without compromising the tracking speed by distillation based multi-exits architecture.

2.2   Multi-exit Architecture

With the development of CNNs, networks with additional exits to anytime prediction have gained more attention. Motivated by the fact that early layers act as the role of feature extractors for later one, Huang *et al.* [14] designed a multi-exit architecture depending on the DenseNet. Recently, Phuong *et al.* [20] proposed the early stopping multi-exit architecture. We take advantage of multi-exit architecture to handle different challenging frames by different exits.

2.3   Knowledge Distillation

The goal of Knowledge Distillation is to transfer knowledge between two models. In an early work, Hinton *et al.* [11] demonstrated that distilling works very well for transferring knowledge from an ensemble or a large highly regularized model into a smaller, distilled model. Subsequently, [22] proposed a training strategy that a model is self-trained by distillation from its previous predictions. Recently, Phuong *et al.* [20] proposed distillation from one part of a model to another part.

3   Proposed Tracking Method

3.1   Identify the Number of Exits

To identify the suitable number of exits, we assume that the frames of a sequence can be classified into three categories: easy (containing small appearance variation), medium (containing medium appearance variation), and difficult (containing large appearance variation) frames in this paper. DMENet with three exits is proposed by the assumption that the target in easy, medium and difficult frames can be located better in the low, medium and high-level feature maps of the network correspondingly. To demonstrate the rationality of our assumption, DMENet without distillation training is tested on the OTB-100 [27] dataset.

The frames of sequences in OTB-100 are classified into three classes: easy, medium and difficult. Different categories of frames are based on the average bounding box overlap ratios among all the 12 compared trackers which are shown in Sect. 5.2. In detail, the medium and difficult categories are composed of the frames that have lower average bounding box overlap ratios than two predefined thresholds 0.7 and 0.5, and that the frames from easy category higher average bounding box overlap ratio than 0.7. As illustrated in Fig. 2a, we show some visual examples of easy, medium, and difficult frames. Three categories of frames are represented with different levels of easy, medium and difficult.

To gather statistics of the real output ratio in each exit, the baseline with 3 exits are trained without knowledge distillation. In each exit, a confidence score is utilized to decide whether to locate the target of the current frame by this exit (high confidence) or leave it to be done by the next exit (low confidence). Hence, only when the confidence score of the current exit is up to a threshold, the target can be located in the current exit. Figure 2b shows that most of the easy/medium/difficult frames come out from the 1st/2nd/3rd exit correspondingly, which demonstrates the rationality of the assumption.

3.2   Network Architecture

As shown in Fig. 1, the proposed network consists of three convolutional layers (conv1-3) that are cloned from the VGG-M network [5]. Two Maxout network In Network (MIN) modules [4] are stacked after the first and second convolutional layer to increase nonlinearity and alleviate the problem of vanishing gradients in ReLU. The three convolutional layers and two MIN modules are used for constructing a shared feature map, which is referred to as the baseline. Sequently, three attention modules are stacked with the baseline network; two residual attention modules and one channel attention module. Each exit shares the same architecture: a ROIAlign layer for extracting features from each RoI, and two convolutional layers (i.e. $Conv\_Exit*\_1$ and $Conv\_Exit*\_2$) for binary classification. The details of the MIN module, attention modules and exit are described in the following sub-sections, receptively.

MIN Module. Despite the advances that have been made in the deep classification-based trackers [15,19,24], some issues remain, including (1) model discriminant ability; (2) the vanishing gradients and saturation of activation units during training. Because the shared feature map [15,19,24] is extracted from a shallow convolutional architecture (*conv1-3*), such shallow architecture can not cope with the nonlinear variations of the target. Besides, the constant 0 will block the gradient flowing through inactivated ReLUs in *conv1-3*, which will result in vanishing gradients. Furthermore, changes in data distribution during the training phases may saturate the activation function, which will slow down the training process (especially for the online updating phases).

To solve these problems, we stack two MIN modules after *conv1* and *conv2*, respectively. As shown in Table 1, a two-layer multi-layer perceptron (MLP) is used to increase the nonlinearity of local patches. After each MLP, there is a maxout unit to overcome the vanishing gradient problem commonly encountered when using ReLU. A maxout unit can be expressed as follows:

$$\mathbf{F}_{i,j,ch} = \max_{m \in [1,k]} (\mathbf{w}_{ch_m}^{\mathrm{T}} \mathbf{x}_{i,j} + b_{ch_m}) \tag{1}$$

where $\mathbf{x}_{i,j}$ is the input centered at the location (i,j), ch is the index of channel in feature, $\mathbf{F}$, which are constructed by taking the maximum across k maxout hidden pieces. The maxout unit acts as a cross-channel max-pooling layer on a convolutional layer, which selects the maximal output to be fed into the next layer. Furthermore, batch normalization $\mathcal{BN}$ is introduced to avoid the covariate shift problem caused by the changes of data distribution. The process in the MIN module can be illustrated as follows:

$$
\begin{aligned}
\mathbf{F}^1_{i,j,ch_1} &= \mathcal{BN}(\mathbf{w}^1_{ch_1}{}^{\mathrm{T}} \mathbf{x}_{i,j} + b^1_{ch_j}), \\
\mathbf{F}^2_{i,j,ch_2} &= \max_{m \in [1,k_1]} (\mathcal{BN}(\mathbf{w}^2_{ch_m}{}^{\mathrm{T}} \mathbf{F}^1_{i,j} + b^2_{ch_m})), \\
\mathbf{F}^3_{i,j,ch_3} &= \max_{m \in [1,k_1]} (\mathcal{BN}(\mathbf{w}^3_{ch_m}{}^{\mathrm{T}} \mathbf{F}^3_{i,j} + b^3_{ch_m})),
\end{aligned}
\tag{2}
$$

Attention Modules. Since our ultimate goal is to stop early on the tracking process by knowledge distillation, the later exits should be more discriminative to guide the early exits. Motivated by this, two residual attention modules are stacked with top and bottom layers of the baseline network to boost the discriminative information at low-level and high-level features, respectively.

The architecture of residual attention modules is shown in Table 1. As shown in Table 1, max pooling is used to enlarge the receptive fields for capturing global features, and the bilinear interpolation operation is used to expand the spatial resolution to original spatial resolution. The operation of the residual module can be expressed as follows:

$$
\begin{aligned}
\mathbf{F}^{\mathbf{R}}(\mathbf{x}) &= (\mathbf{A}^{\mathbf{R}}_{\mathbf{s}}(\mathbf{x}) \otimes \mathbf{F}(\mathbf{x})) \oplus \mathbf{F}(\mathbf{x}), \\
\mathbf{A}^{\mathbf{R}}_{\mathbf{s}}(\mathbf{x}) &= \frac{1}{1 + e^{-\mathbf{A}^{\mathbf{R}}(\mathbf{x})}},
\end{aligned}
\tag{3}
$$

where $\mathbf{A_s^R}(\mathbf{x})$ is activated by a sigmoid function, $\mathbf{F^R}(\mathbf{x})$ is the residual attention feature. $\otimes$ and $\oplus$ indicate the element-wise multiplication and addition,
respectively.

After the second residual attention module, we add a channel attention module [12] to increase the sensitivity of the most discriminative channels for separating the target from the background. The channel attention module takes
the feature F as its input and removes the spatial information of $\mathbf{F}$ by a global
pooling operation. Then, the channel-wise dependencies are obtained by channel
dimension reduction and increasing with two fully-connected layers. After that,
a sigmoid function is used to obtain the channel-wise weights $\mathbf{w_c}$. The output
$\mathbf{F^C}(\mathbf{x})$ is obtained by weighting with $\mathbf{w_c}$:

$$\mathbf{F^C}(\mathbf{x}) = \mathbf{w_c} \cdot \mathbf{F}(\mathbf{x}), \tag{4}$$

Exit. In our multi-exit architecture, each exit consists of an RoIAlign layer
and a two-layer FCN with two output nodes corresponding to the target and
the background, i.e. the exit acts as a binary classifier. The output size of each
$Conv\_Exit*\_1$ is $3 \times 3 \times 128$. After that, The output size of each $Conv\_Exit*\_2$
is $1 \times 1 \times 2$. The confidence score of an exit is utilized to decide whether to locate
the target with high confidence on this exit or continue processing the next exit.

RT-MDNet [15] achieves competitive speed, the RoI features generated from
the coarsest convolutional layer ($conv3$) are limited to locate the target at the
high precision area. To enhance the representation of RoI features, we propose
to combine the RoI features from all early exits to current exit by exploiting
the finer and richer information from multiple convolutional layers than that
available from a single layer. In detail, the RoI feature is a multidimensional
array of size $3 \times 3 \times ch$, where ch is the number of channels. At the current exit,
we fuse the RoI features across all previous exits by concatenation operation.

3.3 Distillation Training for Multi-exit

The goal of our algorithm is to train the multi-exit architecture by knowledge
distillation. We denote a teacher classifier $\mathbf{t}$ and a student classifier $\mathbf{s}$ which is
learning from $\mathbf{t}$. This can be optimized by minimizing the cross-entropy of their
output:

$$\ell^{temp}(\mathbf{t},\mathbf{s}) = -temp^2 \sum_{c=1}^{C} [\mathbf{t}^{1/temp}(\mathbf{x})]_c \log [\mathbf{s}^{1/temp}(\mathbf{x})]_c,$$
$$[\mathbf{s}^{1/temp}(\mathbf{x})]_c = softmax(\mathbf{s}(\mathbf{x})/temp), \tag{5}$$
$$[\mathbf{t}^{1/temp}(\mathbf{x})]_c = softmax(\mathbf{t}(\mathbf{x})/temp),$$

where $\mathbf{t}(\mathbf{x})$ and $\mathbf{s}(\mathbf{x})$ denote the prediction of $\mathbf{t}$ and $\mathbf{s}$, respectively. $temp$ is the
temperature parameter, which controls the softness of the prediction in teacher $\mathbf{t}$.
$[\mathbf{t}^{1/temp}(\mathbf{x})]_c$ and $[\mathbf{s}^{1/temp}(\mathbf{x})]_c$ denote the soft prediction of $\mathbf{t}$ and $\mathbf{s}$, respectively.

C is the number of classes. Then we define the distillation loss in our multi-exit architecture as:

$$\mathcal{L}_{dis}(\mathbf{x}) = \frac{1}{Ex}\sum_{e=1}^{Ex}\frac{1}{|\mathcal{T}(e)|}\sum_{t\in\mathcal{T}(e)}\ell^{temp}(\mathbf{cf}_t(\mathbf{x}),\mathbf{cf}_e(\mathbf{x})) \tag{6}$$

where Ex is the number of exits. $\mathcal{T}(e) \in Ex$ is the set of teacher exits. Here, we set all exits to learn only from the last one. $\mathbf{cf}(\cdot)$ is the classifier at each exit. Then, our network minimizes a loss $\mathcal{L}$ by a combination of a classification loss $\mathcal{L}_{cls}$ and a distillation loss $\mathcal{L}_{dis}$, which is given by

$$\mathcal{L} = \mathcal{L}_{cls} + \alpha\mathcal{L}_{dis} \tag{7}$$

where α is a hyper-parameter that controls the balance between the two-loss terms. In all experiments, we set $\alpha = 1$.

4 Online Tracking Algorithm

4.1 Main Process of Tracking

Given a frame, we totally sample $N = 256$ target candidates $\{c_i\}_{i=1}^{N}$. We evaluate each candidate c_i by their positive scores $p(c_i)$. The positive score $p(c_i)$ reflects the probability of the current candidate belonging to the target class. The tracking result O_l of exit l is chosen by the highest positive score as follow:

$$O_l = \arg\max_{c_i} \ p(c_i) \tag{8}$$

In detail, the positive scores $p(O_l)$ are calculated from the low-level exit to high-level exit. When the $p(O_l)$ of the current exit exceeds the confidence score threshold θ_c, the tracking process is stopped and O_l is taken as the tracking result. Otherwise, continue feeding forward and processing the next exit. If the positive confidence score of the last exit is less than θ_c, the target tracking in the current frame is failed. If no exit yields a sufficiently confident decision for a frame, the one with the highest score in exit 3 is taken as the result.

4.2 Optimization and Settings

Offline Training. For offline pre-training during each iteration, a minibatch is constructed with samples collected from each video. We randomly select 8 frames in each video, and draw 64 positive and 192 negative samples from each frame. The positive bounding boxes have overlap larger than 0.7 with ground-truths by IoU measure while the negative samples have less than 0.5 IoU. We train our models on ImageNet-Vid [23] and training split of LaSOT [9].

Online Training. To catch up with the appearance variations of the specific target, we have to fine-tune the pre-trained model with the first frame of each video. We draw 500 positive and 5000 negative samples based on the same IoU criteria for initializing. Depending on the estimated location, the trackers gather 96 positive and 192 negative examples. Here, we update our tracker in an efficient way which is almost identical to [19].

Optimization Setting. The proposed algorithm is implemented in PyTorch. For offline training, we train the network for 1000 epochs with a learning rate 0.0001. For online updates, the number of iterations for fine-tuning is 15 and the learning rate is set to 0.0003. The weight decay and momentum are fixed to 0.0005 and 0.9, respectively.

Table 1. Ablation on loss $\mathcal{L}_{dis}$.

$\mathcal{L}_{dis}$	E1	E2	E3	Succ (%)	Prec (%)
	✓			63.0	86.2
✓	✓			69.8(+6.8)	93.6(+7.4)
		✓		69.4	93.6
✓		✓		71.4(+2.0)	95.3(+1.7)
			✓	70.8	94.7
✓			✓	71.7(+0.9)	95.6(+0.9)
	✓	✓	✓	69.6	94.1
✓	✓	✓	✓	71.3(+1.7)	95.2(+1.1)

Table 2. Attention modules and RoI fusion.

θ_c	0.5	0.6	0.7	0.8	0.9	0.95
Succ (%)	70.4	71.0	71.2	71.3	71.3	71.4
FPS	45	42	41	36	16	5

(a) θ_c.

Baseline	RA	CA	RoI_f	Succ (%)	Prec (%)
✓				68.1	90.9
✓	✓			68.8	92.3
✓		✓		68.5	91.2
✓	✓	✓		69.5	94.1
✓	✓	✓	✓	71.3	95.2

(b) Attention modules and RoI fusion.

5   Experimental Results and Analysis

To validate the settings of our architecture, we first give a sequence of ablation studies on OTB-50 dataset [26]. Then we verify the robustness of our tracker on four standard datasets. Finally, analysis is also provided to help the understanding of our architecture.

5.1   Ablation Study

Ablation on Performance of Each Exit with $\mathcal{L}_{dis}$. We explore the performance of each exit in Table 1. As the number of exit increases, the accuracy gain is consistent. It suggests that the architecture learns suitable representation for each exit and the highest accuracy are achieved. More interesting, we find that the accuracy of early exits with $\mathcal{L}_{dis}$ (gray lines) improves a lot.

Ablation on Confidence Threshold θ_c. We explore the impact of θ_c in each exit with the success score in Table 2a. From Table 2a, the accuracy of early exits is relatively high. Hence, we test the θ_c from a relatively low value: 0.5. When the

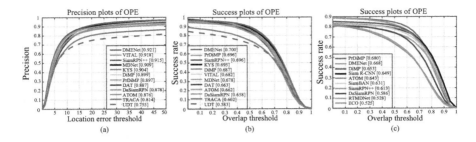

Fig. 3. (a) and (b) are the precision and success plots on OTB-100, respectively.

Table 3. The success scores (%) on the LaSOT dataset. (The **first** and *second* best results are shown in bold and italic), the same below.

	DMENet	PrDiMP	KYS	MDNet	DiMP	ATOM	SiamRPN++	VITAL
Succ (%)	*56.9*	**59.8**	-	39.7	*56.9*	51.5	49.6	39.0

Table 4. The comparisons on the VOT2018 dataset.

	DMENet	SiamRPN++	ATOM	UPDT	RCO	DRT	DaSiamRPN	LADCF
EAO	**0.417**	*0.414*	0.401	0.378	0.376	0.356	0.383	0.389
Ar	**0.605**	*0.600*	0.590	0.536	0.507	0.519	0.586	0.503
Rr	1.69	0.234	0.204	0.184	**0.155**	0.201	0.276	*0.159*

thresholds increase, the success score increases. Considering both the efficiency and accuracy, we set the value of θ_c as 0.8 for $E1$ and $E2$.

Ablation on Attention Modules and RoI Fusion. We explored the effect of the attention modules. The results are shown in Table 2. We can observe that the residual attention and channel attention modules bring more gain in accuracy, The effectiveness of RoI fusion is achieved analogously.

5.2   Comparison with the State-of-the-Art

OTB-100 [27]. We compare the proposed DMENet tracker on OTB-100 dataset [27] with the following recent published 11 trackers: VITAL [24], SiamRPN++ [16], MDNet [19], KYS [3], DiMP [2], PrDiMP [8], DAT [21], DaSiamRPN [28], ATOM [7], TRACA [6], and UDT [25]. The results are shown in Fig. 3(a, b). According to Fig. 3(a, b), our DMENet tracker achieves the best success score among all the state-of-the-art trackers. Our tracker also surpasses the best classification-based tracker VITAL both in precision and success evaluations with a 20 times speed-up.

UAV123 [18]. UAV123 [18] whose characteristics inherently differ from another dataset such as OTB-100. DMENet is compared with some state-of-the-art methods in Fig. 3(c). Figure 3(c) shows that our DMENet achieves competitive results among all the compared trackers.

LaSOT [9]. The LaSOT [9] dataset is a large-scale dataset consisting of 1400 sequences. In LaSOT, trackers are evaluated based on three criteria including precision, normalized precision, and success. Our approach is evaluated on the test set of 280 videos. We show the success score in Table 3. From Table 3, our tracker outperforms the classification-based trackers (i.e., MDNet [19] and VITAL [24]) by a large margin.

VOT2018 [1]. The VOT2018 [1] dataset contains 60 videos. The criteria including accuracy rank(Ar), robustness rank (Rr) and the expected average overlap (EAO). As illustrated in Table 4, our DMENet achieves competitive results with higher ranking within all the compared trackers.

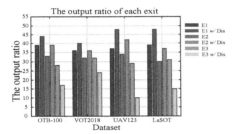

Fig. 4. The output ratio of each exit. E∗ _w/_ indicates E∗ with distillation training.

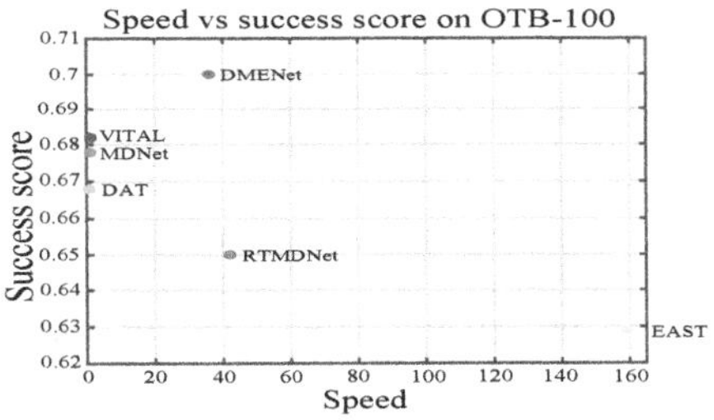

Fig. 5. A comparison of the quality and the speed of state-of-the-art tracking methods on OTB-100.

5.3 Discussion

The proposed DMENet is a fully convolutional network based on multi-exit architecture, which contains a feature extraction part and a classification part. The feature extraction part can extract powerful features at different layers, which guarantees the accuracy of the DMENet. The classification part is constituted by three very simple exits, the knowledge distillation encourages frames to come out from early exit while maintaining higher accuracy. Figure 4 shows that knowledge distillation greatly improves the output ratio of the early exit. As illustrated in Fig. 5, the EAST method [13] is a mixed architecture, whose layers are not online updated except the first two layers for efficiency, which affects its accuracy. Even by online updating, the gained improvement in accuracy is very limited. The AUC of the EAST and DMENet is 0.629 and 0.700 on OTB-100. MDNet processes all the frames through the entire network, which is suboptimal and time-consuming. The single ROI pooling in RT-MDNet degrades its accuracy while increasing speed.

6   Conclusions

An end-to-end multi-exit architecture based on knowledge distillation, dubbed as DMENet, is proposed in this paper. Our DMENet tracker improves speed by encouraging early exit to mimic later, more accurate exits by matching their output. Three MIN and two attention modules at different levels of representation further guarantee the robustness of DMENet. Thus, our DMENet can speed up the classification-based tracking framework, and get state-of-the-art performance. The proposed tracker achieves outstanding performance in four large public tracking benchmarks with 36 FPS.

References

1. Kristan, M., et al.: The sixth visual object tracking VOT2018 challenge results. In: Leal-Taixé, L., Roth, S. (eds.) ECCV 2018. LNCS, vol. 11129, pp. 3–53. Springer, Cham (2019). https://doi.org/10.1007/978-3-030-11009-3_1
2. Bhat, G., Danelljan, M., Van Gool, L., Timofte, R.: Learning discriminative model prediction for tracking. arXiv (2019)
3. Bhat, G., Danelljan, M., Van Gool, L., Timofte, R.: Know your surroundings: exploiting scene information for object tracking. In: Vedaldi, A., Bischof, H., Brox, T., Frahm, J.-M. (eds.) ECCV 2020. LNCS, vol. 12368, pp. 205–221. Springer, Cham (2020). https://doi.org/10.1007/978-3-030-58592-1_13
4. Chang, J.R., Chen, Y.S.: Batch-normalized maxout network in network. Computer Science (2015)
5. Chatfield, K., Simonyan, K., Vedaldi, A., Zisserman, A.: Return of the devil in the details: delving deep into convolutional nets. Computer Science (2014)
6. Choi, J., et al.: Context-aware deep feature compression for high-speed visual tracking. In: CVPR (2018)
7. Danelljan, M., Bhat, G., Khan, F.S., Felsberg, M.: Atom: accurate tracking by overlap maximization. In: CVPR (2019)
8. Danelljan, M., Gool, L.V., Timofte, R.: Probabilistic regression for visual tracking. In: CVPR (2020)
9. Fan, H., et al.: LaSOT: a high-quality benchmark for large-scale single object tracking. In: CVPR (2019)
10. Girshick, R.: Fast R-CNN. In: ICCV (2015)
11. Hinton, G., Vinyals, O., Dean, J.: Distilling the knowledge in a neural network. arXiv (2015)
12. Hu, J., Shen, L., Sun, G.: Squeeze-and-excitation networks. In: CVPR (2018)
13. Huang, C., Lucey, S., Ramanan, D.: Learning policies for adaptive tracking with deep feature cascades. In: ICCV (2017)
14. Huang, G., Chen, D., Li, T., Wu, F., van der Maaten, L., Weinberger, K.Q.: Multi-scale dense networks for resource efficient image classification. arXiv (2017)
15. Jung, I., Son, J., Baek, M., Han, B.: Real-time MDNet (2018)
16. Li, B., Wu, W., Wang, Q., Zhang, F., Xing, J., Yan, J.: SiamRPN++: evolution of Siamese visual tracking with very deep networks. arXiv (2018)
17. Li, X., Ma, C., Wu, B., He, Z., Yang, M.H.: Target-aware deep tracking. In: CVPR (2019)

18. Mueller, M., Smith, N., Ghanem, B.: A benchmark and simulator for UAV track-ing. In: Leibe, B., Matas, J., Sebe, N., Welling, M. (eds.) ECCV 2016. LNCS, vol. 9905, pp. 445–461. Springer, Cham (2016). https://doi.org/10.1007/978-3-319-46448-0_27

19. Nam, H., Han, B.: Learning multi-domain convolutional neural networks for visual tracking. In: CVPR (2016)

20. Phuong, M., Lampert, C.H.: Distillation-based training for multi-exit architectures. In: ICCV (2019)

21. Pu, S., Song, Y., Ma, C., Zhang, H., Yang, M.H.: Deep attentive tracking via reciprocative learning. In: NIPS (2018)

22. Radosavovic, I., Dollár, P., Girshick, R., Gkioxari, G., He, K.: Data distillation: towards omni-supervised learning. In: CVPR (2018)

23. Russakovsky, O., et al.: ImageNet large scale visual recognition challenge. Int. J. Comput. Vis. **115**(3), 211–252 (2015). https://doi.org/10.1007/s11263-015-0816-y

24. Song, Y., et al.: Vital: Visual tracking via adversarial learning. In: CVPR (2018)

25. Wang, N., Song, Y., Ma, C., Zhou, W., Liu, W., Li, H.: Unsupervised deep tracking. In: CVPR (2019)

26. Wu, Y., Lim, J., Yang, M.H.: Online object tracking: a benchmark. In: CVPR (2013)

27. Wu, Y., Lim, J., Yang, M.H.: Object tracking benchmark. TPAMI **37**, 1834–1848 (2015)

28. Zhu, Z., Wang, Q., Li, B., Wu, W., Yan, J., Hu, W.: Distractor-aware Siamese networks for visual object tracking. In: Ferrari, V., Hebert, M., Sminchisescu, C., Weiss, Y. (eds.) ECCV 2018. LNCS, vol. 11213, pp. 103–119. Springer, Cham (2018). https://doi.org/10.1007/978-3-030-01240-3_7

Handwriting Trajectory Reconstruction Using Spatial-Temporal Encoder-Decoder Network

Feilong Wei⬵ and Yuanping Zhu(✉)⬵

Tianjin Normal University, No. 393 Binshuixi Road, Xiqing District, Tianjin, China
zhuyuanping@tjnu.edu.cn

Abstract. Chinese handwriting characters have complex strokes and various writing styles, which makes it difficult to reconstruct handwriting. Aiming at this problem, we propose a handwriting reconstruction method based on a spatial-temporal encoder-decoder network with constrains. Different from other models that generate trajectory coordinates through a fully connected network, the method proposed in this paper outputs heat map sequence. The model is consists of three modules: key point detection module, spatial encoder-decoder module and reconstruction constraint module. The key point detector module and the spatial encoder part of encoder-decoder module are composed of a full convolutional network. The former generates heat maps of all key points which is a branch of the spatial encoder, and the mainly encoding the spatial information of each position on the offline image. The temporal decoder module is composed of a GRU network and an MLP network. Finally, we combine temporal information and reconstruction constraints to generate the final sequence. At each time, the features encoding by the spatial encoder module are combined with the features at the previous time that generate a corresponding heat map. The main contribution of the work of this paper is to propose a method that more suitable for handwriting reconstruction of Chinese handwritten characters. Experimental results show that the CT [6] accuracy of our method has already reached 87.6% on OLHWDB1.1 dataset.

Keywords: Handwriting trajectory reconstruction · Full convolutional network · Encoder-decoder network · Deep learning.

1 Introduction

Handwritten text analysis and recognition [20] has always been an important field of OCR [9], and it has also been the focus of research by scientists in the past decade [14]. Handwriting analysis has been studied for a long time. From the initial rule-based method to the current deep learning network-based methods, the accuracy of recognition has been continuously improved. According to different representations, handwritten text recognition is divided into online

H. Ma et al. (Eds.): PRCV 2021, LNCS 13019, pp. 342–354, 2021.
https://doi.org/10.1007/978-3-030-88004-0_28

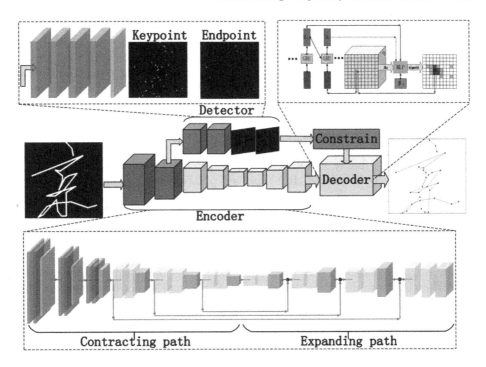

Fig. 1. The framework of handwriting reconstruction method.

handwriting recognition and offline handwriting recognition. Offline characters are represented by two-dimensional static images, while online characters are represented by a continuous coordinate sequence. Online characters also covers the trajectory, speed and angle of the handwriting during writing. Therefore, compared to offline characters, the accuracy of online handwriting recognition is usually higher than offline handwriting recognition. However, offline text collection is more convenient, more suitable for actual application scenarios, and its applications are more extensive. If the dynamic information of the text can be recovered from the two-dimensional static image, the static and dynamic information can be combined to further improve the accuracy of recognition. Moreover, handwriting reconstruction is widely used in smart writing and handwriting identification [7].

Currently, character handwriting reconstruction methods include graph search, template matching and writing rules, as well as deep learning based methods [6,20]. The method based on graph search [17] is to find a path with the least cost according to the minimum energy cost criterion. It is only suitable for the restoration of the writing order of numbers and alphabets. The method based on template matching [11] needs to build a stroke template library, and restore the handwriting by comparing the input image with the template. This method has a wider application range and higher accuracy, but it is too complicated to calculate the best path in the matching process. The method [2] based on writing rules uses the structural characteristics of characters to express the relationship between character strokes, and then uses rules to restore their order. Its disadvantage is that it cannot adapt to changes in writing styles and cannot handle text with broken pens. The method [19] based on deep learning performs a series of preprocessing on the image, and finally performs the arrangement prediction of the order relationship of each pixel through the network. It has poor adaptability to the complicated text with many strokes. Other methods based on deep learning like [6,6] extracted the feature sequence of the two-bit static image, and finally the handwriting sequence is generated through RNN and fully connected network. It is not very adaptable to samples which have complex font and a wide range of stroke's number.

When a person is writing, the visual attention will move with the movement of the handwriting. In machine vision, according to this feature, we express it as the response probability of corresponding position at different times, and should be concentrated in a certain area or point. Therefore, this paper proposes a handwriting reconstruction method based on spatial-temporal encoder-decoder network, which simulates the process of human visual attention movement [18] by predicting the probability of each point on the image at different times.

The rest of this article is organized as follows. The second section introduces Spatial-Temporal Encoder-Decoder Network model proposed in this article in detail. The third section explains in detail the reconstruction constraint proposed in this paper. The fourth section introduces the composition of the loss function in detail. The fifth section is the detail of experiment and results. The last section is the conclusions.

2 Spatial-Temporal Encoder-Decoder Network

In this section, we introduce in detail how our proposed method generates online handwriting sequences based on offline pictures. As mentioned earlier, we did not directly output, but output the absolute position of the maximum probability point at different temporal steps. The Spatial-Temporal Encoder-Decoder Network is divided into three modules to generate handwriting sequence:key point detector module, spatial encoder module, and temporal decoder module. The spatial encoder module is the backbone network of the model, which is essentially a variant FCN network and outputs the spatial features of each position of the offline image. Figure 3 shows its structure. The key point detector module is a branch of the backbone network, which outputs and classifies all key points of the font. Recurrent neural network GRU [3] and Multi-layer perceptron MLP form temporal decoder, which combines spatial features to output the heat map sequence. The overall framework as show as Fig. 1.

2.1 Key Point Detector

The key point detector [1,5,16] module is to return the position of each candidate point through the FCN network [13]. Fully convolutional networks have better spatial generalization capabilities than fully connected networks. This module detects all key points and divides them into two categories: end points and connection points. And then provide this information to the reconstruction constraint module. Since the full convolutional network FCN is more stable than the fully connected network in terms of position regression, more and more people use the full convolutional network FCN when studying key point detection [1]. Fully convolutional network only contains convolutional layer, pooling layer and activation layer. Specific parts are selected through the information connection between parts. This method is very meaningful for target detection (Fig. 2).

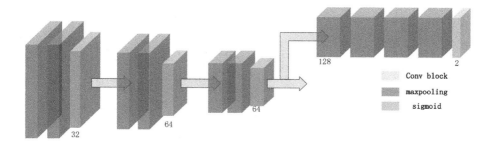

Fig. 2. The structure diagram of key point detector network

The detection network can identify the overall frame of the font and filter out the key parts. It has a certain sensitivity to the turning points of the line segments. Even a curve with a small curvature can identify subtle turning points, which are finally reflected in the output heat map. While detecting the key points, the detection network also determines the length of the output coordinate sequence, and realizes the self-variable length sequence generation.

2.2 Spatial Encoder Network

The key point detector network can find the key points of the font, but only extracts the position information, and cannot analyze the relationship between them (the feature map size is the same as the output image size, and the receptive field is limited, and deeper sequential features are not extracted). FCN cannot extract deeper-scale features well without changing the size of the feature map, so this work will be completed by the spatial encoder network. The spatial encoder network is a special full convolutional network. In order to extract the deep-level visual features of the image and obtain a larger receptive field while keeping the feature map at a certain size. So this part is composed of FCN and U-Net [12]. FCN [13] made a brief introduction in the front. And U-Net [12] is an FCN network with a special structure. U-Net [12] consists of two parts: one is the contracting path and the other is the expanding path. The contraction path can obtain contextual information of different scales, and the expansion path can supplement some of the deep-level information of the image. But the supplement of information is definitely incomplete, so skip connect is needed to combine the higher resolution pictures on the contraction path. Since the method proposed in this paper is to generate a heat map of handwriting points through FCN. It limits the output size of FCN must take into account the size ratio of handwriting points in the original image. And the addition of U-Net [12] is to maintain a certain size scale feature map while obtaining the deeper features of the image. Meanwhile, the stroke texture information of the image can be obtained by expanding the receptive field.

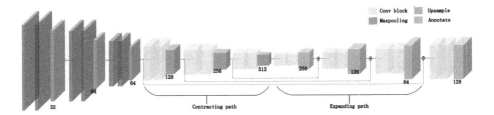

Fig. 3. The structure of spatial encoder network as showing as picture.

The specific structure of spatial encoder network shown in Fig. 3. The image through the spatial encoder network, is encoded as a tensor of size $d \times H' \times W'$. We denote these coding features as Eq. 1,

$$a = \{a_1, a_2, a_3, ..., a_n\}, a_i \in R^d, L = H' \times W' \tag{1}$$

where d is the dimension of a_i.

2.3   Temporal Decoder Network

The Temporal Decoder Network is essentially a candidate determiner composed of MLP and GRU [3]. In order to link the offline image with the variable-length output sequence, the paper [18] calculates an intermediate vector to provide a regional feature filter for subsequent recognition and classification. But we use this intermediate vector to output the coordinate points we need in the image heatmap. Figure 4 shows the work flow of the temporal decoder network, where MLP is an multi-layer perceptron composed of multiple fully connected networks and is the output layer of the temporal decoder network. GRU [3] is an improved version of cyclic neural network RNN, which solves the problem of gradient disappearance or gradient explosion during RNN training, and the space occupancy rate is much smaller than LSTM [4] while achieving the same effect. The hidden state calculation equation of GRU [3] see Eq. 2 ~ Eq. 5.

$$z_t = \sigma \left(W_{hz} H_{t-1} + U_{cz} C_t + b_z \right) \tag{2}$$

$$r_t = \sigma \left(W_{hr} H_{t-1} + U_{cr} C_t + b_r \right) \tag{3}$$

$$\widetilde{H_t} = \tanh\left(W_h\left(H_{t-1} \otimes r_t\right) + U_h C_t + b_h\right) \tag{4}$$

$$H_t = (1 - z_t) \otimes H_{t-1} + z_t \otimes \widetilde{H_t} \tag{5}$$

Among them,$\sigma\left(\cdot\right)$ is the sigmoid function, $z_t, r_t, \widetilde{H_t}$ which is the update gate, reset gate and candidate state. When the temporal decoder network predicts the position of the handwriting point at each time step, it outputs the probability of each area, so the output only needs to maximize the probability of the candidate area. The temporal decoder network combines spatial features a_i and the hidden state H_{t-1} of the current GRU to calculate the maximum probability position of the handwriting point at the current moment (see Eq. 6~Eq. 7),

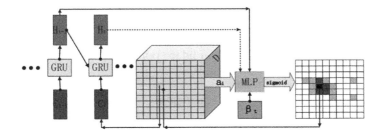

Fig. 4. The calculation process of the temporal decoder network.

$$e_{ti} = v_a^T \tanh\left(W_a H_{t-1} + U_a a_i\right) \tag{6}$$

$$p_{ti} = \frac{\exp\left(e_{ti}\right)}{\sum_{i=0}^{L} \exp\left(e_{ti}\right)} \tag{7}$$

where $v_a \in R^n, W_a \in R^{n' \times n}, U_a \in R^d$. And then we will the most probable point as the C_t to strengthen the relationship between points, like Eq. 8.

$$C_t = a\left[\max\left(p\right)\right] \tag{8}$$

In order to strengthen the path information, this article proposes a handwriting trend feature. The handwriting trend feature is a blank graph (β) whose size is the output scale, and the corresponding position is marked at each time step. Then trend feature are extracted by convolution and be send to the MLP. The final calculation formula is as shown as Eq. 9 $\sim$ Eq. 12.

$$\beta = (0) \in F^{H' \times W'} \tag{9}$$

$$\beta_t = (1_{ij}) \in F^{H' \times W'}, i \in (0, W'), j \in (0, H') \tag{10}$$

$$F_t = f\left(\beta_t\right) \tag{11}$$

$$e_{ti} = v_a^T \tanh\left(W_a H_{t-1} + U_a a_i + U_f f_t\right), f_t \in F_t \tag{12}$$

where β_t is the trajectory picture in the time step t, and $f\left(\cdot\right)$ is a convolution module. Finally the output e_{ti} will be used in Eq. 11 which represents the response of each position of the corresponding time step t.

3 Handwriting Reconstruction Constraints

Although the Spatial-Temporal Encoder-Decoder Network has a certain adaptability to the reconstruction of handwriting Chinese characters with complex fonts and broken pens, the handwriting point probability based on the whole image will be chaotic when faced with such samples. In order to constrain the chaos of handwriting, we designed a connection rule based on different handwriting points, as shown in Fig. 5.

In the key point detection module, we divide all points into connection points and end points. So, we defined two rules:

Theorem 1. *The starting point of the line segment must be the end point*

Theorem 2. *There must be a solid line in the line segment.*

In practical applications, we select candidate points based on the rules and the output of the model (See Eq. 13~ Eq. 15),

$$
p_{ti} = \begin{cases} \frac{\exp(e_{ti}) \times d_i}{\sum_{i=0}^{L} \exp(e_{ti}) \times d_i}, & if\ last\ point \in endpoints, \\ \frac{\exp(e_{ti}) \times k_i}{\sum_{i=0}^{L} \exp(e_{ti}) \times k_i}, & if\ last\ point \in connection points. \end{cases} \tag{13}
$$

$$
l = \frac{1}{n} \phi\,(last\ point, candidate\ point) \tag{14}
$$

$$
P_{ti} = \begin{cases} \max(p_{ti}), & if\ last\ point \in endpoints, \\ \max(p_{ti} \times l). & if\ last\ point \in connection points. \end{cases} \tag{15}
$$

where $\phi\,(l,c)$ in Eq. 14 means interpolating sampling between two points in the original image and n is the number of samples. In addition, $k_i \in key\ point\ map$ and $d_i \in end\ point\ map$. P_{ti} is the final predicted value.

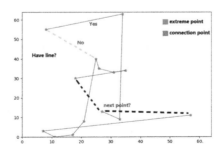

Fig. 5. Example of the situation in the process of reconstruction

4   Loss Function

Since the Spatial-Temporal Encoder-Decoder Network needs to learn key point detection and key point sorting, these two tasks are different. So we define the final loss function as Eq. 16 like [8], where L_{det} represents the loss in the key point detection task and L_{sq} represents the loss in the key point sorting task.

$$L = L_{det} + L_{sq} \tag{16}$$

In order to measure the gap between the predicted map and the label and balance the quantitative relationship between key points and the background, focal loss [10] is used as the loss function, as shown in the Eq. 17.

$$L_{det} = \frac{-1}{N} \sum_{c=1}^{C} \sum_{h=1}^{H} \sum_{w=1}^{W} \begin{cases} \beta \left(1 - p_{cij}\right)^{\alpha} \log\left(p_{cij}\right), & if \ y_{cij} = 1, \\ \left(1 - \beta\right) p_{cij}^{\alpha} \log\left(1 - p_{cij}\right), & otherwise. \end{cases} \tag{17}$$

Different from the traditional sorting loss function, our sorting task is to maximize the probability of the label point at each time, so we directly adopt the cross-entropy loss function.(L_{sq} see Eq. 18)

$$L_{sq} = \frac{-1}{N} \sum_{t=1}^{N} \log\left(p_{t,label[t]}\right) \tag{18}$$

5   Experiment

In order to verify the effectiveness of the proposed method in handwriting reconstruction, this chapter conducts ablation experiments and comparative experiments.

5.1   Dataset Processing

OLHWDB1.1 and Tamil dataset are used in experiment.OLHWDB1.1 includes 3755 types of Chinese characters, which is written on a separate page. The stroke coordinates of the pen tip are recorded. Tamil dataset is from paper [6], which can explore the reconstruction effect on other languages, is a dataset of HP Company's compete. We should convert it to offline form because all of them are trajectory sequence.

Different from offline handwriting characters saved in the form of static images, online handwriting characters retain richer dynamic information when writing in the form of handwriting point sequences. We save the original data in the form of a formula [20] like Eq. 19,

$$[[x_1, y_1, s_1], [x_2, y_2, s_2], ..., [x_n, y_n, s_n]] \tag{19}$$

where x_i and y_i is the coordinate,s_i is the point state. And then, we convert the data set into a specific form for training.

We think that the points that are too dense or the intermediate points on the same line are redundant points. In order to filter excess points, we set two conditions [20] as shown in Eq. 20 and Eq. 21,

$$\sqrt{(x_i - x_{i-1})^2 + (y_i - y_{i-1})^2} \leq T \tag{20}$$

$$\frac{\Delta x_{i-1} \Delta x_i + \Delta y_{i-1} \Delta y_i}{\sqrt{(\Delta x_{i-1}^2 + \Delta x_i^2) \cdot (y_{i-1}^2 + y_i^2)}} \geq C \tag{21}$$

where T is the threshold to filter out points with too dense distance and C is the threshold to filter out the middle point in the same straight line. In order to protect the starting point and end point of the strokes from being screened out, the screening operation is carried out when $s_i = s_{i-1} = s_{i+1}$.

Offline Character Generation. In order to make the key point heat map and label correspond to the offline image, map the preprocessed handwriting point sequence coordinates to the image whose size is $H' \times W'$. Then we resize the image to $H \times W$. (See (a) in Fig. 6) According to the corresponding label, generate a heatmap of key point [15] based on the Gaussian distribution on the image.(See (b) in Fig. 6)

5.2   Implementation Details

The network model in this article is built under the pytorch framework, and the GPU model used by the platform is NVIDIA 1080Ti, which runs on a 64-bit Linux system. In the data preprocessing in this paper, the parameter T in Eq. 20 is $0.05 \times \max(H, W)$ and the parameter C in Eq. 21 is -0.9. The size of image is $H \times W = 512 \times 512$, the heat map and the output scale are $H' \times W' = 64 \times 64$. The dimension of the output a_i of the spatial encoder network is $d = 128$. The hidden state H_t of GRU is a 64-dimensional tensor. Finally, we use the optimizer Adam to set the initial learning rate $lr = 0.001$ and decay to 0.1 every 10 rounds.

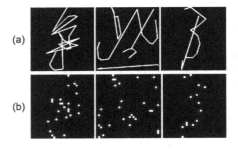

Fig. 6. Offline characters (a) and corresponding heatmap labels (b).

5.3   Evaluation Metrics

At present, there is no unified standard for the evaluation of online handwriting generation problems, such as paper [6,20]. This is also due to the large differences in the methods of generating handwriting.

Due to the particularity of the method proposed in this article. We use the average probability of the corresponding position of each handwriting point of the character as the criterion for the quality of the model.(See Eq. 22)

$$meanP = \frac{1}{K} \sum_{t=1}^{K} p_{t,indice} \qquad (22)$$

where K is the number of trajectory points. Although $meanP$ cannot fully represent the recovery degree of a font handwriting, it can reflect the response degree of the model to handwriting points.

In addition, in order to facilitate the comparison with the paper [6], we also adopted their evaluation index(See Eq. 23 ~ Eq. 24),

$$Starting\ Point\ Accuracy = \frac{Number\ of\ correct\ SP}{Total\ number\ of\ test\ images} \qquad (23)$$

$$Junction\ Point\ Accuracy = \frac{Number\ of\ correct\ JP}{Total\ number\ JP\ points\ in\ test\ data} \qquad (24)$$

when complete trajectory (CT) of an offline character image is perfectly retrieved along with the correct starting point,we evaluate this metric as a positive result.

5.4   Experiment and Result Analysis

In order to verify the necessity of the handwriting trend characteristics (TC) in Eq. 11 and reconstruction constraints (RC) in the model proposed in this article, we conducted corresponding ablation experiments with $meanP$ as the evaluation index. We randomly select 5000 samples from the test set for evaluation (see Table 1) The results of the Table 1 are predictable. The handwriting trend characteristics provides model with the features formed by the handwriting points of all previous moments, and provides enlightening information for the next moment. And reconstruction constraints can properly correct its errors and provide more accurate information for the next moment.

Table 1. The meanP of each method combination

Method	meanP
Decoder	0.559
Decoder +TC	0.723
Decoder +RC	0.753
Decoder +TC+RC	**0.796**

Table 2. Stroke recovery accuracy

	Evaluation Metrics	Tamil	OLHWDB1.1
Bhunia et al. [6].	Starting Point (SP)Accuracy	98.12%	85.00%
	Junction Points (JP) accuracy	**97.02%**	70.40%
	Complete trajectory (CT) retrieval accuracy	95.54%	64.30%
Ours.	Starting point (SP) accuracy	**99.10%**	**93.80%**
	Junction points (JP) accuracy	97.00%	**89.30%**
	Complete trajectory (CT) retrieval accuracy	**95.60%**	**87.60%**

We also conducted a comparative experiment with the paper [6] in Tamil dataset and OLHWDB1.1. The results are from 1,000 randomly selected samples. (see Table 2)

We have compared the accuracy of our proposed method with the method from [6]. We selected it because it is the latest methods in this field as comparison objects and implemented them on the dataset. That ours model are more suitable for trajectory reconstruction of Chinese characters. A few qualitative results of methods are shown in Fig. 7 and Fig. 8.

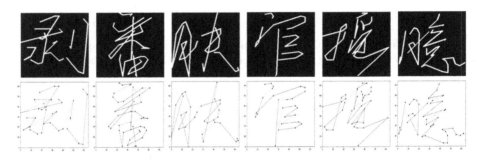

Fig. 7. Examples of the recovery trajectory from offline character on OLHWDB dataset.

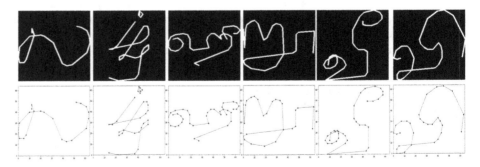

Fig. 8. Examples of the recovery trajectory from offline character on Tamil dataset.

6   Conclusion

This paper proposes a method of regression trajectory sequence that generates heatmap base on Spatial-Temporal Encoder-Decoder Network. The reconstruction results is better than the method [6] on OLHWDB. The coordinates generated in this way cannot be directly trained in the network, and the gradient must also be faulted by generating heat map labels. In future work, we will focus on solving this problem and combine it with GAN to generate a more complete trajectory sequence. In addition, whether the model can completely recover the handwriting of characters that have not been touched before will also be a future research direction.

Acknowledgement. This work was supported by the Natural Science Foundation of Tianjin(Grant No.18JCYBJC85000)

References

1. Cao, Z., Hidalgo, G., Simon, T., Wei, S.E., Sheikh, Y.: Openpose: Realtime multi-person 2d pose estimation using part affinity fields. IEEE Trans. Pattern Anal. Mach. Intell. **43**(1), 172–186 (2021)
2. Cao, Z., Su, Z., Wang, Y.: An offline handwritten chinese character writing sequence recovery model. J. Image Graphics **10**(1), 2074–2081 (2009)
3. Cho, K., et al.: Learning phrase representations using rnn encoder-decoder for statistical machine translation. In: Proceedings of the Empiricial Methods in Natural Language Processing (EMLP) (2014)
4. Hochreiter, S., Schmidhuber, J.: Long short-term memory. Neural Comput. **9**(8), 1735–1780 (1997)
5. Insafutdinov, E., Pishchulin, L., Andres, B., Andriluka, M., Schiele, B.: Deepercut: a deeper, stronger, and faster multi-person pose estimation model. In: Proceeding of European Conference on Computer Vision, pp. 34–50 (2016)
6. Kumar Bhunia, A., et al.: Handwriting trajectory recovery using end-to-end deep encoder-decoder network. In: 2018 24th International Conference on Pattern Recognition (ICPR), pp. 3639–3644 (2018)
7. Lai, S., Jin, L., Zhu, Y., Li, Z., Lin, L.: Synsig2vec: Forgery-free learning of dynamic signature representations by sigma lognormal-based synthesis. IEEE Trans. Pattern Anal. Mach. Intell. **8**(1), 99–112 (2021)
8. Law, H., Deng, J.: Cornernet: detecting objects as paired keypoints. Int. J. Comput. Vis. **128**(3), 642–656 (2020)
9. Li, L., Gao, F., Bu, J., Wang, Y., Yu, Z., Zheng, Q.: An End-to-End OCR Text Re-organization Sequence Learning for Rich-Text Detail Image Comprehension. In: Vedaldi, A., Bischof, H., Brox, T., Frahm, J.-M. (eds.) ECCV 2020. LNCS, vol. 12370, pp. 85–100. Springer, Cham (2020). https://doi.org/10.1007/978-3-030-58595-2_6
10. Papandreou, G., et al.: Towards accurate multi-person pose estimation in the wild. In: 2017 IEEE Conference on Computer Vision and Pattern Recognition (CVPR) (2017)
11. Qiao, Y., Yasuhara, M.: Recover writing trajectory from multiple stroked image using bidirectional dynamic search. In: Proceedings of the 18th International Conference on Pattern Recognition (ICPR), pp. 970–973 (2006)

12. Ronneberger, O., Fischer, P., Brox, T.: U-net: Convolutional networks for biomedical image segmentation. In: Medical Image Computing and Computer-Assisted Intervention – MICCAI 2015, pp. 234–241 (2015)
13. Schwing, A.G., Urtasun, R.: Fully connected deep structured networks. In arXiv:1503.02351. (2015)
14. Shi, B., Xiang, B., Cong, Y.: An end-to-end trainable neural network for image-based sequence recognition and its application to scene text recognition. IEEE Trans. Pattern Anal. Mach. Intell. **39**(11), 2298–2304 (2016)
15. Shi, X., et al.: Deep learning for precipitation nowcasting: A benchmark and a new model. In: 31st Annual Conference on Neural Information Processing Systems, NIPS, pp. 5617–5627 (2017)
16. Wei, S.E., Ramakrishna, V., Kanade, T., Sheikh, Y.: Convolutional pose machines. In: 2016 IEEE Conference on Computer Vision and Pattern Recognition (CVPR), pp. 4724–4732 (2016)
17. Yu, Q., Yasuhara, M.: Recovering drawing order from offline handwritten image using direction context and optimal euler path. In: IEEE International Conference on Acoustics, pp. 765–768 (2006)
18. Zhang, J., et al.: Watch, attend and parse: An end-to-end neural network based approach to handwritten mathematical expression recognition. Pattern Recognit. **71**(11), 196–206 (2017)
19. Zhang, R., Zhan, Y., yang, M.: A method for restoring handwritten strokes based on endpoint sequence prediction. Comput. Sci. **046**(4), 264–267 (2019)
20. Zhang, X.Y., Yin, F., Zhang, Y.M., Liu, C.L., Bengio, Y.: Drawing and recognizing chinese characters with recurrent neural network. IEEE Trans. Pattern Anal. Mach. Intell. **40**(4), 849–862 (2018)

Scene Semantic Guidance for Object Detection

Zhuo Liu, Xuemei Xie$^{(\boxtimes)}$, and Xuyang Li

School of Artificial Intelligence, Xidian University, Xi'an 710071, China
xmxie@mail.xidian.edu.cn

Abstract. In the real world, objects often have ambiguous or indistinctive appearances but they tend to co-vary with other objects and particular scene. In this paper, we exploit scene context to provide a global semantic guidance for object detection. We present a simple but effective Scene Semantic Guidance (SSG) framework, which can be applied as a plug-and-play component, to facilitate the classification ability of detectors. Specifically, to explicitly model scene semantic context, we propose a scene semantic embedding module which leverages an auxiliary task of multi-label classification to learn object-level scene concept. Further, to adaptively incorporate the scene semantic context into the object feature, we propose a semantic consistency guidance module which can strengthen the discrimination of the object feature. Comprehensive experiments on MS-COCO benchmark demonstrate that the proposed SSG framework is effective and generalizable, leading to consistent improvements upon typical detectors, including Faster R-CNN, RetinaNet, and FCOS.

Keywords: Object detection · Scene context · Semantic guidance

1 Introduction

Object detection is one of the fundamental and challenging computer vision problems that aiming to detect objects of predefined categories in digital images. With the rise of deep convolutional neural networks, research in object detection has seen unprecedented progress. Current state-of-the-art detectors can be divided into two categories, anchor-based detectors [12,13,18,23] and anchor-free ones [10,19,22,24].

Either anchor-based or anchor-free detectors typically classify objects using their interior features while ignoring global semantic information, which makes them difficult to recognize objects possessing ambiguous or indistinctive appearances. For examples in Fig. 1, detected by Faster R-CNN [18]. These examples illustrate that ignoring such scene information inevitably places constraints on the accuracy of detection. In the real world, objects tend to co-vary with other objects and particular scene, providing a rich source of contextual associations to be exploited by the visual system. We are thus motivated to explicitly model

© Springer Nature Switzerland AG 2021
H. Ma et al. (Eds.): PRCV 2021, LNCS 13019, pp. 355–365, 2021.
https://doi.org/10.1007/978-3-030-88004-0_29

(a)   (b)

Fig. 1. Some typical detection errors. (a) The ambiguous *snowboard* is false detected and the extremely small *traffic light* is undetected. (b) The indistinctive *tennis racket* is undetected.

the scene context with semantic information, to help recognizing objects that have ambiguous or no distinctive appearances and suppress the false positive detections.

The key computational question is how to represent the scene context in a semantical and informative form. For early times object detectors, a common way of integrating scene context is to integrate a statistical summary of the elements that comprise the scene, like Gist [7]. For modern deep learning based detectors, there are some methods to extract scene context. [3,11] employ recurrent neural networks to encode scene contextual information. [15] extracts the whole image feature as the scene feature by global pooling operation. Although these methods consider the scene context, they do not explicitly utilize the semantic information of scene category as lacking of effective supervised label. To handle this issue, we assume that scene category imposes tight distribution over the kind of objects that might appear in the scene. Consequently, scene category can be defined in terms of the objects which are present in the image. We regard scene category as a combination of objects' categories and propose a scene semantic embedding module that explicitly models scene semantic context by an auxiliary task of multi-label classification.

Further, we need to effectively make use of the scene context information, which is still an exploratory problem in object detection. Recent works [4,5,11, 15,16] have made some efforts to leverage the scene context to improve object detection. However, these methods take consider scene context into all objects rather than as an adaptive manner. Different from them, we observed that the scene context plays a significant effect on those objects processing ambiguous or indistinctive appearances. Based on this observation, we assume that those semantically ambiguous objects have low consistency with the scene, and should consider more scene information for them. Therefore, we propose a semantic consistency guidance module that adaptively incorporates the scene semantic feature into the object feature. Specifically, we take a similarity function to measure the semantical compatibility score of the scene and the object. And then we employ a gated fusion function to fuse both features under the compatibility control. This novel fusion strategy can adaptively adjust the object feature according to the scene semantic, which is helpful to strengthen the scene-aware ability of the object.

Combination of the above two modules is the proposed Scene Semantic Guidance (SSG) framework. It is worth noting that the proposed SSG framework is not limited to a specific detector. Naturally, it can be applied as a plug-and-play component, to facilitate the classification ability of mainstream detectors. In summary, this work mainly has three following contributions:

- To explicitly model scene context with semantic information, we propose a scene semantic embedding module which leverages an auxiliary task of multi-label classification to learn object-level scene concept.
- To adaptively incorporate the scene semantic context into the object feature, we propose a semantic consistency guidance module which can strengthen the discrimination of the object feature.
- We combine the proposed two modules into a framework named Scene Semantic Guidance (SSG). The experimental results on the challenging benchmark MS-COCO demonstrate that the proposed SSG framework can boost three typical anchor-based and anchor-free detectors, which proves its effectiveness and generalization ability.

2   Related Work

2.1   General Object Detection

Current state-of-the-art detectors can be divided into two categories, anchor-based detectors and anchor-free ones. Anchor-based detectors can be generally divided into one-stage methods [13,23] and two-stage methods [12,18]. Both of them first tile a large number of preset anchors on the image, then predict the category and refine the coordinates of these anchors by one or several times, finally output these refined anchors as detection results. More recently, anchor-free detectors have attracted substantial attention due to their novelty and simplicity. One kind of them formulates the object detection problem as a key-point or a semantic-point detection problem, including CornerNet [10], RepPoints [20]. Another type of anchor-free detectors classify each point on the feature pyramids [12] into foreground classes or background, and directly predict the distances from the foreground point to the four sides of the ground-truth bounding box. The popular methods include FCOS [19], ATSS [22], and SPAD [24].

Complementary to these works, in this paper, we focus on how to exploit scene semantic information to improve the classification ability of general detectors including both anchor-based and anchor-free methods.

2.2   Scene Context for Object Detection

In this work, we focus on scene context, which refers to how the objects are related to the environment surrounding them [17,21]. Recent works have leveraged the scene context to improve object detection. Ouyang et al. [16] used the image classification scores as contextual features and concatenated it with the object detection scores to improve detection results. Li et al. [11] exploited stacked Long Short-Term Memory (LSTM) layers to recurrently generates an attention map for an input image to highlight useful global contextual locations. Liu et al. [15] proposed Structure Inference Network that reasons object state in a graph under the guidance of scene context. Cao et al. [1] proposed a Attention-guided Context Feature Pyramid Network to capture effective contextual dependencies from multiple receptive fields. Chen et al. [4] proposed HCE framework that generates hierarchical contextual RoI features by fusing both instance-level and global-level information.

Similar to [4], we extract scene context with semantic information by an auxiliary task of multi-label classification. Different from these methods that treat each object equally when exploiting the scene context, our method takes consider scene context as an adaptive manner. Motivated by [8], which obtains pixel-aware demanding extent by measuring the similarity between the global feature and per-pixel feature, we introduce a semantic consistency guidance module to adaptively incorporate the scene context into the object feature.

3   Approach

3.1   Overview of Scene Semantic Guidance

The overview of Scene Semantic Guidance (SSG) framework is shown in Fig. 2. Since the proposed framework is independent with specific detector, for the sake of simplicity, the method for extracting the objects features is omitted in Fig. 2. The proposed SSG framework aims to strengthen the scene-aware ability of the object feature by the guidance of scene semantic context. Firstly, the *scene semantic embedding* module upon the backbone extracts the scene semantic feature by an auxiliary task of multi-label classification (MLC). Then, the *semantic consistency guidance* module uses a similarity function to calculate the semantic compatibility score θ of the scene feature $\mathbf{s}$ and the object feature $\mathbf{x}$. Further, the scene-aware object feature $\mathbf{x}'$ is generated by a gated fusion of $\mathbf{s}$ and $\mathbf{x}$. Finally, the updated $\mathbf{x}'$ is fed into regression and classification branch of the detector's head, respectively. The whole network is trained end-to-end, and the overall loss is the sum of original detection losses (classification and regression) and the multi-label loss (described in 3.2).

The proposed SSG framework is flexible and generalizable, as it can be applied as a plug-and-play component for mainstream object detectors. In the following subsections, we will elaborate the designing details of the SSE module and the SCG module.

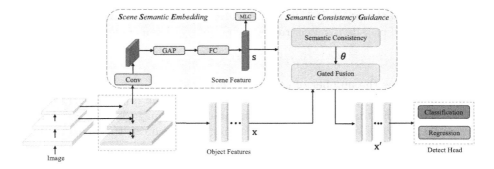

Fig. 2. Overview of the proposed Scene Semantic Guidance (SSG) framework. Firstly, the *scene semantic embedding* module upon the backbone extracts the scene semantic feature by an auxiliary task of multi-label classification (MLC). Then, the *semantic consistency guidance* module uses a similarity function to calculate the semantic compatibility score θ of the scene semantic feature **s** and the object feature **x**. Further, the scene-aware object feature **x**$'$ is generated by a gated fusion of **s** and **x**. Finally, the updated **x**$'$ is fed into regression and classification branch of the detector's head, respectively.

3.2 Scene Sematic Embedding

As aforementioned, conventional detectors may ignore scene semantic information, which is crucial for distinguishing the context-dependent objects. To break this limitation, we exploit scene semantic embedding (SSE) upon the detection backbone, enabling the backbone to learn object-level scene semantic concept.

To obtain the scene semantic concept, a straightforward approach is to generate scene labels (e.g., bedroom, school, or office) on the image-level. However, it requires additional annotations. Moreover, unlike object categories, scene categories are often ambiguous. We argue that scene can be defined in terms of the objects which are present in the image and then regard scene category as a combination of objects' categories. Therefore, the proposed SSE module takes the image-level multi-class labels, which can be conveniently obtained by collecting all instance-level categories in an image.

In practice, the proposed SSE module is based on a multi-label classifier. As shown in Fig. 2, we first apply a convolution layer on the topmost output of FPN [12] to obtain the input feature map, and then employ global average pooling (GAP) for feature aggregation. Here, the additional convolution layer aims to learn high-level scene semantic feature. Formally, let $\mathbf{X} \in \mathbb{R}^{c \times h \times w}$ denotes the topmost output feature map of FPN, where c is the channel dimensionality, h and w are the height and width, respectively. Then the scene semantic feature **s** is extracted as following:

$$\mathbf{s} = f_{gap}(f_{conv}(\mathbf{X})) \in \mathbb{R}^d, \tag{1}$$

where $f_{conv}(\cdot)$ is implemented by 3×3 convolutional layer with stride 1 followed by batch normalization and ReLU in order, which also increases the channel of the output feature map of FPN from 256 to 1024 for higher semantic abstraction. The multi-label classifier is constructed by C binary classifiers for all categories:

$$\hat{\mathbf{y}}_{cls} = f_{cls}(\mathbf{s}) \in \mathbb{R}^C, \qquad (2)$$

where C denotes the number of categories, each element of $\hat{\mathbf{y}}_{cls}$ is a confidence score (logits), and f_{cls} is binary classifier modeled as one fully-connected (FC) layer. We assume that the ground truth label of an image is $\mathbf{y} \in \mathbb{R}^C$, where $y^i = \{0, 1\}$ denotes whether object of category i appears in the image or not. The multi-label loss can be formulated as follows:

$$\mathcal{L}_{mll} = -\sum_{c=1}^{C} y^c \log(\sigma(\hat{y}_{cls}^c)) + (1 - y^c) \log(1 - \sigma(\hat{y}_{cls}^c)), \qquad (3)$$

where $\sigma(\cdot)$ is the sigmoid function.

In this way, the obtained scene semantic feature conveys whole scene context for learning all object categories that appear in the image. Consequently, it can advance the feature learning of the objects that are highly dependent on context clues, which demonstrated later by experiments (see Table 2).

3.3 Semantic Consistency Guidance

Scene context can provide global semantic guidance for overall images, thus may help inferring context-dependent objects in the scene and rectifying misclassification detection results. However, most methods [4,15,16] just apply a simple concatenation fusion when exploiting the scene context, rather than considering the consistency relationship between the object and the scene. We observed that the scene context does not help objects equally. For those objects with ambiguous appearances, the scene context plays a significant effect. Inspired by this, we propose a semantic consistency guidance module (SCG) which adaptively strengthen the discrimination of the object feature.

As shown in Fig. 2, the proposed SCG module is composed of two functions: 1) semantic consistency function, and 2) gated fusion function. More specifically, the semantic consistency function is used to calculate the compatibility score of the scene and the object in the same domain, which represents the consistency that the object exists in the scene. We denote the scene semantic feature as $\mathbf{s} \in \mathbb{R}^d$ and the object feature as $\mathbf{x} \in \mathbb{R}^c$, the dimension c varies with different detectors. In Faster R-CNN [18], the $\mathbf{x}$ is obtained by applying two FC layers with dimension 1024 on the RoI aligned feature. In RetinaNet [13] and FCOS [19], the $\mathbf{x}$ is obtained by applying several convolutional layers with dimension 256 on the output feature map from FPN. To align these two features into a shared space $\mathbb{R}^k$, we apply two individual FC layers with dimension k followed by ReLU to learn the projection for $\mathbf{s}$ and $\mathbf{x}$, respectively. Then we measure the similarity of these two features in the shared space by calculating their cosine distance. Formally, the semantic consistency function can be formulated as:

$$\theta = sim(g_s(\mathbf{s}), g_x(\mathbf{x})) \in [0, 1], \qquad (4)$$

where θ is the compatibility score within $[0, 1]$ since the projected space is positive, $sim(\cdot)$ denotes the similarity function implemented by cosine distance, $g_s(\cdot)$ and $g_x(\cdot)$ are the nonlinear projection functions implemented by one FC layer.

The other one is the gated fusion function which adaptively fuses the scene semantic feature and the object feature guided by their compatibility score. Intuitively, we should consider more scene information for those objects that have low compatibility with the scene. Hence, we implement the gated fusion function by a weighted sum of the original object feature and the transformed scene-aware object feature. Given the scene feature $\mathbf{s}$, the object feature $\mathbf{x}$ can be updated as following:

$$\mathbf{x}' = \theta\mathbf{x} + (1 - \theta)\phi(\mathbf{W}\mathbf{s} + \mathbf{U}\mathbf{x}), \tag{5}$$

where θ denotes the compatibility score calculated by Eq. 4, ϕ denotes $tanh$ activate function, $\mathbf{W}$ and $\mathbf{U}$ are linear transformer implemented by one FC layer. From Eq. 5 we can see that the smaller the ϕ, the object feature $\mathbf{x}$ will incorporate more information from the scene feature $\mathbf{s}$. Finally, the updated feature $\mathbf{x}'$ is fed into the regression and classification branch of the detection head.

4 Experiments

4.1 Dataset and Evaluation Metrics

We conduct extensive experiments on MS-COCO 2017 dataset [14] to demonstrate the effectiveness and generalization ability of the proposed Scene Semantic Guidance framework. MS-COCO 2017 is the most popular and challenging benchmark for general object detection, which contains 80 common object categories, 118K images for training, 5K images for validation and 20K for testing. Following the common practice [13,18,19], we use the `train2017` split for training, report ablation results on the `val2017` split. We adopt the standard COCO-style Average Precision (AP) as the evaluation metrics.

4.2 Implementation Details

We implement the proposed method and all baseline methods based on MMDetection v2.4 codebase [2]. The implementations of the baselines and our method strictly follow the default settings of MMDetection. Images are resized such that the short edge has 800 pixels while the long edge has less than 1333 pixels. We use no data augmentation except horizontal flipping for training. The ResNet [9] is exploited as backbone, which is pre-trained on ImageNet [6]. We use 2 NVIDIA TITAN RTX GPUs for training with a total batch size of 16 (8 images per GPU) in both ablation study and performance comparison. We train all models with SGD optimizer for 12 epochs in the total, with the initial learning rate as 0.01 and decreased by a factor of 0.1 at 8th epoch and 11th epoch. Weight decay and momentum are set as 0.0001 and 0.9, respectively. We also adopt the linear warming up strategy to begin the training of our model.

Table 1. Compared with baselines (Faster R-CNN [18], RetinaNet [13] and FCOS [19]) on MS-COCO val2017. "SSG" denotes the proposed Scene Semantic Guidance framework.

Backbone	Method	SSG	AP	AP^{50}	AP^{75}	AP^S	AP^M	AP^L
ResNet-50	Faster R-CNN		37.4	58.1	40.4	21.2	41.0	48.1
		√	**38.5**	**59.7**	**41.5**	**23.1**	**42.3**	**49.0**
	RetinaNet		36.5	55.4	39.1	20.4	40.3	48.1
		√	**37.7**	**58.0**	**40.0**	**22.5**	**41.6**	**48.3**
	FCOS		36.6	55.7	38.8	20.7	40.1	47.4
		√	**37.5**	**57.1**	**39.8**	**22.8**	**41.1**	**48.5**
ResNet-101	Faster R-CNN		39.4	60.1	43.1	22.4	43.7	51.1
		√	**40.1**	**61.4**	**43.5**	**22.9**	**43.9**	**52.7**
	RetinaNet		38.5	57.6	41.0	21.7	42.8	50.4
		√	**39.7**	**59.9**	**42.3**	**23.8**	**44.1**	**51.1**
	FCOS		39.2	58.8	42.1	22.9	42.8	51.6
		√	**40.0**	**60.0**	**42.8**	**23.2**	**44.1**	**52.7**

4.3 Comparisons with Baselines

To demonstrate the generality of the proposed SSG framework on both anchor-based and anchor-free detectors, we consider three well-known object detectors as our baseline systems, including Faster R-CNN [18], RetinaNet [13], and FCOS [19]. All detectors are instantiated with two different backbones, i.e., ResNet-50 and ResNet-101 with FPN [12]. Comparison results are shown in Table 1. The proposed SSG framework achieves consistent accuracy gains overall all baseline detectors on all evaluation metrics. Specifically, our method improves about 1.2% AP for three baselines with ResNet-50 backbone. Particularly, the AP^S of MS-COCO depicted in Table 1 which represents the performance of small objects also gets significantly improved about 2% AP compared with baselines. These results demonstrate that the proposed SSG framework is more competitive in small object detection where objects usually possess no distinctive appearances.

For visualization purpose to compare with baselines, several examples of detection results are given in Fig. 3. Following a common threshold of 0.5 used for visualizing detected objects, we only illustrate a detection when its score is higher than the threshold. As we can see, some ambiguous objects (e.g., *snowboard*, *refrigerator*) are mislabeled, and some indistinctive objects (e.g., *traffic light*, *surfboard*) are undetected, where the scene is obvious and useful cues. These results indicate that our method is able to help recognizing objects processing ambiguous or indistinctive appearances by exploiting scene semantic information.

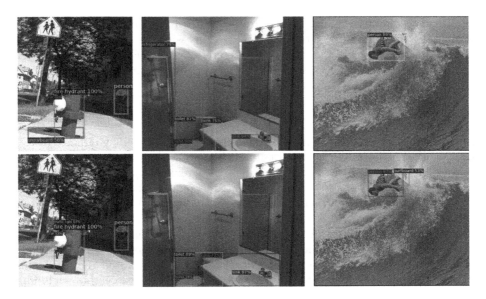

Fig. 3. Comparison of detection results. The first row is the results of each baseline (from left to right: Faster R-CNN, RetinaNet, and FCOS), and the second row is the results of each baseline with the proposed SSG framework.

4.4   Individual Module Effect

We study the impact of the individual module of the proposed SSG framework for different detectors and results are shown in Table 2. Solely applying the SSE module on the detection backbone gives about 0.4% AP improvement for three baselines. This verifies that SSE module advances the feature learning for context-dependent object categories while extracting the scene semantic feature. By adding the SCG module, the performance is further boosted since introducing the scene semantic information into the objects. These results suggest that these two modules in the SSG framework are complementary with each other.

Table 2. Effect of each module on MS-COCO `val2017` with ResNet-50-FPN.

Method	SSE	SCG	AP	Params(M)	Flops(G)
Faster R-CNN			37.4	41.53	207.07
	√		37.8	43.97	207.68
	√	√	**38.5**	46.07	209.25
RetinaNet			36.5	37.74	239.32
	√		36.8	40.19	239.48
	√	√	**37.7**	40.62	241.33
FCOS			36.6	32.02	200.5
	√		37.1	34.46	200.62
	√	√	**37.5**	34.89	202.47

We also compare the number of parameters with baselines, it shows that the proposed SSG framework gets considerable gains with a small relative increase in the number of parameters (<8%) and inference cost of FLOPs (<1%). These results suggest that the proposed SSG framework is effectiveness and efficiency.

5 Conclusions

In this paper, we proposed a novel and effective Scene Semantic Guidance (SSG) framework that leverages scene semantic information to provide a global semantic guidance for the object feature, thus help recognizing objects processing indistinctive or ambiguous appearances. By introducing an auxiliary task of multi-label classification to extract the scene semantic feature, the proposed scene semantic embedding module can advance the feature learning of the objects. And by measuring the semantical compatibility of the object and the scene, the proposed semantic guidance module can adaptively strengthen the discrimination of the object feature. Comprehensive experiments demonstrated the consistent outperforming accuracy on three different types well-known detectors, including Faster R-CNN, RetinaNet and FCOS. We believe it can be easily applied to other similar detectors as a plug-and-play component.

References

1. Cao, J., Chen, Q., Guo, J., et al.: Attention-guided context feature pyramid network for object detection. arXiv preprint arXiv:2005.11475 (2020)
2. Chen, K., Wang, J., Pang, J., et al.: MMDetection: Open mmlab detection toolbox and benchmark. arXiv preprint arXiv:1906.07155 (2019)
3. Chen, X., Gupta, A.: Spatial memory for context reasoning in object detection. In: ICCV (2017)
4. Chen, Z.M., Jin, X., Zhao, B., et al.: Hierarchical context embedding for region-based object detection. In: ECCV (2020)
5. Chen, Z., Huang, S., Tao, D.: Context refinement for object detection. In: ECCV (2018)
6. Deng, J., Dong, W., Socher, R., et al.: Imagenet: a large-scale hierarchical image database. In: CVPR (2009)
7. Divvala, S.K., Hoiem, D., Hays, J.H., et al.: An empirical study of context in object detection. In: CVPR (2009)
8. Fu, J., Liu, J., Wang, Y., et al.: Adaptive context network for scene parsing. In: ICCV (2019)
9. He, K., Zhang, X., Ren, S., et al.: Deep residual learning for image recognition. In: CVPR (2016)
10. Law, H., Deng, J.: Cornernet: detecting objects as paired keypoints. In: ECCV (2018)
11. Li, J., Wei, Y., Liang, X., et al.: Attentive contexts for object detection. IEEE TMM **19**(5), 944–954 (2016)
12. Lin, T.Y., Dollár, P., Girshick, et al.: Feature pyramid networks for object detection. In: CVPR (2017)

13. Lin, T.Y., Goyal, P., Girshick, R., et al.: Focal loss for dense object detection. In: ICCV (2017)
14. Lin, T.Y., Maire, M., Belongie, S., et al.: Microsoft coco: Common objects in context. In: ECCV (2014)
15. Liu, Y., Wang, R., Shan, S., et al.: Structure inference net: object detection using scene-level context and instance-level relationships. In: CVPR (2018)
16. Ouyang, W., Wang, X., Zeng, X., et al.: Deepid-net: deformable deep convolutional neural networks for object detection. In: CVPR (2015)
17. Qiao, X., Zheng, Q., Cao, Y., et al.: Tell me where i am: object-level scene context prediction. In: CVPR (2019)
18. Ren, S., He, K., Girshick, R., et al.: Faster r-cnn: towards real-time object detection with region proposal networks. IEEE TPAMI **39**(6), 1137 (2017)
19. Tian, Z., Shen, C., Chen, H., et al.: Fcos: fully convolutional one-stage object detection. in: ICCV (2019)
20. Yang, Z., Liu, S., Hu, H., et al.: Reppoints: point set representation for object detection. In: ICCV (2019)
21. Yuan, Y., Chen, X., Wang, J.: Object-contextual representations for semantic segmentation. In: European Conference on Computer Vision (2020)
22. Zhang, S., Chi, C., Yao, Y., et al.: Bridging the gap between anchor-based and anchor-free detection via adaptive training sample selection. In: CVPR (2020)
23. Zhang, X., Wan, F., Liu, C., et al.: Freeanchor: Learning to match anchors for visual object detection. In: NeurIPS (2019)
24. Zhu, C., Chen, F., Shen, Z., et al.: Soft anchor-point object detection. In: ECCV (2020)

Training Person Re-identification Networks with Transferred Images

Junkai Deng[1], Zhanxiang Feng[1], Peijia Chen[1], and Jianhuang Lai[1,2,3(✉)]

[1] School of Computer Science and Engineering Sun Yat-sen University, Guangzhou, Guangdong, China
dengjk6@mail2.sysu.edu.cn, stsljh@mail.sysu.edu.cn
[2] Guangdong Province Key Laboratory of Information Security Technology, Guangzhou, China
[3] Key Laboratory of Machine Intelligence and Advanced Computing, Ministry of Education, Guangzhou, China

Abstract. Recently, researchers begin using generated as well as transferred images to expand the training dataset of person re-identification. It expands the training dataset's sample space, which contributes to learning more robust features for the Re-ID network. However, these works only transfer a certain kind of feature between images. There is no known attempt to transfer multiple features across multiple images. By transferring multiple features, the sample space expands further, allowing Re-ID networks to learn more robust features. In this paper, we propose a unified framework and pipeline to integrate multiple feature transfer networks. Users are also free to determine how to transfer these features. Based on the above framework and pipeline, we create a large amount of transferred images from the Market-1501 dataset and create an expanded dataset named Market1501-EX. Further more, we propose a corresponding person identification labeling method for identity loss, using the generation information provided by the dataset. We use this dataset to train a Re-ID network, and the training result in this paper rests at mid-high level of related works which also utilize generated images.

Keywords: Person re-identification · Feature transfer · Framework · Pipeline

1 Introduction

Person Re-identification (Re-ID) has gained extensive public interest, with its potential application in public security. Particularly, images and videos captured from surveillance cameras could rely on Re-ID to quickly identify a particular suspect from a large crowd of people within a quite-long time span. During the COVID-19 outbreak, this technology could also be used to track down an infector's trace, making it easier to lock on other potential infectors.

This project is supported by the NSFC (62076258, 61902444).

H. Ma et al. (Eds.): PRCV 2021, LNCS 13019, pp. 366–378, 2021.
https://doi.org/10.1007/978-3-030-88004-0_30

Currently, person re-identification is facing a problem that the features extracted by the network rely too much on cues which are not directly related to the person's identity, such as people's clothes and poses. This has caused problem, as a person's cloth could change between different images, and a person's pose is almost never consistent. The limited number of training data is making person re-identification networks easy to overfit. Some recent works are trying to provide solutions to this problem, with new network structures, new datasets which deliberately provides images in scenes of the same person wearing different clothes across different images, and so on.

With the research into generative adversarial networks (GANs) gaining promising results, the idea of using generated images for person re-identification comes around in 2017, when Zheng et al. [30] try to generated a large amount of images using a generative adversarial network and then using these images as the training data of a person re-identification network. Since then, the idea of expanding current person re-identification dataset has gone through more exploration. Different feature transfer networks based on different generative adversarial networks have been proposed [1,4,14]. These works help to expand the sample space of the training data. However, these works mostly focus on one particular type of image feature, leaving other features untouched. This inspires us to break up an image into different kinds of features, and combining them more freely to generate even more generated images. This paper provides an attempt on this idea, whose main contributions are:

1. We propose a unified framework which aims to combine different kinds of feature transfer networks into a big generation module, and a corresponding pipeline for it.
2. Based on the framework and pipeline, we take our step to expand the Market-1501 dataset into a new, greatly expanded dataset "Market1501-EX".
3. We propose a person labeling method utilizing the details of the image generation process.
4. The experimental results show that the overall result lied on par with most related works, and the labeling method outperforming some other frequently used labeling strategies.

2   Related Works

2.1   Person Re-identification

In the early age of person re-identification, researchers mainly use the color and outline cues of a person in an image as a feature of this person, without any use of neural networks [6]. Since the introduction of CNN [9] into computer vision-related fields such as AlexNet [8], researchers begin using deep learning into person re-identification problems [10,27].

Recent person re-identification networks are mainly built upon various ResNet neural networks, often making changes to the popular ResNet-50 variant [3,21,29]. The attention mechanism is also integrated into the person re-identification networks in [11,13,20]. In Chen et al. [3], they use two types of

368   J. Deng et al.

additional structures to force the orthogonality of weights and features, making
the network less likely to extract features with high correlation.

It is also worth mentioning [26]. In their work, they propose a network which
extracts features entirely based on the outline of a person, without any use of
color. They conclude that this would contribute to the learning of robust features
with little amount of change in clothing.

Recently, researchers has become interested in cross-domain training
[22,23]. They train a person re-identification network using one dataset, but test
it against another dataset. Cross-domain training helps the network to learn
more robust features.

2.2 Image Generation and Person Re-identification

The idea of using generated images in person re-identification is first introduced
by Zheng et al. [30] in 2017. They use a GAN to learn from the training dataset,
and generate new images in addition to the original real images, forming a new
expanded dataset. This new dataset is then used as the training dataset of
a person re-identification network. The overall architecture of this method is
shown in Fig. 1. Subsequent works [14] have a similar architecture, using a GAN
to generate image, and a person Re-ID network to be trained. Recently, some
researchers [29] begin to integrate the person re-identification network directly
into the GAN, by sharing the weights learned by encoders.

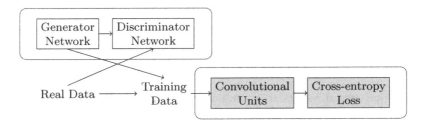

Fig. 1. The overall architecture proposed by Zheng et al. Images are first used to
train a GAN, then mixed with the generation results as a new training set for person
re-identification module.

Human Pose Transfer. Human pose transfer (HPT) is a relatively new problem. It is first proposed in 2017 by Ma et al. [15]. Liu et al. [14] have managed
to separate a person's pose feature from its appearance features. Prior to them,
Siarohin et al. [19] have applied a human pose transfer network on the Market-
1501 dataset, but without testing its effects on the person re-identification task.
Recent approaches to human pose transferring include Neverova et al. [16], Zhu
et al. [32], Pan et al. [17]. and Ge et al. [5].

Human Cloth Transfer. There are also attempts to change a person's clothing instead of changing pose. Among them the most remarkable result is DG-Net, proposed by Zheng et al. [29]. Deng et al. [4] utilize another network called CycleGAN proposed by Zhu et al. [31] to achieve the changing of a person's cloth.

Wang et al. [24] use 3D modeling to create a large amount of clothing textures. They also emulate different data collecting environments to expand the dataset. With the advance of image rendering and ray tracing, it would be more common to see expanded dataset using emulation techniques.

2.3   Loss Functions

There are three most-used loss functions when training a person re-identification network. These three frequently-used loss functions are triplet loss, verification loss and identity loss. Identity loss relies on the labeling of an image. It sees the person re-identification as an image classification problem. A common technique is label smoothing which provides some weight on other classes instead of using one-hot hard labeling. Smooth labeling prevents the network from too confident in classification. Labeling generated images proves to be a problem because the absence of ground truths is unable to obtain the one-hot labels. Researchers have found various ways to assign a label for them. Zheng et al. [30] propose a method named "LSRO". They use one-hot hard labeling with label smoothing for real images, and assign the same weight to every class when the image is a generated one. There are other labeling methods too. The "all in one" method stuffs every generated image into a new class label. And the "pseudo label" method, using current network's prediction's maximum value as the assumed label, forcing subsequent training epochs to move closer to this assumption [30]. Liu et al. [14] have another approach. They use hard labeling on real images, and apply label smoothing on generated labeling, as generated images are not as real as real images. Zheng et al. [29] use teacher-student model to determine a generated image's labeling, with the student model trying to fit the result by a pre-trained teacher model. The other two loss functions, verification loss and triplet loss, are not involved with labeling. Verification loss penalizes the network when it fails to predict whether the two given images are of the same person. Triplet loss encourages the network to move the features of the same person's images closer, while moving them further away from other people's features. Usually, a person re-identification network uses multiple loss functions during its training process.

3   A Unified Generation Framework and Pipeline

In this section, the details of the unified framework is briefly outlined. In this paper, the features that could be used by Re-ID networks are roughly put into four classes. The first of them are **environmental features**, which include the information of the position of the camera, the illumination status, the camera status, etc. There are also three person-related features: **pose features** which include the person's pose in the image; **cloth features** which include the person's clothing in the image, such as the cloth they wear, whether they are carrying handbags, etc.; and **identity features** which directly link to the person's identity. Identity features include the person's sex, body figure, skin color, etc. Among these four types of feature in an image, the former three types of features would be volatile, but the last type of feature would be consistent throughout different images. The ultimate goal of a person re-identification network would be extracting the last type of feature from an image. By mixing up these four types of features from different images, a GAN would generate a lot more images. The maximum number of images possible is s^4, where s is the total number of real images.

3.1   The Framework and Pipeline

The overall structure of the framework is shown in Fig. 2. The structure takes its resemblance from [30], which separates the generation module away from the person re-identification module. This is the most flexible design, as every part of the framework could be easily swapped out and substituted. Furthermore, every part of the framework needs not to be loaded into memory all at once, which greatly reduces the hardware environment requirements.

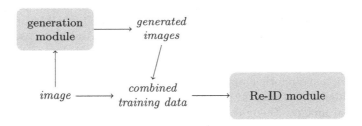

Fig. 2. The overall structure of the framework, the generated images are used as the input of the person re-identification network.

The image generation module's inside structure is shown in Fig. 3. It mainly uses a "generation cycle" to generate new images. From an image's perspective, it is like cascade feature transfer. An image goes through one of the transfer networks to produce a generated image with different clothes or poses. Then, the generated image could either exit the loop, or go back to the beginning to have another type of feature changed. How an image gets into and out of the loop is customizable. Here, two types of transfer networks are installed: the pose transfer network, and the cloth transfer network. This provides the transfer of two types of image features together. The pose transfer network we use is the network proposed by Isola et al. [7]. It requires the human pose estimation data of every image in the dataset, which could be obtained by a pre-trained network by Cao et al. [2, 25]. As for the cloth transfer network, we use the generation part of DG-Net [29]. DG-Net has its person re-identification embedded, but only its related generation portion is used.

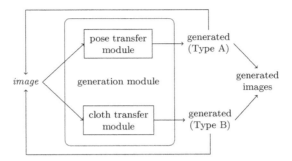

Fig. 3. The structure and pipeline of the image generation module. An image would go through this module several times before it is used as the final, generated image.

This approach does not require an embedded person re-identification network into the framework. Instead, a full Re-ID network will be appended to the end of it. This is because we use several distinct transfer generation networks in a cascading style, we could not find a suitable location to embed the person re-identification network. Shall we put the network into the pose transfer network, we risk having the extracted feature from the embedded network to include cloth information. The same applies to the cloth transfer network.

The core of the whole generation process is a file called "generation guide file". This file acts as a record of how an image should gone through the generation loop. Given n different types of transfer networks present in the generation module, then every record in the generation guide file contains $2n + 1$ fields. The first $n + 1$ fields represent the source real images used as the material of feature transferring, and the latter n fields represent the output file of each generation loop. The first field always represents the image's identity feature, and the following n images each provides a type of irrelevant feature.

3.2   Market1501-EX

Using the aforementioned framework and pipeline, we expand the Market-1501 dataset into an expanded dataset named "Market1501-EX" (Refer to Fig. 4). The Market1501-EX dataset uses four different generation schemes with one generation guide file: pose-only, cloth-only, pose first then cloth, cloth first then pose. The number of each of these four set of generated images is 140k, random sampled before mixing to form a new training dataset. It also comes with the raw generation guide file, thus downstream applications could trace back on every generated image on how this image is generated.

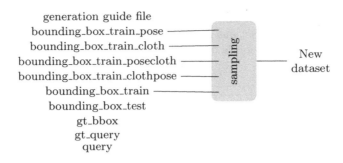

generation guide file
bounding_box_train_pose
bounding_box_train_cloth
bounding_box_train_posecloth
bounding_box_train_clothpose
bounding_box_train
bounding_box_test
gt_bbox
gt_query
query

sampling

New
dataset

Fig. 4. Arrangement of Market1501-EX.

3.3   Person Re-identification and Labeling

The person re-identification network we use is ABD-Net, proposed by Chen et al. [3]. We use identity loss as the loss function of the person Re-ID network.

We use a new labeling strategy, which utilizes the generation guide file. It takes a step back from the LSRO proposed in [30]. It follows the following rules:

- If the image is a real image, then it is identical to one-hot hard labeling. (the corresponding class label is set to 1, all other class labels are set to 0)
- If the image only goes through one generation cycle, then set every class label as 0, and for both real images, their respective class label is added by 1/2.
- If the image only goes through two generation cycles, then set every class label as 0, and for all real images, their respective class label is added by 1/3.

This idea could be expanded to multiple cycles. Suppose image x is generated from $x_0, x_1, \cdots, x_n$ after n cycles of generation loops, where $x_0, x_1, \cdots, x_n$ are related to k different class labels $y_1, y_2, \cdots, y_k$. For each class label, there are $m_1, m_2, \cdots, m_k$ images in set $\{x_0, x_1, \cdots, x_n\}$. Then the label for x is shown in the following equation. After that, label smoothing may be applied.

$$y_i = \begin{cases} \frac{y_i}{n+1} & i \in \{y_1, y_2, \cdots, y_k\} \\ 0 & \text{otherwise} \end{cases}, \text{for } i = 1, \cdots, K \qquad (1)$$

This labeling method is supposed to be more specific than LSRO, because every class label that has clearly no relevance to the generated image is set to 0 (if label smoothing is not applied) or a very small value (if label smoothing is applied). Any class label that has relevance to the generated image has a relatively higher weight than the evenly-distributed LSRO. This would help the network to learn more efficiently. Also, without one-hot hard labeling on generated images, this method takes the generated images' quality into account, and avoids making the network over-confident on generated images.

4   Experiment

4.1   Generated Images

Fig. 5. A collection of generated images. Four groups of results are put side-to-side. For each group, the first three columns are real images from Market-1501 dataset which provides identity feature, pose feature and cloth feature respectively. The fourth column shows the pose-only images. The fifth column shows the cloth-only images. The sixth column shows the pose-then-cloth images and the last column shows the cloth-then-pose images.

Figure 5 illustrates some of the generation results. It is clearly shown that most of the image generation tasks are successfully completed, preserving the identity information of the original image in the first column. Different types of transferred images help expand the size of the training data, which helps the person re-identification network to extract identity-related features.

There is a small fraction of generated images that are not ideal. Both pose transferring and cloth transferring could fail. And their failed results are shown in Fig. 6. There are two typical types of pose transfer failures - no person at all, or unrecognizable limbs. Cloth transferring failures are mostly the confusion in coloring the images.

Fig. 6. Left: Some examples of failed pose transfer generation. The first row contains no person, while the second row has unrecognizable limbs. Right: Some examples of failed cloth transfer generation. They are hard to recognize the shape of a human, and sometimes mistaking other elements as clothes.

4.2   Re-ID Results

Comparison with Some Related Methods. Table 1 shows the result of ours against several other generation-based approaches. It is shown that our result is better than other approaches which separates image generation and person re-identification. It sits in a relatively good position among the approaches that embeds person Re-ID network into the generation network. We speculate that embedding the person re-identification network into the generation network has better results, because it directly uses the extracted features during transfer generation. Converting features back into images could be lossy, and this would affect the accuracy of the person re-identification network.

Table 1. Comparison with some related works.

	Embedded Re-ID?	Rank-1	mAP
Li et al. [12]	Yes	66.1	32.7%
Zeng et al. [28]	Yes	88.45	71.46%
Bhuiyan et al. [1]	Yes	88.51	74.55%
Qian et al. [18]	Yes	89.43	72.58%
Ge et al. [5]	Yes	90.5	77.7%
Zheng et al. [29]	Yes	**94.8**	**86.0%**
Zheng et al. [30]	No	83.97	66.07%
Liu et al. [14]	No	87.65	68.92%
Ours	No	**90.41**	**72.88%**

Table 2. Result on different expansion sets.

Base	Rank-1	Rank-5	Rank-10	mAP
	88.66	95.31	96.94	**72.73%**
base + pose	87.71	95.37	97.00	70.53%
base + pose-cloth	88.48	95.04	96.79	71.97%
base + cloth-pose	88.69	95.34	96.82	72.59%
base + cloth	**89.25**	**95.81**	**97.18**	71.38%

Ablation Studies. Table 2 shows the training result on individual expanded image sets. It is clear that going through two-phase image generation achieves better results than only going through pose transferring. However, they perform less ideal than going through cloth transferring alone. We specially observe that when using the pose transferred images, the result is even worse than the baseline. Part of the reason is that the failed generation images given by pose transfer network are worse than the failed images by cloth transfer network. Failed pose transfer images often miss some human parts, or completely unrecognizable. Compared to failed cloth transfer images which only fail in coloring, they are less likely useful to person re-identification training. And the generation error would accumulate through multiple generation cycles. Another possible cause is that the Re-ID network we use is specially aimed to combat the issue of extracting correlated features caused by the pattern on clothes. It is not very good at distinguishing different poses. And the large variety of poses confuses the Re-ID network. Nevertheless, they achieve better results than the baseline, which doesn't use any expanded dataset.

Table 3. Test on different labeling methods.

	Rank-1	Rank-5	Rank-10	mAP
One-hot	85.81	94.42	96.50	66.62%
LSRO	88.33	94.71	96.38	**72.12%**
Ours	**89.25**	**95.81**	**97.18**	71.38%

Table 3 shows the training result using different labeling methods. This paper mainly compares it with one-hot hard labeling and LSRO labeling proposed by [30]. It is shown that the labeling method in we proposed has an obviously better result under the Rank-k criteria, although it is slightly less ideal under the mAP criteria.

Finally, Table 4 shows that different number of generated images would have a different impact on the overall performance of the Re-ID network. When the number of generated images is below around 20,000, more images would yield better results. However, when more generated images are added to the training

Table 4. Impact of different number of generated images on the Re-ID result.

0	Rank-1	Rank-5	Rank-10	mAP
	88.66	95.31	96.94	72.73%
4000	89.13	95.37	97.18	71.28%
8000	89.28	95.84	97.54	71.50%
12000	89.46	95.75	97.65	71.98%
16000	89.58	95.90	97.71	72.02%
20000	**90.41**	**96.17**	97.48	**72.88%**
24000	89.52	96.08	**97.77**	72.22%
30000	89.67	96.08	97.54	71.50%

set, the result took a downturn. With the best overall performance sitting at around 20,000 generated images. When there are not many generated images, the added images indeed expand the training set's sampling space, attributing to the learning of better-quality features. However, when the number is too large, the generated images, which obviously have lower quality than real images, come to overwhelm the real images, confusing the network. Also, too many generated images would make the training process too long.

5   Conclusion

In this paper, a unified framework which aimed to provide a common proto-col to all existing image generation networks for person re-identification is pro-posed. Furthermore, an expansion based on the Market-1501 dataset, named Market1501-EX dataset, is provided, which provides flexibility and all sorts of details to its users. Finally, utilizing the generation details provided by Market1501-EX dataset, we propose a method for person identity labeling. Experimental results show that the training results from the generated images which have gone through more than one round of image generation, are better than the result from images which have gone through pose transferring. This shows that going through multiple rounds of generation could be useful to some extent. As for the labeling scheme, we show that its performance is better than other common techniques.

References

1. Bhuiyan, A., Liu, Y., Siva, P., Javan, M., Ayed, I.B., Granger, E.: Pose guided gated fusion for person re-identification. In: WACV, pp. 2675–2684 (2020)
2. Cao, Z., Simon, T., Wei, S.E., Sheikh, Y.: Realtime multi-person 2d pose estimation using part affinity fields. In: CVPR (2017)
3. Chen, T., et al.: Abd-net: Attentive but diverse person re-identification. In: ICCV, pp. 8351–8361 (2019)

4. Deng, W., Zheng, L., Ye, Q., Kang, G., Yang, Y., Jiao, J.: Image-image domain adaptation with preserved self-similarity and domain-dissimilarity for person re-identification. In: CVPR, pp. 994–1003 (2018)
5. Ge, Y., et al.: Fd-gan: Pose-guided feature distilling gan for robust person re-identification. arXiv preprint arXiv:1810.02936 (2018)
6. Gheissari, N., Sebastian, T.B., Hartley, R.: Person reidentification using spatiotemporal appearance. In: CVPR, vol. 2, pp. 1528–1535. IEEE (2006)
7. Isola, P., Zhu, J.Y., Zhou, T., Efros, A.A.: Image-to-image translation with conditional adversarial networks. In: CVPR, pp. 1125–1134 (2017)
8. Krizhevsky, A., Sutskever, I., Hinton, G.E.: Imagenet classification with deep convolutional neural networks. NIPS **25**, 1097–1105 (2012)
9. LeCun, Y., Bottou, L., Bengio, Y., Haffner, P.: Gradient-based learning applied to document recognition. IEEE **86**(11), 2278–2324 (1998)
10. Li, W., Zhao, R., Xiao, T., Wang, X.: Deepreid: deep filter pairing neural network for person re-identification. In: CVPR, pp. 152–159 (2014)
11. Li, W., Zhu, X., Gong, S.: Harmonious attention network for person re-identification. In: CVPR, pp. 2285–2294 (2018)
12. Li, Y., Wang, T., Liu, L.: Random style transfer for person re-identification with one example. AIMS Math. **6**(5), 4715–4733 (2021)
13. Liu, H., Feng, J., Qi, M., Jiang, J., Yan, S.: End-to-end comparative attention networks for person re-identification. TIP **26**(7), 3492–3506 (2017)
14. Liu, J., Ni, B., Yan, Y., Zhou, P., Cheng, S., Hu, J.: Pose transferrable person re-identification. In: CVPR, pp. 4099–4108 (2018)
15. Ma, L., Jia, X., Sun, Q., Schiele, B., Tuytelaars, T., Van Gool, L.: Pose guided person image generation. arXiv preprint arXiv:1705.09368 (2017)
16. Neverova, N., Guler, R.A., Kokkinos, I.: Dense pose transfer. In: ECCV (September 2018)
17. Pan, H., Cao, X.: Pose transfer based on generative adversarial networks. In: Su, R. (ed.) 2020 International Conference on Image, Video Processing and Artificial Intelligence, vol. 11584, pp. 239–244. International Society for Optics and Photonics, SPIE (2020)
18. Qian, X., et al.: Pose-normalized image generation for person re-identification. In: ECCV (September 2018)
19. Siarohin, A., Sangineto, E., Lathuiliere, S., Sebe, N.: Deformable gans for pose-based human image generation. In: CVPR, pp. 3408–3416 (2018)
20. Sun, J., Li, Y., Chen, H., Zhang, B., Zhu, J.: Memf: multi-level-attention embedding and multi-layer-feature fusion model for person re-identification. PR **116**, 107937 (2021)
21. Sun, Y., Zheng, L., Yang, Y., Tian, Q., Wang, S.: Beyond part models: person retrieval with refined part pooling (and a strong convolutional baseline). In: ECCV, pp. 480–496 (2018)
22. Tang, Y., Xi, Y., Wang, N., Song, B., Gao, X.: Cgan-tm: a novel domain-to-domain transferring method for person re-identification. TIP **29**, 5641–5651 (2020)
23. Wang, H., Hu, J., Zhang, G.: Multi-source transfer network for cross domain person re-identification. IEEE Access **8**, 83265–83275 (2020)
24. Wang, Y., Liao, S., Shao, L.: Surpassing real-world source training data: Random 3d characters for generalizable person re-identification. In: ACMMM, pp. 3422–3430 (2020)
25. Wei, S.E., Ramakrishna, V., Kanade, T., Sheikh, Y.: Convolutional pose machines. In: CVPR (2016)

26. Yang, Q., Wu, A., Zheng, W.S.: Person re-identification by contour sketch under moderate clothing change. In: TPAMI (2020)
27. Yi, D., Lei, Z., Liao, S., Li, S.Z.: Deep metric learning for person re-identification. In: ICPR, pp. 34–39. IEEE (2014)
28. Zeng, Z., Wang, Z., Wang, Z., Zheng, Y., Chuang, Y.Y., Satoh, S.: Illumination-adaptive person re-identification. TMM **22**(12), 3064–3074 (2020)
29. Zheng, Z., Yang, X., Yu, Z., Zheng, L., Yang, Y., Kautz, J.: Joint discriminative and generative learning for person re-identification. In: CVPR, pp. 2138–2147 (2019)
30. Zheng, Z., Zheng, L., Yang, Y.: Unlabeled samples generated by GAN improve the person re-identification baseline in vitro. In: ICCV, pp. 3754–3762 (2017)
31. Zhu, J.Y., Park, T., Isola, P., Efros, A.A.: Unpaired image-to-image translation using cycle-consistent adversarial networks. In: ICCV, pp. 2223–2232 (2017)
32. Zhu, Z., Huang, T., Shi, B., Yu, M., Wang, B., Bai, X.: Progressive pose attention transfer for person image generation. In: CVPR (June 2019)

ACFIM: Adaptively Cyclic Feature Information-Interaction Model for Object Detection

Chen Song, Xu Cheng[⊠], Lihua Liu, and Daqiu Li

School of Computer and Software, Nanjing University of Information Science and Technology, Nanjing 210044, China
xcheng@nuist.edu.cn

Abstract. Object detection is one of the most fundamental tasks toward image content understanding due to their wide applications in real-world. Although numerous algorithms have been proposed, implementing effective and efficient object detection is still very challenging for now, especially for the challenges in restricted situations of multi-size objects and weak semantic information. In this paper, we propose a feature information-interaction visual attention model for multi-layer feature fusion and enhancement, which utilizes channel information to weight self-attentive feature maps, completing extraction, fusion and enhancement of global semantic feature with local contextual information of the object. Additionally, we also propose an adaptively cyclic feature information-interaction model, which adopts branch prediction to decide the number of visual attention, accomplishing adaptive fusion of global semantic feature and local fine-grained information. Numerous experiments on the benchmark dataset PASCAL VOC and MS COCO show that our method effectively achieves significant improvements over baseline model.

Keywords: Computer vision · Object detection · Convolutional neural network

1 Introduction

Object detection in an image [1] includes category recognition and position prediction, which is the underlying task in computer vision. It provides a prerequisite of feature extraction for the high-level visual tasks [2–4] such as video object detection, salient object detection, object tracking, etc. With the rapid development of deep learning technology recently, object detectors have been developed and achieved high efficiency, which are now widely utilized in many fields [5–7] such as intelligent video surveillance, robot vision, autonomous driving, etc.

Before the prosperity of deep learning, traditional algorithms [8] mainly concentrated on feature extraction and region classification with appropriate detection accuracy on

This work is supported in part by the National Natural Science Foundation of China (Grant No. 61802058, 61911530397); and in part by the Project funded by the China Postdoctoral Science Foundation (Grant No. 2019M651650).

H. Ma et al. (Eds.): PRCV 2021, LNCS 13019, pp. 379–391, 2021.
https://doi.org/10.1007/978-3-030-88004-0_31

the benchmark dataset of PASCAL VOC [9]. During this time, the drawbacks of these algorithms are revealed. Firstly, most models adopted the sliding window scheme to generate region proposals, which resulted in redundant proposals and imbalanced samples with positive and negative regions. Secondly, feature descriptors (Haar-like features [10], Scale Invariant Feature Transform [11], Histogram of Gradients [12], etc.) failed to fully capture the high-level semantic and low-level fine-grained object information. Finally, the overall structure lacked global optimization strategy.

Recently, deep learning has experienced an unprecedented development boom and performed the promising results in many visual tasks. The existing object detection pipelines based on deep learning can be classified into: anchor-based object detection pipelines [14–24] and anchor-free object detection pipelines [25–29]. Nevertheless, these models exist the problem of inaccurate location prediction, the major reason is inadequate feature scheme of global and local feature of the object.

To solve the above problems, the contributions of this work are summarized as follows:

(1) We propose a feature information-interaction attention model under visual perception to enrich semantic feature of the objects, which accomplishes multi-layer information extraction, feature response maps fusion and enhancement of global feature with the object contextual information.
(2) We also develop an adaptively cyclic feature information-interaction module, which is integrated into out model to achieve the adaptive fusion of semantic features and local fine-grained information, better distinguishing the similarities and differences between objects.
(3) Experimental results on the PASCAL VOC and MS COCO benchmarks show that our proposed method gains significant improvements over the baseline method.

2   Related Works

2.1   Anchor-Based Object Detectors

The anchor-based object detectors divide the detection task into three steps: region proposal generation, feature extraction and classification prediction. Girshick et.al [14] proposed R-CNN (Regions with CNN features), which utilized Selective Search algorithm to generate region proposals, and then predicted the coordinate location of the objects. Faster RCNN [15] is trained in end-to-end manner by introducing RPN module, which generated anchors containing three scales and aspect ratios for each pixel of the feature map. Additionally, SSD [16], an one-stage object detector, was proposed to predict multi-scale objects through predefined anchors. After that, DSSD [17], STDN [18] and RefineDet [19] were proposed to fuse multi-layer low-level and high-level information through feature fusion and enhancement. Recently, some other algorithms [20–24], such as neural architecture search, contextual information fusion, transformer etc., is developed to finetune baseline detectors.

2.2 Anchor-Free Object Detectors

The anchor-free object detection pipelines predict border coordinates by combining keypoints pairs learned from feature maps, which eliminate the requirement to design anchors and reduce the model parameters. Law et al. [25] proposed CornerNet based on paired keypoints (the top-left and the bottom-right corner), which utilized Hourglass [26] to predict the heatmap of corner locations. Additionally, corner pooling was developed to better localize keypoints. To better localize corner points, central keypoint detection was introduced to by CenterNet [27] to assist in localizing objects, which proposed both central pooling and cascaded corner pooling to enrich object boundaries and internal information. Tian et al. [28] proposed FCOS (Fully Convolutional One-stage Object Detection), which translates object detection task into a pixel-level prediction fashion. Zhang et al. [29] proposed RepPoints, a new finer representation of objects as a set of sample points, which modeled fine-grained localization information and identified local region features.

2.3 Visual Attention Models

In the computer vision field, visual attention models are utilized to highlight the edge, texture, visual change, etc. SENet [30], the winner of ImageNet2017 Classification Competition, learnt the relevance and difference between channels and then weighted them. Afterward, CBAM [31] and BAM [32] were presented to integrate channel and spatial attention modules into tandem and parallel modes, which enhanced feature information flow between channels. Currently, the subsequent attention pipelines [33, 34] take self-attentive feature fusion, adaptive convolution kernel selection and multistage feature fusion to enhance information flow between feature maps. Additionally, self-attentive models have also developed to integrate spatial information into feature maps. Inspired by non-local similarity of image denoise, Wang et.al [35] proposed Non-local Network for exploring spatial correlation between feature maps, which developed non-local blocks to capture long-range dependencies. Nevertheless, it greatly increased the computational complexity.

3 The Proposed Method

3.1 Problems and Motivations

Single Shot MultiBox Detector (SSD) is one of the best detectors with respect to speed and accuracy trade-off. Nevertheless, some drawbacks limit the accuracy of the algorithm. Firstly, the semantic information of shallow layers is weak, and it fails to capture global dependent information to predict small and dense clusters of objects. Secondly, the feature maps from medium layers exist the problem of feature confusion, which make it difficult to accurately regress bounding boxes. Finally, the deep layers have less object contextual information, making it fail to predict large objects confidently.

Inspired by information guidance between self-attentive models [35], we propose a feature information-interaction model, which introduces feature map channel weights on the self-attentive module and takes a weighted mechanism to focus on the regional block.

On this basis, an adaptively cyclic information-interaction visual model is developed to solve the problem of insufficient feature fusion, which concentrates on the feature map more than once, better distinguishing the similarities and differences between objects.

3.2 Feature Information-Interaction Model

As mentioned above, the existing self-attentive models associate the internal information of feature maps and concentrate on the local information of object, ignoring the inter-channel feature information association, i.e., the global semantic feature information of the object. To tackle this problem effectively, we propose a feature information-interaction model (FIM), where weighted channels of feature maps are proposed to perceive the global semantic and local fine-grained feature of the object. The overall structure of FIM is shown in Fig. 1.

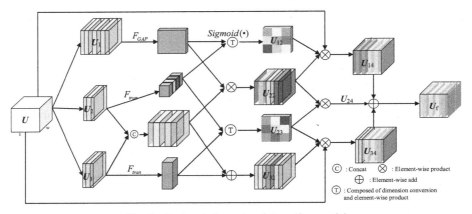

Fig. 1. Feature information-interaction model

Given the feature map $U \in \mathbf{R}^{H \times W \times C}$, U_1, U_2, U_3 are obtained through convolution operator respectively, where $U_1 \in \mathbf{R}^{H \times W \times C}, U_2 \in \mathbf{R}^{H \times W \times C/8}$ and $U_3 \in \mathbf{R}^{H \times W \times C/8}$. Noticed that the channel attention and self-attentive module are achieved by these feature maps. For the channel attention, global average pooling and sigmoid are utilized to get feature map with weighted channel, and the resulting feature map $U_{12} \in \mathbf{R}^{1 \times 1 \times C}$. For the self-attention module, we utilize dimension transformation to obtain the intermediate feature map. Then, weighted channel supervision information is utilized to assist the intermediate self-attention feature map. Afterwards, U_{22}, U_{23} and U_{32} are obtained to represent enhanced feature maps with global semantic features and local fine-grained information respectively, where $U_{22} \in \mathbf{R}^{H \times W \times C}, U_{23} \in \mathbf{R}^{1 \times 1 \times C}$ and $U_{32} \in \mathbf{R}^{H \times W \times C}$. Additionally, we merge the original feature map U into enhanced feature maps for enriching semantic feature information. Therefore, the adaptively weighted attention information can be obtained through threshold multiply and add operations through Eq. (1).

$$U_{14} = (1 + \alpha)(U_{12} \otimes U_{22}) \oplus (1 - \alpha)U \tag{1}$$

where U_{14} is the enhanced feature map, U_{12}, U_{22} and U are the intermediate layer information, respectively. And α denotes the predicted threshold through convolution, $\otimes$ and $\oplus$ represent element-multiply and element-add operations, respectively. And then, self-attentive feature maps can be obtained in spatial dimension through Eq. (2). and Eq. (3), respectively.

$$U_{24} = \alpha U_{22} \oplus \beta U_{23} \tag{2}$$

$$U_{34} = (1 + \alpha)(U_{23} \otimes U_{32}) \oplus (1 - \alpha)U \tag{3}$$

where U_{24} and U_{34} are the enhanced feature maps, U, U_{22}, U_{23} and U_{32} are the intermediate layer information, respectively. α and β are the predicted threshold and we set $\alpha + \beta = 1$ empirically. Finally, the threshold weighting method is adopted to generate resulting feature map through Eq. (1).

$$U_f = \alpha(U_{14}) \oplus \beta(U_{24}) \oplus \gamma(U_{34}) \tag{4}$$

where U_f is the resulting feature map, U_{14}, U_{24}, and U_{34} are the intermediate enhanced features, respectively, α, β and γ are adaptive threshold.

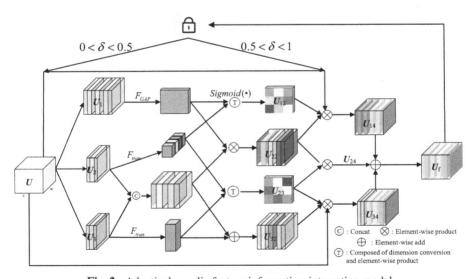

Fig. 2. Adaptively cyclic feature information-interaction model

3.3 Adaptively Cyclic Feature Information-interaction Model

To enrich global semantic feature and local contextual information of the object, we also propose an adaptively cyclic information-interaction model (ACFIM) to strengthen the ability of feature extraction. Concretely speaking, the convolutional prediction module is developed to control the location and number of visual attention, adaptively concentrates on the feature map more than once, better distinguishing the similarities and differences between objects.

Based on our knowledge and experience, feature maps of shallow layer have enriched local fine-grained information of the object, and feature maps of deep layer contain abundant global semantic feature of the object. Therefore, different choices of cycle times and location of feature fusion between shallow-medium-deep layers are different. For feature maps of shallow layer, we set three times default for cyclic model to better enrich and represent global feature flow of the object. For feature maps of medium and deep layer, we set two times default for cyclic model for finetuning information flow between high-level semantic feature and low-level fine-grained information. Additionally, we set an intermediate threshold δ for deciding cyclic location to better adapt the enhanced feature map. If δ less than predefined threshold (we set 0.5 default), we merge the enhanced feature map into the original feature map for loop initialization. Otherwise, we merge the enhanced feature map into the intermediate feature map for feature fusion and initialization. The overall structure of ACFIM is shown in Fig. 2.

3.4 Loss Function

The overall objective loss function of SSD is a weighted sum of the localization loss and the confidence loss, more detail information can be referred in [16]. In our model, we adopt Focal loss for confidence loss to address the problem of class imbalance. Additionally, we slightly adjust the parameters between the default anchors and the ground-truth boxes, as shown in Eq. (5).

$$L_{loc}(x,p,g,d) = \sum_{i \in pos}^{N} \sum_{m \in \{cx,cy,w,h\}} x_{ij}^k L_1(p_i^m - k_j^m)$$

$$k_j^{cx} = (g_j^{cx} - d_i^{cx})/d_i^w/\text{var}^1 \quad k_j^{cy} = (g_j^{cy} - d_i^{cy})/d_i^h/\text{var}^1$$

$$k_j^w = \log(\frac{g_j^w}{d_i^w} + \text{var}^2) \quad k_j^h = \log(\frac{g_j^h}{d_i^h} + \text{var}^2) \tag{5}$$

where p, g, and d denote the predicted value, ground-truth bounding boxes parameters and the default anchors parameters, respectively. Let $x_{ij}^k = \{0, 1\}$ be an indicator for matching the i-th default box to j-th ground truth box of category k.var^1 and var^2 denote the variance of object weight and height, respectively. We set var^1 and var^2 to 1.05 and 0.2 empirically.

4 Experimental Results

In order to compare the proposed method with the baseline model SSD [16] fairly, our experimental results are all based on backbone network VGG [36]. We experiment on the PASCAL VOC [9] and MS COCO [13], two mainstream object detection benchmark datasets, and ablation studies are tested on PASCAL VOC2007. The metric to evaluate detection performance are Average Precision (AP), mean Average Precision (mAP) and Frame Per Second (FPS).

4.1 Datasets

We evaluate our model on object detection benchmark datasets, namely PASCAL VOC and MS COCO. For the PASCAL VOC2007, it is divided into trainval and test sets, and the former contains 5011 images and the latter contains 4952 images annotated with category and bounding box of objects. For the MS COCO dataset, the trainval35k with 119297 images is utilized to train and the standard test-dev 2015 split with 20288 images is used to test. Additionally, more fine-grained metrics are adopted by the dataset to evaluate the model comprehensively. MS COCO defines that the object is small (S $< 32^2$), medium ($32^2 < S < 96^2$), large (S $> 96^2$), where S is measured as the number of pixels in the image, AP_S, AP_M, AP_L denote mAP of model where objects are small, medium, and large.

4.2 Experimental Setup

We implement the proposed method on the framework of Pytorch, ands utilize SSD as our baseline model. The proposed visual attention model is embedded in the extras four prediction modules. During the training stage, without bells and whistles, we follow the original SSD strategies, including data augmentation, backbone network, scale and aspect ratios for predefined anchors, etc. While we slightly change our learning rate scheduling for better accommodation of our model. All new conv-layers are initialized with the MSRA method [37].

4.3 Pascal Voc2007

We train our model on the benchmark dataset PASCAL VOC2007 and VOC2012 trainval set, and test the model on VOC2007 test set. We utilize 10^{-3} learning rate for the first 80k iterations, then decrease it to 10^{-4} for the next 20k iterations and 10^{-5} for the rest 20k iterations. Additionally, we adopt a "warmup" strategy that gradually ramps up the learning rate, which contributes to stabilizing the training process. The momentum and weight decay are set to 0.9 and 0.0005, respectively.

Table 1 and Table 2 show experimental results on the PASCAL VOC test set, and All the methods are trained on VOC2007 and VOC2012 trainval set. Without bells and whistles, our method achieves 79.7% mAP with 300 × 300 input images and 82.1% mAP with 512 × 512 input images, exceeding the latest SSD300* (the latest version updated by the authors with) by 2.2 points and SSD512* (the latest version model with 300 × 300 input images) by 2.6 points. For some categories such as aero, bike, horse, person and train which occupy the major area in an image, our proposed method gains a remarkable improvement of 2–7 points than baseline model of SSD300*, outperforming most of the other detectors. For the medium objects like boat, chair and table, our detector with 300 × 300 input gains a moderate improvement of 2–5 points than SSD300*. For some small objects like bottle, chair and plant, Our our proposed method gains a remarkable improvement of 3–7 points of SSD300*.

Table 1. Detection results on PASCAL VOC2007 test set, and the bold red and bold blue fonts denote the best and the second best results for single category, respectively.

Method	mAP	Aero	Bike	Bird	Boat	Bottle	Bus	Car	Cat	Chair	Cow
Region-based object detectors											
Faster [15]	73.2	76.5	79.0	70.9	65.6	52.1	83.1	84.7	96.4	52.0	81.9
ION [20]	75.6	79.2	83.1	77.6	65.6	54.9	85.4	85.1	87.0	54.4	80.6
Faster [15]	76.4	79.8	80.7	76.2	68.3	55.9	85.1	85.3	89.8	56.7	87.8
MR-CNN [21]	78.2	80.3	84.1	78.5	70.8	68.5	88.0	85.9	87.8	60.3	85.2
R-FCN [22]	80.5	79.9	87.2	81.5	72.0	69.8	86.8	88.5	89.8	67.0	88.1
R-DAD [23]	80.2	90.0	86.6	81.3	71.2	66.0	83.4	83.7	94.5	63.2	84.0
Single-stage object detectors											
SSD300* [16]	77.5	79.5	83.9	76.0	69.6	50.5	87.0	85.7	88.1	60.3	81.5
SSD512* [16]	79.5	84.8	85.1	81.5	73.0	57.8	87.8	88.3	87.4	63.5	85.4
DSSD321 [17]	78.6	81.9	84.9	80.5	68.4	53.9	85.6	86.2	88.9	61.1	83.5
DSSD512 [17]	81.5	86.6	86.2	82.6	74.9	62.5	89.0	88.7	88.8	65.2	87.0
ESSD300 [38]	79.4	82.6	86.1	79.8	72.2	54.7	86.8	86.9	88.2	62.8	85.2
Feature-fused SSD [39]	78.9	82.0	86.5	78.0	71.7	52.9	86.6	86.9	88.3	63.2	83.0
STDN300 [18]	78.1	81.1	86.9	76.4	69.2	52.4	87.7	84.2	88.3	60.2	81.3
STDN513 [18]	80.9	86.1	89.3	79.5	74.3	61.9	88.5	88.3	89.4	67.4	86.5
MDSSD300 [40]	78.6	86.5	87.6	78.9	70.6	55.0	86.9	87.0	88.1	58.5	84.8
MDSSD512 [40]	80.3	88.8	88.7	83.2	73.7	58.3	88.2	89.3	87.4	62.4	85.1
Ours	79.7	86.3	86.9	77.8	74.0	56.9	87.2	86.5	88.1	65.0	84.8
Ours (512)	82.1	87.8	88.7	82.6	75.2	64.4	88.8	88.6	88.7	67.5	88.1

Table 2. Detection results on PASCAL VOC2007 test set, and the bold red and bold blue fonts denote the best and the second best results for single category, respectively.

Method	Table	Dog	Horse	Mbike	Person	Plant	Sheep	Sofa	Train	Tv
Region-based object detectors										
Faster [15]	65.7	84.8	84.6	77.5	76.7	38.8	73.6	73.9	83.0	72.6
ION [20]	73.8	85.3	82.2	82.2	74.4	47.1	75.8	72.7	84.2	80.4
Faster [15]	69.4	88.3	88.9	80.9	78.4	41.7	78.6	79.8	85.3	72.0
MR-CNN [21]	73.7	87.2	86.5	85.0	76.4	48.5	76.3	75.5	85.0	81.0
R-FCN [22]	74.5	89.8	90.6	79.9	81.2	53.7	81.8	81.5	85.9	79.9

(continued)

Table 2. (*continued*)

Method	Table	Dog	Horse	Mbike	Person	Plant	Sheep	Sofa	Train	Tv
R-DAD [23]	64.2	**92.8**	**90.1**	**88.6**	**87.3**	**62.2**	82.8	70.9	**88.8**	77.6
Single-stage object detectors										
SSD300* [16]	77.0	86.1	87.5	84.0	79.4	52.3	77.9	79.5	87.6	76.8
SSD512* [16]	73.2	86.2	86.7	83.9	82.5	55.6	81.7	79.0	86.6	80.0
DSSD321 [17]	78.7	86.7	88.7	86.7	79.7	51.7	78.0	80.9	87.2	79.4
DSSD513 [17]	78.7	88.2	89.0	87.5	83.7	51.1	**86.3**	**81.6**	85.7	**83.7**
ESSD300 [38]	78.2	87.5	88.0	87.0	80.0	56.1	80.2	80.4	**887**	78.1
Feature-fused SSD [39]	76.8	86.1	88.5	87.5	80.4	53.9	80.6	79.5	88.2	77.9
STDN300 [18]	77.6	86.6	88.9	87.8	76.8	51.8	78.4	81.3	87.5	77.8
STDN513 [18]	79.5	86.4	89.2	88.5	79.3	53.0	77.9	81.4	86.6	**85.5**
MDSSD300 [40]	73.4	84.8	89.2	88.1	78.0	52.3	78.6	74.5	86.8	80.7
MDSSD512 [40]	75.1	84.7	89.7	88.3	83.2	56.7	**84.0**	77.4	83.9	77.6
Ours	**79.8**	86.2	88.1	86.0	79.9	55.3	79.8	81.2	86.5	77.8
Ours (512)	**80.2**	87.5	88.2	**88.9**	**84.0**	**56.9**	83.4	**82.2**	88.4	82.0

Table 3. Comparison of mAP for different models on PASCAL VOC2007 test set.

Attention model	Data	Feature fusion	mAP
SENet [30]	07 + 12	Element-wise sum	77.93
Non-local network [35]	07 + 12	Element-wise sum	78.12
GCNet [41]	07 + 12	Element-wise sum	78.11
BAM [32]	07 + 12	Element-wise sum	77.61
CBAM [31]	07 + 12	Element-wise sum	77.90
SKNet [42]	07 + 12	Element-wise sum	78.16
SKNet [42]	07 + 12	Element-wise product	78.02
FIM (Ours)	07 + 12	Element-wise sum	79.48
FIM (Ours)	07 + 12	Element-wise product	79.30
ACFIM	07 + 12	Element-wise sum	79.70
ACFIM (Ours)	07 + 12	Element-wise product	79.62

4.4   Ablation Study on PASCAL VOC2007

To further validate the effectiveness of the proposed method, we study some significant design factors for the proposed modules. In these experiments, the model is trained with the combined dataset from VOC2007 trainval set and VOC2012 trainval, and tested on the VOC2007 test with 300×300 input images.

The Effectiveness of Visual Attention Models: In Table 3, we compare proposed method with other visual attention models under same baseline model. For fairness and simplicity, we simply replace our module with other visual attention models. Besides our method, the baseline SSD model with SKNet achieves the best performance. Among these models, channel-wise attention of SKNet is efficient in excavating the internal relationship between channels. Our model considers object feature information-interaction under visual perception, which contributes to excavating the similarity and difference of feature maps and achieves the best performance among all models.

FIM vs. ACFIM: To validate the effectiveness of our proposed cyclic feature-interaction method and other design factors, we experiment on the same baseline configuration by Sect. 4.2. In Table 3, the baseline model with our proposed ACFIM achieves the best performance, which exceeds SKNet and FIM by 1.54 and 0.22 points. Although FIM incorporates semantic and local detailed information to concentrate on important features of the object, we believe that multiple attention contributes to learning and distinguishing objects more finely. We also adopt different feature fusion methods to explore the best fusion mode of feature maps, as shown in Table 3, the element-wise sum of feature maps is better than the element-wise product.

4.5   Ms Coco

To further validate the effectiveness of proposed method, we train our model on MS COCO [31]. We utilize the trainval35k (118287 images) for training and evaluate the results on the minival. We train the model with 10^{-3} for the first 280k iterations, then 10^{-4} and 10^{-5} for another 120k and 40k iterations. In Table 4, we observe that our method achieves 27.6% AP@[0.5:0.95], 46.8% AP@0.5, and 28.7% AP@0.75, which improves the baseline model SSD300* by 2.5, 3.7, and 2.9 points respectively. Our model with 512×512 input images also outperforms the baseline SSD512* by 2.1, 2.5, and 2.5 points respectively. It is noticeable that our model with 300×300 and 512×512 input images achieve 8.9% AP and 13.0% AP for small objects ($S < 32^2$) respectively, which improves SSD (2.3% and 2.1%) with a large margin and proves that our proposed method

Table 4. Detection results on MS COCO dataset.

Method	Data	Network	Avg.Precision, IOU:			Avg.Precision, area:		
			0.5:0.95	0.5	0.75	S	M	L
SSD300*	trainval	VGG	25.1	43.1	25.8	6.6	25.9	41.4
SSD512*	trainval	VGG	28.8	48.8	30.3	10.9	31.8	43.5
Ours(300)	trainval	VGG	27.6	46.8	28.7	8.9	29.6	42.8
Ours(512)	trainval	VGG	30.9	51.3	32.8	13.0	35.2	46.4

is more powerful on detection of small objects. For the medium ($32^2 < S < 96^2$), our method with 512×512 input images achieve 35.2% AP, which improves SSD (3.4%) with a large margin.

5 Conclusion

In this paper, we propose a feature information-interaction model under visual perception, which adopts channel information of feature map to weight self-attention feature maps, completing the extraction, fusion and enhancement of global semantic feature with object contextual information. Then we also propose an adaptively cyclic feature information-interaction model, which adopts a branch prediction mechanism to decide the number of visual attention, accomplishing adaptive fusion of global semantic features and local detailed information repeatedly. Experiments show that our proposed method significantly improves the accuracy of the baseline model SSD. In future work, we plan to incorporate visual attention module and Transformer representation in the baseline model to further improve its effectiveness.

References

1. Fischler, M., Elschlager, R.: The representation and matching of pictorial structures. IEEE Trans. Comput. **100**(1), 67–92 (1973)
2. Borji, A., Cheng, M.-M., Hou, Q., Jiang, H., Li, J.: Salient object detection: a survey. Computat. Vis. Media **5**(2), 117–150 (2019). https://doi.org/10.1007/s41095-019-0149-9
3. Chen, Y., Cao, Y., Hu, H., et al.: Memory enhanced global-local aggregation for video object detection. In: Proceedings of the IEEE Conference on Computer Vision and Pattern Recognition, Piscataway, NJ, pp. 10337–10346. IEEE (2020)
4. Zheng, Y., Liu, X., Cheng, X., et al.: Multi-task deep dual correlation filters for visual tracking. IEEE Trans. Image Process. **29**, 9614–9626 (2020)
5. Fu, Z., Chen, Y., Yong, H., et al.: Foreground gating and background refining network for surveillance object detection. IEEE Trans. Image Process. **28**(12), 6077–6090 (2019)
6. Dai, X.: Hybridnet: a fast vehicle detection system for autonomous driving. Sig. Process.: Image Commun. **70**, 79–88 (2019)
7. Du, G., Wang, K., Lian, S.: Vision-based robotic grasping from object localization, pose estimation, grasp detection to motion planning: a review. arXiv preprint arXiv:1905.06658 (2019)
8. Felzenszwalb, P.F., Girshick, R.B., McAllester, D., et al.: Cascade object detection with deformable part models. In: Proceedings of the IEEE Conference on Computer Vision and Pattern Recognition, Piscataway, NJ, pp. 2241–2248. IEEE (2010)
9. Everingham, M., Van Gool, L., Williams, C.K., et al.: The pascal visual object classes (VOC) challenge. Int. J. Comput. Vision **88**(2), 303–338 (2009)
10. Rätsch, M., Romdhani, S., Vetter, T.: Efficient face detection by a cascaded support vector machine using haar-like features. Joint Pattern Recog. Symp. **3175**, 62–70 (2004)
11. Lowe, D.G.: Object recognition from local scale-invariant features. In: Proceedings of the IEEE International Conference on Computer Vision, Piscataway, NJ, pp. 1150–1157 IEEE (1999)
12. Dalal, N., Triggs, B.: Histograms of oriented gradients for human detection. In: Proceedings of the IEEE Conference on Computer Vision and Pattern Recognition, Piscataway, NJ, pp. 886–893. IEEE (2005)

13. Lin, T.Y., Maire, M., Belongie, S., Hays, et al.: Microsoft coco: Common objects in context, In: Proceedings of the IEEE International Conference on Computer Vision, Piscataway, NJ, pp.740–755. IEEE (2014)
14. Girshick, R., Donahue, J., Darrell, et al.: Rich feature hierarchies for accurate object detection and semantic segmentation. In: Proceedings of the IEEE Conference on Computer Vision and Pattern Recognition, Piscataway, NJ, pp. 580–587. IEEE (2014)
15. Ren, S., He, K., Girshick, R., et al.: Faster R-CNN: towards real time object detection with region proposal networks. arXiv preprint arXiv:1506.01497 (2015)
16. Liu, W., et al.: SSD: Single shot multibox detector. In: Leibe, B., Matas, J., Sebe, N., Welling, M. (eds.) ECCV 2016. LNCS, vol. 9905, pp. 21–37. Springer, Cham (2016). https://doi.org/10.1007/978-3-319-46448-0_2
17. Fu, C.Y., Liu, W., Ranga, A., et al.: DSSD: deconvolutional single shot detector. arXiv preprint arXiv:1701.06659 (2017)
18. Zhou, P., Ni, B., Geng, C., et al.: Scale-transferrable object detection. In: Proceedings of the IEEE Conference on Computer Vision and Pattern Recognition, Piscataway, NJ, pp. 528–537. IEEE (2018)
19. Zhang, S., Wen, L., Bian, X., et al.: Single-shot refinement neural network for object detection. In: Proceedings of the IEEE Conference on Computer Vision and Pattern Recognition, Piscataway, NJ, pp. 4203–4212. IEEE (2018)
20. Bell, S., Zitnick, C.L., Bala, K., et al.: Inside-outside net: detecting objects in context with skip pooling and recurrent neural networks. In: Proceedings of the IEEE Conference on Computer Vision and Pattern Recognition, Piscataway, NJ, pp. 2874–2883. IEEE (2016)
21. Zagoruyko, S., Lerer, A., Lin, T.Y., et al.: A multipath network for object detection. arXiv preprint arXiv:1604.02135 (2016)
22. Dai, J., Li, Y., He, K., et al.: R-fcn: object detection via region-based fully convolutional networks. arXiv preprint arXiv:06409 (2016)
23. Bae, S.H.: Object detection based on region decomposition and assembly. In: Proceedings of the AAAI Conference on Artificial Intelligence, pp. 8094–8101 (2019)
24. Beery, S., Wu, G., et al.: Context r-cnn: Long term temporal context for per-camera object detection. In: Proceedings of the IEEE Conference on Computer Vision and Pattern Recognition, Piscataway, NJ, pp. 13075–13085. IEEE (2020)
25. Law, H., Deng, J.: Cornernet: detecting objects as paired keypoints. In: Agapito, L., Bronstein, M.M., Rother, C. (eds.) ECCV 2018. LNCS, pp. 734–750. Springer, Cham (2018). https://doi.org/10.1007/978-3-030-30952-7
26. Newell, A., Yang, K., Deng, J.: Stacked hourglass networks for human pose estimation. In: Leibe, B., Matas, J., Sebe, N., Welling, M. (eds.) ECCV 2016. LNCS, vol. 9912, pp. 483–499. Springer, Cham (2016). https://doi.org/10.1007/978-3-319-46484-8_29
27. Duan, K., Bai, S., Xie, L., et al.: Centernet: keypoint triplets for object detection. In: Proceedings of the IEEE International Conference on Computer Vision, Piscataway, NJ, pp. 6569–6578. IEEE (2019)
28. Tian, Z., Shen, C., Chen, H., et al.: FCOS: fully convolutional one-stage object detection. In: Proceedings of the IEEE Conference on Computer Vision and Pattern Recognition, Piscataway, NJ, pp. 9627–9636. IEEE (2019)
29. Yang, Z., Liu, S., Hu, H., et al.: Reppoints: point set representation for object detection. In: Proceedings of the IEEE International Conference on Computer Vision, Piscataway, NJ, pp. 9656–9665, IEEE (2019)
30. Hu, J., Shen, L., Sun, G.: Squeeze-and-excitation networks. arXiv preprint arXiv:1709.01507 (2017)
31. Woo, S., Park, J., Lee, J.-Y., Kweon, I.S.: CBAM: convolutional block attention module. In: Ferrari, V., Hebert, M., Sminchisescu, C., Weiss, Y. (eds.) ECCV 2018. LNCS, vol. 11211, pp. 3–19. Springer, Cham (2018). https://doi.org/10.1007/978-3-030-01234-2_1

32. Park, J., Woo, S., Lee, J.-Y., et al.: BAM: bottleneck attention module. arXiv preprint arXiv: 1807.06514 (2018)
33. Liu, J.J., Hou, Q., et al.: Improving convolutional networks with self-calibrated convolutions. In: Proceedings of the IEEE Conference on Computer Vision and Pattern Recognition, Piscataway, NJ, pp. 10096–10105. IEEE (2020)
34. Yang, Q.L., Zhang, Y.B.: SA-Net: shuffle attention for deep convolutional neural networks. arXiv preprint arXiv:2102.00240 (2021)
35. Wang, X., Girshick, R., Gupta, A., et al.: Non-local neural networks. In: Proceedings of the IEEE Conference on Computer Vision and Pattern Recognition, Piscataway, NJ, pp. 7794–7803. IEEE (2018)
36. Simonyan, K., Zisserman, A.: Very deep convolutional networks for large-scale image recognition. arXiv preprint arXiv:1409.1556 (2014)
37. He, K., Zhang, X., Ren, S., et al.: Delving deep into rectifiers: surpassing human-level performance on imagenet classification. In: Proceedings of the IEEE International Conference on Computer Vision, Piscataway, NJ,pp. 1026–1034. IEEE (2015)
38. Zheng, L., Fu, C., Zhao, Y.: Extend the shallow part of single shot multibox detector viconvolutional neural network. In: Tenth International Conference on Digital Image Processing. International Society for Optics and Photonics, pp. 10806–1080613. (2018)
39. Cao, G., Xie, X., Yang, W., et al.: Feature-fused SSD: fast detection for small objects. In: Ninth International Conference on Graphic and Image Processing, pp. 10615–106151E (2018)
40. Cui, L., Ma, R., Lv, P., et al.: MDSSD: multi-scale deconvolutional single shot detector for small objects. arXiv preprint arXiv:1805.07009 (2018)
41. Cao, Y., Xu, J.R., Lin, S. et al.: Gcnet: Non-local networks meet squeeze-excitation networks and beyond. In: Proceedings of the IEEE/CVF International Conference on Computer Vision Workshops. Piscataway, NJ. IEEE(2019)
42. Li, X., Wang, W., Hu, X., et al.: Selective kernel networks. In: Proceedings of the IEEE/CVF Conference on Computer Vision and Pattern Recognition. Piscataway, NJ, pp. 510–519. IEEE(2019)

Research of Robust Video Object Tracking Algorithm Based on Jetson Nano Embedded Platform

Xiangyang Luo[✉], Chao Zhang, and Zhao Lv

Anhui University, Hefei, China
luoxiangyang12@163.com

Abstract. The majority of recent work has tended to improve the overall tracking capability of trackers. Despite the success of these methods, their inherent complexity has limited their scope of application. In contrast, it is more practical to improve the performance on embedded platforms without increasing the complexity of tracking algorithms. Limited by the computing power of the embedded system, this paper proposed an interesting tracker based on the Kernel correlation filter (KCF). To tackle the occlusion and disappearance problem during object tracking, we suggest a lightweight object detection network to relocate the target by using color features and then reinitialize the tracker. In addition, we avoid the problem of scale variation in object tracking by adaptively cross-linking the KCF algorithm with the embedded platform. The proposed method improves the performance of KCF tracker on embedded platform without increasing its complexity and achieves robust tracking of video targets. We have captured a lot of videos and done a lot of experiments to prove the effectiveness of our method.

Keywords: Object detection · Embedded systems · Kernel correlation filter

1 Introduction

With the continuous development of computer technology, the computing power of PC is constantly improving, making the use of computers to implement human vision functions one of the hottest topics at present. As one of the fundamental research problem in computer vision community, object tracking [1] has received much attention from scholars in related fields. Most of the recent work tends to improve the overall tracking capability by optimizing tracking algorithms in high-performance computing platforms. Compared with general-purpose computer systems that focus on high-speed and massive numerical computing power, embedded platforms have lower arithmetic power and limited processing power to ensure that tracking algorithms achieve the same tracking speed on embedded platforms as on PCs. However, the embedded platform has lower cost, lower power consumption and more flexible application. It will be more practical if the video targets tracking can be achieved robustly and efficiently on embedded platforms.

© Springer Nature Switzerland AG 2021
H. Ma et al. (Eds.): PRCV 2021, LNCS 13019, pp. 392–403, 2021.
https://doi.org/10.1007/978-3-030-88004-0_32

In recent years, breakthroughs have been made in target tracking algorithms, especially the introduction of correlation filtering, which greatly increases the calculation speed by converting the operation process from time domain to frequency domain. KCF [2] introduces a cyclic matrix based on correlation filtering and adds a Gaussian kernel to ridge regression, which substantially improves the performance of the tracker and makes it possible to work on real-time systems. With the successful application of deep learning in computer vision tasks, it has also started to be applied in target tracking algorithms, and a large number of research results have been achieved. Many scholars have also tried to integrate correlation filtering methods with deep learning methods to improve the tracker's capabilities by leveraging the strengths of each. In [3], Jack Valmadre proposed to integrate correlation filtering with CNN to better utilize the advantages of both through end-to-end training.

Lightweight platform based object tracking has become increasingly popular due to its low cost and flexibility of sensing without powerful computing power. Especially in the field of deep learning, training network parameters online with large amounts of data can greatly affect tracking speed [4–7]. An increasing number of scholars tend to improve the tracking speed by sacrificing a portion of accuracy. In [8], Luca Bertinetto proposes SiamFC, a tracking method based on an offline end-to-end trained fully convolutional twin network, its tracking speed can reach 86 frames per second, and its performance exceeds most real-time trackers. In [9], David Held uses a twin network-based regression approach to learn the changing relationship between target appearance and motion, which can track at speeds of 100 frames per second. However, in terms of real-time performance, the correlation filter-based algorithm still has a great advantage. Although the tracking accuracy of KCF is slightly lower, its tracking speed can reach 172 frames per second. Our intention is to improve the performance of KCF tracker on embedded platform without increasing its complexity.

2 Integration of KCF Algorithm with Color Features

In this paper, to ensure the implementation of object tracking on embedded platform, we choose the fastest lightweight neural network SSD Mobilenet-V2 (300*300) for object detection and an improved KCF tracker which has better real-time performance for object tracking. The primary process of object tracking in this paper is as follows: 1) Detect the target (pedestrian) based on video stream and extract its color histogram features. 2) Initialize the KCF tracker by target position to achieve object tracking. 3) If the target disappears temporarily due to some reasons such as occlusion, the detector restarts working and continuously detects the same type of target (pedestrians) in the next frame. The candidate region with the highest correlation with the original feature and whose correlation exceeds a preset threshold is selected as the target. 4) By using the Proportion-Integral-Differential (PID) controller, the distance between the car and the target is generally constant.

KCF algorithm replaces single-channel grayscale features with multi-channel HOG features based on CSK [10] algorithm to improve tracker performance. Furthermore, KCF introduces the circular matrix and diagonalize it with the Discrete Fourier Transform, which increases the computational speed by several orders of magnitude. This ensures that KCF has a decent accuracy while maintaining a relatively high tracking speed.

The objective of KCF algorithm is to solve the ridge regression problem. Its objective function is defined as

$$f(z) = w^T z. \tag{1}$$

The kernel function is $\varphi^T(x)\varphi\left(x'\right) = k\left(x, x'\right)$, so the Objective function is defined as

$$f(z) = w^T z = \sum_{i=1}^{n} \alpha_i k(z, x_i). \tag{2}$$

Then convert the problem of solving w from the original space to solving α in the dual space, and translate it to the Fourier domain:

$$\hat{\alpha} = \frac{\hat{y}}{\hat{k}^{xx} + \lambda}, \tag{3}$$

Where $\hat{k}^{xx}$ denotes the first row of the kernel matrix K and $\hat{\ }$ denotes the discrete Fourier transform of the vector.

The response of each test sample is calculated. And the one with largest correlation is selected as the tracking result.

$$\hat{f}(z) = \hat{k}^{xz} \odot \hat{\alpha}, \tag{4}$$

Where $\hat{k}^{xz}$ refers to the first row of the kernel matrix $\hat{k}^z$ of the test and training samples on the kernel space.

The selection of target features is crucial to the performance of trackers, we found that the tracker using multi-feature fusion [11–13] has a significantly higher tracking accuracy than the tracker using a single feature [10, 14], but its tracking speed is also inevitably decreased. This complex tracker is more suitable for high performance computers, and it is difficult to achieve the real-time requirements on low performance platforms. The limited computing power of Jetson Nano is difficult to guarantee its computational power requirements. In the field of computer vision, the Histogram of Oriented Gradient (HOG) is used as a common feature descriptor which reflects the surface texture features and contour shape of the target. One of the major advantages of the HOG feature is its simplicity, which can be easily implemented on a lightweight computing platform [15].

The purpose of fusing multiple valid features is to improve the tracking accuracy of the tracker. It reduces the possibility of tracking failure when the target disappears briefly due to occlusion or other reasons, which is done at the expense of tracker speed. But is it also possible to improve the tracker accuracy while ensuring the tracking speed? We chose an alternative approach using multiple features. To avoid increasing the complexity of the tracker, we additionally extract other features of the target and use this feature together with the tracker to find the target after it disappears. In this way, the real-time performance of the tracker is still guaranteed and the accuracy of the tracker is also further improved.

Among the image features such as color, shape, and texture. Color features are the most significant visual features with high stability and reliability. In pedestrian tracking, choosing color features with scale invariance and rotation invariance to work together with HOG features can improve the efficiency and accuracy of the system.

We used correlation as a measure of histogram similarity. The correlation comparison formula comes from the statistical term correlation coefficient, which is a quantity that studies the degree of linear correlation between variables. If two images are extremely similar, then to a certain extent they can be considered as the same. If H_1, H_2 represent the color histograms of the two images, and $d(H_1, H_2)$ represents the correlation between these two images, then the correlation can be derived from Eq. (5).

$$d(H_1, H_2) = \frac{\sum_I (H_1(I) - \overline{H}_1)(H_2(I) - \overline{H}_2)}{\sqrt{\sum_I (H_1(I) - \overline{H}_1)^2 \sum_I (H_2(I) - \overline{H}_2)^2}} \tag{5}$$

In which:

$$\overline{H}_k = \frac{1}{N} \sum_J H_k(J) \tag{6}$$

When the target disappears, we detect similar objects in the next frame, find the region that has the greatest correlation with the target, and then compare the correlation with the threshold value, if it is greater than the threshold, the target has reappeared. Meanwhile, the tracker is also looking for a target, and the two compete with each other, and we pick the faster one. Since the selected target detection algorithm and target tracking algorithm both have good real-time performance, and the feature extraction method also has high extraction efficiency, thus the tracking speed is ensured on the embedded platform.

3 Adaptive Cross-Linking of KCF Algorithm with Electronic Control Platform

The tracking box of KCF algorithm is fixed. When the scale of the target changes, the tracker cannot track in an adaptive way. As the interference information in the tracking box increases, the problem of tracking failure cannot be avoided. It becomes a challenging issue to implement robust scale estimation in visual object tracking. The follow-up algorithm [16–19] solves the scale variation problem mainly based on the following three methods, which are scale pool-based method, chunk-based method, and feature point-based method.

The main purpose of adding a scale adaptive algorithm to the tracker is to reduce the possibility of tracking failure. However, just as a tracker using multi-feature fusion is slower than a KCF tracker using a single feature, a correlation filter-like algorithm with scale adaptation is slower than a KCF algorithm with a fixed tracking box. In this paper, we propose the method of adaptive cross-linking KCF algorithm with the electronic control platform to ensure the proportion of valid information in the tracking box, which is achieved by ensuring that the car maintains a relatively fixed distance from the target (pedestrian). The aim is to improve the performance of the KCF tracker on the Jetson Nano platform without increasing its complexity.

X. Luo et al.

The PID algorithm is the classical algorithm in automatic control algorithms, based on the proportional (P), integral (I) and differential (D) of deviations to achieve automatic control. The objective function is defined as:

$$u(t) = K_p \left[e(t) + \frac{1}{T_i} \int_0^t e(t)dt + T_d \frac{de(t)}{d_t} \right] \tag{7}$$

In the PID controller, the role of the proportional segment is to respond to the deviation instantaneously; the role of the integral segment is to eliminate the steady-state error that occurs in the proportional segment; and the differential segment achieves anticipatory control based on the trend of the deviation to reduce oscillations in the control process (Fig. 1).

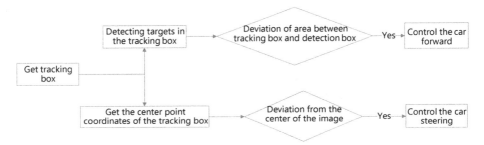

Fig. 1. Object tracking flow chart

At the beginning, we obtain the size of the initial tracking box. Then, we perform target detection in the subsequent tracking box to get the area occupied by the valid information. Finally, we use PID algorithm to control the car forward according to the deviation of the area between the detection box and the tracking box. Meanwhile, we regard the horizontal distance between the center of the tracking box and the center of the image as a deviation to control the car steering by PID algorithm. The process of obtaining the valid information in the tracking box is as follows:

1. Get the position of the tracking box in the current frame.
2. Use the mask to extract the tracking box to reduce the computational effort.
3. Perform target detection in the tracking box to obtain the area occupied by valid information.

4 Experiments

4.1 Comparison with KCF Tracker

Figure 2 shows part of the process from the disappearance of the tracking target to its relocation. It can be seen that in the third image (frame 379 in the original video), another pedestrian appears to the right of the tracking target. In the fifth image (frame 386 in

Fig. 2. Example of object stabilization tracking

the original video), the tracking target has been obscured by the pedestrian, causing it to disappear. In the sixth image (frame 401 in the original video), the target is repositioned, and it takes only 15 frames from the disappearance of the target to its relocation. It can be seen from the images that the target is relocated once it reappears, which can fully meet the real-time requirements of the system, and this is thanks to the selection of the threshold value. Selecting the correct threshold to relocate to the target as soon as it appears is also an important task.

We have captured 60 videos under different illumination conditions and scenarios for testing the performance of our proposed method compared to the conventional KCF algorithm on the Jetson Nano. During the shooting of these videos, the target is kept at a relatively fixed distance from the camera to avoid tracking failure caused by scale changes. In each video, the target is passed behind by other pedestrians during walking, causing the tracking target to disappear briefly. And the duration of each video is kept at 15–25 s. We use three metrics to measure the performance of the algorithms. The first is precision, which represents the number of successfully tracked videos against the total number of videos. The second is the average relocation time (ART), which represents the number of frames elapsed from the disappearance of the target to its rediscovery. The last one is the tracking speed, which is expressed using frames per second (FPS). In the experiments, the tracking was successful if the tracker could relocate the target even after its transient disappearance, otherwise the tracking failed (Table 1).

The experimental results show that our approach improves the tracking accuracy without reducing the tracking speed compared with the conventional KCF algorithm. Moreover, the ART is reduced from 26 to 19. This shows that when the target disappears transiently, using a tracker and a detector together can find the target faster than using

Table 1. Summary of experimental results on the 60 videos dataset

Algorithm	Mean precision	ART	Mean FPS
KCF	50%	26	32
KCF + color feature	66.7%	19	33

the tracker alone. Moreover, we certainly want to find it again as soon as possible to ensure the timeliness of tracking.

In the KCF tracker, the size of the tracking box is fixed. However, when the target reappears, its distance from the camera has a higher probability of changing compared to the original position, leading to a change in the scale of the target. In this case, it is difficult to find the target again using the KCF tracker alone. In contrast, finding the target again by object detection and comparing the color feature correlation between each candidate can reinitialize the tracker and improve the robustness of tracking.

SSD Mobilenet-V2 (300*300) can reach a detection speed of 39 FPS in Jetson Nano, which is faster than KCF's tracking speed that can completely guarantee the real-time performance. And we found that the tracking speed is relatively faster when the tracking frame match the size of the target. The combination of these reasons leads to a slightly faster tracking speed of our approach than the traditional KCF algorithm. We also took an additional 20 videos in which the target was not obscured by other pedestrians. The experimental results show that the accuracy of both methods is close to 100%.

The KCF algorithm can achieve a processing speed of 172 FPS on a high-performance PC, but only 32 FPS on an embedded platform with limited computing power. We also found that when using the traditional KCF algorithm for target tracking on the embedded platform, even if the tracking target is not obscured, the KCF tracking frame sometimes slips due to the different video backgrounds, which means the target deviates from the center of the tracking box and thus leads to tracking failure. However, by combining the KCF algorithm with color features, the target can be quickly relocated when tracking fails (Fig. 3).

4.2   The Selection of Thresholds

When using color features to relocate the targets, it is necessary to determine the effect of different threshold values on tracking accuracy in order to select the appropriate threshold. We use the horizontal coordinate to represent the correlation threshold and the vertical coordinate to show the tracking accuracy. As we can see in Fig. 4, the probability of tracking failure increases when the threshold value is chosen below 0.85 or above 0.92. The reason for this is that the tracker is attracted to similar objects when the threshold is low, while when the threshold is high, although the tracked targets do not change, their backgrounds are always changing because the targets are constantly moving, which makes it difficult for them to achieve such a high correlation, leading to tracking failure. The experimental results show that the highest tracking accuracy, nearly 0.64, can be achieved when the threshold value is taken in the range of 0.85 to

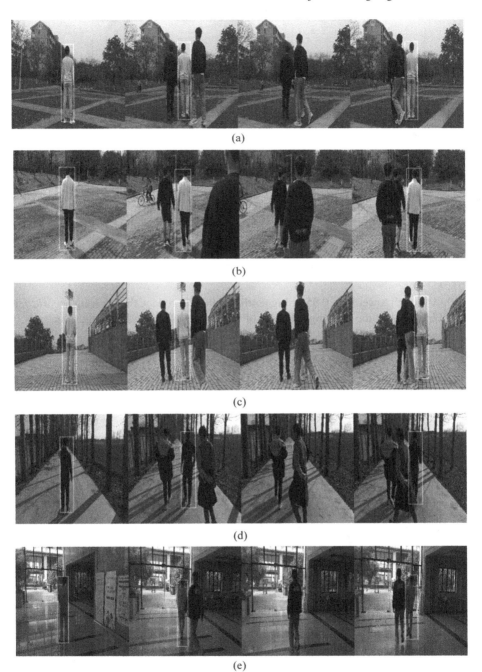

(a)

(b)

(c)

(d)

(e)

Fig. 3. Experimental results for several different occasions and illumination conditions. (a), (b) are the results under different illumination conditions and (b)–(e) are the results under different scenes.

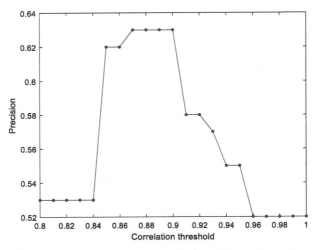

Fig. 4. Accuracy rates corresponding to different thresholds

0.90. Although lower than 0.85 or higher than 0.90 will reduce the tracking accuracy, it is still higher than the tracking accuracy of the tracker using the KCF algorithm only.

4.3 Pedestrian Tracking by Adding Object Detection

Fig. 5. Traditional KCF algorithm for pedestrian tracking

Figure 5 shows the performance of the KCF algorithm for pedestrian tracking on the Jetson Nano, and the top left corner shows the real-time frame rate. The results show that the effective information in the tracking box is gradually decreasing as the distance between the pedestrian and the camera increases. Figure 6 shows the results of object detection in the tracking box, where the green rectangular box is the KCF tracking box and the blue rectangular box is the object detection box. The results show that the proportion of the valid information in the target detection box is more than that in the KCF tracking box. Moreover, we can find that adding object detection to the object

Fig. 6. Pedestrian tracking by adding object detection

tracking process does not reduce the speed of object tracking, because the use of mask images greatly reduces the computational load of object detection.

Fig. 7. Example of embedded car

Figure 7 shows the embedded car used in our experiments. In the experiment, we control the car forward according to the deviation of the area between the detection box and the tracking box. Meanwhile, we control the car steering by the size of the deviation of the tracking box from the image center. And the experimental results show that object detection in the tracking box does not degrade the real-time performance of the algorithm.

5 Conclusions

We select the KCF algorithm, which is faster and relatively more accurate, as the basic tracker. When the embedded car follows pedestrian, if the pedestrian disappear transiently due to occlusion and other reasons, the proposed algorithm can find the target again by object detection and comparing the correlation between the color features of each candidate and the tracking target, and then use it to reinitialize the tracker. We also use the adaptive cross-linking of KCF algorithm with the electronic control platform to keep the tracker at a relatively fixed distance from the tracked pedestrian to provide a more stable tracking environment. The experimental results demonstrate that our algorithm achieved better performance than the traditional KCF algorithm on embedded platforms.

Acknowledgements. The research work described in this paper is supported by Anhui Province Natural Science Foundation (No.1908085MF203) and Anhui provincial natural science research project of colleges and Universities (No.KJ2020A0034). The authors would like to thank all members of Intelligent Video Research Group from IIP-HCI lab of Anhui University for their valuable suggestions and assistance in preparing this paper.

References

1. Smeulders, W.A., Chu, et al.: Visual tracking: an experimental survey. Pattern Anal. Mach. Intell. **36**(7), 1442–1468 (2014)
2. Henriques, J.F., Caseiro, R., Martins, P., et al.: High-speed tracking with kernelized correlation filters. IEEE Trans. Pattern Anal. Mach. Intell. **37**(3), 583–596 (2015)
3. Valmadre, J., Bertinetto, L., Henriques, J.F., et al.: End-to-end representation learning for correlation filter based tracking. IEEE, 5000–5008 (2017)
4. Li, H., Li, Y., Porikli, F.: DeepTrack: learning discriminative feature representations online for robust visual tracking. IEEE Trans. Image Process. **25**(4), 1834–1848 (2016)
5. Danelljan, M., Robinson, A., Shahbaz Khan, F., Felsberg, M.: Beyond correlation filters: learning continuous convolution operators for visual tracking. In: Leibe, B., Matas, J., Sebe, N., Welling, M. (eds.) ECCV 2016. LNCS, vol. 9909, pp. 472–488. Springer, Cham (2016). https://doi.org/10.1007/978-3-319-46454-1_29
6. Danelljan, M., Bhat, G., Khan, F.S., et al.: ECO: efficient convolution operators for tracking. IEEE Comput. Soc., 6931–6939 (2016)
7. Nam, H., Han, B.: Learning multi-domain convolutional neural networks for visual tracking. IEEE, 4293–4302 (2016)
8. Bertinetto, L., Valmadre, J., Henriques, J.F., Vedaldi, A., Torr, P.H.S.: Fully-convolutional siamese networks for object tracking. In: Hua, G., Jégou, H. (eds.) ECCV 2016. LNCS, vol. 9914, pp. 850–865. Springer, Cham (2016). https://doi.org/10.1007/978-3-319-48881-3_56
9. Held, D., Thrun, S., Savarese, S.: Learning to track at 100 FPS with deep regression networks. In: Leibe, B., Matas, J., Sebe, N., Welling, M. (eds.) ECCV 2016. LNCS, vol. 9905, pp. 749–765. Springer, Cham (2016). https://doi.org/10.1007/978-3-319-46448-0_45
10. Henriques, J.F., Caseiro, R., Martins, P., Batista, J.: Exploiting the circulant structure of tracking-by-detection with kernels. In: Fitzgibbon, A., Lazebnik, S., Perona, P., Sato, Y., Schmid, C. (eds.) ECCV 2012. LNCS, vol. 7575, pp. 702–715. Springer, Heidelberg (2012). https://doi.org/10.1007/978-3-642-33765-9_50
11. Li, Y., Zhu, J.: A scale adaptive kernel correlation filter tracker with feature integration. In: Agapito, L., Bronstein, M.M., Rother, C. (eds.) ECCV 2014. LNCS, vol. 8926, pp. 254–265. Springer, Cham (2015). https://doi.org/10.1007/978-3-319-16181-5_18
12. Bertinetto, L., Valmadre, J., Golodetz, S., et al.: Staple: complementary learners for real-time tracking. Comput. Vis. Pattern Recogn. IEEE, 1401–1409 (2016)
13. Zhu, G.B., Wang, J.Q., Wu, Y., Zhang, X.Y., Lu, H.Q.: MC-HOG correlation tracking with saliency proposal. In: Proceedings of the 13th AAAI Conference on Artificial Intelligence, AAAI Press, Phoenix, USA, pp. 3690–3696 (2016)

14. Bolme, D.S., Beveridge, J.R., Draper, B.A., et al.: Visual object tracking using adaptive correlation filters. In: The Twenty-Third IEEE Conference on Computer Vision and Pattern Recognition, CVPR 2010, San Francisco, CA, USA, pp. 13–18, June 2010. IEEE (2010)
15. Dalal, N., Triggs, B.: Histograms of oriented gradients for human detection. In: IEEE Computer Society Conference on Computer Vision and Pattern Recognition. IEEE (2005)
16. Danelljan, M., Häger, G., Khan, F.S., et al.: Accurate scale estimation for robust visual tracking. Br. Mach. Vis. Conf., 1–11 (2014)
17. Xu, Y., Wang, J., Hang, L., et al.: Patch-based scale calculation for real-time visual tracking. IEEE Signal Process. Lett. **23**(1), 40–44 (2015)
18. Montero, A.S., Lang, J., Laganiere, R.: Scalable kernel correlation filter with sparse feature integration. In: IEEE International Conference on Computer Vision Workshop. IEEE Computer Society (2016)
19. Liu, T., Gang, W., Yang, Q.: Real-time part-based visual tracking via adaptive correlation filters. In: 2015 IEEE Conference on Computer Vision and Pattern Recognition (CVPR). IEEE (2015)

Classification-IoU Joint Label Assignment for End-to-End Object Detection

Xiaolin Gu, Min Yang[✉], Ke Liu, and Yi Zhang

Beijing SunWise Space Technology Ltd., Beijing, China
yangmin@sunwisespace.com

Abstract. Using prediction-aware label assignment to choose positive samples has proven to be the key to remove Non-maximum Suppression (NMS) and reach end-to-end object detection. However, existing prediction-aware label assignment methods combine the classification cost and localization cost simply by multiplication or addition. It might produce unstable assignment results since the classification cost and localization cost are used independently to assign labels, and thus degrades the detection accuracy. In this paper, we propose a classification-IoU joint label assignment that takes advantage of the correlation between classification and localization to choose positive samples. In particular, we add category and localization consistency constraints to the cost function for label assignment, and improve the stability of label assignment. Besides, a classification-IoU joint representation is applied to improve the correlation between classification and localization in the process of training. It is trained by the loss function that considers classification and location quality simultaneously. Extensive experiments demonstrate the effectiveness of our method. On COCO dataset, the end-to-end object detector with the techniques we proposed, achieves competitive performance against the state-of-the-art object detectors.

Keywords: End-to-end object detector · Prediction-aware label assignment · Classification-IoU joint label assignment · Classification-IoU joint representation

1 Introduction

Object detection is one of the fundamental tasks in computer vision and enables numerous downstream applications. One of the challenging topic for current object detectors is label assignment. Specifically, how to define the positive samples for each object and negative samples for the background has always been an open problem. Most object detectors [1,5,6,11,16,18–21,24,25,30] use many-to-one label assignment to choose positive samples. They pre-define thousands of candidate boxes on the image and choose the box candidates whose location cost is smaller than the threshold as positive samples. For one target, there is more than one positive sample. This rule provides adequate foreground samples

© Springer Nature Switzerland AG 2021
H. Ma et al. (Eds.): PRCV 2021, LNCS 13019, pp. 404–415, 2021.
https://doi.org/10.1007/978-3-030-88004-0_33

to obtain a strong and robust feature representation. However, redundant and near-duplicate results are produced during inference, thus making non-maximum suppression(NMS) [17] necessary postprocessing.

Recently, end-to-end object detectors [2,23,27,31,33] achieve great success. Prediction-aware label assignment that dynamically assign the foreground samples according to the cost of classification and regression simultaneously, is the key to remove Non-maximum Suppression (NMS) and reach end-to-end. OneNet [23] propose the Minimum Cost Label Assignment to choose positive samples, where the cost is the summation of classification cost and location cost. For each target, only one sample of minimum cost among all samples is assigned as a positive sample, others are all negative samples. DeFCN [27] propose a prediction-aware one-to-one label assignment, where the function used to select positive sample is the product of category score and IoU score between the predict and target box.

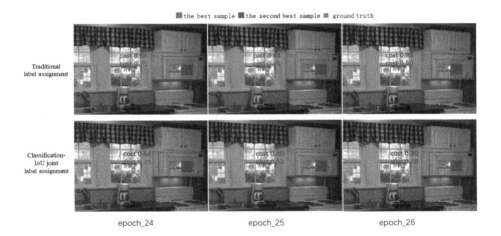

Fig. 1. Positive samples chosen by traditional prediction-aware label assignment and classification-IoU joint label assignment. For one target, the red and blue bounding boxes indicate respectively the best and second best samples chosen by minimizing the label assignment cost. The costs are also shown in the corresponding images. The green bounding boxes indicate the ground truth of the target. The first row shows the assignment results by using traditional prediction-aware label assignment, where the positive sample changes obviously during continuous epochs. The bottom row shows the assignment results for classification-IoU joint label assignment, which are more stable.

However, these prediction-aware label assignment methods simply combine the classification cost and location cost by multiplication or addition, which produce unstable assignment results. Most object detectors use two branches to predict classification and location respectively, and training them by different loss function. Similar with the process of training, the classification and localization cost are used independently in the prediction-aware label assignment, and simply combined by multiplication or addition. It leads to the ambiguity of label assignment, and might produce unstable assignment results, as shown in the first rows in Fig. 1. For one target, the positive sample changes obviously during continuous epochs. In order to improve the stability of label assignment, we propose a classification-IoU joint label assignment that utilizes the correlation between classification and localization. In particular, we add category and localization consistency constraints to the cost function for label assignment, and use Hungarian algorithm to choose the positive samples. It improves the stability of label assignment, as shown in the bottom rows in Fig. 1. For one target, the positive sample is a stable point during continuous epochs.

Furthermore, a classification-IoU joint representation is used to improve the correlation between classification and localization in the process of training. It is trained by the loss function that considers classification and localization quality simultaneously. On COCO dataset [15], our end-to-end detector with the proposed techniques achieves competitive performance against state-of-the-art detectors.

Some dense object detectors also use the classification-IoU joint representations to rank the prediction results. Generalized Focal Loss (GFL) [12,13] merge the quality estimation into the class prediction vector to form a joint representation of localization quality and classification, and use the generalized focal loss for training. Different with GFL, we weights the positive and negative examples asymmetrically, whereas GFL deals with them equally. VarifocalNet [29] proposed a IoU-aware classification scores that simultaneously represent the object presence confidence and localization accuracy, to produce a more accurate rank of detections in dense object detectors. It use the varifocal loss as classification loss for training. Varifocal loss use the IoU score to weight the positive samples. If a positive sample has a high gt-IoU, its contribution to the loss will thus be relatively big, which is more suitable for many-to-one label assignment. Different with varifocal loss, we treat all positive samples equally for training.

2   Related Work

2.1   Prediction-Aware Label Assignment

To achieve end-to-end detection, many approaches are explored in the previous literature. DETR [33] is proposed to directly output the predictions without any hand-crafted label assignment and post-processing procedure. It treats the label assignment as a bipartite matching problem by using the foreground loss [14] in training as the matching cost, which can be rapidly solved by the Hungarian algorithm [22]. OneNet [23] propose minimum cost assignment to assign labels. The cost is the summation of classification cost and location cost between sample and ground-truth. For each ground-truth box, only one sample of minimum cost among all samples is assigned to a positive sample, others are all negative samples. DeFCN [27] proposes a prediction-aware one-to-one label assignment to enable end-to-end detection, where the function used to select positive sample is the product of category score and IoU score between the predict and target box. All of them use one-to-one label assignment. For each ground-truth, only one sample is chosen as a positive sample; others are all negative ones. However, all of the above prediction-aware label assignment use the classification and localization cost independent, and simply combine them by multiplication or addition, which produce unstable assignment results during training. In order to improve the stability of assignment results, we propose a classification-IoU joint label assignment, which considers not only the cost of classification and localization, but also the correlation between them.

2.2 Classification-IoU Joint Representation

In order to get a more accurate ranking score, Fitness NMS [26], IoU-Net [9], IoU-aware RetinaNet [28] and FCOS [25] utilized a separate branch to perform localization quality estimation in a form of IoU or centerness score. Then they multiply the localization quality scores and the classification scores as the ranking scores during inference. However, this separate formulation causes the inconsistency between training and test as well as unreliable quality predictions. Because the localization quality and classification score are usually trained independently but compositely utilized during inference.

In order to solve this problem, some detectors propose the classification-IoU joint representations to present the prediction of classification and localization quality, and train it by the loss function that considers the correlation between classification and localization. Generalized Focal Loss [12,13] merge the quality estimation into the class prediction vector to form a joint representation of localization quality and classification, and use a vector to represent arbitrary distribution of box locations. It use Generalized Focal Loss to train the classification. Similarity, VarifocalNet [29] propose to learn IoU-aware classification scores that simultaneously represent the object presence confidence and localization accuracy, to produce a more accurate rank of detections in dense object detectors. It use the varifcoal loss as the classification loss for training. Although good results have been obtained, both of them are designed for dense object detectors, which use many-to-one label assignment and need to use NMS to remove the redundant results.

Zhou [31] propose a compact Positive Sample Selector (PSS) head for automatic selection of the best bounding box for each target, which is a separate branch of the localization, and trained independently by a stop-gradient operation. Different with [31], we first apply the classification-IoU joint scores as the ranking scores in end-to-end object detection.

3   Methodology

3.1   Classification-IoU Joint Label Assignment

The cost function of label assignment plays a crucial role for choosing suitable positive samples in end-to-end object detectors. Most of them use foreground loss [23] as the matching cost to assign labels, which is the sum of category loss and location loss,

$$C_f = \lambda_{cls}C_{cls} + \lambda_{loc}C_{loc}, \tag{1}$$

where λ_{cls} and λ_{loc} is the weight for classification cost and location cost respectively. Classification cost C_{cls} is computed by focal loss [14] as follow,

$$C_{cls} = -\alpha\left(1 - p_j\right)^\gamma log\left(p_j\right) - \sum_{i=0,i\neq j}^{n}\left(1 - \alpha\right)\left(1 - p_i\right)^\gamma log\left(1 - p_i\right), \tag{2}$$

where α and $(1 - \alpha)$ are the weighting factor for the target category and other categories. j is the index for target category and i is the index for other category. p is the prediction of classification. γ is the factor for down-weights to target category and other categories. Location cost C_{loc} is composed by point distance and box IoU loss, that is defined as follow,

$$C_{loc} = \lambda_{IoU}C_{IoU} + \lambda_{L1}C_{L1}, \tag{3}$$

where C_{IoU} and C_{L1} are box IoU loss and L1 loss between ground-truth and samples respectively. λ_{IoU} and λ_{L1} are coefficients. In this paper, we set the $\lambda_{L1} = 0$, and just use the C_{IoU} as the location cost.

However, most object detectors use two branches to predict classification and localization respectively, and training them by different loss function. Generally, focal loss is used to train classification, and IoU loss or L1 loss is applied to train localization. Similar with the process of training, foreground loss uses the cost of classification and localization independently, and simply combines them by the addition. It leads to the ambiguity of label assignment, and get unstable assign results during training, as shown in Fig. 1. In order to improve the stability of label assignment, we propose a classification-IoU joint label assignment that takes the advantage of the correlation between classification and localization to choose positive samples, and add the classification and localization consistency constraint in the cost function. Final cost function is defined as follow,

$$C = \lambda_{con}C_{con} + \lambda_{cls}C_{cls} + \lambda_{loc}C_{loc}, \tag{4}$$

where λ_{con}, λ_{cls} and λ_{loc} are the coefficients of C_{con}, C_{cls} and C_{loc} respectively. C_{con} is the cross entropy cost between classification and localization, which is defined as follow,

$$C_{con} = -\left(q log\left(p_j\right) + \left(1 - q\right)log\left(1 - p_j\right)\right), \tag{5}$$

where q is set as the IoU scores between the generated bounding box and the ground-truth box. C_{con} is a constraint that indicates the consistency between category and IOU score. When candidate samples have similar classification and localization costs, it is more likely to choose samples with similar category and IOU score as positive samples. C_{loc} and C_{cls} is the same as in foreground loss as proposed in Eq. 1. We use one-to-one label assignment to find a suitable positive sample for each ground-truth box, which solved by Hungarian algorithm and using C as the matching cost.

3.2   Classification-IoU Joint Representation

Modern end-to-end object detectors use the focal loss [14] to train the classification, which is proposed to address the extreme imbalance problem between foreground and background classes. It is defined as follow,

$$
\begin{cases}
-\alpha \left(1 - p\right)^{\gamma} log\left(p\right) & y = 1 \\
-\left(1 - \alpha\right) p^{\gamma} log\left(1 - p\right) & y = 0,
\end{cases}
\tag{6}
$$

where y specifies the ground-truth class and p is the predicted probability for foreground class. While weighting factor $(\alpha, (1 - \alpha))$ is used to balance the importance of positive and negative examples, and the modulating factor $((1 - p)^{\gamma}, p^{\gamma})$ automatically down-weights the contribution of easy examples during training and rapidly focuses the model on hard examples.

In order to get a classification-IoU joint representation to improve the correlation between classification and localization, we use the location quality (e.g. IoU scores) q instead the one-hot category label y in focal loss, which follows the conventional definition between $[0, 1]$. Specifically, $q = 0$ denotes the negative samples with 0 quality score, and $q > 0$ stands for the positive samples with target IoU score. It is defined as follow,

$$
\begin{cases}
-\left(q log\left(p\right) + \left(1 - q\right) log\left(1 - p\right)\right) & q > 0 \\
-\alpha p^{\gamma} log\left(1 - p\right) & q = 0
\end{cases}
\tag{7}
$$

Because positive samples are extremely rare compared with negative in end-to-end object detectors using one-to-one label assignment, our loss only reduces the loss contribution from negative examples by scaling their loss with a factor of p^{γ}, and does not down-weight positive examples in the same way.

4   Experiments

4.1   Implementation Details

In this section, we construct several experiments and visualizations on COCO [15] datasets. We use the same head network as FCOS [25], and the default backbone is ResNet-50 [7]. The backbone is initialized with the pre-trained weights on ImageNet [4] with frozen batch normalization [8]. In the training phase, input images are reshaped so that their shorter side is 800 pixes. All the experiments are trained on 4 GPUs with 4 images per GPU for 36 epochs. The initial learning rate is set to 0.01 and then decreased by 10 at epoch 24 and 33. We use Synchronized SGD [10] to optimize all the models with a weight decay of 0.0001 and a momentum of 0.9. All the training hyper-parameters are identical to the 2x schedule in MMDetection [3]. We set $\lambda_{cls} = 1.0$, $\lambda_{loc} = 1.0$, $\lambda_{iou} = 2.0$, $\lambda_{L1} = 0.0$ and $\lambda_{con} = 0.1$ for training. In inference, top-100 scoring boxes are selected as the final output.

Table 1. The effect of Classification-IoU label assignment and Calssification-IoU joint representation on COCO val set. The baseline detector uses the foreground loss for label assignment and focal loss to train classification.

Model	Classification-IoU joint label assignment	Classification-IoU joint score	mAP	AP_{50}	AP_{75}
Ours			38.7	57.4	41.8
	✓		39.2	57.4	42.6
		✓	38.9	57.6	41.9
	✓	✓	**39.7**	**59.8**	**43.1**

4.2 Ablation Study

We first take some experiments to study the impact of our propose method, as shown in Table 1. The baseline end-to-end detector is based on the framework of FCOS without the branch of centerness. In the baseline object detector, the cost function for label assignment is foreground loss as defined in Eq. 1 and the classification loss used in training is focal loss [14]. The first row shows the performance of it and 38.7% mAP is acquired. Replacing the label assignment by classification-IoU joint label assignment defined in Eq. 4, the performance is improve to 39.2% mAP, as shown in the second rows. We find that using classification-IoU joint label assignment can get more stable assign results as shown in Fig. 1, which achieves 0.5% mAP absolute gains than the label assignment based on foreground loss. The third row shows the performance of the object detector use the classification-IoU joint score and 38.9% mAP is acquired. By using the loss function proposed in Eq. 7 to train the classification-IoU joint score that considers the correlation between classification and localization in training, a more accurate rank of detections is obtained and the performance of detector is improved. The bottom row shows the performance of the object detector use both classification-IoU joint label assignment and the classification-IoU joint scores and 39.7% mAP is acquired. It shows that these two methods used together can achieve a better result.

We investigate the effect of the hyper-parameter λ_{con} of classification-IoU joint label assignment on detection performance. λ_{con} is the weight of cost C_{con}. We show the performance in Table 2. When set λ_{con} to a large value, such as 1.0 or 0.6, the performance of detector is getting worse. In the initial stage, IoU scores are small. If use a high weight factor for C_{con}, it will tend to choose the samples with both small classification and location scores as the positive samples, which is not the correct positive samples for training the detector. When set λ_{con} with a small value, such as 0.01, the cost of C_{con} has no effect, so the performance is similar to baseline detector. Among those, $\lambda_{con} = 0.1$ work best (39.2% mAP), and we adopt this value for all the following experiments.

Table 2. The effect of hyper-parameter λ_{con} on COCO val set. The other hyper-parameters λ_{loc} and λ_{cls} is set to 2.0 and 1.0 respectively as DeFCN.

λ_{con}	mAP	AP_{50}	AP_{75}
1.0	37.4	56.1	40.6
0.6	38.4	56.2	41.5
0.2	38.9	**57.6**	41.7
0.1	**39.2**	57.4	**42.6**
0.01	38.7	57.4	41.8

4.3   Comparison with State-of-the-art

We also compare our end-to-end object detector, which is built on FCOS framework, with some state-of-the-art object detectors as shown in Table 3 and Some qualitative results of our detector are shown in Fig. 2. For the dense object detectors, FCOS [25] is an anchor free object detector, that use many-to-one label assignment to choose the positive samples, and predicts centerness scores to suppress the low-quality detections. VarifocalNet [29] is based on ATSS [30], and use the varifocal loss for training. CenterNet [32] is an anchor free object detector, which use the one-to-one label assignment, and only choose the grid point which is nearest to the targets center as positive sample and perform classification and regression on these foreground samples.

Table 3. Comparison of state-of-art object detectors on COCO dataset. We compare our object detectors with dense object detectors and end-to-end object detectors. For the dense object detectors, NMS is used to remove redundant results. For end-to-end object detectors, top-100 scoring boxes are selected as the final output. The inference speed of all detectors in terms of frame per second (FPS) is tested on a single TITAN RTX GPU.

Method	Backbone	mAP	AP_{50}	AP_{75}	Times (ms)
CenterNet	ResNet-50	35.0	53.5	37.2	33.3
FCOS	ResNet-50	38.7	57.2	41.7	48.0
VarifocalNet	ResNet-50	**44.6**	**62.5**	**48.1**	54.3
PSS	ResNet-50	**42.3**	**60.2**	**46.0**	-
$DeFCN_{POTO}$	ResNet-50	39.2	56.6	42.7	44.3
$DeFCN_{POTO+3DMF}$	ResNet-50	40.6	58.0	44.7	52.3
DeFCN	ResNet-50	41.4	59.5	45.6	53.2
Ours	ResNet-50	39.7	59.8	43.1	36.3

For end-to-end object detectors, we compare our detector with the DeFCN [27] and PSS [31]. DeFCN is based on FCOS, and propose to use the POTO label assignment to choose positive samples. What's more, it propose a 3D max filter to utilize the multi-scale features and improve the discriminability of convolutions in the local region. Besides, in order to provide adequate supervision, it use the auxiliary loss based on many-to-one label assignment for training. PSS [31] propose a compact Positive Sample Selector (PSS) head for automatic selection of the best bounding box for each target, which is a separate branch of the localization, and trained independently by a stop-gradient operation.

Fig. 2. Detection examples of applying our end-to-end detector on COCO val-dev. The score threshold is 0.3 for visualization.

Compared with the CenterNet and FCOS, our detector achieves 3.0% mAP and 1.0% mAP gaps by using the same backbone respectively. However, the results of our detector is still lower than VarifocalNet. Both of FCOS and VarifocalNet use the many-to-one label assignment. The adequate foreground samples lead to a strong and robust feature representation. However, FCOS choose all the points in the targets as positive samples, that contains a lot of low quality samples. Our detector use one-to-one label assignment, for one target , there is only one positive samples that is the highest quality samples for target, so we achives 1.0% mAP gaps with FCOS. For VarifocalNet, varifocal loss weight the positive samples with the training target q. If a positive example has a high gt-IoU, its contribution to the loss will thus be relatively big. It proposes a star-shaped bounding box feature representation to refine coarse bounding boxes. Compared with it, our detector has the more simpler one-to-one label assignment strategy and only one box regression head to regress the bounding box.

Our end-to-end object detector get a better performance than DeFCN without 3D max filter and auxiliary loss. But the performance is still lower than the full DeFCN and PSS. For DeFCN, it propose the 3D max filter to utilize the multi-scale features, and add an independent branch from box regression head to compute the localization quality. What's more, it use the auxiliary loss based on many-to-one label assignment in training to provide adequate supervision. PSS learn objective involves both many-to-one and one-to-one label assignments, which use the many-to-one label assignment to choose dense samples for training a baseline detector. Then use the one-to-one label assignment to choose samples for training the PSS head. Compared with these detectors, our end-to-end object detector is very simple that we just change the cost function for label assignment and the classification loss for training, which has a faster inference speed, which are shown in the last column of Table 3.

5   Conclusion

In this work, we have propose a classification-IoU joint label assignment to choose more stable positive samples, which considers not only the cost of classification and localization, but also the consistency between them. Moreover, we first apply the classification-IoU joint scores in end-to-end object detectors. With the proposed techniques, our end-to-end object detector achieves competitive performance against the state-of-the-art detectors.

References

1. Cai, Z., Vasconcelos, N.: Cascade R-CNN: delving into high quality object detection. In: Proceedings of the IEEE Conference on Computer Vision and Pattern Recognition, pp. 6154–6162 (2018)
2. Carion, N., Massa, F., Synnaeve, G., Usunier, N., Kirillov, A., Zagoruyko, S.: End-to-end object detection with transformers. In: Vedaldi, A., Bischof, H., Brox, T., Frahm, J.-M. (eds.) ECCV 2020. LNCS, vol. 12346, pp. 213–229. Springer, Cham (2020). https://doi.org/10.1007/978-3-030-58452-8_13
3. Chen, K., et al.: MMdetection: OpenMMLab detection toolbox and benchmark. arXiv preprint arXiv:1906.07155 (2019)

4. Deng, J., Dong, W., Socher, R., Li, L.J., Li, K., Fei-Fei, L.: ImageNet: a large-scale hierarchical image database. In: 2009 IEEE Conference on Computer Vision and Pattern Recognition, pp. 248–255. IEEE (2009)

5. Girshick, R.: Fast R-CNN. In: Proceedings of the IEEE International Conference on Computer Vision, pp. 1440–1448 (2015)

6. He, K., Gkioxari, G., Dollár, P., Girshick, R.: Mask R-CNN. In: Proceedings of the IEEE International Conference on Computer Vision, pp. 2961–2969 (2017)

7. He, K., Zhang, X., Ren, S., Sun, J.: Deep residual learning for image recognition. In: Proceedings of the IEEE Conference on Computer Vision and Pattern Recognition, pp. 770–778 (2016)

8. Ioffe, S., Szegedy, C.: Batch normalization: accelerating deep network training by reducing internal covariate shift. In: International Conference on Machine Learning, pp. 448–456. PMLR (2015)

9. Jiang, B., Luo, R., Mao, J., Xiao, T., Jiang, Y.: Acquisition of localization confidence for accurate object detection. In: Proceedings of the European Conference on Computer Vision (ECCV), pp. 784–799 (2018)

10. Krizhevsky, A., Sutskever, I., Hinton, G.E.: ImageNet classification with deep convolutional neural networks. Adv. Neural Inf. Process. Syst. **25**, 1097–1105 (2012)

11. Law, H., Teng, Y., Russakovsky, O., Deng, J.: CornerNet-lite: efficient keypoint based object detection. arXiv preprint arXiv:1904.08900 (2019)

12. Li, X., Wang, W., Hu, X., Li, J., Tang, J., Yang, J.: Generalized focal loss v2: learning reliable localization quality estimation for dense object detection. In: Proceedings of the IEEE/CVF Conference on Computer Vision and Pattern Recognition, pp. 11632–11641 (2021)

13. Li, X., et al.: Generalized focal loss: Learning qualified and distributed bounding boxes for dense object detection. In: NeurIPS (2020)

14. Lin, T.Y., Goyal, P., Girshick, R., He, K., Dollár, P.: Focal loss for dense object detection. In: Proceedings of the IEEE International Conference on Computer Vision, pp. 2980–2988 (2017)

15. Lin, T.-Y., et al.: Microsoft COCO: common objects in context. In: Fleet, D., Pajdla, T., Schiele, B., Tuytelaars, T. (eds.) ECCV 2014. LNCS, vol. 8693, pp. 740–755. Springer, Cham (2014). https://doi.org/10.1007/978-3-319-10602-1_48

16. Liu, W., et al.: SSD: single shot multibox detector. In: Leibe, B., Matas, J., Sebe, N., Welling, M. (eds.) ECCV 2016. LNCS, vol. 9905, pp. 21–37. Springer, Cham (2016). https://doi.org/10.1007/978-3-319-46448-0_2

17. Neubeck, A., Van Gool, L.: Efficient non-maximum suppression. In: 18th International Conference on Pattern Recognition (ICPR 2006), vol. 3, pp. 850–855. IEEE (2006)

18. Pang, J., Chen, K., Shi, J., Feng, H., Ouyang, W., Lin, D.: Libra R-CNN: towards balanced learning for object detection. In: Proceedings of the IEEE/CVF Conference on Computer Vision and Pattern Recognition, pp. 821–830 (2019)

19. Redmon, J., Divvala, S., Girshick, R., Farhadi, A.: You only look once: Unified, real-time object detection. In: Proceedings of the IEEE Conference on Computer Vision and Pattern Recognition, pp. 779–788 (2016)

20. Redmon, J., Farhadi, A.: YOLOv3: an incremental improvement. arXiv preprint arXiv:1804.02767 (2018)

21. Ren, S., He, K., Girshick, R., Sun, J.: Faster R-CNN: towards real-time object detection with region proposal networks. In: Advances in Neural Information Processing Systems (NIPS) (2015)

22. Stewart, R., Andriluka, M., Ng, A.Y.: End-to-end people detection in crowded scenes. In: Proceedings of the IEEE Conference on Computer Vision and Pattern Recognition, pp. 2325–2333 (2016)
23. Sun, P., Jiang, Y., Xie, E., Yuan, Z., Wang, C., Luo, P.: OneNet: towards end-to-end one-stage object detection. arXiv preprint arXiv:2012.05780 (2020)
24. Tan, M., Pang, R., Le, Q.V.: EfficientDet: scalable and efficient object detection. In: Proceedings of the IEEE/CVF Conference on Computer Vision and Pattern Recognition, pp. 10781–10790 (2020)
25. Tian, Z., Shen, C., Chen, H., He, T.: FCOS: fully convolutional one-stage object detection. In: Proceedings of the IEEE/CVF International Conference on Computer Vision, pp. 9627–9636 (2019)
26. Tychsen-Smith, L., Petersson, L.: Improving object localization with fitness NMS and bounded IoU loss. In: Proceedings of the IEEE Conference on Computer Vision and Pattern Recognition, pp. 6877–6885 (2018)
27. Wang, J., Song, L., Li, Z., Sun, H., Sun, J., Zheng, N.: End-to-end object detection with fully convolutional network. In: Proceedings of the IEEE/CVF Conference on Computer Vision and Pattern Recognition, pp. 15849–15858 (2021)
28. Wu, S., Li, X., Wang, X.: IoU-aware single-stage object detector for accurate localization. Image and Vision Computing **97**, 103911 (2020)
29. Zhang, H., Wang, Y., Dayoub, F., Sunderhauf, N.: VarifocalNet: an IoU-aware dense object detector. In: Proceedings of the IEEE/CVF Conference on Computer Vision and Pattern Recognition, pp. 8514–8523 (2021)
30. Zhang, S., Chi, C., Yao, Y., Lei, Z., Li, S.Z.: Bridging the gap between anchor-based and anchor-free detection via adaptive training sample selection. In: Proceedings of the IEEE/CVF Conference on Computer Vision and Pattern Recognition, pp. 9759–9768 (2020)
31. Zhou, Q., Yu, C., Shen, C., Wang, Z., Li, H.: Object detection made simpler by eliminating heuristic NMS. arXiv preprint arXiv:2101.11782 (2021)
32. Zhou, X., Wang, D., Krähenbühl, P.: Objects as points. arXiv preprint arXiv:1904.07850 (2019)
33. Zhu, X., Su, W., Lu, L., Li, B., Wang, X., Dai, J.: Deformable DETR: deformable transformers for end-to-end object detection. arXiv preprint arXiv:2010.04159 (2020)

Joint Learning Appearance and Motion Models for Visual Tracking

Wenmei Xu[1], Hongyuan Yu[3], Wei Wang[3], Chenglong Li[1,2(✉)],
and Liang Wang[3]

[1] Anhui Provincial Key Laboratory of Multimodal Cognitive Computation,
School of Computer Science and Technology, Anhui University, Hefei, China
[2] Institute of Physical Science and Information Technology, Anhui University, Hefei,
China
[3] Center for Research on Intelligent Perception and Computing,
Institute of Automation, Chinese Academy of Sciences, Beijing, China

Abstract. Motion information is a key characteristic in the description
of target objects in visual tracking. However, seldom of existing works
consider the motion features and tracking performance is thus easily
affected when appearance features are not reliable in challenging scenarios. In this work, we propose to leverage motion cues in a novel deep
network for visual tracking. In particular, we employ the optical flow to
effectively model motion cues and reduce background interferences. With
a modest impact on efficiency, both appearance and motion features are
used to significantly improve tracking accuracy and robustness. At the
same time, we use a few strategies to update our tracker online so that
we can avoid error accumulation. Extensive experiments validate that
our method achieves better results against state-of-the-art methods on
several public datasets, while operating at a real-time speed.

Keywords: Visual tracking · Discriminative online learning · Optical
flow · Motion information

1 Introduction

Visual tracking is a popular but challenging research topic in computer vision.
According to the bounding box of the first frame of an arbitrary object in a given
video, visual tracking is to track it in subsequent video frames. It suffers from
many challenges in real-world scenarios, such as background clutter, occlusion,
deformation, illumination variation, fast motion and low resolution. At the same
time, visual tracking is an open and desirable research field with a wide range
of applications, including autonomous driving, video surveillance and human-
machine interaction, to name a few.

Recently, visual tracking based on Siamese network [1,3,8,11,15,16,20,23,28]
has attracted much attention, mainly because the schemes of offline learning well
balance tracking accuracy and speed. But these methods lack a model update

H. Ma et al. (Eds.): PRCV 2021, LNCS 13019, pp. 416–428, 2021.
https://doi.org/10.1007/978-3-030-88004-0_34

mechanism, which makes them still have a large accuracy gap comparing to online update trackers [2,4,5,13,19,26]. The ATOM [4] is proposed to divide the tracking problem into a classification task and an estimation task, which highly improves the tracking accuracy. It updates the classification task online for adapting to the tracking object changes and determines the rough position of an object. To make the classification task learn faster, it adopts a Conjugate-Gradient-based optimization strategy. And then it estimates an object's bounding box by iteratively optimizing multiple initial bounding boxes. The estimation task is trained offline on large-scale datasets. However, the ATOM only considers appearance features of the current frame and does not take advantage of motion information between frames in videos. It will cause trackers to drift when the object encounters partial occlusion or deformation, as shown in Fig. 1. The images on the left show that trackers can accurately locate the object, but ATOM will drift when it encounters occlusion or interference. As a result, tracking fails and it is difficult to keep up with the object again.

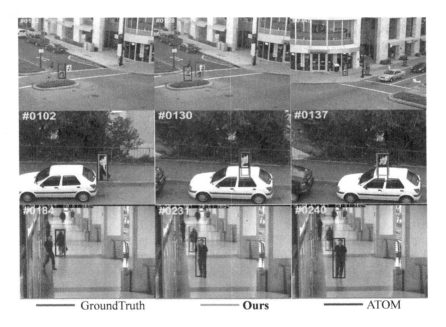

GroundTruth   **Ours**   ATOM

Fig. 1. A comparison of our approach with ATOM [4]. The ATOM tracker will drift when it encounters occlusion or interference. Our approach employs accurate optical flow and considers the motion information between frames in videos. Thus, we can obtain the magnitude and direction of the object motion and avoid drifting. The second row shows that not only can we distinguish between the target and the background correctly but also the resulting bounding box is more compact.

In this work, we explore motion information in a novel deep network for visual tracking, instead of complex and time-consuming LSTM networks and

3D Convolutional Neural Networks, and only precise optical flow is used to improve robustness and accuracy of visual tracking. Specifically, for the classification task, we use appearance and motion information of the object to improve the discriminative ability of the classifier. Unlike using a pre-trained model to extract fixed optical flow, we can achieve task-specific optical flow by online updating the optical flow estimation network a few times. Moreover, our optical flow estimation network has fewer parameters and the error rate of optical flow is rather low, we further obtain accurate optical flow directly. For the estimation task, we follow the ATOM tracker's offline training of the target state estimation module to predict the Intersection over Union (IoU) overlap between the target ground truth and an estimated bounding box. The effectiveness of our tracker is shown in Fig. 1.

We carry out comprehensive experiments on several challenging benchmarks, including OTB2015 [24], VOT2018 [14] and LaSOT [7]. The experiments prove that our method is effective under the challenges of occlusion, deformation, fast motion and low resolution, Table 2 shows the comparison results. Our tracking approach achieves the best performance on these tracking benchmarks, while we can run at frame-rates beyond real-time. Furthermore, detailed comparative experiments and analyses are provided to demonstrate the potential of our approach.

2   Related Work

The tracking methods based on deep learning improve the performance of visual tracking by using more discriminative deep features. The Siamese network treats tracking problems as similarity learning, given a pair of object template and search region, the feature is extracted using Siamese networks of shared parameters, and then the similarity map is calculated using cross-correlation operation. It overcomes the limitation of pre-trained deep CNN models and conducts an end-to-end training. A series of improvements have been made in subsequent methods. For instance, SiamRPN [16] utilizes the region proposal network including classification branch and regression branch and formulates tracking task as a one-shot local detection task; DaSiamRPN [28] learns distractor-aware Siamese network for robust and long-term tracking; SiamRPN++ [15] explores deeper backbone networks such as ResNet or Inception and solves the effect of padding on translation invariance; SiamMask [23] combines visual tracking and segmentation, and a three-branch structure is used to estimate the location of object; SiamFC++ [25] puts forward a simple and effective framework of anchor-free, which reduces the number of hyper-parameters; ATOM [4] adopts classification module of online learning and estimation module of offline training for accurate tracking. Nevertheless, these methods do not take into account the motion information between frames in videos, and simply use appearance and semantic information of the current frame.

In addition to utilizing appearance and semantic information about the object, motion information is also crucial for visual tracking. The TSN tracker [21] uses tuple learning modules to store historical samples of the object, and temporal convolutional networks learn the similarity between current candidate proposals and historical samples; The DSTN tracker [22] improves the TSN tracker using static memory, which uses LSTM networks to dynamically model historical information of the object, but its speed can only reach 1fps. Quite a few trackers also use optical flow to obtain motion information. For instance, SINT [20] uses optical flow to filter candidate samples with inconsistent motion, which employs the variational approach for optical flow estimation, and all parameters are set manually without any learning; Gladh et al. [10] combines handcrafted and deep appearance features with deep motion features in a DCF-based tracking-by-detection framework, and it uses pre-trained optical flow networks for action recognition to extract features from the off-the-shelf optical flow images, and its speed is only 0.0659fps; FlowTrack [29] warps historical frames to specified frame guided by optical flow information that is estimated by a deep flow network [6], which can enhance appearance features because different frames provide a variety of viewpoints, deformations and illuminations. But its optical flow estimation network parameters are 100 times more than ours and optical flow error rate is also higher than ours. In contrast to these methods, not only can we learn the task-specific optical flow in an end-to-end manner but also we do not need to pre-compute and store optical flow.

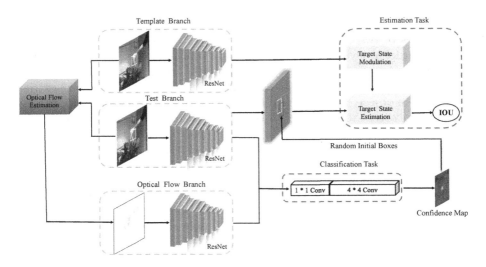

Fig. 2. An overview of our tracking framework. The network framework consists of three branches, namely the template branch, the test branch and the optical flow branch. The optical flow branch is obtained from an optical flow estimation network. We input three branches into backbone network ResNet to extract features firstly, and then carry out classification and estimation tasks to obtain the final tracking result.

3   Proposed Method

In this section, we describe the proposed tracking framework in detail. As shown in Fig. 2, the framework consists of three branches. The optical flow branch is obtained by inputting two adjacent frames in videos into optical flow estimation networks. Our feature extraction network is ResNet pre-trained on ImageNet and will not be fine-tuned. We also divide tracking into two tasks, a classification task in (Sect. 3.1) and an estimation task in (Sect. 3.2). The first task uses appearance and motion information of an object to distinguish between object and background, which can provide an initial bounding box of the object, and this is where the confidence map has the highest response. The appearance and motion information of the object corresponds to the test branch and optical flow branch respectively. The second task firstly uses the first frame of a video, namely template branch, to initialize the target state modulation module, and we can obtain modulation vectors as priori information of the object. Then, a number of bounding boxes are randomly sampled on an initial bounding box obtained by classification task, and we can find out bounding boxes with three largest IoU values by the target state estimation module. Finally, we average bounding boxes of the three largest IoU values as the final tracking result.

3.1   Classification Task

For accurate classification and real-time performance, our online classification task consists of two convolutional layers, a 1×1 convolutional layer and a 4×4 convolutional layer. Firstly, the template branch is cropped and does data augmentation at a given bounding box ground truth, thus there is more data to initialize the classification task. And then we use ResNet networks to extract features of an object, input Block 4 features into classification convolutional layers, and apply least-square loss and the optimization strategy of the ATOM [4] to learn promptly. Secondly, two adjacent frames in videos, namely template branch and test branch, are cropped according to the bounding box of the previous frame, and then we input them into optical flow estimation networks to obtain the optical flow branch. The Block 4 features of the test branch and optical flow branch are fused and input into the already initialized classification task. Finally, we can predict the rough location of the object by the highest response position of the confidence map. We find that if only features of the current frame are used for classification, the tracker will drift in the case of interference, partial occlusion and fast motion. If the optical flow branch union is used, the motion information such as magnitude and direction of motion between frames can be provided, therefore making the classification task more accurate and robust than our baseline approach.

The optical flow can capture pixel displacement between two adjacent frames. And optical flow methods are based on the assumption that brightness is constant, and it can detect changes of image pixel intensity with time and infer movement magnitude and direction of an object. Inspired by TVNet [9], which replaces the traditional TV-L1 [27] method with convolutional layers to extract optical flow, we update this smart and fast network for optical flow estimation. Specifically, in the tracking process, we input two adjacent frames and use formula Eq. (2) as loss function. It is optimized 5 times in the first frame to learn task-specific features and then it's fixed. Normally, we set the initial value of optical flow to be zero, but we can choose a better optical flow initializer by training the network several times. In addition, experimental results show that it has low optical flow errors and reduces irrelevant noise when we update optical flow estimation networks. In order to meet requirements of real-time speed, we crop two adjacent frames according to bounding box of the object, then perform estimation network training and optical flow estimation. After comprehensive experiments and balancing speed and performance, we can achieve the best results by optimizing optical flow estimation networks for 5 times. Our loss function is derived as follows.

An image at a point (x, y) at time t has moved $(\Delta x, \Delta y)$ through time Δt, we can formulate that: $I_0(x, y, t) = I_1(x + \Delta x, y + \Delta y, t + \Delta t)$ because of brightness is unchanged. Assuming that displacement between two frames is small, this can be approximated according to the Taylor Series: $I_1(x + \Delta x, y + \Delta y, t + \Delta t) = I_0(x, y, t) + \frac{\partial I}{\partial x}\Delta x + \frac{\partial I}{\partial y}\Delta y + \frac{\partial I}{\partial t}\Delta t$. Then, the following formula can be obtained: $\frac{\partial I}{\partial x}\frac{\Delta x}{\Delta t} + \frac{\partial I}{\partial y}\frac{\Delta y}{\Delta t} + \frac{\partial I}{\partial t} = 0$, i.e., $I_x u_1 + I_y u_2 + I_t = 0$, where $\boldsymbol{u} = [u_1, u_2]$ is the optical flow. Its nonlinear form is: $I_1(x + \boldsymbol{u}) - I_0(x)$, which can be linearized using Taylor expansions, leading to:

$$\rho(\boldsymbol{u}) = \nabla I_1(x + \boldsymbol{u}^0)(\boldsymbol{u} - \boldsymbol{u}^0) + I_1(x + \boldsymbol{u}^0) - I_0(x) \tag{1}$$

where $\boldsymbol{u}^0$ is initialization of $\boldsymbol{u}$. Accordingly the main optimization equation of loss function is:

$$min_{u(x), x \in \Omega} \sum_{x \in \Omega} (|\nabla u_1(x)| + |\nabla u_2(x)|) + \lambda |\rho(\boldsymbol{u})| \tag{2}$$

where the first term $|\nabla u_1(x)| + |\nabla u_2(x)|$ represents smoothing constraint, while the second term $\rho(\boldsymbol{u})$ represents the assumption that brightness is constant, λ is typically set to 0.15.

3.2   Estimation Task

According to our careful observation, some trackers can track the object correctly, but the resulting bounding box is not accurate, resulting in lower accuracy. Therefore, a well-designed tracker not only requires a powerful classifier but also can produce an accurate bounding box. And using a multi-scale search to estimate the target's bounding box is not accurate and expensive, because it cannot adapt to the scale change of an object. In addition, the introduction of region proposal network (RPN) and anchor box setting undoubtedly improve the accuracy of tracking, but it involves a lot of hyper-parameters. The tracking performance is particularly sensitive to these hyper-parameters and we need to design carefully. Thus, we use the maximum overlap strategy as [4] for our estimation task of offline training. Firstly, we sample template images and test images in the datasets, and input them into the backbone network ResNet without any fine-tuning, specifically, we use Block3 and Block4 features. Secondly, we input features obtained by the template branch and bounding boxes of template images into the target state modulation module for generating modulation vectors. Thirdly, modulation vectors and features obtained from the test branch and disturbed bounding boxes of test images are input into the target state estimation module for learning as much as possible about the Intersection over Union (IoU) overlap. The bounding box's predictive IoU formula is:

$$IoU(B) = p(z(x_0, B_0) \cdot t(x, B)) \tag{3}$$

where p represents the IoU predictor module, it consists of three fully connected layers; z represents modulation vectors, x_0 and B_0 represent the template frame and corresponding bounding boxes, respectively; t represents features obtained by test frames, x and B represent test frames and corresponding disturbed bounding boxes, respectively. We train the target state estimation module by minimizing errors in predicting the IoU.

4   Experiments

4.1   Implementation Details

Our tracker is implemented in Python, using PyTorch. It runs at about 25 FPS on a GeForce GTX TITAN X GPU. It is due to the following two aspects. Firstly, we improve the running speed by image processing, image normalization and multi-processing. Secondly, our optical flow estimation model has few parameters, which can be proved by calculating the FLOPs and parameters of our model. The speed comparison results of the trackers are shown in Table 1. We use recently proposed Large-scale Single Object Tracking (LaSOT) [7] and TrackingNet datasets [18] to train our estimation task. The COCO dataset [17] is used to enhance our training datasets. We crop input images by 5 times in accordance with bounding boxes and resize them to 288 × 288. Our batch size is set to 64, and we use mean square loss and the ADAM optimizer to optimize network parameters of the estimation task for 50 epochs. The initial learning rate is set at 10^{-3}. In the tracking stage, the estimation task will not be updated. The classification task will be updated every 10 frames. And some flags will be set to mark whether it is a hard negative sample or an uncertain sample. The classification task is also updated and the learning rate is doubled when it comes to hard negative samples.

4.2   Comparison Results

We compare our proposed approach with state-of-the-art methods on three challenging tracking benchmarks. Because the backbone network used in our baseline method [4] is ResNet-18, therefore we have two versions, Ours-18 and Ours-50

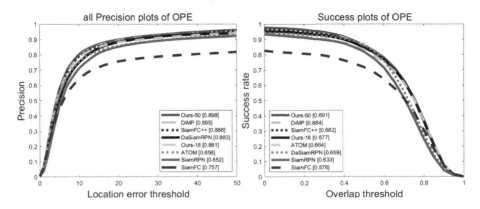

Fig. 3. Precision and success plots show a comparison of our approach with the state-of-the-art approach on the OTB2015 dataset.

Table 1. Comparison of the tracking speed

Trackers	ATOM [4]	PrDiMP [5]	ROAM [26]	FlowTrack [29]	MDNet [19]	DSTN [22]	TSN [21]	[10]	Ours
Speed (fps)	30	30	13	12	1	1	1	0.0659	25

employ ResNet-18 and ResNet-50 as backbone network respectively. It is worth noting that we do not use the GOT10k [12] joint to train our model as DiMP [2], only using three datasets. And the backbone network used by the DiMP tracker is ResNet-50.

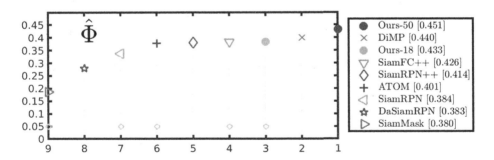

Fig. 4. Expected averaged overlap performance on the VOT2018 dataset.

OTB2015: The OTB2015 dataset contains 100 video sequences, including 11 challenging attributes, such as FM (fast motion), BC (background clutters), MB (motion blur), SV (scale variation), DEF (deformation), IPR (in-plane rotation), OPR (out-of-plane rotation), IV (illumination variation), OV (out-of-view), OCC (occlusion), LR (low resolution). According to the evaluation standard of the OTB2015 dataset, we use precision plot and success plot to show them. For precision plot, it is considered successful if the distance between the centers of the predicted box and ground truth box is under 20 pixels. For success plot, it is considered successful if the intersection-over-union overlap of predicted result and ground truth is more than a given threshold. We compare our tracker on the OTB2015 dataset with state-of-the-art trackers [1, 2, 4, 16, 25, 28]. Figure 3 shows that our tracker has superior results. Our ResNet-18 version outperforms ATOM with a relative gain of 1.3% in the success plot and 2.3% in the precision plot, and our ResNet-50 version achieves the best success score of 0.691.

Table 2. Precision/Success rate based on different challenge attributes

Challenges	Trackers					
	SiamRPN	DaSiamRPN	ATOM	SiamFC++	DiMP	**Ours-50**
FM	0.797/0.605	0.824/0.624	0.866/0.687	0.872/0.687	0.873/0.683	**0.874/0.687**
BC	0.809/0.598	0.862/0.646	0.796/0.620	0.843/0.657	0.823/0.635	0.842/0.652
MB	0.826/0.629	0.827/0.629	0.863/0.703	0.880/0.688	0.893/0.698	**0.878/0.694**
SV	0.844/0.619	0.856/0.639	0.887/0.691	0.904/0.701	0.906/0.697	**0.914/0.709**
DEF	0.829/0.620	0.878/0.645	0.827/0.619	0.862/0.651	0.868/0.653	**0.881/0.668**
IPR	0.861/0.633	0.893/0.654	0.901/0.680	0.878/0.661	0.902/0.672	0.902/0.677
OPR	0.854/0.628	0.878/0.644	0.853/0.647	0.878/0.663	0.881/0.660	**0.890/0.671**
IV	0.856/0.650	0.861/0.652	0.854/0.663	0.879/0.684	0.898/0.690	0.885/0.684
OV	0.742/0.552	0.734/0.545	0.856/0.651	0.864/0.670	0.882/0.674	0.864/0.670
OCC	0.787/0.590	0.816/0.614	0.796/0.618	0.852/0.661	0.849/0.656	**0.869/0.677**
LR	0.982/0.642	0.942/0.636	0.985/0.691	0.997/0.709	0.996/0.707	**0.993/0.708**
ALL	0.852/0.633	0.883/0.659	0.858/0.664	0.886/0.682	0.895/0.684	**0.898/0.691**

VOT2018: The VOT2018 dataset contains 60 video sequences, and the performance of trackers is generally evaluated using robustness, accuracy and Expected Average Overlap (EAO). Robustness is the number of failures, accuracy is the average overlap rate of successful tracking, and EAO integrates the above two evaluation standards to evaluate the performance of trackers more comprehensively. Our ResNet-18 version outperforms ATOM with a relative gain of 3.2%, our ResNet-50 version achieves the best EAO score of 0.451, the result shown in Fig. 4 is an average of 5 runs to reduce randomness.

LaSOT: The LaSOT dataset is a long-term tracking dataset in which objects may disappear and then come into view again. It contains 1400 video sequences, each video has an average of 2512 frames. Its evaluation standard is One-Pass-Evaluation (OPE) according to OTB2015. Figure 5 shows the performance of our tracker on the LaSOT test set consisting of 280 videos. For the sake of fair comparisons, we only train with the official training set. In particular, in a long-term tracking dataset, the performance of modules with re-detection objects is generally better. However, our approach outperforms ATOM with a relative gain of 1.9%, which reflects the importance of motion information.

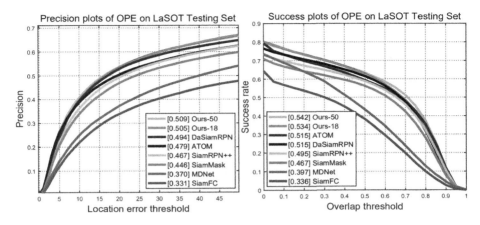

Fig. 5. Precision and success plots show a comparison of our approach with state-of-the-art methods on the LaSOT dataset.

4.3 Ablation Study

To prove the validity of our proposed optical flow branch, we perform an ablation study. As shown in Fig. 6, the first line is video sequence Human4, in which optical flow information will suppress static background information (e.g. traffic signs) and provide the direction information. Therefore, our tracker does not drift, which is also proved by our confidence map. But our baseline method appears to drift in the case of partial occlusion. The second line is video sequence Woman, in which optical flow information also inhibits static background information (e.g. vehicles). Although the vehicle covers legs of the pedestrian, our tracker is not disturbed at all, the classification is more accurate, and the resulting bounding box is more compact. The third line is video sequence Walking, in which the woman is walking forward, and when a similar distractor suddenly appears and walks backward, our baseline method will drift. However, after adding the optical flow branch, the tracker will follow the established direction, thus we can avoid drifting. This fully demonstrates the importance of motion information in videos and the effectiveness of our method. On the other hand, we also try to perform tracking using only the optical flow branch without the test branch, but the results are not satisfactory. Consequently, appearance features are important, and motion information can be used as a supplement to improve the robustness of visual tracking.

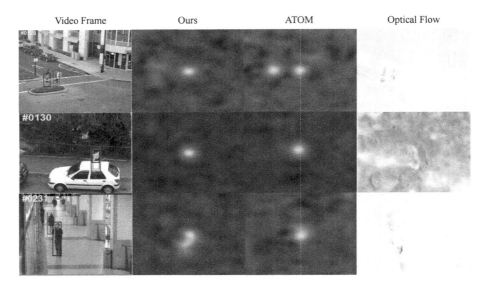

Video Frame   Ours   ATOM   Optical Flow

Fig. 6. During the tracking process, optical flow and confidence maps are visualized on the OTB2015 dataset. The red bounding box is ground truth, the green bounding box is our tracking method, and the blue bounding box is our baseline method. (Color figure online)

5   Conclusion

In this paper, we not only use appearance information of the current frame to distinguish the object and background during tracking but also explore whether motion information between video frames can make classifiers more intelligent, especially under partial occlusion, fast motion, low resolution and other challenges. And our optical flow branch does not need to use any pre-trained model and optical flow ground truth, what is more, it can learn the optical flow of task-specific. Through a large number of experiments and theory analyses, we find the significance of motion information for visual tracking. While improving the performance of visual tracking, we use several methods to ensure real-time speed, including image cropping, image normalization and multi-processing. Finally, after ablation experiments and comparison among three challenging datasets, our tracker obtains favorable performance.

References

1. Bertinetto, L., Valmadre, J., Henriques, J.F., Vedaldi, A., Torr, P.H.: Fully-convolutional Siamese networks for object tracking. In: European Conference on Computer Vision, pp. 850–865 (2016)
2. Bhat, G., Danelljan, M., Gool, L.V., Timofte, R.: Learning discriminative model prediction for tracking. In: Proceedings of the IEEE International Conference on Computer Vision, pp. 6182–6191 (2019)

3. Chen, Z., Zhong, B., Li, G., Zhang, S., Ji, R.: Siamese box adaptive network for visual tracking. In: Proceedings of the IEEE Conference on Computer Vision and Pattern Recognition, pp. 6667–6676 (2020)
4. Danelljan, M., Bhat, G., Khan, F.S., Felsberg, M.: Atom: accurate tracking by overlap maximization. In: Proceedings of the IEEE Conference on Computer Vision and Pattern Recognition, pp. 4660–4669 (2019)
5. Danelljan, M., Gool, L.V., Timofte, R.: Probabilistic regression for visual tracking. In: Proceedings of the IEEE Conference on Computer Vision and Pattern Recognition, pp. 7181–7190 (2020)
6. Dosovitskiy, A., et al.: FlowNet: learning optical flow with convolutional networks. In: Proceedings of the IEEE International Conference on Computer Vision, pp. 2758–2766 (2015)
7. Fan, H., et al.: LaSOT: a high-quality benchmark for large-scale single object tracking. In: Proceedings of the IEEE Conference on Computer Vision and Pattern Recognition, pp. 5374–5383 (2019)
8. Fan, H., Ling, H.: Siamese cascaded region proposal networks for real-time visual tracking. In: Proceedings of the IEEE Conference on Computer Vision and Pattern Recognition, pp. 7952–7961 (2019)
9. Fan, L., Huang, W., Gan, C., Ermon, S., Gong, B., Huang, J.: End-to-end learning of motion representation for video understanding. In: Proceedings of the IEEE Conference on Computer Vision and Pattern Recognition, pp. 6016–6025 (2018)
10. Gladh, S., Danelljan, M., Khan, F.S., Felsberg, M.: Deep motion features for visual tracking. In: International Conference on Pattern Recognition, pp. 1243–1248 (2016)
11. Guo, D., Wang, J., Cui, Y., Wang, Z., Chen, S.: SiamCAR: siamese fully convolutional classification and regression for visual tracking. In: Proceedings of the IEEE Conference on Computer Vision and Pattern Recognition, pp. 6268–6276 (2020)
12. Huang, L., Zhao, X., Huang, K.: GOT-10k: a large high-diversity benchmark for generic object tracking in the wild. IEEE Trans. Patt. Anal. Mach. Intell. (01), 1 (2019)
13. Jung, I., Son, J., Baek, M., Han, B.: Real-time MDNet. In: Proceedings of the European Conference on Computer Vision, pp. 83–98 (2018)
14. Kristan, M., et al.: The sixth visual object tracking vot2018 challenge results. In: Proceedings of the European Conference on Computer Vision (2018)
15. Li, B., Wu, W., Wang, Q., Zhang, F., Xing, J., Yan, J.: SiamRPN++: evolution of Siamese visual tracking with very deep networks. In: Proceedings of the IEEE Conference on Computer Vision and Pattern Recognition, pp. 4282–4291 (2019)
16. Li, B., Yan, J., Wu, W., Zhu, Z., Hu, X.: High performance visual tracking with Siamese region proposal network. In: Proceedings of the IEEE Conference on Computer Vision and Pattern Recognition, pp. 8971–8980 (2018)
17. Lin, T.Y., et al.: Microsoft COCO: common objects in context. In: European Conference on Computer Vision, pp. 740–755 (2014)
18. Muller, M., Bibi, A., Giancola, S., Alsubaihi, S., Ghanem, B.: TrackingNet: a large-scale dataset and benchmark for object tracking in the wild. In: Proceedings of the European Conference on Computer Vision, pp. 300–317 (2018)
19. Nam, H., Han, B.: Learning multi-domain convolutional neural networks for visual tracking. In: Proceedings of the IEEE Conference on Computer Vision and Pattern Recognition, pp. 4293–4302 (2016)
20. Tao, R., Gavves, E., Smeulders, A.W.: Siamese instance search for tracking. In: Proceedings of the IEEE Conference on Computer Vision and Pattern Recognition, pp. 1420–1429 (2016)

21. Teng, Z., Xing, J., Wang, Q., Lang, C., Feng, S., Jin, Y.: Robust object tracking based on temporal and spatial deep networks. In: Proceedings of the IEEE International Conference on Computer Vision, pp. 1144–1153 (2017)
22. Teng, Z., Xing, J., Wang, Q., Zhang, B., Fan, J.: Deep spatial and temporal network for robust visual object tracking. IEEE Trans. Image Process. **29**, 1762–1775 (2019)
23. Wang, Q., Zhang, L., Bertinetto, L., Hu, W., Torr, P.H.: Fast online object tracking and segmentation: a unifying approach. In: Proceedings of the IEEE Conference on Computer Vision and Pattern Recognition, pp. 1328–1338 (2019)
24. Wu, Y., Lim, J., Yang, M.H.: Online object tracking: a benchmark. In: Proceedings of the IEEE Conference on Computer Vision and Pattern Recognition, pp. 2411–2418 (2013)
25. Xu, Y., Wang, Z., Li, Z., Yuan, Y., Yu, G.: SiamFC++: towards robust and accurate visual tracking with target estimation guidelines. In: Proceedings of the AAAI Conference on Artificial Intelligence, pp. 12549–12556 (2020)
26. Yang, T., Xu, P., Hu, R., Chai, H., Chan, A.B.: ROAM: recurrently optimizing tracking model. In: Proceedings of the IEEE Conference on Computer Vision and Pattern Recognition, pp. 6717–6726 (2020)
27. Zach, C., Pock, T., Bischof, H.: A duality based approach for realtime TV-L 1 optical flow. In: Joint Pattern Recognition Symposium, pp. 214–223 (2007)
28. Zhu, Z., Wang, Q., Li, B., Wu, W., Yan, J., Hu, W.: Distractor-aware Siamese networks for visual object tracking. In: Proceedings of the European Conference on Computer Vision, pp. 101–117 (2018)
29. Zhu, Z., Wu, W., Zou, W., Yan, J.: End-to-end flow correlation tracking with spatial-temporal attention. In: Proceedings of the IEEE Conference on Computer Vision and Pattern Recognition, pp. 548–557 (2018)

ReFlowNet: Revisiting Coarse-to-fine Learning of Optical Flow

Leyang Xu and Zongqing Lu[✉]

Tsinghua Shenzhen International Graduate School, Shenzhen, China
`luzq@sz.tsinghua.edu.cn`

Abstract. Recent work on optical flow estimation has been recognized as a computer vision task based on multi-resolution learning. However, existing methods only extract features by using plain convolution, while feature representation in the network affects the accuracy of optical flow estimation. To solve this problem, we propose a symmetrical factorization convolution block to extracting features for rich semantic information, increasing the depth and nonlinearity of the network. In addition, warping operation is often employed in optical flow estimation networks, but it brings low confidence information in computation. A confidence map module is proposed in order to compensate for the warping operation. In this paper, we propose a new optical flow estimation network called ReFlowNet. Experiments demonstrate that our network achieves competitive performance with other methods on the MPI-Sintel dataset.

Keywords: Optical flow · Symmetrical factorization convolution block · Confidence map

1 Introduction

Optical flow estimation is a key instrument in computer vision, which describes the pixel displacement between two consecutive images [3]. Recent researches towards optical flow estimation show three essential modules in their neural network architecture: feature pyramid, warping, and cost volume [5,7,18]. To address the large displacement problem in optical flow estimation, the coarse-to-fine strategy is frequently adopted. This paper aims to analyze the problems caused by the network structure of the coarse-to-fine strategy for optical flow estimation and to improve the accuracy of optical flow estimation from the coarse-to-fine strategy.

First, we analyze the implementation of the feature extraction network adopted by the general optical flow estimation network [5,18], which only uses strided convolution to extract features roughly and cannot capture longer range dependencies well between two raw images in large motion situations. To address this issue, simply increasing the number of convolution blocks may be a solution, but it is not elegant. Therefore, we propose a symmetrical factorization convolution block to extract complex features in pyramid levels. Specifically, we use

H. Ma et al. (Eds.): PRCV 2021, LNCS 13019, pp. 429–442, 2021.
https://doi.org/10.1007/978-3-030-88004-0_35

1D kernel convolution in symmetrical factorization convolution block to collect short-range feature information and 2D kernel convolution in the block to catch long-range features. At the tail of the block, we use the channel shuffle technique to fuse long-range and short-range features. The advantage of this block is that it is effective to produce a more accurate estimation of the flow without introducing many parameters.

Secondly, warping is a common operation in optical flow estimation, as shown in Fig. 1, we find that the warping operation brings low confidence information. To address this issue, we tailor a confidence map module to focus on the warped features that are of high confidence.

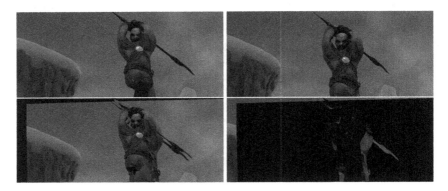

Fig. 1. Warping operation brings low confidence information. The top row: the first frame, the second frame, from left to right. The bottom row: the second frame is warped with the optical flow, the error map between the first frame and the warped map, from left to right.

Thirdly, regarding the correlation layers, [24] constructs a 4D convolutional matching module for volumetric correspondence processing, and [12] deforms the cost volume to alleviate the warping artifacts. [9,17] suggest that correlation normalization improves convergence and the final performance in unsupervised learning of optical flow. Hence, we introduce correlation normalization in our network.

Therefore, based on our three contributions, we propose ReFlowNet, and our contributions are accordingly three-fold.

1) We propose symmetrical factorization convolution blocks to build the feature pyramid by merging long-range and short-range features more appropriately.
2) We propose confidence map modules to focus on the warped features that are of high confidence.
3) We introduce correlation normalization into supervised learning of optical flow.

2 Related Work

Optical Flow Estimation. Classic variational approaches have laid the foundation in optical flow estimation since the research of Horn and Schunck [3], which is based on brightness constancy and spatial smoothness assumption. For large displacement issues, from coarse to fine is a commonly adopted strategy. Later work [23] combined CNN feature extractor and variational approach for flow field refinement. However, they sacrificed huge computation costs to maintain optical flow estimation accuracy, so when in the real-time environment they are impractical.

FlowNet [2] opens the door to optical flow research which is training end-to-end CNNs on a synthetic dataset to estimate optical flow. They attempted to build two CNN architectures FlowNetS and FlowNetC, the correlation layer in FlowNetC has evolved into an important part in optical flow estimation and disparity estimation network architecture design. The successive work FlowNet2 [8] stacks a few FlowNetS and FlowNetC, introducing warping operation which turns out to be an indispensable component in later networks. However, FlowNet2 is too heavy and is not suited for mobile devices.

SpyNet [15] addresses the model size issue by only taking spatial pyramid and warping principles without correlation layer. PWC-Net [18] and Lite-FlowNet [5] explores a lightweight network associated feature pyramid, warping, and cost volume, they achieve excellent performance and keep network architecture light. VCN [24] explores the cost volume module redesign that constructs a 4D convolutional matching module for volumetric correspondence processing. Devon [12] deforms the cost volume to alleviate the warping artifacts. IRR [7]proposes an iterative residual refinement scheme that takes the output from the previous iteration as input and iteratively refines it, HD3 [25] treats cost volume as a probabilistic distribution and applies uncertainty estimation for optical flow estimation.

Occlusion Estimation in Optical Flow. Occlusion estimation is an ill-posed problem that optical flow has been got stuck for a long time. Several papers suggest that occlusion handling approaches in optical flow roughly into these points. 1) Occlusion handling in loss function design, occlusion constrained by the photometric loss [26] occluded pixels excluded from the loss computation [14,21], 2) utilizing forward-backward consistency and occlusion-disocclusion symmetry [6] 3) using multiple frames which benefit from long-term temporal cues.[11,13,16], 4) DDflow [10] presents a data distillation approach to learn the optical flow of occluded pixels, which works particularly well for near image boundaries. 5) Layered optical flow estimation [27] combines a layered motion representation with a soft-mask module, MaskFlowNet [29] puts forward learnable occlusion mask filters occluded areas immediately after feature warping without any explicit supervision and deformable convolution layer, but it consists of two stages and the overall architecture is huge.

Attention Mechanism. Attention mechanism can be viewed as a weighted assignment. Spatial attention and channel attention are two important attention modules [22]. Channel attention module [4] is widely applied in computer vision tasks, such as semantic segmentation and super resolution [28]. Spatial attention module or self-attention mechanism [20] has excellent performance capabilities. Our work is closely related to the family of attention mechanism computer vision applications.

3   Proposed Method: ReFlowNet

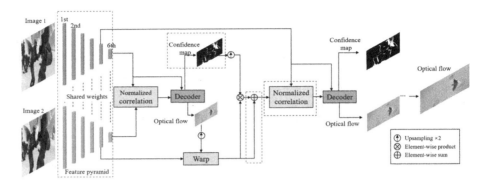

Fig. 2. Overall network architecture. For clear display, we only draw connections for the 5th and 6th levels, with the context network omitted. Other pyramidal levels have the same structure. In this work, we propose three contributions. Our first contribution (in the red dashed box) is the symmetrical factorization convolution block used to extract features for each layer of the feature pyramid (see Fig. 4 for details). Our second contribution (in the two green dashed boxes) is to generate a confidence map to better exploit the warped features. Our third contribution (in the blue dashed box) is to leverage the normalized correlation. (Color figure online)

3.1   Network Structure

Figure 2 illustrates the overall structure of our ReFlowNet. Given two consecutive images I_1 and I_2, our task is to estimate the pixel displacement φ representing optical flow from I_1 to I_2.

First, we put the image pair into a feature pyramid to extract the corresponding feature maps F_1, F_2:

$$(F_1, F_2) = f(I_1, I_2), \tag{1}$$

where f represents the feature extraction network.

Secondly, for the top level of the feature pyramid (the 6th level shown in Fig. 2), we use normalized correlation on the two feature maps, and then through a multi-layer CNN decoder module, we obtain a coarse optical flow map and a confidence map. The confidence map shows where the coarse optical flow map is of high confidence so that we can transfer the high-confidence optical flow to the lower level. The remaining levels have a similar operation, so we present the expression for the 6th level as an example:

$$c^6 = \text{corr}(F_1^6, F_2^6), \tag{2}$$

$$(\varphi^6, \theta^6) = \text{decoder}(c^6), \tag{3}$$

where c denotes the correlation of two feature maps, φ means the flow map, and θ is the confidence map, with superscript for the pyramid level.

Thirdly, we upsample two times the upper-level flow and confidence maps to get the right scale, and warp the lth level feature map of the second frame by using the upsampled flow from $(l+1)$th level, to obtain warped feature map ω^l. Then we calculate the element-wise product $\otimes$ of ω^l and the confidence map θ^l to attain the compensated warped feature γ^l:

$$\omega^l = \text{warp}(F_2^l, \text{up}_2(\varphi^{l+1})), \tag{4}$$

$$\gamma^l = \omega^l \otimes \theta^l + \omega^l, \tag{5}$$

and then using γ^l, we update the flow and confidence maps:

$$c^l = \text{corr}(\gamma^l, F_1^l), \tag{6}$$

$$(\varphi^l, \theta^l) = \text{decoder}(c^l). \tag{7}$$

Finally, we make flow predictions from the 6th level to the 2nd level in a coarse-to-fine manner. The final flow displacement at the 2nd level is 4-time upsampled to the same size as the input image by bilinear interpolation. The final flow displacement passing through a context network (same as PWC-Net [18]) produces the final refined flow prediction.

Next, we will explain our three contributions in detail.

3.2 Symmetrical Factorization Convolution Block

As raw images provide less feature information, many networks of optical flow estimation replace the fixed image pyramid with learnable feature pyramids. We observe that their simple convolution can only extract short-range features. For the issue of large displacement in optical flow estimation, we argue that enlarging the respective field is also important. Therefore, to achieve our aim of obtaining long-range and short-range features simultaneously by using simple operations without introducing many parameters or computations, we propose symmetrical factorization convolution blocks into the feature pyramid.

To make the parameters of the basic building block not increase, it is necessary to determine the setting of the number of input and output channels. Assuming that the number of input channels is C_1 and the number of output channels is C_2, the required parameters after passing through the convolution block shown in Fig. 3(a) are

$$C_1 \times 3 \times 3 \times C_1 = 9C_1C_2, \tag{8}$$

after the convolution block shown in Fig. 3(b), the required parameters are:

$$\left(\frac{C_1}{2} \times 1 \times 3 \times \frac{C_2}{2} + \frac{C_2}{2} \times 3 \times 1 \times \frac{C_2}{2}\right) \times 2 = \frac{3}{2}C_1C_2 + \frac{3}{2}C_2^2, \qquad (9)$$

if we want the parameters of Fig. 3(b) to not increase compared to the parameters of Fig. 3(a), the following condition should be met:

$$\left(\frac{3}{2}C_1C_2 + \frac{3}{2}C_2^2\right) \le 9C_1C_2, \qquad (10)$$

simplified to get $C_2 \le 5C_1$, in the experiment, the setting of the number of input and output channels only needs to ensure $C_2 \le 5C_1$.

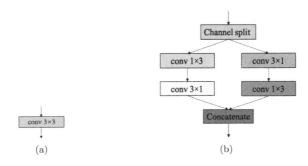

(a)   (b)

Fig. 3. Comparison of a common building block with symmetrical factorization convolution building block. (a) The implementation of convolution layers in the feature pyramid level in a general optical flow estimation network. (b) The proposed symmetrical factorization convolution block.

A layer implementation of the feature pyramid in a general optical flow estimation network is shown in Fig. 4(a). We use symmetrical factorization convolution building block to replace the second 3×3 convolution kernel in Fig. 4(a) with a 3×1 convolution kernel and 1×3 convolution kernel to obtain rich feature semantic information, and then the symmetrical factorization convolution block is symmetrical up and down, left and right. The fusion of feature information between channels is accomplished by a 1×1 convolution and channel shuffle.

3.3 Confidence Map Module

Feature warping could alleviate complex and large motion issues. The process can be formulated as

$$\text{warp}(F_2, \varphi(x)) = F_2(x + \varphi(x)), \qquad (11)$$

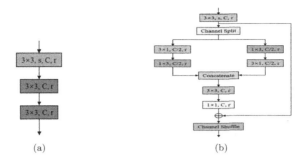

(a)   (b)

Fig. 4. Comparison of common convolution block with symmetrical factorization convolution block. (a) The implementation of convolution layers in the feature pyramid level in the general network. (b) The proposed symmetrical factorization convolution block. The notation 'r' is the dilation of the kernel, 'C' represents the number of output channels, and 's' is the stride of the convolution. We replace the second convolution kernel in (a) with a 3×1 convolution kernel and 1×3 convolution kernel and add a 1×1 convolution kernel and channel shuffle at the end. Before the channel shuffle, there is a skip connection.

where x is the pixel position and $\varphi(x)$ is the estimated pixel displacement. As $\varphi(x)$ records the pixel displacement from the first frame to the second frame, ideally, we can restore the first frame by warping the second frame and the flow map. But in fact, the warping operation is easy to bring low confidence information in the computation on the image and flow.

To improve the confidence of the warping operation, we introduce a confidence map module generated by the decoder, which is a multi-layer CNN that has dense connections between convolutional layers. The number of filters at each convolutional layer in the decoder is the same as PWC-Net. The difference is that we add a convolutional layer, and then activate the output by a sigmoid function. Under the guidance of the attention mechanism, use the broadcast mechanism in TensorFlow to make the warped feature map and the confidence map into the Hadamard product.

3.4 Normalized Correlation Layer

Normalizing correlation is a popular practice to find correspondences between two feature maps, formulated as

$$
\mathcal{C}^L = \sum_{d=1}^{D} \left(\frac{F_{x,d}^{(1,L)} - \mu^{(1,L)}}{\sigma^{(1,L)}} \right) \left(\frac{F_{x+\varphi(x),d}^{(2,L)} - \mu^{(2,L)}}{\sigma^{(2,L)}} \right), \tag{12}
$$

where $F^{(i,L)} \in \mathbb{R}^{\frac{H}{2^L} \times \frac{W}{2^L} \times D}$ denotes the feature map for image i at level L, with H, W and D denoting the height, weight and dimension of the feature map, respectively; x means the pixel position in image 1, $\varphi(x)$ is the pixel shift, and

$\mu^{(i,L)}$ and $\sigma^{(i,L)}$ are the sample mean and standard deviation of $F^{(i,L)}$ over its spatial and feature dimensions.

3.5   Training Loss

We use the multi-scale loss function proposed in [19]:

$$\mathcal{L}(\Theta) = \sum_{l=\hat{L}}^{L} \alpha_l \sum_x \|\varphi_\Theta^l(x) - \varphi_{gt}^l(x)\|_2 + \lambda\|\Theta\|_2^2, \qquad (13)$$

where Θ denotes the set of all the learnable parameters in our network, $\|\cdot\|_2$ denotes L2 norm to regularize parameters of the network; α_l and $\varphi_\Theta^l(x)$ are the weight and the predicted flow at the lth pyramid level, respectively; λ denotes the trade-off weight; and $\hat{L}$ represents the desired level that we want to output (level 2 in the experiment), out of totally L levels. The loss function is optimized by using the Adam optimizer.

For fine-tuning, we use a robust version of training loss:

$$\mathcal{L}(\Theta) = \sum_{l=\hat{L}}^{L} \alpha_l \sum_x \left(\|\varphi_\Theta^l(x) - \varphi_{gt}^l(x)\|_1 + \epsilon\right)^q + \lambda\|\Theta\|_2^2, \qquad (14)$$

where $\|\cdot\|_1$ denotes the L1 norm, $q < 1$ gives less penalty to outliers, and ϵ is a small constant.

4   Experiments

We evaluate our method on the standard optical flow benchmark dataset MPI-Sintel [1]. Following previous works, we pretrain our network on the FlyingChairs and FlyingThings. We compare our results to FlowNet [2], FlowNet2 [8], SpyNet [15], LiteFlowNet [5] and PWC-Net [18]. We consider average endpoint error (EPE) as the evaluation metric, where EPE is the ranking metric on the MPI Sintel benchmark.

4.1   Implementation Details

Data Augmentation. We expose augmentations on 50% of images randomly, including geometric transformations: flipping, translation, scaling. Specifically, we horizontally and vertically flip images; translate images by a range $[-5\%, 5\%]$ of the original image size on x and y axis independently; scale from $[0.95, 1.05]$ of original image size.

Training Procedure. We first train the modules using the FlyingChairs dataset in TensorFlow. Each image is cropped to 448×384 patches during data augmentation and a batch size of 8. We use the learning rate schedule starting from 0.0001 and reduce the learning rate from the 200K iteration by half at every

$100K$ iteration. Our iterations are $700K$ in total. The parameters in training loss (13) are set to be $\alpha = (0.0025, 0.005, 0.01, 0.02, 0.08, 0.32)$, $\lambda = 0.0004$. Then we fine-tune the models on the FlyingThings where samples with extremely difficult data were omitted. The cropped image size is 768×384 and the batch size is 4. Finally, we fine-tune the models on the Sintel training sets. The cropped image size is 768×384 and the batch size is 4. We use robust loss function (14) with $\epsilon = 0.01$ and $q = 0.4$ (Figs. 5, 6).

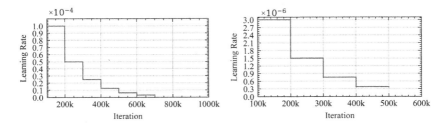

Fig. 5. Left: Learning rate used on FlyingChairs dataset. Right: Learning rate used on FlyingThings dataset.

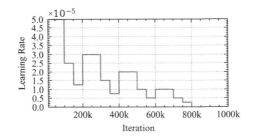

Fig. 6. Learning rate used on MPI-Sintel dataset.

All the experiments are conducted on an NVIDIA 1080 Ti GPU.

4.2   Results and Analyses

Ablation Study. To understand how the three new components affect the model performance, we perform an ablation study, by training different variants of ReFlowNet on the FlyingChairs dataset (splitting 3% for validation) and testing them on the FlyingChairs validation set and the Sintel training sets.

From the results in Table 1, we can observe the followings.

First, among the three new components, when only a single one is adopted, the symmetrical factorization convolution block produces the most improvement,

Table 1. Component ablation (SFCB: Symmetrical factorization convolution block; CM: confidence map; NC: normalized correlation). The model has been trained on the FlyingChairs and finetuned on the FlyingThings, and we test different components using the FlyingChairs validation set and the Sintel training sets. The smaller the value in the table, the better the performance.

SFCB	CM	NC	Trained on chairs			Finetuned on FlyingThings		
			Chairs	Clean	Final	Chairs	Clean	Final
×	×	×	1.32	4.08	4.81	1.84	3.20	4.44
✓			1.15	3.74	4.61	1.63	2.82	4.25
	✓		1.15	4.06	4.94	1.69	3.27	4.47
		✓	1.17	4.28	5.19	1.66	2.89	4.44
✓	✓		**1.11**	3.97	4.78	**1.45**	2.73	4.15
✓		✓	1.17	3.66	4.67	1.67	**2.61**	4.23
	✓	✓	1.24	4.04	4.85	1.66	2.91	4.16
✓	✓	✓	1.27	**3.61**	**4.58**	1.72	2.70	**4.14**

and it actually can improve the performance in every case from the baseline. Secondly, when only a combination of two new components is adopted, it generally prefers the combinations with the symmetrical factorization convolution block involved. Thirdly, when all three new components are jointly adopted, although the performance does not improve on FlyingChairs, it mostly performs the best on Sintel. In short, every component contributes to the overall performance, especially the symmetrical factorization convolution block.

Comparison with Closely-Related Work. Table 2 shows the results of our model (ReFlowNet) and other closely-related coarse-to-fine networks on the Sintel datasets.

Table 2. Average EPE results of different methods on the MPI Sintel datasets. The smaller the value, the better the performance.

Methods	Sintel clean		Sintel final		Time (s)
	Train	Test	Train	Test	
FlowNetS [2]	4.50	6.96	5.45	7.76	0.01
FlowNetC [2]	4.31	6.85	5.87	8.51	0.05
FlowNet2 [8]	2.02	3.96	3.14	6.02	0.12
SpyNet [15]	4.12	6.64	5.57	8.36	0.16
LiteFlowNet [5]	2.52	4.86	4.05	6.09	0.09
PWC-Net-Base	1.07	6.47	2.22	7.36	0.08
ReFlowNet (Ours)	1.31	5.32	1.94	6.19	0.08

From Table 2, we can observe the following patterns. First, our ReFlowNet performs better than PWC-Net [18] in all but one case. Secondly, FlowNet2 [8] achieves impressive performance by stacking several basic models into a large-capacity model, but it takes 0.12 s while our ReFlowNet only takes 0.08 s. Thirdly, our ReFlowNet performs better than SpyNet [15]. SpyNet uses image pyramids while ReFlowNet learns feature pyramids. SpyNet feeds CNNs with images, while ReFlowNet feeds a correlation result of image features. This indicates that using the feature pyramid is better than using the image pyramid, and it is even better to use the symmetrical factorization convolution block in the feature pyramid. Finally, LiteFlowNet [5] furthers incorporates a flow regularization layer to deal with outliers, but it requires stage-wise training. In terms of running time, LiteFlowNet takes 0.09 s, while ReFlowNet only takes 0.08 s. In summary, we can see that our method works quite well.

4.3   MPI-Sintel Test Results

It can be seen from Fig. 7 and Fig. 8 that the predicted optical flow of ReFlowNet in the dashed box is better than that of the baseline network. The difference between the raw images of Fig. 7 and Fig. 8 is that the images of the Clean are clean while the images of the Final are motion-blurred.

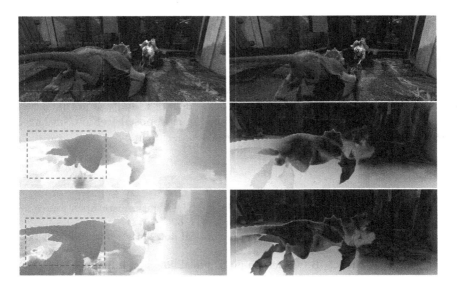

Fig. 7. Qualitative results on the Sintel-Clean test set. The top row: the pair of raw images. The middle row: flow map of baseline network, error map of baseline, from left to right. The bottom row: flow map of ReFlowNet, error map of ReFlowNet, from left to right. (Error maps are normalized to be in the range between 0 and 1.)

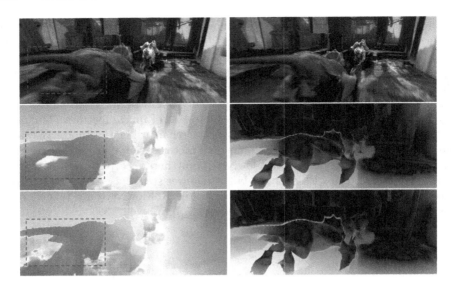

Fig. 8. Qualitative results on the Sintel-Final test set. The top row: the pair of raw images. The middle row: flow map of baseline network, error map of baseline, from left to right. The bottom row: flow map of ReFlowNet, error map of ReFlowNet, from left to right. (error maps are normalized to be in the range between 0 and 1.)

5 Conclusions

We propose a new optical flow estimation network ReFlowNet to produce better optical flow estimation: we introduce symmetrical factorization convolution blocks, add confidence map modules into the network, and use normalized correlation. Based on these three contributions, our proposed ReFlowNet can outperform the baseline, as validated in various experiments.

References

1. Butler, D.J., Wulff, J., Stanley, G.B., Black, M.J.: A naturalistic open source movie for optical flow evaluation. In: ECCV, pp. 611–625 (2012)
2. Dosovitskiy, A., et al.: FlowNet: learning optical flow with convolutional networks. In: ICCV, pp. 2758–2766 (2015)
3. Horn, B.K., Schunck, B.G.: Determining optical flow. Artif. Intell. **17**(1), 185–203 (1981)
4. Hu, J., Shen, L., Sun, G.: Squeeze-and-excitation networks. In: CVPR, pp. 7132–7141 (2018)
5. Hui, T.W., Tang, X., Loy, C.C.: LiteFlowNet: A lightweight convolutional neural network for optical flow estimation. In: CVPR, pp. 8981–8989 (2018)
6. Hur, J., Roth, S.: MirrorFlow: exploiting symmetries in joint optical flow and occlusion estimation. In: ICCV, pp. 312–321 (2017)

7. Hur, J., Roth, S.: Iterative residual refinement for joint optical flow and occlusion estimation. In: CVPR, pp. 5747–5756 (2019)
8. Ilg, E., Mayer, N., Saikia, T., Keuper, M., Dosovitskiy, A., Brox, T.: FlowNet 2.0: evolution of optical flow estimation with deep networks. In: CVPR, pp. 2462–2470 (2017)
9. Jonschkowski, R., Stone, A., Barron, J.T., Gordon, A., Konolige, K., Angelova, A.: What matters in unsupervised optical flow. arXiv preprint arXiv:2006.04902 (2020)
10. Liu, P., King, I., Lyu, M.R., Xu, J.: DDFlow: learning optical flow with unlabeled data distillation. In: AAAI, pp. 8770–8777 (2019)
11. Liu, P., Lyu, M., King, I., Xu, J.: SelFlow: self-supervised learning of optical flow. In: CVPR, pp. 4571–4580 (2019)
12. Lu, Y., Valmadre, J., Wang, H., Kannala, J., Harandi, M., Torr, P.: Devon: deformable volume network for learning optical flow. In: WACV, pp. 2705–2713 (2020)
13. Maurer, D., Bruhn, A.: ProFlow: learning to predict optical flow. arXiv preprint arXiv:1806.00800 (2018)
14. Meister, S., Hur, J., Roth, S.: UnFlow: unsupervised learning of optical flow with a bidirectional census loss. In: AAAI, pp. 7251–7259 (2018)
15. Ranjan, A., Black, M.J.: Optical flow estimation using a spatial pyramid network. In: CVPR, pp. 4161–4170 (2017)
16. Ren, Z., Gallo, O., Sun, D., Yang, M.H., Sudderth, E.B., Kautz, J.: A fusion approach for multi-frame optical flow estimation. In: WACV, pp. 2077–2086 (2019)
17. Rocco, I., Arandjelovic, R., Sivic, J.: Convolutional neural network architecture for geometric matching. IEEE Trans. Pattern Anal. Mach. Intell. **41**(11), 2553–2567 (2019)
18. Sun, D., Yang, X., Liu, M.Y., Kautz, J.: PWC-Net: CNNs for optical flow using pyramid, warping, and cost volume. In: CVPR, pp. 8934–8943 (2018)
19. Sun, D., Yang, X., Liu, M.Y., Kautz, J.: Models matter, so does training: an empirical study of CNNs for optical flow estimation. IEEE Trans. Patt. Anal. Mach. Intell. **42**, 1408–1423 (2020)
20. Wang, X., Girshick, R.B., Gupta, A., He, K.: Non-local neural networks. arXiv preprint arXiv:1711.07971 (2017)
21. Wang, Y., Yang, Y., Yang, Z., Zhao, L., Wang, P., Xu, W.: Occlusion aware unsupervised learning of optical flow. In: CVPR, pp. 4884–4893 (2018)
22. Woo, S., Park, J., Lee, J.Y., Kweon, I.S.: CBAM: convolutional block attention module. In: ECCV, pp. 3–19 (2018)
23. Xu, J., Ranftl, R., Koltun, V.: Accurate optical flow via direct cost volume processing. In: CVPR, pp. 5807–5815 (2017)
24. Yang, G., Ramanan, D.: Volumetric correspondence networks for optical flow. In: NeurIPS, pp. 793–803 (2019)
25. Yin, Z., Darrell, T., Yu, F.: Hierarchical discrete distribution decomposition for match density estimation. In: CVPR, pp. 6037–6046 (2019)
26. Yu, J.J., Harley, A.W., Derpanis, K.G.: Back to basics: unsupervised learning of optical flow via brightness constancy and motion smoothness. In: Hua, G., Jégou, H. (eds.) ECCV 2016. LNCS, vol. 9915, pp. 3–10. Springer, Cham (2016). https://doi.org/10.1007/978-3-319-49409-8_1
27. Zhang, X., Ma, D., Ouyang, X., Jiang, S., Gan, L., Agam, G.: Layered optical flow estimation using a deep neural network with a soft mask. In: IJCAI, pp. 1170–1176 (2018)

28. Zhang, Y., Li, K., Li, K., Wang, L., Zhong, B., Fu, Y.: Image super-resolution using very deep residual channel attention networks. In: Ferrari, V., Hebert, M., Sminchisescu, C., Weiss, Y. (eds.) ECCV, pp. 294–310 (2018)
29. Zhao, S., Sheng, Y., Dong, Y., Chang, E.I.C., Xu, Y.: MaskFlownet: asymmetric feature matching with learnable occlusion mask. In: CVPR, pp. 6277–6286 (2020)

Local Mutual Metric Network for Few-Shot Image Classification

Yaohui Li, Huaxiong Li$^{(\boxtimes)}$, Haoxing Chen, and Chunlin Chen

Department of Control and Systems Engineering, Nanjing University,
Nanjing, China
{yaohuili,haoxingchen}@smail.nju.edu.cn,
{huaxiongli,clchen}@nju.edu.cn

Abstract. Few-shot image classification aims to recognize unseen categories with only a few labeled training samples. Recent metric-based approaches tend to represent each sample with a high-level semantic representation and make decisions according to the similarities between the query sample and support categories. However, high-level concepts are identified to be poor at generalizing to novel concepts that differ from previous seen concepts due to domain shifts. Moreover, most existing methods conduct one-way instance-level metric without involving more discriminative local relations. In this paper, we propose a *Local Mutual Metric Network (LM2N)*, which combines low-level structural representations with high-level semantic representations by unifying all abstraction levels of the embedding network to achieve a balance between discrimination and generalization ability. We also propose a novel local mutual metric strategy to collect and reweight local relations in a bidirectional manner. Extensive experiments on five benchmark datasets (i.e. *mini*ImageNet, *tiered*ImageNet and three fine-grained datasets) show the superiority of our proposed method.

Keywords: Few-shot learning · Metric learning · Attention · Local representation

1 Introduction

Despite that deep learning [10,18] has made tremendous advances in many machine learning tasks [37], its data-driven nature restricts its performance and efficiency in practical applications where collecting enough samples and correctly labeling them are expensive. Therefore, Few-shot Learning (FSL) [7,8,16] is widely studied recently, which aims to generalize to novel unseen categories with scarce data after training on seen categories with sufficient data. Few-shot Image Classification (FSIC) [12,23] is one of the well-studied fields in FSL, which aims at making classification on unseen categories with a small number of training samples.

© Springer Nature Switzerland AG 2021
H. Ma et al. (Eds.): PRCV 2021, LNCS 13019, pp. 443–454, 2021.
https://doi.org/10.1007/978-3-030-88004-0_36

Recently, many meta-learning based approaches have been proposed to solve the FSIC problem, which can be divided into two main streams: optimization-based approaches [8,31] and metric-based approaches [20,29]. Specifically, approaches based on optimization target at learning a suitable initialization or updating strategy for the base model, which helps the base model to converge on novel tasks with a few steps of gradient descent. Metric-based approaches aim to learn a transferable feature space, in which the homogeneous samples are close to each other while the heterogeneous samples are far away and the query samples are categorized by measuring their similarities with the labeled support categories.

Metric-based methods [9,21,22,30,32,34] have attracted a lot of attention due to their efficiency and simplicity on solving the FSIC problem. However, there are still some limitations remain to be solved.

First, most metric-based approaches adopt the high-level semantic representation [3,30,32] to describe each sample. Although high-level representations contain rich semantic information with high degree of discrimination, they are not shared between different categories compared with low-level features [19]. Meanwhile, recent metric-based approaches mainly follow the paradigm of meta-learning, which is inherently designed to enhance the generalization ability of the base model by learning inductive bias (meta knowledge) [11]. However, several recent works [5,33] show that the classic pretraining combined with fine tuning beats the advanced meta-learning methods, which indicates that generalizing just by meta knowledge is far from enough. Second, recent methods mainly conduct instance-level metric [22,30,32] while ignore the local relations between two feature maps, which can not reveal the true similarities and is obviously inefficient in practice.

To solve the above limitations, in this paper, we first propose a local representation fusion strategy to combine low-level representations with high-level representations. We further propose a non-parametric Local Mutual Metric Module (LM3), which comprehensively compares two feature maps by collecting and reweighting their local relations.

Our main contributions are as follows:

- We propose a representation fusion strategy to make use of both low-level and high-level features by a learned fusion layer, which aims to achieve a balance between discrimination and generalization ability.
- We design a novel local mutual metric strategy for the FSIC problem, which explores the pairwise local relations between two feature maps in a bidirectional manner.
- We adopt a Convolutional Block Attention Module (CBAM) [36] to select and highlight discriminative semantic regions.

2   Related Works

Recent meta-learning based FSIC approaches can be broadly divided into two branches: optimization-based and metric-based approaches.

2.1   Optimization-Based Approaches

Optimization-based approaches [2,8,31] utilize the meta learner to learn optimal initialization parameters for the base learner, which is sensitive to novel samples so that the base model can fast adapt to unseen categories with only a few steps of gradient descent. MAML [8] is a typical optimization-based method for the FSIC, which concludes parameterized meta knowledge from a series of episodes (tasks) to boost model-agnostic generalization ability. Later optimization-based approaches mainly follow the ideology of MAML.

However, optimization-based approaches suffer from the computation of high-order gradient [25], which is out of question in our metric-based LM2N.

2.2   Metric-Based Approaches

Metric-based approaches [4,9,21,22,30,32,34] aim to learn a transferable embedding space (feature extractor), in which to map labeled support samples and unlabeled query samples. Final classification is completed by comparing the distances between query samples and support categories. Obviously, the key points of metric-based approaches are transferable feature representations and an effective metric strategy.

Specifically, Matching Network [34] utilizes a LSTM for feature embedding and compares the cosine similarities between feature maps for classification. Note that the Matching Network also proposes the widely used episodic training mechanism. Prototypical Network [30] utilizes the center of the support class as the class representation (prototype) and makes classification by computing the Euclidean distances between query samples and class prototypes. Relation Network [32] follows the representation strategy of [30] and proposes a novel MLP-based metric, which aims to learn a transferable and effective nonlinear metric function.

Unlike the above approaches, CovaMNet [22] represents images by local representations, which treats a feature map (a $h \times w \times d$ tensor) as $h \times w$ d-dimension local representation vectors. CovaMNet makes classification according to the distribution consistency between query samples and support categories. DN4 [21] and SAML [9] also follow the local representation strategy.

Based on the local representation mechanism, our LM2N unifies low-level and high-level features for a fused local representation. Unlike the existing instance-level metric, we propose a novel local mutual metric strategy, which explores the local relations between two feature maps bilaterally to reveal their local correlations comprehensively. Note that some methods have explored the local metric in FSIC, i.e. DN4 [21] and SAML [9]. DN4 utilizes a unidirectional Nearest Neighbor searching to acquire the local relations of two feature maps and makes classification according to the summation of the pairwise local similarities, which shows promising performance. SAML uses a relation matrix to collect and select useful pairwise local vectors, followed by a MLP-based nonlinear metric network [32].

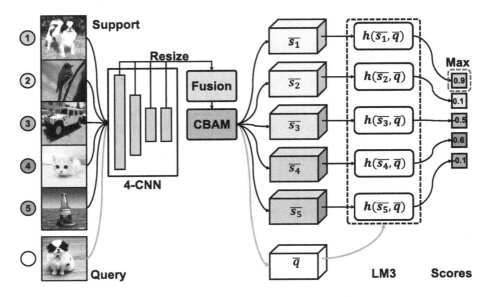

Fig. 1. The overview of our proposed LM2N under the 5-way 1-shot setting.

3 Methodology

3.1 Problem Formulation

In this paper, we follow the classic episodic training [34] mechanism. Specifically, we utilize a series of episodes (tasks) on the training set to train the model and evaluate the model with a series of episodes on the test set.

In each episode, we randomly sample a support set with N categories (K samples per category) $\mathcal{S} = \{\{s_1^1, y_1\}, \{s_1^2, y_1\}, \cdots, \{s_N^K, y_N\}\}$ and an unlabeled query set with NM samples $\mathcal{Q} = \{q_1, q_2, \cdots, q_{NM}\}$. Notice that $\mathcal{S}$ and $\mathcal{Q}$ share the same label space. Finally, each episode can be viewed as a N-way K-shot task, which aims to categorize the NM query samples into the N categories with only K labeled samples per category.

3.2 Overview

The overview of our proposed LM2N is shown in Fig. 1. First, we feed the support set (5-way 1-shot setting) and a query sample into the embedding network $\mathcal{F}_\theta$. Then, a fusion layer $\mathcal{I}_\phi$ is applied to fuse the outputs of all layers of the feature extractor, which automatically learns the weight of each layer to balance the discrimination and generalization. After the representation fusion, an attention module $\mathcal{A}_\omega$ based on the CBAM [36] is utilized to highlight the semantic regions. Finally, we measure the similarities between the query feature map with five support categories through the LM3 and obtain the final similarity scores.

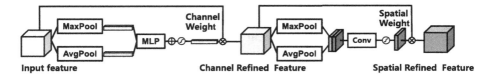

Fig. 2. The structure of the CBAM.

3.3 Local Representation Fusion Layer

Representation fusion has achieved promising performance in object detection and image segmentation [24,38] while it is rarely used in the field of FSIC. In this paper, we introduce a representation fusion strategy into FSIC.

Specifically, given a n-layer embedding network, we aim to unify the outputs of these n layers (n representation maps). In practice, there is no enough prior knowledge to give appropriate weights to these n feature maps. Therefore, a fusion layer $\mathcal{I}_\phi$ is utilized to learn the fusion weights under the meta-learning framework. In practice, we resize these posterior feature maps to the same size and combine them by concatenation before putting into the fusion layer. After the representation fusion, we can obtain a balanced feature map:

$$\hat{x} = \mathcal{I}_\phi(\mathcal{F}_\theta(x)) \in \mathbb{R}^{h \times w \times d}. \tag{1}$$

Where $\mathcal{F}_\theta(x)$ denotes the outputs of the meaningful layers of the backbone $\mathcal{F}_\theta$. Following [9], we view the 3D vector $\hat{x}$ as $h \times w$ d-dimension local vectors: $\hat{x} = \{v_1, v_2, \cdots, v_{hw}\}$ and each local vector v_i represents the information of the corresponding region of the feature map.

3.4 Attention Module

To help the model capture the semantic regions, we adopt the widely adopted attention mechanism CBAM [36,40] in our model (see in Fig. 2).

Generally, CBAM consists of two parts, i.e. channel attention and spatial attention. Specifically, channel attention first squeezes the feature map along the spatial dimension by global average pooling and global max pooling to obtain two weight tensors. Then the two weight tensors are combined to generate the channel attention weight, which is further used to refine the feature map. Similar to the channel attention, spatial attention squeezes the feature map along the channel dimension and generates the spatial attention weight, which expands along the channel dimension to refine the feature.

Through the attention module $\mathcal{A}_\omega$, the semantic objects are effectively highlighted, which contributes to the further metric (see Table 3).

$$\bar{x} = \mathcal{A}_\omega(\hat{x}). \tag{2}$$

3.5   Local Mutual Metric Module (LM3)

In this paper, we propose a novel local mutual metric strategy to explore the local relations between two feature maps in a bidirectional manner. Given a support class $S_n = \{s_n^1, s_n^2, \cdots, s_n^K\}$ and a query instance q, following the above procedures, we adopt the central class strategy [30] here for concise expression:

$$\bar{s}_n = \frac{1}{|S_n|} \sum_{(s_n^i, y_n) \in S_n} \mathcal{A}_\omega(\mathcal{I}_\phi(\mathcal{F}_\theta(s_n^i))). \tag{3}$$

We rewrite them by local representations:

$$\bar{s}_n = \{t_1, t_2, \cdots, t_{hw}\}, \bar{q} = \{l_1, l_2, \cdots, l_{hw}\}. \tag{4}$$

Then a LM3 is applied to calculate the similarity between $\bar{s}_n$ and $\bar{q}$. We first define a local compare function $g(,)$ [9] to calculate the similarity between two local vectors. Mathematically, the local mutual metric function is defined as:

$$h(\bar{s}_n, \bar{q}) = \sum_{i=1}^{hw} \sum_{j=1}^{hw} \alpha_{i,j} g(t_i, l_j) + \sum_{j=1}^{hw} \sum_{i=1}^{hw} \beta_{i,j} g(t_i, l_j), \tag{5}$$

$$\alpha_{i,j} = \frac{e^{[g(t_i, l_j)/T]}}{\sum_{k=1}^{hw} e^{[g(t_i, l_k)/T]}}, \beta_{i,j} = \frac{e^{[g(t_i, l_j)/T]}}{\sum_{k=1}^{hw} e^{[g(t_k, l_j)/T]}}. \tag{6}$$

Here, T denotes the influence coefficient of the similarity, which gives the similar local vector pairs much higher weights while gives the dissimilar local vector pairs much lower weights. The value of the local mutual metric function $h(,)$ represents the similarity score between the query sample q and the n-th support class S_n.

Given an episode, there are N categories in the support set $\{S_1, S_2, \cdots, S_N\}$ and NM unlabeled samples in the query set $\{q_1, q_2, \cdots, q_{NM}\}$. Therefore, for each query sample there are N similarity scores involving N support classes. We utilize the softmax function to compute the probabilities and the final loss function of this episode is defined as:

$$Loss = -\frac{1}{NM} \sum_{i=1}^{NM} \sum_{j=1}^{N} \varphi(\hat{y}_i = y_j) log \frac{e^{h(\bar{s}_j, \bar{q}_i)}}{\sum_{k=1}^{N} e^{h(\bar{s}_k, \bar{q}_i)}}, \tag{7}$$

in which $\hat{y}_i$ represents the predicted label for q_i and $\varphi()$ is a flag function that equals one if its argument is true and zero otherwise.

4   Experiments

4.1   Datasets

We evaluate our method on five benchmark datasets, and all images are resized to 84×84 in our model.

*mini*ImageNet. *mini*ImageNet [34] is sampled from the ImageNet [6], which consists of 100 categories and 600 samples per category. We split it into 64, 16 and 20 categories for training, validation and test by convention, respectively [22].

*tiered*ImageNet. *tiered*ImageNet [27] is also a subset of the ImageNet [6]. There are 608 categories and 779, 165 images in it. We split the dataset into 351, 97 and 160 categories for training, validation and test, respectively.

Stanford Dogs, Stanford Cars and CUB-200-2011. We utilize three fine-grained datasets to evaluate the fine-grained FSIC performance of our method. For *Stanford Dogs* [14], we divide it into 70, 20 and 30 classes. For *Stanford Cars* [17], we split it into 130, 17 and 49 categories while for *CUB-200-2011* [35], we choose 100 classes for training and both 50 classes for validation and test [22].

4.2 Network Architecture

In our model, the main network architecture consists of the embedding network, the representation fusion layer and the attention module.

Specifically, the embedding network is a 4-layer CNN [21,22] with four convolutional blocks and each convolutional block contains a convolutional layer with 64 3×3 filters, a batch normalization layer and a Leaky ReLU layer. In addition, we add 2×2 max pooling after the first two convolutional blocks. The representation fusion layer is actually a 1×1 convolutional layer.

The attention module consists of a 2-layer MLP to learn the channel attention weight and a 1×1 convolutional layer to learn the spatial attention weight [40].

4.3 Experimental Settings

We implement our model under the framework of Pytorch [26]. Specifically, under the setting of the meta-learning (episodic training), we train, validate and test our model on a series of N-way K-shot episodes (tasks) randomly sampled from the training set, validation set and test set respectively. Specifically, for the end-to-end training stage, we randomly construct 200, 000 episodes from the training set and 20, 000 episodes from the validation set. When $N = 5, K = 1$, a 5-class episode contains 1 support sample and 10 query samples per calss; when $N = 5, K = 5$, a 5-class episode contains 5 support samples and 10 query samples per class.

Notice that we initialize the learning rate to 0.005 and halve it per 40, 000 episodes. A widely adopted Adam [15] optimizer is applied to optimize our model during the training procedure. During the test stage, we randomly construct 600 episodes to evaluate the performance of our model on novel concepts.

4.4 Comparisons with Other Methods

Experiments on the Routine FSIC Datasets. We compare our LM2N with both optimization-based and metric-based SOTA methods on *mini*ImageNet and

Table 1. Comparisons with the state-of-the-art methods on *mini*ImageNet and *tiered*ImageNet with 95% confidence intervals. The best and the second best results of each column are shown in red and blue respectively.

Model	Type	*mini*ImageNet		*tiered*ImageNet	
		1-shot(%)	5-shot(%)	1-shot(%)	5-shot(%)
MAML [8]	Optimization	48.70 ± 1.84	63.11 ± 0.92	51.67 ± 1.81	70.30 ± 1.75
MTL [31]	Optimization	45.60 ± 1.84	61.20 ± 0.90	–	–
MAML++ [2]	Optimization	52.15 ± 0.26	68.32 ± 0.44	–	–
MatchingNet [34]	Metric	43.56 ± 0.84	55.31 ± 0.73	–	–
IMP [1]	Metric	49.60 ± 0.80	68.10 ± 0.80	–	–
SAML [9]	Metric	52.64 ± 0.56	67.32 ± 0.75	–	–
DSN [29]	Metric	51.78 ± 0.96	68.99 ± 0.69	–	–
TNet [39]	Metric	52.39	67.89	–	–
GNN [28]	Metric	50.33 ± 0.36	66.41 ± 0.63	–	–
ProtoNet [30]	Metric	49.42 ± 0.78	68.20 ± 0.66	48.58 ± 0.87	69.57 ± 0.75
RelationNet [32]	Metric	50.44 ± 0.82	65.32 ± 0.70	54.48 ± 0.93	71.31 ± 0.78
CovaMNet [22]	Metric	51.19 ± 0.76	67.65 ± 0.63	54.98 ± 0.90	71.51 ± 0.75
DN4 [21]	Metric	51.24 ± 0.74	71.02 ± 0.64	53.37 ± 0.86	74.45 ± 0.70
LM2N(Ours)	Metric	53.22 ± 0.76	71.98 ± 0.91	54.66 ± 0.79	73.31 ± 0.85

Table 2. Experiments on three fine-grained datasets. The best and the second best results of each column are shown in red and blue respectively.

Model	5-Way accuracy(%)					
	Stanford dogs		Stanford Cars		CUB-200-2011	
	1-shot	5-shot	1-shot	5-shot	1-shot	5-shot
MatchingNet [34]	35.80	47.50	34.80	44.70	61.16	72.86
ProtoNet [30]	37.59	48.19	40.90	52.93	51.31	70.77
MAML [8]	45.81	60.01	48.17	61.85	55.92	72.09
RelationNet [32]	44.49	56.35	48.59	60.98	62.45	76.11
CovaMNet [22]	49.10	63.04	56.65	71.33	60.58	74.24
DN4 [21]	45.41	63.51	59.84	88.65	52.79	81.45
LRPABN [13]	46.17	59.11	56.31	70.23	63.63	76.06
LM2N(Ours)	51.15	67.63	59.98	84.65	64.95	79.82

*tiered*ImageNet. As is shown in Table 1, our LM2N achieves promising performance on both datasets under both settings. Specifically, on the *mini*ImageNet, our LM2N achieves the best results under both 1-shot and 5-shot settings with 0.58% and 0.96% improvements respectively. Compared with our base model Prototypical Network, our method achieves 3.80% and 3.78% improvements. On the *tiered*ImageNet, our LM2N also achieves competitive performance compared with the other methods.

The outstanding performances indicate the effectiveness of our method, which explores the mutual local relations between the query samples and the support categories. Most previous methods compare the query samples and the support

Table 3. Ablation study on *mini*ImageNet.

Ablation models			5-way accuracy (%)	
Fusion	Attention	LM3	1-shot	5-shot
			49.24	68.70
	✓		50.42	68.96
✓			50.16	68.03
		✓	51.63	70.13
✓	✓		50.91	69.65
	✓	✓	52.46	71.01
✓		✓	51.93	71.15
✓	✓	✓	53.22	71.98

categories on the instance-level while DN4 conducts a local metric by searching out the local nearest neighbors. Unlike these methods, our L2MN takes a mutual view to collect local relations and takes a whole picture of the feature maps. Besides, the representation fusion strategy enhances the generalization ability to novel concepts and the attention mechanism highlights the discriminative features.

Experiments on the Fine-Grained FSIC Datasets. We also compare our method with the SOTA method on three fine-grained FSIC datasets. As is shown in Table 2, LM2N achieves competitive performances. Specifically, our LM2N achieve the best performances on four columns with 0.14%–4.12% improvement while on the other two columns LM2N both achieves the second best results.

Our LM2N is naturally suitable for fine-grained problems. For one thing, LM2N utilizes the local metric strategy to explore the local relations between two feature maps, for another, the attention mechanism helps to highlight the discriminative local regions, which is obviously effective on fine-grained FSIC.

5   Discussion

5.1   Ablation Study

In this paper we introduce the representation fusion strategy to FSIC tasks and propose a novel local mutual metric mechanism. Moreover, we adopt CBAM as the attention module to highlight the discriminative local regions [40]. In general, these three parts achieve promising performance together while we also want to know their respective contributions to the model. Therefore, we design the ablation study to reveal the contributions of each part. Specifically, we step-by-step remove one or several parts and make comparisons with the original model (see Table 3). Experimental results show that each component more or less contributes to the model.

5.2   Cross-Domain FSIC Analysis

To further evaluate the robustness of our LM2N when facing severe domain-shift, we conduct cross-domain experiments by training on the *mini*ImageNet and testing on the *CUB-200-2011* [5]. Experimental results are shown in Table 4. From the table we can see that our LM2N shows better generalization ability than most previous methods when facing large domain shift.

Table 4. Cross-domain performance (*mini***ImageNet**→ *CUB*). We adopt the results of the Resnet-18 reported by [5].

Model	Embedding	5-way 5-shot(%)
Baseline [5]	Resnet-18	65.57 ± 0.70
MatchingNet [34]	Resnet-18	53.07 ± 0.74
ProtoNet [30]	Resnet-18	62.02 ± 0.70
MAML [8]	Resnet-18	51.34 ± 0.72
RelationNet [32]	Resnet-18	57.71 ± 0.73
CovaMNet [22]	Conv-64F	63.21 ± 0.68
DN4 [21]	Conv-64F	63.42 ± 0.70
QPN (Ours)	Conv-64F	64.40 ± 0.79

Table 5. The performance with different similarity functions on *mini*ImageNet.

Metric	5-way 1-shot	5-way 5-shot
Gaussian	51.88 ± 0.72	70.83 ± 0.85
Euclidean	51.08 ± 0.87	70.63 ± 0.82
Cosine	53.22 ± 0.76	71.98 ± 0.91

5.3   Influence of Different Local Compare Functions

In the local mutual metric module, we utilize the local compare function $g(,)$ to measure the similarity between two local vectors. Certainly, the local metric function has many choices, e.g. cosine similarity, Euclidean similarity and Gaussian similarity [9].

Cosine similarity:

$$g(a,b) = \frac{a^T b}{||a|| \cdot ||b||}. \tag{8}$$

Euclidean similarity:

$$g(a,b) = \frac{1}{e^{||a-b||^2}}. \tag{9}$$

Gaussian similarity:

$$g(a,b) = e^{a \cdot b}. \tag{10}$$

Obviously, as is shown in Table 5, cosine similarity achieves the best performance under both settings.

6   Conclusion

In this paper, we propose a Local Mutual Metric Network for the FSIC. We propose a novel metric strategy LM3 to explore the local relations between two feature maps in a bidirectional manner for a more effective metric. We also achieves a balance between discrimination and generalization ability by multi-level representation fusion. An attention module is adopted to highlight the semantic objects. Experimental results on five FSIC benchmark datasets show the superiority of our method. Ablation study and the experiments on the cross-domain FSIC task demonstrate the effectiveness and the robustness of our method.

Acknowledgement. This work was partially supported by National Natural Science Foundation of China (Nos. 62176116, 71732003, 62073160, and 71671086) and the National Key Research and Development Program of China (Nos. 2018YFB1402600).

References

1. Allen, K., Shelhamer, E., Shin, H., Tenenbaum, J.: Infinite mixture prototypes for few-shot learning. In: ICML, pp. 232–241 (2019)
2. Antoniou, A., Edwards, H., Storkey, A.: How to train your maml. In: ICLR (2018)
3. Chen, H., Li, H., Li, Y., Chen, C.: Multi-scale adaptive task attention network for few-shot learning. arXiv preprint arXiv:2011.14479 (2020)
4. Chen, H., Li, H., Li, Y., Chen, C.: Multi-level metric learning for few-shot image recognition. arXiv preprint arXiv:2103.11383 (2021)
5. Chen, W.Y., Liu, Y.C., Kira, Z., Wang, Y.C.F., Huang, J.B.: A closer look at few-shot classification. In: ICLR (2019)
6. Deng, J., Dong, W., Socher, R., Li, L.J., Li, K., Fei-Fei, L.: Imagenet: a large-scale hierarchical image database. In: CVPR, pp. 248–255 (2009)
7. Fe-Fei, L., et al.: A bayesian approach to unsupervised one-shot learning of object categories. In: ICCV, pp. 1134–1141 (2003)
8. Finn, C., Abbeel, P., Levine, S.: Model-agnostic meta-learning for fast adaptation of deep networks. In: ICML, pp. 1126–1135 (2017)
9. Hao, F., He, F., Cheng, J., Wang, L., Cao, J., Tao, D.: Collect and select: semantic alignment metric learning for few-shot learning. In: ICCV, pp. 8460–8469 (2019)
10. He, K., Zhang, X., Ren, S., Sun, J.: Deep residual learning for image recognition. In: CVPR, pp. 770–778 (2016)
11. Hochreiter, S., Younger, A.S., Conwell, P.R.: Learning to learn using gradient descent. In: International Conference on Artificial Neural Networks, pp. 87–94 (2001)
12. Hou, R., Chang, H., Ma, B., Shan, S., Chen, X.: Cross attention network for few-shot classification. In: NeurIPS, pp. 4003–4014 (2019)
13. Huang, H., Zhang, J., Zhang, J., Xu, J., Wu, Q.: Low-rank pairwise alignment bilinear network for few-shot fine-grained image classification. IEEE Trans. Multimedia **23**, 1666–1680 (2020)
14. Khosla, A., Jayadevaprakash, N., Yao, B., Li, F.F.: Novel dataset for fine-grained image categorization: stanford dogs. In: CVPR Workshop on FGVC, vol. 2 (2011)
15. Kingma, D.P., Ba, J.: Adam: a method for stochastic optimization. In: ICLR (2015)
16. Koch, G., Zemel, R., Salakhutdinov, R., et al.: Siamese neural networks for one-shot image recognition. In: ICML Deep Learning Workshop, vol. 2 (2015)

17. Krause, J., Stark, M., Deng, J., Fei-Fei, L.: 3D object representations for fine-grained categorization. In: ICCV Workshop, pp. 554–561 (2013)
18. Krizhevsky, A., Sutskever, I., Hinton, G.E.: Imagenet classification with deep convolutional neural networks. In: NeurIPS, vol. 25, pp. 1097–1105 (2012)
19. LeCun, Y., Bengio, Y., Hinton, G.: Deep learning. Nature **521**(7553), 436–444 (2015)
20. Li, W., Wang, L., Huo, J., Shi, Y., Gao, Y., Luo, J.: Asymmetric distribution measure for few-shot learning, pp. 2957–2963 (2020)
21. Li, W., Wang, L., Xu, J., Huo, J., Gao, Y., Luo, J.: Revisiting local descriptor based image-to-class measure for few-shot learning. In: CVPR, pp. 7260–7268 (2019)
22. Li, W., Xu, J., Huo, J., Wang, L., Gao, Y., Luo, J.: Distribution consistency based covariance metric networks for few-shot learning. In: AAAI, pp. 8642–8649 (2019)
23. Li, Y., Li, H., Chen, H., Chen, C.: Hierarchical representation based query-specific prototypical network for few-shot image classification. arXiv preprint arXiv:2103.11384 (2021)
24. Lin, T.Y., Dollár, P., Girshick, R., He, K., Hariharan, B., Belongie, S.: Feature pyramid networks for object detection. In: CVPR, pp. 2117–2125 (2017)
25. Mishra, N., Rohaninejad, M., Chen, X., Abbeel, P.: A simple neural attentive meta-learner. In: ICLR (2018)
26. Paszke, A., et al.: Pytorch: an imperative style, high-performance deep learning library. In: NeurIPS, vol. 32, pp. 8026–8037 (2019)
27. Ren, M., et al.: Meta-learning for semi-supervised few-shot classification. In: ICLR (2018)
28. Satorras, V.G., Estrach, J.B.: Few-shot learning with graph neural networks. In: ICLR (2018)
29. Simon, C., Koniusz, P., Nock, R., Harandi, M.: Adaptive subspaces for few-shot learning. In: CVPR, pp. 4136–4145 (2020)
30. Snell, J., Swersky, K., Zemel, R.: Prototypical networks for few-shot learning. In: NeurIPS, pp. 4080–4090 (2017)
31. Sun, Q., Liu, Y., Chua, T.S., Schiele, B.: Meta-transfer learning for few-shot learning. In: CVPR, pp. 403–412 (2019)
32. Sung, F., Yang, Y., Zhang, L., Xiang, T., Torr, P.H., Hospedales, T.M.: Learning to compare: relation network for few-shot learning. In: CVPR, pp. 1199–1208 (2018)
33. Tian, Y., Wang, Y., Krishnan, D., Tenenbaum, J.B., Isola, P.: Rethinking few-shot image classification: a good embedding is all you need? In: ECCV, pp. 266–282 (2020)
34. Vinyals, O., Blundell, C., Lillicrap, T., Kavukcuoglu, K., Wierstra, D.: Matching networks for one shot learning. In: NeurIPS, pp. 3637–3645 (2016)
35. Wah, C., Branson, S., Welinder, P., Perona, P., Belongie, S.: The caltech-ucsd birds-200-2011 dataset (2011)
36. Woo, S., Park, J., Lee, J.Y., Kweon, I.S.: Cbam: convolutional block attention module. In: ECCV, pp. 3–19 (2018)
37. Zhang, C., Li, H., Chen, C., Qian, Y., Zhou, X.: Enhanced group sparse regularized nonconvex regression for face recognition. IEEE Trans. Pattern Anal. Mach. Intell. (2020). https://doi.org/10.1109/TPAMI.2020.3033994
38. Zhao, H., Shi, J., Qi, X., Wang, X., Jia, J.: Pyramid scene parsing network. In: CVPR, pp. 2881–2890 (2017)
39. Zhu, W., Li, W., Liao, H., Luo, J.: Temperature network for few-shot learning with distribution-aware large-margin metric. Pattern Recogn. **112**, 107797 (2021)
40. Zhu, Y., Liu, C., Jiang, S.: Multi-attention meta learning for few-shot fine-grained image recognition. In: IJCAI, pp. 1090–1096 (2020)

SimplePose V2: Greedy Offset-Guided Keypoint Grouping for Human Pose Estimation

Jia Li[1], Linhua Xiang[1], Jiwei Chen[1,2], and Zengfu Wang[1,2(✉)]

[1] Department of Automation, University of Science and Technology of China, Hefei, China
{jialee,xlh1995,cjwbdw6}@mail.ustc.edu.cn,zfwang@ustc.edu.cn
[2] Institute of Intelligent Machines, Chinese Academy of Sciences, Hefei, China

Abstract. We propose a simple yet reliable bottom-up approach with a good trade-off between accuracy and efficiency for the problem of multi-person pose estimation. To encode the multi-person pose information in the image, we employ Gaussian response heatmaps to encode the keypoint position information of all persons. And we propose a set of guiding offsets to encode the pairing information between keypoints belonging to the same individuals. We use an Hourglass Network to infer the said heatmaps and guiding offsets simultaneously. During testing, we greedily assign the detected keypoints to different individuals according to the guiding offsets. Besides, we introduce a peak regularization into the pixel-wise L_2 loss for keypoint heatmap regression, improving the precision of keypoint localization. Experiments validate the effectiveness of the introduced components. Our approach is comparable to the state of the art on the challenging COCO dataset under fair conditions.

Keywords: Bottom-up · Human pose · Guiding offset · Heatmap

1 Introduction

The problem of multi-person pose estimation is to localize the 2D skeleton keypoints (body joints) for all persons given a single image [9,15,17]. It is a fundamental task with many applications. Some recent work can address the problem of single-person pose estimation extremely well thanks to the development of convolutional neural networks (CNNs) specially designed for human pose estimation such as Hourglass Network [8,9,13] and HRNet [18]. But the challenge becomes much more difficult when multiple persons appear in the scene at the same time. And it has not been solved well so far considering accuracy, speed and approach simplicity.

Student paper.

© Springer Nature Switzerland AG 2021
H. Ma et al. (Eds.): PRCV 2021, LNCS 13019, pp. 455–467, 2021.
https://doi.org/10.1007/978-3-030-88004-0_37

Approach Taxonomy. Existing approaches for this problem can be roughly divided into two categories: ***top-down*** and ***bottom-up***. Some recent top-down approaches [2,4,16,18,21] have achieved high accuracy. However, most of them have low prediction efficiency especially when many people exist in the scene, for they must rely on an advanced person detector to detect all the persons and estimate all the single-person poses one by one. The bottom-up approaches [1,3,7,9,12] instead infer the keypoint positions and corresponding keypoint grouping cues of all persons indiscriminately using a feed-forward network (here, we mainly discuss CNN-based approaches) and then group the detected keypoints into individual human poses. The run time of the grouping process may be very fast and nearly constant regardless of the person number in the image. Therefore, developing bottom-up approaches is challenging but attractive for a better trade-off between accuracy and efficiency. In this paper, we focus on the bottom-up approaches.

Keypoint Coordinate Encoding. Here, we only review the most important work. With the arrival of the deep learning era, DeepPose [20] for the first time uses CNNs to regress the Cartesian coordinates of a fixed number of person keypoints directly. As a result, this approach cannot handle the situation of multiple persons. PersonLab [15], CenterNet [22] and PifPaf [7] decompose the task of keypoint localization into two subproblems at each position: binary classification to discriminate whether this current pixel is a keypoint, and offset regression to the ground-truth keypoint position. However, special techniques are essential to make their approaches work well, e.g., leveraging focal loss [10] for the classification task and Laplace-based L_1 loss [6] for the offset regression task [7]. By contrast, Tompson et al. [19] and many later researchers use CNNs to predict the Gaussian response heatmaps of person keypoints and then obtain the keypoint coordinates by finding the local maximum responses in the heatmaps. But high-res feature maps are required to alleviate the precision decline of keypoint localization. Zhang et al. [21] design a novel distribution-aware coordinate representation of keypoint (DARK) for the top-down approaches.

Keypoint Grouping Encoding. The encoding of keypoint grouping (or association) information is critical for the post-processing in bottom-up approaches. Here, we conclude the existing approaches for keypoint grouping into two categories: ***global grouping*** and ***greedy grouping***. Prior work, such as DeeperCut [5], Associative Embedding [12] and CenterNet [22], presents different approaches to encode the global keypoint grouping information. For instance, Associative Embedding employs Hourglass Network [13] to infer the identity tags for all detected keypoints. These tags are grouping cues to cluster keypoints into individual human poses. And CenterNet [22] proposes center offsets away from person centers as keypoint grouping cues. By contrast, Part Affinity Fields [1], PersonLab [15] and PIFPAF [7] are greedy grouping approaches, associating adjacent nodes in a human skeleton tree independently rather than finding the global solution of a global graph matching problem.

Some state-of-the-art (SOTA) bottom-up approaches are remarkable in estimation accuracy but have so complicated structure and many hyperparameters affecting results that we cannot clearly figure out the contributions of introduced components or compare these approaches equally. In this work, we propose a bottom-up approach, which is easy to follow and works well in terms of accuracy, speed and clarity. We use Hourglass-104 [8] as the inference model and select CenterNet [22] as the baseline approach. In this paper: (1) we present a greedy keypoint grouping method, which we refer to as *greedy offset-guided keypoint grouping* (GOG). The adjacent keypoints of each person are associated by a set of guiding offsets; (2) we rethink the existing Gaussian heatmap encoding-decoding method for keypoint coordinates and present a peak regularization for the heatmap regression task to improve the precision of keypoint localization.

2   METHOD

2.1   Preliminary

Keypoint grouping information is essential for multi-person pose estimation. Here, we propose a novel grouping encoding for person keypoints as the form of **guiding offsets** to greedily "connect" the adjacent keypoints belonging to the same persons. A guiding offset illustrated in Fig. 1 is the displacement vector starting at a keypoint J_{from} of a person and pointing to his/her next keypoint J_{to}. An overview of our approach is shown in Fig. 2, which is mainly inspired by SimplePose [9] and CenterNet [22] but different from them or other bottom-up approaches [3,7,12,15] based on offsets.

PersonLab [15] and PIFPAF [7] detect an arbitrary root keypoint of a person and use bidirectional offsets to generate a tree-structured kinematic graph of that person. Merely the root keypoint coordinate is obtained in the confidence maps of classification while the subsequent keypoint coordinate is regressed directly on the basis of the offset fields and the former keypoint location. Abundant root keypoints are detected as the seeds of the kinematic graphs to ensure they can detect as many human poses as possible. Hence, they have to apply non-maximum suppression (NMS) to remove redundant person poses. CenterNet [22] detects the center point of each person instance as the root "node" and localizes keypoints using classification heatmaps and refinement offsets. CenterNet only associates the center point with the keypoints which are within the same inferred person bounding box according to center offsets.

However, both precise offset regression to keypoint locations and localization of person center points are vague visual tasks. Previous CNN-based work such as DeepPose [20] and Tompson et al. [19] has already suggested that inferring keypoint heatmaps is easier than inferring keypoint Cartesian coordinates. The vague ground-truth keypoint of a person is different from his/her bounding box or segmentation boundary with clear visual concepts. As a result, we localize all keypoints using Gaussian response heatmaps. As for person keypoint grouping, we infer a set of guiding offsets at the positions close to every keypoint pointing to its adjacent keypoint.

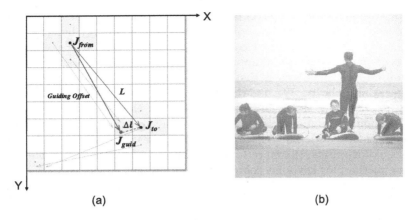

(a)　　　　　　　　　　(b)

Fig. 1. Definition of guiding offset. (a) The ground-truth guiding offset placed at the position of keypoint J_{from} points to its adjacent keypoint J_{to}. Note that J_{from} and J_{to} are in different heatmap channels, the figure is just for description. In practice, the inferred guiding offset points to the floating-point position J_{guid}. (b) The inferred guiding offsets around the "right hip" keypoints of all persons. They guide to the corresponding "right ankle" keypoints of the same individuals.

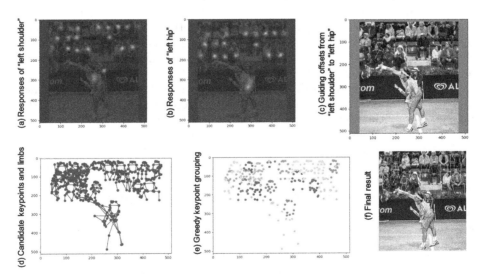

Fig. 2. Overview of the proposed approach. (a), (b) and (c): we infer the keypoints and the guiding offsets simultaneously using Hourglass-104 [8,13,22]. There are 17 types of keypoint heatmaps and 19 types of guiding-offset feature maps. (d): next, the top k ($k = 32$ in all our experiments) scoring keypoints and limbs (connections of paired keypoints) of each type and with high confidence are collected. (e): subsequently, they are grouped into individual human poses greedily using our GOG algorithm. (f): finally, we measure the confidence of each human pose by averaging the Gaussian response values of its keypoints and filter out the human poses with low confidence.

2.2   Definition of Gaussian Responses

We use the most popular coordinate encoding-decoding method for keypoint localization, i.e., Gaussian response heatmaps. Assuming $I \in \mathbb{R}^{W \times H \times 3}$ is the input image of width W and height H, the ground-truth keypoint heatmaps for the network output are $S^* = (S_1^*, S_2^*, \cdots, S_C^*) \in [0,1]^{\frac{W}{R} \times \frac{H}{R} \times C}$, in which R is the network output stride and C is the number of person keypoint types. In our approach, $R = 4$ for Hourglass Network [8,9,13,22] and $C = 17$ for COCO dataset [11]. Considering a pixel value at $g(u, v, j)$ in the j-th ground-truth keypoint heatmap, before generating the ground truth at g, we map g to the position $\tilde{g}(x, y, j)$ in the input space $\mathbb{R}^{W \times H \times C}$ using this [1,9] transformation:

$$\tilde{g}(x, y, j) = g(u \cdot R + R/2 - 0.5, v \cdot R + R/2 - 0.5, j). \tag{1}$$

If the ground-truth keypoint at the position $\tilde{p}$ is the nearest keypoint away from $\tilde{g}$, we generate the ground-truth Gaussian response value at $g(u, v, j)$ as:

$$S^*(u, v, j) = \exp\left[-\frac{(\tilde{g}_x - \tilde{p}_x)^2 + (\tilde{g}_y - \tilde{p}_y)^2}{2\sigma_k^2} \right], \tag{2}$$

where $\sigma_k = 7$ is a fixed number for different keypoint types for simplicity in all our experiments. The coordinate transformation from g to $\tilde{g}$ is critical for keypoint localization precision because we hold the perspective everywhere in this work that each pixel in the image occupies a 1×1 "area" and that pixel value lies exactly at the center of the pixel cell.

During decoding (testing), we correspondingly upsample the final heatmaps $S = (S_1, S_2, \cdots, S_C)$ inferred by the network R times and then find the local maximums in the enlarged heatmaps as the detected keypoints. Essentially, when we employ networks to infer the said keypoint heatmaps, our final goal is *ranking* the heatmap pixels in a **correct order** according to their Gaussian responses rather than regressing the heatmap pixels' exact responses. If the inferred Gaussian response at the ground-truth keypoint position is higher than those of the nearby pixels in the enlarged heatmaps, we can detect the keypoint successfully.

Here, we provided a detailed description of this decoding process for keypoint localization. The bilinear or bicubic interpolation approach is usually used in resizing the heatmaps. Please refer to Fig. 3, the interpolation approach for enlarging the Gaussian response heatmaps determines whether we can restore and find the accurate keypoint positions. Let us consider the most serious situation of the precision loss in keypoint heatmap representation. Supposing a ground-truth keypoint of type j lies at the position $(3.5, 3.5)$ in the input image, let us see if we can distinguish it from the false positives at $(4, 4)$ and $(4, 5)$.

Owing to the design of our encoding (refer to Eq. 2), the inferred responses at $(4, 4)$ and $(4, 5)$ are both high. If the interpolated response at $(3.5, 3.5)$ is higher, then we can still obtain the precise keypoint position in theory[1]. Otherwise, a

---

[1] We can theoretically estimate the floating-point keypoint position according to the Gaussian kernel with the fixed standard deviation σ_k used in our encoding.

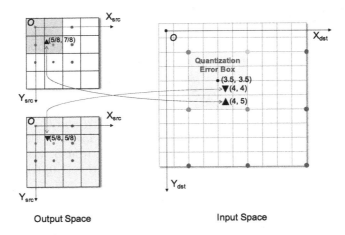

Fig. 3. Enlarging the resolution $\frac{W}{R} \times \frac{H}{R}$ of the inferred heatmaps using the bilinear interpolation approach (**left-top**) or bicubic interpolation approach (**left-bottom**) to the spatial extent $W \times H$ of the original input (**right**). In this example, the interpolation stride $R' = R = 1/4$. R is the output stride of the network. The pixels in the blue cells and green cells are used to interpolate the response values at the positions of $(4,4)$ and $(4,5)$ in the enlarged heatmaps at the resolution of the input image respectively. (Color figure online)

localization error is inevitable. Here, we assume that the network infers heatmaps perfectly. If we use the bilinear interpolation approach, any interpolated response $\tilde{s}(x,y)$ at $\tilde{g}(x,y)$ in the quantization error box (yellow area in the **right** of Fig. 3) is mapped from the response $s(u,v)$ at $g(u,v)$ in the inferred heatmaps and has the constant value:

$$s(u,v) = \begin{bmatrix} 1-u & u \end{bmatrix} \begin{bmatrix} s(0,0) & s(0,1) \\ s(1,0) & s(1,1) \end{bmatrix} \begin{bmatrix} 1-v \\ v \end{bmatrix}$$
$$= s(0,0) = s(0,1) = s(1,0) = s(1,1). \tag{3}$$

Therefore, we cannot distinguish the true keypoint from the nearby false positives in theory. That means the keypoint localization error occurs.

By contrast, if we use the bicubic interpolation approach, the computed response at $\tilde{g}(x,y)$ in the quantization error box is based on the sixteen response values at $g(0,0)$, $g(0,1)$, ..., $g(3,3)$. It is easy to know that the interpolated response value $\tilde{s}(3.5, 3.5)$ is higher than $\tilde{s}(4,4)$ and $\tilde{s}(4,5)$ in the enlarged heatmaps $\tilde{S}_j \in \mathbb{R}^{W \times H}$. Thus, we can estimate the ground-truth keypoint position in theory. In practice, we simply regard the integer coordinates of the local maximums in the enlarged heatmaps of size $W \times H$ as the detected keypoint. Note that the ground-truth keypoint coordinates are integers in the COCO dataset [11]. And in this case, original keypoint locations can be obtained using our heatmap encoding-decoding method for keypoint coordinates.

2.3 Definition of Guiding Offsets

As illustrated in Fig. 1 (a), the ground-truth guiding offset originated from key-point J_{from}, which points to keypoint J_{to}, is a displacement vector calculated as:

$$\{J_{to}(x) - J_{from}(x), J_{to}(y) - J_{from}(y)\}. \tag{4}$$

We place a set of ground-truth guiding offsets around the "start" keypoint J_{from}. A human pose skeleton with 19 "limbs" in PIFPAF [7] is used (refer to Fig. 4). A limb in the human pose skeleton is a "connection" between two adjacent keypoints. Consequently, we use 19 types of guiding offsets to associate keypoints unless mentioned otherwise. The labeled area around the "start" keypoint is set to 7×7 when we train the network. If different guiding offsets overlap at the same position, we keep the guiding offset with the smallest Euclidean norm.

During decoding, we only use the guiding offsets predicted at the local maximum positions in the keypoint heatmaps as the keypoint grouping cues. Supposing we have two keypoint candidates J'_{from} and J'_{to} in the inferred heatmaps to be paired and they have Gaussian response values $J'_{from}(s)$ and $J'_{to}(s)$ respectively. The guiding offset at J'_{from} points to the position J'_{guid}. We measure the connection score between them as:

$$S(J'_{from}, J'_{to}) = J'_{from}(s) \cdot J'_{to}(s) \cdot \exp\left(-\frac{\Delta l}{L}\right) \cdot \mathbb{1}(\Delta l \leq d_0), \tag{5}$$

where L is the length between J'_{from} and J'_{to} in Euclidean space, Δl is the "error" distance between J'_{guid} and J'_{to} in Euclidean space, and $\mathbb{1}$ is the indicator function. If Δl is bigger than the hyperparameter d_0, the pairing is refused. We set $d_0 = 40$ for all kinds of guiding offsets for simplicity in our experiments.

Equation 5 takes into account the "start" keypoint's Gaussian response, the "end" keypoint's Gaussian response, and the "guiding" error normalized by the scale of the current human pose instance. Here, we refer to a connection of the paired keypoints with high connection confidence as a "limb" candidate. All limb candidates are sorted according to their connection scores.

2.4 Post-processing

After network inference, we decode the network outputs, i.e., keypoint heatmaps and guiding offsets, into multi-person pose results. We slide a 3×3 window in the enlarged keypoint heatmaps and find the local maximums as the detected keypoints. Assuming we have detected a keypoint J_{from} and a keypoint J_{to} of the adjacent types, they are paired and connected as described in Sect. 2.3. We independently pair the adjacent keypoints in the human pose skeleton. After that, We need to group the paired keypoints into individuals, which is shown in Fig. 2(d)~(e).

To solve the keypoint assignment problem, we employ a greedy offset-guided keypoint grouping (GOG) algorithm, which is similar to the keypoint association algorithm in SimplePose [9] and CenterNet [22]. The adjacent limbs in the human pose skeleton are assembled greedily. And the keypoints are grouped into individuals accordingly. The GOG algorithm is greedy and fast.

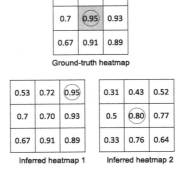

Fig. 4. Human pose skeleton with 19 limbs to associate the adjacent keypoints [7].

Fig. 5. Keypoint localization in the inferred heatmaps. We find the local maximums in the red circles as the detected keypoints. (Color figure online)

2.5 Network Structure and Losses

Following CenterNet [22], we employ Hourglass-104 as the backbone network that consists of two stacked hourglass modules [13] and has 104 convolutional layers. We only add 1×1 convolutional layers into the backbone to regress the desired outputs. The focal L_2 loss [9] is applied to supervise the regression of keypoint heatmaps while only the guiding offsets close to "start" keypoints are supervised by the L_1 loss and normalized by individual scales.

2.6 Peak Regularization for Heatmap Regression

The L_2 loss measures the pixel-wise distances between the inferred heatmaps and the ground-truth heatmaps. However, a low L_2 loss cannot guarantee the keypoint localization precision [14]. Please refer to Fig. 5, the L_2 loss of the inferred heatmap 1 is lower than that of the inferred heatmap 2 but the keypoint localization in heatmap 2 is more accurate. This example suggests that there is a gap between the loss function and the evaluation metric. To alleviate this gap, we introduce the peak regularization for keypoint heatmap regression.

As demonstrated in Sect. 2.2, we can detect the keypoint successfully if the inferred Gaussian response at the ground-truth keypoint position is higher than those of the nearby pixels. Therefore, in Fig. 5, the peak point of Gaussian response in the inferred heatmap 2 is correct while the peak point in the inferred heatmap 1 deviates from the ground-truth position. Let us say a ground-truth keypoint of type j lies at the position $p(u, v)$. Assuming that the predicted Gaussian response by the network at $p(u, v)$ in the j-th inferred heatmap

$S_j \in [0,1]^{\frac{W}{R} \times \frac{H}{R}}$ is $S_j(u,v)$. We propose the peak regularization into the Gaussian response regression at the keypoint position $p(u,v)$ as:

$$\sum_{m=-k}^{k} \sum_{n=-k}^{k} \max\left\{0, S_j(u+m, v+n) - S_j(u,v) + \Delta\right\}, \tag{6}$$

in which we consider the $(2k+1) \times (2k+1)$ area around the ground-truth Gaussian peak point, and Δ is a hyperparameter. We set $k=2$ and $\Delta = 0.02$ in our experiments. All the Gaussian responses of the ground-truth keypoints are supervised by the L_2 loss with the peak regularization during network training.

Please refer to Fig. 5 again, the peak regularization loss of the inferred heatmap 2 is less than that of the inferred heatmap 1, which is different from the L_2 loss. The introduced peak regularization makes the network infer the correct Gaussian peak point.

Consequently, the total loss of the network during training includes: (1) the L_2 loss of the keypoint heatmaps, (2) the peak regularization loss of the Gaussian responses, and (3) the L_1 loss of the guiding offsets.

3   Experiments

3.1   Implementation Details

Our system is implemented using Python and Pytorch. We present experiments on the COCO dataset [11]. It consists of the training set, the test-dev set, and the validation set. COCO dataset uses OKS-based average precision (AP) to evaluate results. Our models are trained on the training set and evaluated on the validation set (excluding images without people) and test-dev set. The training images are cropped to 512×512 patches with random scaling, rotation, flipping, and translation. We fine-tune the pre-trained Hourglass-104 in [22] using the Adam optimizer with the batch size of 32, the initial learning rate of $1e-3$ for less than 150 epochs. For evaluation we use single-scale inference, keeping the same as CenterNet [22], PersonLab [15], and PIFPAF [7] for fair comparisons.

3.2   Error Analyses

It is essential to learn the upper limit of our system. We replace the network outputs with the ground truth and obtain the theoretical upper bound: 86.6% AP on the COCO validation set. If we define more "limbs" in the human pose skeleton, let us say 44 limbs, to connect adjacent keypoints, the upper bound reaches 91.0% AP. It further increases to 95.5% AP when output stride $R = 1$.

3.3   Ablation Studies

We validate different encoding-decoding methods for multi-person keypoint coordinates in Table 1, in which, "qnt" represents quantifying the numerical keypoint

Table 1. Ablation studies of the encoding-decoding methods for keypoint coordinates on the COCO validation set.Here, flip augmentation is not used during testing.

Method	DARK [21]	qnt	qnt + ro	Ours1	Ours1 + ro	Ours1 w/$B.I.$	Ours1 w/$P.R.$
AP	54.5	56.2	62.8	64.4	64.1	62.3	65.8

Table 2. Results of accuracy and speed (single-scale inference) on the COCO validation set. We use the same computer with a 2080 Ti GPU to test speed equally. The faster speed in parenthesis is obtained using FP16 inference provided by Nvidia Apex.

Method	Backbone	P.R	w/Flip	AP	FPS (FP16)	Input length
CenterNet (**baseline**) [22]	Hourglass-104		✓	64.0	8.3	~640
SimplePose [9]	IM-Hourglass		✓	66.3	0.7	768
PersonLab [15]	RestNet-101			61.2	–	801
PIFPAF [7]	RestNet-101			65.7	4.0	641
Ours1	Hourglass-104			64.4	27.8 (36.6)	640
Ours2	Hourglass-104	✓		65.8	27.8 (36.6)	640
Ours3	Hourglass-104		✓	66.3	14.7 (21.4)	640
Ours4 (**final**)	Hourglass-104	✓	✓	67.5	14.7 (21.4)	640

to an integer coordinate before generating the Gaussian distribution, "ro" is short for refinement offsets [7,8,15,22] improving localization precision. "$B.I.$" represents bilinear interpolation and "$P.R.$" represents the proposed peak regularization. DARK [21] works pretty well in some top-down approaches, and we have modified and applied it to our bottom-up approach.

Experiments in Table 1 indicate that our encoding-decoding method is better. DARK assumes that the predicted keypoint responses obey a Gaussian distribution, but the network outputs usually do not meet this assumption in bottom-up approaches, in which inferred Gaussian responses are often far from ground-truth values. Refinement offsets failed in improving our result (-0.3% AP, see "Ours1 +ro" VS "Ours1"), suggesting that naive offset regression to precise keypoint positions is not accurate enough. As expected in Sect. 2.2, using bilinear interpolation in heatmap upsampling leads to a large accuracy drop of **2.1%** AP ("Ours1 w/ $B.I.$" vs "Ours1"). The introduced peak regularization brings about a significant increase of **1.4%** AP ("Ours1 w/ $P.R.$" vs "Ours1").

3.4   Results

We compare our approach with the reproducible SOTA bottom-up approaches on the COCO dataset [11]. Please refer to Table 2, both the baseline CenterNet and our approach are based on the Hourglass-104 backbone [8]. The long side of the input image is resized to a fixed number. Our approach (Ours4) significantly exceeds the baseline in precision ($+3.5\%$ AP) on the COCO validation set under equal conditions. In addition, the results in Table 2 suggest that our approach has obtained a satisfying trade-off between accuracy and speed.

Table 3. Results on the COCO test-dev set (single-scale inference). Entries marked with "*" are produced by corresponding official source code and models. AP^M and AP^L are the evaluation metrics for medium objects and large objects respectively.

Method	Backbone	AP	AP^M	AP^L	Input Length
CenterNet* [22]	Hourglass-104	63.0	58.9	70.4	~640
SimplePose* [9]	IM-Hourglass	64.6	60.3	71.6	768
PersonLab [15]	RestNet-101	65.5	61.3	71.5	1401
PIFPAF* [7]	RestNet-101	64.9	60.6	71.2	641
HigherHRNet [3]	HRNet-W32	64.1	57.4	73.9	>640
Ours (**final**)	Hourglass-104	66.1	64.2	69.7	640

According to the results in Table 2 and Table 3. Our preliminary system has obvious advantages taking into account accuracy, speed, and simplicity.

4   Conclusions and Future Work

We have presented a novel bottom-up approach for the problem of multi-person pose estimation, which is simple enough yet achieves a respectable trade-off between accuracy and efficiency. Specifically, we rethink encoding-decoding methods for multi-person keypoint coordinates and improve keypoint heatmap regression with peak regularization, leading to more accurate localization. And we propose the guiding offsets between adjacent keypoints as the keypoint grouping cues and assemble human skeletons greedily. Experiments on the challenging COCO dataset have demonstrated the advantages and much room for accuracy improvement of our preliminary working system. We will try more advanced networks designed specially for human pose estimation to improve our results.

References

1. Cao, Z., Simon, T., Wei, S.E., Sheikh, Y.: Realtime multi-person 2D pose estimation using part affinity fields. In: IEEE Conference on Computer Vision and Pattern Recognition, pp. 1302–1310 (2017)
2. Chen, Y., Wang, Z., Peng, Y., Zhang, Z., Yu, G., Sun, J.: Cascaded pyramid network for multi-person pose estimation. In: IEEE Conference on Computer Vision and Pattern Recognition, pp. 7103–7112. IEEE (2018)
3. Cheng, B., Xiao, B., Wang, J., Shi, H., Huang, T.S., Zhang, L.: HigherHRNet: scale-aware representation learning for bottom-up human pose estimation. In: Proceedings of the IEEE Conference on Computer Vision and Pattern Recognition, pp. 5386–5395 (2020)
4. He, K., Gkioxari, G., Dollár, P., Girshick, R.: Mask R-CNN. In: 2017 IEEE International Conference on Computer Vision, pp. 2980–2988. IEEE (2017)

5. Insafutdinov, E., Pishchulin, L., Andres, B., Andriluka, M., Schiele, B.: DeeperCut: a deeper, stronger, and faster multi-person pose estimation model. In: Leibe, B., Matas, J., Sebe, N., Welling, M. (eds.) ECCV 2016. LNCS, vol. 9910, pp. 34–50. Springer, Cham (2016). https://doi.org/10.1007/978-3-319-46466-4_3

6. Kendall, A., Gal, Y.: What uncertainties do we need in Bayesian deep learning for computer vision? In: Advances in Neural Information Processing Systems, pp. 5574–5584 (2017)

7. Kreiss, S., Bertoni, L., Alahi, A.: PifPaf: composite fields for human pose estimation. In: Proceedings of the IEEE Conference on Computer Vision and Pattern Recognition, pp. 11977–11986 (2019)

8. Law, H., Deng, J.: CornerNet: detecting objects as paired keypoints. In: Ferrari, V., Hebert, M., Sminchisescu, C., Weiss, Y. (eds.) Computer Vision – ECCV 2018. LNCS, vol. 11218, pp. 765–781. Springer, Cham (2018). https://doi.org/10.1007/978-3-030-01264-9_45

9. Li, J., Su, W., Wang, Z.: Simple pose: rethinking and improving a bottom-up approach for multi-person pose estimation. In: Proceedings of the AAAI Conference on Artificial Intelligence, vol. 34, pp. 11354–11361 (2020)

10. Lin, T.Y., Goyal, P., Girshick, R., He, K., Dollar, P.: Focal loss for dense object detection. In: 2017 IEEE International Conference on Computer Vision, pp. 2999–3007. IEEE (2017)

11. Lin, T.-Y., et al.: Microsoft COCO: common objects in context. In: Fleet, D., Pajdla, T., Schiele, B., Tuytelaars, T. (eds.) ECCV 2014. LNCS, vol. 8693, pp. 740–755. Springer, Cham (2014). https://doi.org/10.1007/978-3-319-10602-1_48

12. Newell, A., Huang, Z., Deng, J.: Associative embedding: end-to-end learning for joint detection and grouping. In: Advances in Neural Information Processing Systems, pp. 2277–2287 (2017)

13. Newell, A., Yang, K., Deng, J.: Stacked hourglass networks for human pose estimation. In: Leibe, B., Matas, J., Sebe, N., Welling, M. (eds.) ECCV 2016. LNCS, vol. 9912, pp. 483–499. Springer, Cham (2016). https://doi.org/10.1007/978-3-319-46484-8_29

14. Nibali, A., He, Z., Morgan, S., Prendergast, L.: Numerical coordinate regression with convolutional neural networks. arXiv preprint arXiv:1801.07372 (2018)

15. Papandreou, G., Zhu, T., Chen, L.-C., Gidaris, S., Tompson, J., Murphy, K.: PersonLab: person pose estimation and instance segmentation with a bottom-up, part-based, geometric embedding model. In: Ferrari, V., Hebert, M., Sminchisescu, C., Weiss, Y. (eds.) Computer Vision – ECCV 2018. LNCS, vol. 11218, pp. 282–299. Springer, Cham (2018). https://doi.org/10.1007/978-3-030-01264-9_17

16. Papandreou, G., et al.: Towards accurate multi-person pose estimation in the wild. In: 2017 IEEE Conference on Computer Vision and Pattern Recognition, pp. 3711–3719. IEEE (2017)

17. Ronchi, M.R., Perona, P.: Benchmarking and error diagnosis in multi-instance pose estimation. In: 2017 IEEE International Conference on Computer Vision, pp. 369–378. IEEE (2017)

18. Sun, K., Xiao, B., Liu, D., Wang, J.: Deep high-resolution representation learning for human pose estimation. In: Proceedings of the IEEE Conference on Computer Vision and Pattern Recognition, pp. 5693–5703 (2019)

19. Tompson, J., Jain, A., LeCun, Y., Bregler, C.: Joint training of a convolutional network and a graphical model for human pose estimation. In: Proceedings of the 27th International Conference on Neural Information Processing Systems-Volume 1, pp. 1799–1807. MIT Press (2014)

20. Toshev, A., Szegedy, C.: DeepPose: human pose estimation via deep neural networks. In: Proceedings of the IEEE Conference on Computer Vision and Pattern Recognition, pp. 1653–1660 (2014)
21. Zhang, F., Zhu, X., Dai, H., Ye, M., Zhu, C.: Distribution-aware coordinate representation for human pose estimation. In: Proceedings of the IEEE/CVF Conference on Computer Vision and Pattern Recognition, pp. 7093–7102 (2020)
22. Zhou, X., Wang, D., Krähenbühl, P.: Objects as points. arXiv preprint arXiv: 1904.07850 (2019)

Control Variates for Similarity Search

Jeremy Chew$^{(\boxtimes)}$ and Keegan Kang

Singapore University of Technology and Design, Singapore, Singapore
{jeremy_chew,keegan_kang}@sutd.edu.sg

Abstract. We present an alternative technique for similarity estimation under locality sensitive hashing (LSH) schemes with discrete output. By utilising control variates and extra information, we are able to achieve better theoretical variance reductions compared to maximum likelihood estimation with extra information. We show that our method obtains equivalent results, but slight modifications can provide better empirical results and stability at lower dimensions. Finally, we compare the various methods' performances on the MNIST and Gisette dataset, and show that our model achieves better accuracy and stability.

Keywords: Similarity search · Control variates · Variance reduction

1 Introduction

Suppose we are given a dataset $X_{n \times p}$ where we want to estimate some similarity measure $\rho(\boldsymbol{x}_i, \boldsymbol{x}_j) := p_{ij}$ between any two observations (thus denoted as row vectors, $\boldsymbol{x}_i$ and $\boldsymbol{x}_j$). Computing all pairwise similarities would take at least $O(n^2 p)$ time, which is costly when n, p are large.

Locality sensitive hashing (LSH) schemes (such as random projections [1,8], sign random projections [3], and minwise hashing [2]) allow for efficient dimensional reduction of the original dataset X from p-dimensional vectors to k-dimensional vectors [16], $k \ll p$. We construct a hash function $h : \mathbb{R}^p \to \mathbb{R}$ (involving random variables) and compute $h(\boldsymbol{x}_i) = v_i, 1 \leq i \leq n$. $\mathbb{P}[v_i = v_j]$ is used to estimate $\rho(\boldsymbol{x}_i, \boldsymbol{x}_j)$. In practice, we hash k times, and compute $\frac{\sum_{s=1}^{k} \mathbb{1}_{\{v_{is} = v_{js}\}}}{k}$ to find an estimate of $\rho(\boldsymbol{x}_i, \boldsymbol{x}_j)$. This lowers the computational time for obtaining an estimate of each pairwise similarity from $O(n^2 p)$ to $O(n^2 k)$, plus the additional pre-processing time required to compute the hashed value of each vector.

Using computational statistics, one can use techniques such as maximum likelihood estimation [13] to estimate pairwise similarities with sign random projections with extra information [11], or control variates to estimate pairwise similarities with random projections [9,10] with extra information. These methods keep to the same order of pre-processing time as the original LSH schemes, but obtain a lower variance than the ordinary estimates using the same number of samples.

© Springer Nature Switzerland AG 2021
H. Ma et al. (Eds.): PRCV 2021, LNCS 13019, pp. 468–479, 2021.
https://doi.org/10.1007/978-3-030-88004-0_38

2   Our Contributions

Both [11,13] use the MLE technique for variance reduction for a class of LSH schemes similar to sign random projections, where the estimator relies on computing some form of $\frac{\sum_{s=1}^{k} 1_{\{v_{is}=v_{js}\}}}{k}$. However, we show that we can adapt the control variate technique in [9,10], using extra vectors to work with these LSH schemes, to come up with a control variate estimator for our similarity estimates. We show that this estimator obtains the same theoretical variance reduction as the MLE with the additional benefit of increased numerical stability. We also provide a generalised framework that one can use to generate estimators for any arbitrary number of control variates, allowing one to generate additional extra vectors for further variance reduction. Finally, we demonstrate our results via empirical simulations on the MNIST [12] and Gisette datasets [6,14].

3   Review of Preliminary Concepts

We first review the control variate technique. Control variates are used for variance reduction in Monte Carlo simulations [15]. By exploiting the difference between a known estimator and its observed result, the original estimator can be corrected to reduce the error between the observed value and the actual value.

Let X, Y be random variables. Suppose we want to estimate $\mu_X = \mathbb{E}[X]$. If we can find a random variable Y such that $\mu_Y = \mathbb{E}[Y]$ can be analytically calculated, then, $Z = X + c(Y - \mu_Y)$ is also an unbiased estimator for X as $\mathbb{E}[Z] = \mathbb{E}[X] + c(\mathbb{E}[Y] - \mu_Y) = \mu_X$ for any choice of the coefficient c. By choosing $c = -\frac{\text{Cov}(X,Y)}{\text{Var}(Y)}$, we get an expression for the variance of Z as $\text{Var}(Z) = \text{Var}(X) - \frac{(\text{Cov}(X,Y))^2}{\text{Var}(Y)}$ which always results in a reduced variance as long as $\text{Cov}(X,Y)$ is non-zero. This choice of c is the optimal coefficient to minimise the variance [15].

We now explain the extra information idea in [10,11] using Fig. 1.

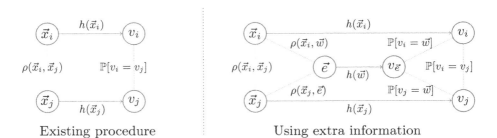

Existing procedure   Using extra information

Fig. 1. Using one extra vector e for which we know relevant information (shown in blue) to estimate unknown $\rho(\boldsymbol{x}_i, \boldsymbol{x}_j)$ (shown in red). (Color figure online)

In the original LSH scheme, we know the hash function h and its relation to $\mathbb{P}[v_i = v_j]$ (blue). We condition on these information to find $\rho(\boldsymbol{x}_i, \boldsymbol{x}_j)$ (red).

In the extra information case, we generate an extra vector $e \in \mathbb{R}^p$, then compute and store $\rho(x_i, e), 1 \leq i \leq n$. This takes $O(np)$ time and requires $O(n)$ space. We then compute the hashed values $v_i, 1 \leq i \leq n$ and v_e. Finally, we condition on all of the known information (blue) to estimate $\rho(x_i, x_j)$ (red).

We now describe what happens in practice where we hash k times. The hashed results are stored in a matrix $Y_{n+1,k}$, where the last row corresponds to the k hashed values $h(e)$. We denote the last row as Y_e.

We remark that the hashed values can be either discrete [2] or binary [3], depending on the LSH scheme. The MLE approach [11] then considers the following sets for binary and discrete-typed hashes where:

$$A := \{s \mid Y_{is} \neq Y_{js}, Y_{js} = Y_{es}\} \quad B := \{s \mid Y_{is} = Y_{js}, Y_{js} \neq Y_{es}\} \tag{1}$$
$$C := \{s \mid Y_{is} = Y_{js}, Y_{js} = Y_{es}\} \quad D := \{s \mid Y_{is} \neq Y_{js}, Y_{js} \neq Y_{es}\} \tag{2}$$

is the collection of indices for binary hashes, $1 \leq s \leq k$ and

$$A := \{s \mid Y_{is} = Y_{js}, Y_{js} = Y_{es}\} \tag{3}$$
$$B := \{s \mid Y_{is} \neq Y_{js}, Y_{is} \neq Y_{es}, Y_{js} = Y_{es}\} \tag{4}$$
$$C := \{s \mid Y_{is} \neq Y_{js}, Y_{is} = Y_{es}, Y_{js} \neq Y_{es}\} \tag{5}$$
$$D := \{s \mid Y_{is} = Y_{js}, Y_{is} \neq Y_{es}, Y_{js} \neq Y_{es}\} \tag{6}$$
$$E := \{s \mid Y_{is} \neq Y_{js}, Y_{is} \neq Y_{es}, Y_{js} \neq Y_{es}\} \tag{7}$$

is the collection of indices for discrete hashes, $1 \leq s \leq k$.

Then, let n_i denote the cardinality of the set i, with $n_A + n_B + n_C + n_D = k$ for binary hashes, and $n_A + n_B + n_C + n_D + n_E = k$ for discrete hashes. Also, let p_i be the probability that an observed element falls in set i.

3.1   Binary Hashes

For any given observed cardinalities n_A, n_B, n_C, n_D, the likelihood of such an event, given parameters p_A, p_B, p_C, p_D, would be

$$\mathcal{L}(p_A, p_B, p_C, p_D) = \frac{k!}{n_A! n_B! n_C! n_D!} p_A^{n_A} p_B^{n_B} p_C^{n_C} p_D^{n_D} \tag{8}$$

However, as $\rho(x_i, e)$ have been stored and computed, then we also have the following constraints

$$p_A + p_C = \rho(x_j, e) := p_{je} \tag{9}$$
$$p_C + p_D = \rho(x_i, e) := p_{ie} \tag{10}$$
$$p_A + p_B + p_C + p_D = 1 \tag{11}$$

Taking the log-likelihood and substituting in the above constraints, we obtain

$$l(p_C) = K + n_A \log(p_{je} - p_C) + n_B \log(1 + p_C - p_{ie} - p_{je}) \\ + n_C \log(p_C) + n_D \log(p_{ie} - p_C) \tag{12}$$

$\hat{p}_C$ can thus be expressed as a root of a cubic, as described in [11], and found via numerical methods.

3.2   Discrete Hashes

The procedure for discrete hashes is similar. We have the same constraints, but since we now have five variables (instead of four in the binary case), we write our log-likelihood in terms of two variables, say p_A and p_D to obtain

$$
l(p_A, p_D) = K + n_A \log(p_A) + n_B \log(p_{je} - p_A) + n_C \log(p_{ie} - p_A) \\
+ n_D \log(p_D) + n_E \log(1 - p_{ie} - p_{je} + p_A - p_D) \tag{13}
$$

By calculus, we can express

$$
p_D = \frac{n_D(1 + p_A - p_{ie} - p_{je})}{n_D + n_E} \tag{14}
$$

and find p_A expressed as a root of a cubic via numerical methods.

However, finding the root of these cubics are not numerically stable for low values of k, which leads us to consider a new estimator.

4   Our Control Variate Estimator: Binary Hashes

Suppose we have generated a random vector $e \in \mathbb{R}^p$, and computed and stored the similarities $\rho(x_i, e)$ for each observation, similar to the setup in [10,11]

Now, suppose we want to estimate $p_{ij} = \rho(x_i, x_j)$ for any pair of observations x_i, x_j, and we have k hashes. We define the following random variables

$$
A := \frac{\sum_{s=1}^{k} 1_{\{Y_{is} = Y_{js}\}}}{k} \quad B := \frac{\sum_{s=1}^{k} 1_{\{Y_{is} = Y_{es}\}}}{k} \quad C := \frac{\sum_s 1_{\{Y_{js} = Y_{es}\}}}{k} \tag{15}
$$

Then, $\mathbb{E}[A] = p_{ij}$, $\mathbb{E}[B] = \rho(x_i, e)$ and $\mathbb{E}[C] = \rho(x_j, e)$. We know $\mathbb{E}[B], \mathbb{E}[C]$ from pre-calculating $\rho(x_i, e)$ and $\rho(x_j, e)$.

We define the following new estimator:

$$
A' = A + c_1(B - \mathbb{E}[B]) + c_2(C - \mathbb{E}[C]) \tag{16}
$$

A' is an unbiased estimator of p_{ij} and guaranteed to have a lower variance than A as long as either (or both of) $\text{Cov}(A, B)$ and $\text{Cov}(A, C)$ are non-zero.

By finding the partial derivative of $\text{Var}(A')$ and solving for optimal $\hat{c}_1, \hat{c}_2$, we can get the optimal control variate corrections. Thus, we have

$$
\text{Var}(A') = \text{Var}(A) + c_1^2 \text{Var}(B) + c_2^2 \text{Var}(C) + 2c_1 \text{Cov}(A, B) + 2c_2 \text{Cov}(A, C) \\
+ 2c_1 c_2 \text{Cov}(B, C) \tag{17}
$$

The partial derivatives are

$$
\frac{\partial \text{Var}(A')}{\partial c_1} = 2c_1 \text{Var}(B) + 2\text{Cov}(A, B) + 2c_2 \text{Cov}(B, C) \tag{18}
$$

$$
\frac{\partial \text{Var}(A')}{\partial c_2} = 2c_2 \text{Var}(C) + 2\text{Cov}(A, C) + 2c_1 \text{Cov}(B, C) \tag{19}
$$

Denoting the covariance matrix between B and C as Σ, and packing the covariances $\mathrm{Cov}(A, B), \mathrm{Cov}(A, C)$ as a column vector Σ_A, we can write the equations forming the optimal values of c_1 and c_2 as follows: $[c_1 \quad c_2]^T = \Sigma^{-1} \Sigma_A$.

Now, the control variate corrections c_1 and c_2 involve the true value of p_{ij} in the calculations of Σ and Σ_A, which we do not have. To get around this, one solution is to obtain an initial estimate for p_{ij} using $\frac{1}{k} \sum_{s=1}^{k} A$ similar to [10]. Another solution involves finding analytic expressions for Σ and Σ_A, collecting the p_{ij} terms on one side, and finally solving the resultant cubic.

Theorem 1. *Suppose we set $g(p_C)$ to be the cubic found after calculus in* (12) *and $f(p_C)$ to be the cubic we find after our control variate approach. We have that $\frac{f(p_A)}{g(p_A)} = -2$.*

Proof. We can write A, B, C in our estimator as

$$A = \frac{n_B + n_C}{k} \qquad B = \frac{n_C + n_D}{k} \qquad C = \frac{n_A + n_C}{k} \tag{20}$$

We set $A' = p_{ij}$, which we rewrite in terms of p_C by using the result presented in (9)-(11). Furthermore, as A, B, C are Bernoulli random variables, we can easily calculate their expectations, variances and covariances as follows

$$\mathbb{E}[A] = p_{ij} = 1 + 2p_C - p_{ie} - p_{je} \tag{21}$$
$$\mathrm{Var}(A) = p_{ij}(1 - p_{ij}) = (1 + 2p_C - p_{ie} - p_{je})(p_{ie} + p_{je} - 2p_C) \tag{22}$$
$$\mathbb{E}[B] = p_{ie} \qquad\qquad\qquad \mathbb{E}[C] = p_{je} \tag{23}$$
$$\mathrm{Var}(B) = p_{ie}(1 - p_{ie}) \qquad\qquad \mathrm{Var}(C) = p_{je}(1 - p_{je}) \tag{24}$$
$$\mathrm{Cov}(A, B) = p_C - p_{ie}(1 + 2p_C - p_{ie} - p_{je}) \quad \mathrm{Cov}(B, C) = p_C - p_{ie}p_{je} \tag{25}$$
$$\mathrm{Cov}(A, C) = p_C - p_{je}(1 + 2p_C - p_{ie} - p_{je}) \tag{26}$$

We can then substitute all these rewritten expressions back into (16). By bringing A' over, and cross-multiplying to remove the denominator, we obtain the following equation, which we call $f(p_C)$:

$$\begin{aligned}
f(p_C) = &- 2n_1 p_C^3 + 4n_1 p_C^2 p_{ie} + 2n_1 p_C^2 p_{je} - 2n_1 p_C^2 - 2n_1 p_C p_{ie}^2 - 2n_1 p_C p_{ie} p_{je} \\
&+ 2n_1 p_C p_{ie} - 2n_2 p_C^3 + 2n_2 p_C^2 p_{ie} + 2n_2 p_C^2 p_{je} - 2n_2 p_C p_{ie} p_{je} - 2n_3 p_C^3 \\
&+ 4n_3 p_C^2 p_{ie} + 4n_3 p_C^2 p_{je} - 2n_3 p_C^2 - 2n_3 p_C p_{ie}^2 - 6n_3 p_C p_{ie} p_{je} + 2n_3 p_C p_{ie} \\
&- 2n_3 p_C p_{je}^2 + 2n_3 p_C p_{je} + 2n_3 p_{ie}^2 p_{je} + 2n_3 p_{ie} p_{je}^2 - 2n_3 p_{ie} p_{je} - 2n_4 p_C^3 \\
&+ 2n_4 p_C^2 p_{ie} + 4n_4 p_C^2 p_{je} - 2n_4 p_C^2 - 2n_4 p_C p_{ie} p_{je} - 2n_4 p_C p_{je}^2 + 2n_4 p_C p_{je}
\end{aligned} \tag{27}$$

which is exactly equivalent to the $\frac{1}{2}$ of cubic in [11].

5 Our Estimator: Discrete Hashes

Unlike the MLE approach for discrete hashes where we define a new collection of sets, our estimator uses the same random variables defined in the binary case.

We find that we require the true value of p_{ij} to calculate c_1 and c_2. As such, we are presented with the same two options as in the binary case. We can either use $\frac{1}{k}\sum_s^k A$ as an initial estimate for p_{ij}, or we can collect the p_{ij} terms on one side, and solve the root for the resultant cubic. Similarly, this result is equivalent to the cubic for discrete hashes.

Theorem 2. *Suppose we set $g(p_A)$ to be the cubic found after calculus in (13) and $f(p_A)$ to be the cubic we find after our control variate approach. We have that $\frac{f(p_A)}{g(p_A)} = -2n_D - n_E$.*

The equivalency of maximum likelihood estimation and control variates have been explored in [4] and [17], where they show how solving the "constrained Monte Carlo" with non-parametric maximum likelihood estimation coincides with the method of control variates. This implies that the variance reduction for our control variate method is asymptotically equivalent to the variance reduction for maximum likelihood estimation method [11].

However, the control variate method allows us to substitute in a proxy for p_{ij} in our coefficients c_1, c_2 (or even compute the empirical covariance), rather than resorting than solving for a root of a cubic. This is less computationally costly than running Newton-Raphson. Moreover, Theorem 2 gives some (theoretical) insight why a naive implementation of the MLE estimator may not be numerically stable at low values of k, due to the values of n_D and n_E, which may be zero for small k.

6   Application to Sign Random Projections

We demonstrate how our estimator can be used for sign random projections (SRP). In SRP, we want an estimate for the angle θ_{ij} between any two vectors $\boldsymbol{x}_i$ and $\boldsymbol{x}_j$ [5]. Given a data matrix $X_{n\times p}$, we generate a random matrix $R_{p\times k}$ with entries i.i.d from $N(0,1)$. We next compute $V = \text{sgn}(XR)$ where we define $\text{sgn}(x) = 1_{\{x\geq 0\}}$. The estimate of θ_{ij} is given by $\hat{\theta} = \sum_{s=1}^{k}\frac{\pi}{k}\cdot 1_{\{V_{is}\neq V_{js}\}}$.

Suppose we now generate an extra vector $\boldsymbol{e}$, similar to [11]. Then we only need to make minor modifications to our estimator to have it calculate estimates for the angles between two vectors. Specifically, we have $\mathbb{E}[A] = 1 - \frac{\theta_{ij}}{\pi}, \mathbb{E}[B] = 1 - \frac{\theta_{ie}}{\pi}$, and $\mathbb{E}[C] = 1 - \frac{\theta_{je}}{\pi}$.

Computing $\mathbb{E}[AB] = \mathbb{E}[AC] = \mathbb{E}[BC]$ involves finding the "three-way" probability of A, B, and C. In the case of estimating angles, this is equivalent to having the three vectors $\boldsymbol{x}_i$, $\boldsymbol{x}_j$, and $\boldsymbol{e}$ fall on one side of some hyperplane. This probability is given by $1 - \frac{\theta_{ij}+\theta_{ie}+\theta je}{2\pi}$, with the proof given in [11].

In general, in order to obtain estimates for any LSH scheme, we need to first obtain an expression for the "three-way" probability of A, B, and C. In other words, we need to be able to calculate an expression for the expectation of the statistic $1_{\{Y_{is}=Y_{js}=Y_{es}\}}$. We discuss more on these in Sect. 8.

With the above modifications, our estimator is able to directly output an estimate of the angle between any two vectors.

7   Experimental Results

We demonstrate our estimators by using the MNIST test dataset [12] and the
Gisette dataset [7,14]. The MNIST test dataset has $n = 10,000$ observations
with $p = 784$ parameters. The Gisette dataset has $n = 13,500$ observations
with $p = 5000$ parameters. We normalise the vectors in both datasets to unit
length. To ensure reproducibility, we use a constant seed for our experiments.
Furthermore, we set the extra vector e to be the mean vector across all observations in each dataset. The simulations are written in Python, and make use of
GPU-accelerated computations through the library PyTorch.

In order to prevent possible biases in picking "good pairs" of vectors, we
choose to calculate all possible pairwise estimates for the full dataset. We then
calculate the mean-squared error between each estimate and the actual value.

We vary k over the range of $\{10, 20, \ldots, 100\}$, and for k, we calculate and
average the results over 100 iterations. For each individual simulation run, we
calculate 4 different estimates, namely CV_SUB, CV_CUBIC, MLE, and SRP. Respectively, they refer to: using an initial estimate for the default control variate procedure, solving the cubic derived from control variates, solving the cubic derived
from MLEs [11], and sign random projections [5].

We compute the average of all mean-squared errors of $49,995,000$ pairwise
angular similarity estimates for the MNIST test dataset, as well as the average
mean-squared errors of $91,118,250$ pairwise angular similarity estimates for the
Gisette dataset. These are displayed in Figs. 2 and 3 respectively.

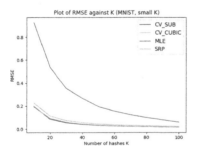

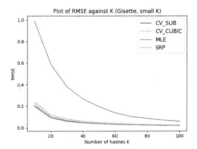

Fig. 2. Averaged MSE for MNIST test dataset

Fig. 3. Averaged MSE for Gisette dataset

In general, we note that CV_SUB and CV_CUBIC both consistently outperform
SRP and MLE. Furthermore, we also note that CV_SUB, that is, simply improving an
initial estimate using control variates, outperforms solving the cubic in CV_CUBIC.
Finally, although CV_CUBIC and MLE have been shown to be equivalent in the
previous sections, empirical results show that the control variates approach offers
greater stability and accuracy compared to the MLE approach.

This is because the calculation for MLE relies on n_A, n_B, n_C, n_D. If any of them are zero due to few observations, this may result in a division-by-zero error. Cross-multiplying can help prevent this error, but the numerical stability would still be affected. On the other hand, CV_SUB uses the empirical value of the control variate coefficient c. This takes the correlation of the already observed vectors into account, thus giving more accurate and stable results.

We also plot the standard deviation of the mean-squared errors in Figs. 4 and 5. We note that CV_SUB tends to have lower standard deviations compared to other estimates. We also note that for larger values of k, we can observe lower variances and higher accuracies compared to the baseline of SRP.

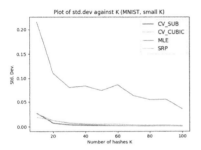

Fig. 4. Standard deviations for MNIST test dataset

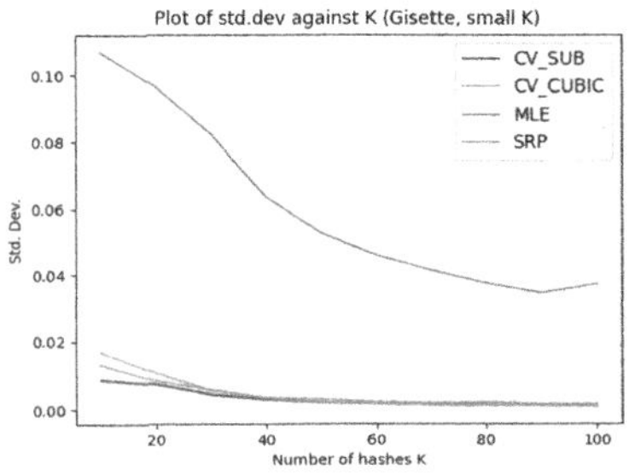

Fig. 5. Standard deviations for Gisette dataset

We also run another set of simulations for a wider range of $k = \{200, 250, \ldots 1000\}$, to show the asymptotic behaviour of the estimators for larger values of k. This time, the results are obtained by averaging over 25 iterations. Again, we calculate these values from all possible pairs of vectors in both the MNIST and Gisette dataset. The results are graphed and displayed in Figs. 6 and 7 for the MNIST and Gisette dataset respectively.

Table 1. Average MSE for $k = 50$ and $k = 1000$

	MNIST ($k = 50$)	Gisette ($k = 50$)	MNIST ($k = 1000$)	Gisette ($k = 1000$)
SRP	0.0448 ± 0.0153	0.0477 ± 0.00867	0.00216 ± 0.000381	0.002400 ± 0.000498
MLE	0.195 ± 0.222	0.195 ± 0.159	0.00165 ± 0.000205	0.00192 ± 0.000107
CV_SUB	$\mathbf{0.0324 \pm 0.00731}$	$\mathbf{0.0372 \pm 0.00578}$	$\mathbf{0.00163 \pm 0.000180}$	$\mathbf{0.00192 \pm 0.000119}$
CV_CUBIC	0.0345 ± 0.00974	0.0393 ± 0.00621	0.00164 ± 0.00974	0.00192 ± 0.000107

We observe that MLE only outperforms SRP for larger values of k, but does not match the performance of control variate. We also observe that as k increases, MLE asymptotically approaches the same performance as CV_CUBIC. This verifies the equivalence of the two approaches as proven in Theorems 1 and 2.

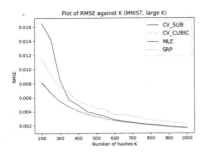

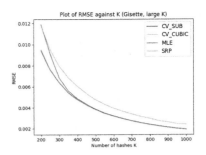

Fig. 6. Averaged MSE for MNIST test dataset

Fig. 7. Averaged MSE for Gisette dataset

Table 1 displays the averaged MSE (with bounds of 3 standard deviations) for $k = 50$ and $k = 1000$ for both the MNIST and Gisette datasets. We can see that for both datasets, the MSE for MLE and CV_CUBIC are approximately equal at $k = 1000$, verifying that our two methods are asymptotically identical.

We note that although large values of k give higher accuracies, it does not make sense to use values of k which are similar in magnitude to the original dimensions of the data, p. This is because the computational complexities of our estimators are in the order of $O(n^2 k)$, while the computational complexity of calculating each pairwise angle is $O(n^2 p)$. Using larger values of k would result in a slower running time, due to the additional overhead of preprocessing the hash values. In this sense, we are only interested in the results for small values of k, but nevertheless, our CV approach still matches/outperforms the MLE approach even for large k.

To verify this, we display the average running time of the various algorithms for $k = \{50, 500, 1000\}$ on the MNIST dataset in Table 2. As expected, for small k, all estimators are faster than directly calculating the angles, but this changes as k increases. We note that SRP is the fastest, with our estimator, CV_SUB coming in close second. There is a tradeoff between accuracy and efficiency, but because both algorithms keep to the same order of complexity, the time tradeoff is not too significant. On the other hand, due to the need to numerically solve a cubic (through Newton-Raphson), CV_CUBIC and MLE have a much more noticeable accuracy-efficiency tradeoff.

Overall, our results shows that our control variate estimator, not only offers an increase in accuracy from the baseline of SRP, but also consistently outperforms the MLE approach proposed by [11].

8   Discussion on Control Variates

An added benefit of using control variates is the ability to easily extend to include additional extra vectors. For example, we can generate further extra vectors $e_1, e_2, \ldots e_j$, which can also be used to obtain even more accurate estimates.

Table 2. Average time taken (in seconds) for $k = \{50, 500, 1000\}$ on MNIST dataset

	$k = 50$	$k = 500$	$k = 1000$
Direct calculation	135.3		
SRP	11.28	138.4	287.4
MLE	99.05	881.5	1805
CV_SUB	15.20	146.7	294.9
CV_CUBIC	47.38	181.1	335.3

For each extra vector e_t, we first pre-compute and store $\rho(x_i, e_t)$ for all observations. Then we have the following random variables for each extra vector:

$$A := \frac{\sum_s 1_{\{Y_{is}=Y_{js}\}}}{k} \qquad B_t := \frac{\sum_s 1_{\{Y_{is}=Y_{e_t s}\}}}{k} \qquad C_t := \frac{\sum_s 1_{\{Y_{js}=Y_{e_t s}\}}}{k} \qquad (28)$$

We can disregard the additional random variables of the form $1_{\{Y_{e_m s}=Y_{e_n s}\}}$ because their covariance with A is 0. In general, for any two random variables to have a non-zero covariance, they must share at least one vector in common. Because we are interested in reducing the variance of A, any random variable we consider should at least share a common vector with A.

Thus, we can define the new estimator of A' of p_{ij} as follows:

$$A' = A + \sum_{t=0}^{j} c_{t,1}(B_t - \mathbb{E}[B_t]) + c_{t,2}(C_t - \mathbb{E}[C_t]) \qquad (29)$$

Minimising $\text{Var}(A')$, we obtain the following system of equations for c:

$$\begin{bmatrix} c_{1,1} & c_{1,2} & \cdots & c_{j,1} & c_{j,2} \end{bmatrix}^T = \Sigma^{-1}\Sigma_A \qquad (30)$$

where Σ refers to the covariance matrix of the random vector $\mathbf{X}$ whose elements are $(B_1, C_1, \ldots B_j, C_j)^T$, and Σ_A refers to the column vector whose elements are the covariances of $\mathbf{X}$ with respect to A.

In order to solve the system of equations above, we would need the covariance matrices. Since $\text{Cov}(X, Y) = \mathbb{E}[XY] - \mathbb{E}[X]\mathbb{E}[Y]$, and we already have $\mathbb{E}[X]$ and $\mathbb{E}[Y]$, we would only need to find expressions for $\mathbb{E}[XY]$.

Now, B_t and C_t are random variables that relate how two vectors are similar to one another. Thus, the expectation $\mathbb{E}[B_i C_j]$ for any i, j rely on which vectors are being referred to. If they share a common vector, such that the three vectors in question are x, y and z, then the product becomes finding the expectation of the random variable $1_{\{Y_{xs}=Y_{ys}=Y_{zs}\}}$. This is the "three-way" similarity between the three vectors. For sign random projections, this is the probability that the three vectors lie on the same side of a hyperplane.

If the random variables do not share any common vectors, e.g. B_1 and C_2, then, their covariance would equal to 0. This would mean that in the covariance matrix Σ above, there are several entries which we would know to be 0.

This makes our covariance matrix sparse, and hence computationally easier to compute the control variate coefficients.

We give an example of the covariance matrix Σ for $j = 4$ extra vectors:

$$
\begin{array}{c}
\begin{array}{cccccccc}
B_1 & C_1 & B_2 & C_2 & B_3 & C_3 & B_4 & C_4
\end{array} \\
\begin{array}{c}
B_1 \\ C_1 \\ B_2 \\ C_2 \\ B_3 \\ C_3 \\ B_4 \\ C_4
\end{array}
\left[
\begin{array}{cc|cc|cc|cc}
v & c & c & 0 & c & 0 & c & 0 \\
c & v & 0 & c & 0 & c & 0 & c \\ \hline
c & 0 & v & c & c & 0 & c & 0 \\
0 & c & c & v & 0 & c & 0 & c \\ \hline
c & 0 & c & 0 & v & c & c & 0 \\
0 & c & 0 & c & c & v & 0 & c \\ \hline
c & 0 & c & 0 & c & 0 & v & c \\
0 & c & 0 & c & 0 & c & c & v
\end{array}
\right]
\end{array}
\tag{31}
$$

where v, c are the respective variances and covariances to be calculated.

9   Conclusion

We have shown how to use control variates to construct estimators for similarity estimation under LSH schemes. We have also demonstrated how our estimator could be used for sign random projections. The empirical results also show that our control variates estimator outperforms other estimators that use extra vectors to improve accuracy. Furthermore, we have shown how our estimator can easily be extended to include greater numbers of extra vectors, which would otherwise require redefining of contingency tables in other approaches.

We believe that this strategy of using control variates can help improve estimates of vector similarities. The stability and accuracy at even low values of k is an added improvement over other similar approaches. Hence, we believe that our framework of using control variates to achieve variance reduction could be beneficial when both fast computation and high accuracy is wanted.

Acknowledgements. This work is funded by the Singapore Ministry of Education Academic Research Fund Tier 2 Grant MOE2018-T2-2-013, as well as with the support of the Singapore University of Technology and Design's Undergraduate Research Opportunities Programme.

The authors also thank the anonymous reviewers for their comments and suggestions for improvement, which has helped to enhance the quality of the paper.

References

1. Achlioptas, D.: Database-friendly random projections: Johnson-Lindenstrauss with binary coins. J. Comput. Syst. Sci. **66**(4), 671–687 (2003)
2. Broder, A.Z.: On the resemblance and containment of documents. In: Compression and Complexity of Sequences 1997, Proceedings, pp. 21–29. IEEE (1997)

3. Charikar, M.S.: Similarity estimation techniques from rounding algorithms. In: Proceedings of the Thirty-Fourth Annual ACM Symposium on Theory of Computing, pp. 380–388. ACM (2002)
4. Glynn, P.W., Szechtman, R.: Some new perspectives on the method of control variates. In: Fang, K.T., Niederreiter, H., Hickernell, F.J. (eds.) Monte Carlo and Quasi-Monte Carlo Methods 2000, pp. 27–49. Springer, Heidelberg (2002). https://doi.org/10.1007/978-3-642-56046-0_3
5. Goemans, M.X., Williamson, D.P.: Improved approximation algorithms for maximum cut and satisfiability problems using semidefinite programming. J. ACM (JACM) **42**(6), 1115–1145 (1995)
6. Guyon, I., Gunn, S., Ben-Hur, A., Dror, G.: Result analysis of the NIPS 2003 feature selection challenge. In: Saul, L.K., Weiss, Y., Bottou, L. (eds.) Advances in Neural Information Processing Systems 17, pp. 545–552. MIT Press (2005)
7. Guyon, I., Gunn, S.R., Ben-Hur, A., Dror, G.: Result analysis of the NIPS 2003 feature selection challenge. In: NIPS, vol. 4, pp. 545–552 (2004)
8. Indyk, P., Motwani, R.: Approximate nearest neighbors: towards removing the curse of dimensionality. In: Proceedings of the Thirtieth Annual ACM Symposium on Theory of Computing, STOC 1998, pp. 604–613. ACM, New York (1998). https://doi.org/10.1145/276698.276876, http://doi.acm.org/10.1145/276698.276876
9. Kang, K.: Using the multivariate normal to improve random projections. In: Yin, H., et al. (eds.) IDEAL 2017. LNCS, vol. 10585, pp. 397–405. Springer, Cham (2017). https://doi.org/10.1007/978-3-319-68935-7_43
10. Kang, K.: Correlations between random projections and the bivariate normal. Data Min. Knowl. Disc. **35**(4), 1622–1653 (2021). https://doi.org/10.1007/s10618-021-00764-6
11. Kang, K., Wong, W.P.: Improving sign random projections with additional information. In: International Conference on Machine Learning, pp. 2479–2487. PMLR (2018)
12. LeCun, Y., Bottou, L., Bengio, Y., Haffner, P.: Gradient-based learning applied to document recognition. Proc. IEEE **86**(11), 2278–2324 (1998)
13. Li, P., Hastie, T.J., Church, K.W.: Improving random projections using marginal information. In: Lugosi, G., Simon, H.U. (eds.) COLT 2006. LNCS (LNAI), vol. 4005, pp. 635–649. Springer, Heidelberg (2006). https://doi.org/10.1007/11776420_46
14. Lichman, M.: UCI Machine Learning Repository (2013). http://archive.ics.uci.edu/ml
15. Rubinstein, R.Y., Marcus, R.: Efficiency of multivariate control variates in Monte Carlo simulation. Oper. Res. **33**(3), 661–677 (1985)
16. Slaney, M., Casey, M.: Locality-sensitive hashing for finding nearest neighbors [lecture notes]. IEEE Signal Process. Mag. **25**(2), 128–131 (2008)
17. Szechtman, R., Glynn, P.W.: Constrained Monte Carlo and the method of control variates. In: Proceeding of the 2001 Winter Simulation Conference (Cat. No. 01CH37304), vol. 1, pp. 394–400. IEEE (2001)

Pyramid Self-attention for Semantic Segmentation

Jiyang Qi[1], Xinggang Wang[1,2(✉)], Yao Hu[3], Xu Tang[3], and Wenyu Liu[1]

[1] School of Electronic Information and Communications, Huazhong University of Science and Technology, Wuhan, China
{jiyangqi,xgwang,liuwy}@hust.edu.cn
[2] Hubei Key Laboratory of Smart Internet Technology, School of Electronic Information and Communications, Huazhong University of Science and Technology, Wuhan, China
[3] Alibaba Group, Hangzhou, China
{yaoohu,buhui.tx}@alibaba-inc.com

Abstract. Self-attention is vital in computer vision since it is the building block of Transformer and can model long-range context for visual recognition. However, computing pairwise self-attention between all pixels for dense prediction tasks (*e.g.*, semantic segmentation) costs high computation. In this paper, we propose a novel pyramid self-attention (PySA) mechanism which can collect global context information far more efficiently. Concretely, the basic module of PySA first divides the whole image into $R \times R$ regions, and then further divides every region into $G \times G$ grids. One feature is extracted for each grid and then self-attention is applied to the grid features within the same region. PySA keeps increasing R (*e.g.*, from 1 to 8) to harvest more local context information and propagate global context to local regions in a parallel/series manner. Since G can be kept as a small value, the computation complexity is low. Experiment results suggest that as compared with the traditional global attention method, PySA can reduce the computational cost greatly while achieving comparable or even better performance on popular semantic segmentation benchmarks (*e.g.*, Cityscapes, ADE20k). The project code is released at https://github.com/hustvl/PySA.

Keywords: Semantic segmentation · Self-attention · Context modeling · Deep learning

1 Introduction

Semantic segmentation is one of the fundamental tasks in computer vision, which requests a model to classify every pixel in the image. Nowadays, semantic segmentation methods based on the fully convolutional network (FCN) have made great progress. However, the convolution operation, designed for modeling local relations, is limited on capturing long-range dependencies.

© Springer Nature Switzerland AG 2021
H. Ma et al. (Eds.): PRCV 2021, LNCS 13019, pp. 480–492, 2021.
https://doi.org/10.1007/978-3-030-88004-0_39

Much of the literature has explored approaches to enhance the long-range correlation modeling ability of convolutional neural networks (CNN). DeepLab-v2 [2] proposes atrous spatial pyramid pooling to enlarge the receptive field and exploit multi-scale features. PSPNet [31] utilizes the pyramid pooling operation to aggregate contextual information. Furthermore, some works leverage the attention mechanism to collect global context information. PSANet [32] predicts an attention map to connect every pixel with all the others. NonLocal [23] introduces self-attention [21] mechanism into CNNs to capture long-range context information by modeling global pixel-wise dependencies. However, the computation complexity of traditional image-level dense pixel-wise self-attention grows geometrically with the resolution of the input feature map. The resulting complexity in time and space are huge, especially for semantic segmentation task in which larger feature maps are required to predict finer segmentation results. To make the attention operation more efficient, CCNet [12] replaces the global attention by two consecutive criss-cross attention operations and achieves even better performance in semantic segmentation and object detection tasks.

Nevertheless, how to effectively and efficiently compute self-attention for visual recognition is still an open problem. We investigate this problem from a classical computer vision perspective - integrating self-attention with spatial pyramid matching [13]. To this end, we propose pyramid self-attention (PySA) which computes multi-scale region-level self-attention and aggregate coarse-scale global context with fine-scale local context. Concretely, the basic module of PySA first divides the whole image into $R \times R$ regions, and then further divides every region into $G \times G$ grids. One feature is extracted for each grid and then self-attention is applied to enhance the grid features within the same region. PySA keeps increasing R (e.g., from 1 to 8) to harvest more local context information and propagate global context to local regions in a parallel/series manner. Since G can be kept as a small value, the computation complexity is low.

We have conducted extensive experiments on popular large-scale datasets (e.g., Cityscapes [5], ADE20k [34]). Compared against the traditional global dense self-attention module, with an input feature map sized 97×97, our PySA can reduce the FLOPs more than one-hundred times while achieving comparable or even better performance.

To summarize, our main contributions are three-fold:

- We introduce the pyramid self-attention (PySA) mechanism which applies multi-scale region-level self-attention operations in a series/parallel manner to approximate the global self-attention and capture long-range context information more efficiently and effectively.
- PySA is a plug-and-play module, which can be applied in many dense prediction vision tasks and is helpful to develop more efficient and effective vision Transformers [8].
- We conduct extensive experiments on popular semantic segmentation datasets. The PySA-based segmentation networks reduce the computation complexity greatly, while achieving comparable or even better performance.

2   Related Work

Semantic Segmentation. Semantic segmentation has been promoted intensely by the development of deep neural networks. FCN [18] firstly replaces all the fully-connected layers in the network with convolution layers, which endows network the ability of pixel-level prediction. To further fuse the long-range semantic information from high-level features and precise localization information from low-level features, U-Net [20], DeconvNet [19], SegNet [1], and Deeplab-v3+ [4] utilize the encoder-decoder structure for multi-level feature fusion. [3,27] leverage dilated convolution layers to enlarge the receptive field and remove the last two downsampling operation to generate denser segmentation results. Deeplab-v2 [2] introduces ASPP module to fuse the contextual information. PSPNet [31] proposes pyramid pooling module to embed multi-scale scenery context features. Several approaches also utilize CRF [3,33] or MRF [17] to capture long-range context information and refine the coarse segmentation results.

Attention Mechanism. Attention mechanism has been effectively used in various tasks. SENet [11] adaptively recalibrates channel-wise feature responses by modeling relationships between channels using channel-wise attention mechanism. Self-attention [6,21,25] has become a popular basic operation in natural language process tasks. NonLocal [23] performs global pixel-wise self-attention which calculates the interactions between any two positions, regardless of their positional distance.

In recent years, many semantic segmentation methods basing on attention mechanism were proposed. PSANet [32] connects every pixel to all the other ones through a self-adaptively learned attention mask. CCNet [12] replaces the global attention by two consecutive criss-cross attention operations where every criss-cross attention captures long-range context information in horizontal and vertical direction. DANet [9] applies self-attention on both spacial dimension and channel dimension. OCNet [29] introduced the interlaced sparse self-attention scheme with two successive attention modules to capture long-range and short-range information respectively.

3   Approach

In this section, we first revisit the general self-attention module briefly, then elaborate the details of basic modules of PySA and how we build our network in series and parallel manners.

3.1   Self-attention Module Revisited

Self-attention mechanism in computer vision tasks is usually used to capture the long-range context dependency information [9,12,23,28,29]. Given an input feature map $\mathbf{X} \in \mathbb{R}^{HW \times C_{in}}$, the self-attention module first generate three feature maps $\{\mathbf{Q}, \mathbf{K}, \mathbf{V}\} \in \mathbb{R}^{HW \times C_{out}}$ through three 1×1 convolution layers respectively.

Then we further calculate the pixel-wise attention map $\mathbf{A} = softmax(\mathbf{Q}\mathbf{K}^T)$, where $\mathbf{A} \in \mathbb{R}^{HW \times HW}$. After that, the output feature can be obtained by $\mathbf{Y} = \mathbf{A}\mathbf{V}$, where $\mathbf{Y} \in \mathbb{R}^{HW \times C_{out}}$. The non-local block [23] also applies a 1×1 convolution layer on $\mathbf{Y}$ to increase the number of channels from C_{out} to C_{in}, and adds the input to it to obtain the final output.

However, when inputting high-resolution feature maps in semantic segmentation task, the attention map $\mathbf{A} \in \mathbb{R}^{HW \times HW}$ will be huge. The resulting complexity in time and space are both $\mathcal{O}(HW \times HW)$, which greatly increases the GPU memory cost and time consumption.

3.2   Pyramid Self-attention Mechanism

To capture long-range context information more efficiently and effectively, we propose the pyramid self-attention mechanism, which computes multi-scale region-level self-attention and aggregate coarse-scale global context with fine-scale local context.

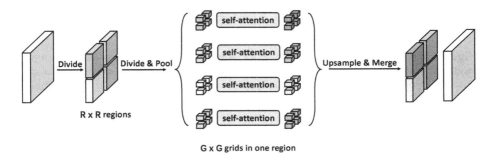

Fig. 1. Overview of one basic module of PySA, where each color denotes one individual region. Both R and G are set to 2. That is to say, the whole image is divided into 2×2 regions and every region is further divided into 2×2 grids. (Color figure online)

Basic Module of Pyramid Self-attention Mechanism. The overall architecture of one basic module of PySA is presented in Fig. 1. Given an input feature map sized $H \times W$, we first divide it into $R \times R$ regions along the spatial dimension. The resolution of every cropped region will be $\frac{H}{R} \times \frac{W}{R}$. Then we further divide each region into $G \times G$ grids and extract one feature for every grid by global average pooling. After that, we only apply the self-attention operation to grid features within the same region.

Thereby, when setting R to 1 and G to be equal to H and W, the basic module of PySA will be degenerated to the typical global pixel-wise self-attention. While when setting R to a larger value, the self-attention operation will be only applied locally within each region individually. And when reducing G, only coarser contextual information from the pooled grid features will be aggregated. Generally, the coarse pooled features still contain enough semantic information for a pixel far away. At last, we directly upsample the enhanced feature map to the same resolution as input via bilinear interpolation. The Fig. 1 shows the procedure when both R and G are set to 2.

Please note that the computation complexity of each basic module of PySA is $\mathcal{O}(R^2 \times (G^2 \times G^2))$ which is *not related to the resolution of input feature maps*. In this way, when the R and G are set small, the entire operation will be very efficient. By applying several basic modules with different R and G, the final output feature will be able to capture information of different granularity from various distances.

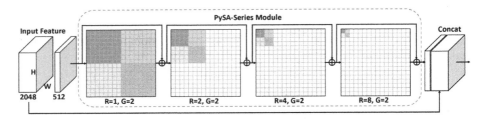

Fig. 2. Illustration of a PySA-Series block which consists of four consecutive basic modules. Note that different from the Fig. 1, each color in this image denotes a grid but not a region. For clarity, the grids in only one region are colored. (Color figure online)

Series Pyramid Self-attention Network. With the basic module proposed above, we build our Series Pyramid Self-Attention Network (PySA-Series) by replacing the traditional non-local attention module in NonLocal Network [23] with our series PySA module, which cascades several basic modules with increasing R and unchanged G, as illustrated in Fig. 2.

Following previous methods [23], given an input feature map produced by ResNet backbone [10], we first apply a 1×1 convolution to reduce the number of channels from 2048 to 512. On top of this feature map with reduced channels, we apply the series PySA module to model the multi-scale context information. For our default setting, we use four consecutive basic modules with R set to 1, 2, 4, 8 respectively and G set to 2. Thereby, the model initially divides the whole image into four grids and fuses the coarse semantic information from them with each other. Then, the same operation is applied in each grid individually and recursively. In this way, the model first aggregates coarse features from the whole image, and then recursively refines the features locally within the divided regions. For the stability of training, we apply residual connection after each basic module.

Finally, we concatenate the enhanced feature map with the input feature map to get the output, which will be directly used to predict the final pixel-level classification result.

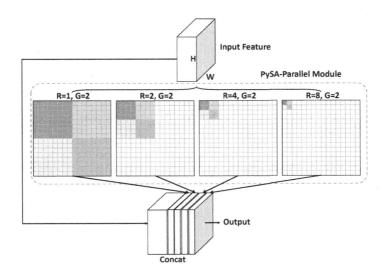

Fig. 3. Illustration of a PySA-Parallel block which applies four basic modules in parallel. For clarity, the grids in only one region are colored. Note that we remove the region dividing mechanism in our default setting of PySA-Parallel as illustrated in Sect. 3.2, which is different from the setting shown in this figure. (Color figure online)

Parallel Pyramid Self-attention Network. Similar to the PySA-Series described above, we further propose the Parallel Pyramid Self-Attention Network (PySA-Parallel) which applies multiple basic modules in parallel (Fig. 3).

Inputting the feature map directly generated by ResNet, four basic modules with different R and G will be applied to the input, where C_{out} of each basic module is 512. The outputted four feature maps will contain contextual information with various granularity and various receptive fields. Then, all the enhanced feature maps will be concatenated with the input feature map, which will be the input of the final pixel-level classification layer.

However, different from the series structure of PySA-Series which allows subsequent modules to leverage and refine the long-range coarse information captured by previous modules, the parallel structure results in that only the local feature from CNN backbone is inputted to the basic sub-modules in parallel PySA module. This leads to that the basic modules with a larger R lost the ability to utilize the long-range context information propagated from the basic modules with a small R. Besides, the fusion of features from different basic modules is only processed by the final classification layer which is not strong enough.

To provide global context information to all basic modules, for the default setting of PySA-Parallel, we remove the region dividing mechanism by setting all the R of basic modules to 1 and further change G to be the product of original R and G (*e.g.*, from $(R = 2, G = 2)$ to $(R = 1, G = 4)$). In this way, all basic modules in PySA-Parallel can capture global semantic information from the whole image individually with only little computational cost.

4   Experiments

We carry out extensive experiments on popular large-scale semantic segmentation datasets (*e.g.*, Cityscapes [5], ADE20k [34]), which reveal the efficiency and effectiveness of PySA.

4.1   Experimental Setup

Datasets and Metrics. Our experiments are mainly conducted on Cityscapes and ADE20K, which are the most commonly used benchmarks for semantic segmentation task.

– Cityscapes focuses on semantic segmentation of urban street scenes. It contains 5,000 high-quality pixel-level annotated frames in addition to 20,000 weakly annotated frames from 50 different cities. In our experiments, we only use the 5,000 finely annotated frames. The number of frames for training, validation, and testing are 2,975, 500, and 1,525 respectively.
– ADE20K contains 150 semantic categories totally, which makes it more challenging than most other semantic segmentation datasets. It provides 20K images for training, 2K images for validation.

For evaluation, we adopt mean IoU (Intersection over Union (IoU) averaged over all classes) as the metric.

Implementation Details. Following previous work [12,31], we adopt the modified ResNet [10] as our backbone for all of our experiments. Specifically, we remove the downsampling operations of the last two stages and apply dilated strategy [3,27] on the subsequent convolution layers. The stride of the final output of CNN backbone is 8. We adopt ResNet50 for ablation study and ResNet101 for comparison with previous methods.

As for training, the input image is augmented by random resizing with a scaling ratio ranging from 0.5 to 2, random cropping, and random horizontal flipping. The input resolution is 769 for Cityscapes and 512 for ADE20K. We employ stochastic gradient descent (SGD) with learning rate = 0.01, a momentum of 0.9, and a weight decay of 0.0005. The learning rate decays following polynomial learning rate policy used in [2,16], where the initial learning rate is multiplied by $1 - (\frac{iter}{iter_{max}})^{power}$ after each iteration with $power = 0.9$. The number of iterations $iter_{max}$ is set to 80K for Cityscapes and 160K for ADE20K.

4.2   Ablation Study

In this section, we study the series PySA module and parallel PySA module with a few alternatives. The results are shown in Table 1, Table 2, and Table 3. Concretely, every pair of brackets in the "Configuration" column of the tables denotes one basic module whose R is set to the first value in the brackets and G is set to the second value. We also present the increased FLOPs of non-local module [23], RCCA module [12], and our series and parallel PySA modules. All of our ablation experiments are conducted on the validation set of Cityscapes.

Table 1. Comparison of different number of basic modules and different dividing granularity. Every pair of brackets in the "Configuration" column of tables denotes one basic module whose R is set to the first value in the brackets and G is set to the second value. In this way, the number of bracket pairs denotes the number of basic modules.

Method	Configuration (R, G)	mIoU	Δ GFLOPs
Baseline	–	74.1	0
+ NonLocal [23]	–	77.3	50.29
+ RCCA [12]	–	78.5	0.17
+ PySA-Series	(1, 4) (4, 4)	78.7	**0.17**
	(1, 2) (2, 2) (4, 2) (8, 2)	**78.8**	0.24
	(1, 8) (2, 8) (4, 8) (12, 8)	**78.8**	5.96
+ PySA-Parallel	(1, 4) (4, 4)	78.1	**0.24**
	(1, 2) (2, 2) (4, 2) (8, 2)	78.8	0.31
	(1, 8) (2, 8) (4, 8) (12, 8)	**79.1**	9.04

Number of Basic Modules. According to the Table 1, using four basic modules can achieve better performance than only using two basic modules, especially for PySA-Parallel which gets a mIoU improvement of 0.7. The deeper or wider architecture increases the capacity of network, which may be the reason for the performance improvement. Enough capacity may be more critical for PySA-Parallel to aggregate the long-range context information effectively.

Dividing Granularity. We also test the effect of finer granularity by enlarging G to 8, as the "(1, 8) (2, 8) (4, 8) (12, 8)" configuration in Table 1, where 12×8 is approximately equal to the edge size of the input feature map sized 97×97. Thereby, the basic module with $R = 12$ and $G = 8$ whose grid size will be 1 can model pixel-level fine-grained dependencies. To align the input feature map and the pooled feature map more precisely, we change the resolution of input image to 768×768 for this setting, and the resulting input feature map of the attention module will be 96×96 which is exactly equal to 12×8. However, the experimental results show that using pixel-level fine-grained information cannot bring much performance improvement but will increase the FLOPs required greatly (more than twenty times larger than our default settings). This may be because that the detailed information of unrelated pixels far away may bring noises, and the local pixel-level information may have been exhaustively modeled by convolutional operations.

Table 2. Influence of removing the region dividing mechansim.

Method	Configuration (R, G)	w/o. Region Dividing	mIoU	Δ GFLOPs
Baseline	–	–	74.1	0
+ NonLocal [23]	–	–	77.3	50.29
+ RCCA [12]	–	–	78.5	0.17
+ PySA-Series	(1, 2) (2, 2) (4, 2) (8, 2)		**78.8**	**0.24**
	(1, 2) (1, 4) (1, 8) (1, 16)	✓	78.7	0.27
+ PySA-Parallel	(1, 2) (2, 2) (4, 2) (8, 2)		78.8	**0.31**
	(1, 2) (1, 4) (1, 8) (1, 16)	✓	**79.3**	0.38

Region Dividing. As stated in Sect. 3.2, we adopt the region dividing mechanism upon PySA-Series to limit the range of self-attention and further reduce the computation cost. While we remove the region dividing mechanism on PySA-Parallel for our default setting to allow every basic module in PySA-Parallel to model long-range context information from the whole image individually. Experiments with and without region dividing are presented in Table 2.

By removing the region dividing, the PySA-Parallel achieves a mIoU improvement of 0.5 with little computation cost, while the performance of PySA-Series is slightly worse than before. Our explanation is that the gain of mIoU on PySA-Parallel is because that the basic modules with larger R in PySA-Parallel can't leverage the long-range context information propagated from the basic modules with a small R, and the final classification layer is not strong enough to fuse the multi-scale features effectively. So, the removal of region dividing (setting R of all basic modules to 1) endows every basic module in PySA Parallel the ability to perceive the whole image. While the series structure of PySA Series allows subsequent basic modules to utilize the long-range information captured by previous basic modules with a small R. Therefore, PySA-Series can't achieve improvement by removing the region dividing mechanism. Besides, the fine-grained detailed feature from unrelated pixels far away may contain much noise, which may be the cause for the slight performance reduction on PySA-Series.

From Global to Local *vs*. From Local to Global. Furthermore, to evaluate the influence of the order of basic modules with different perception ranges, we reverse the order of basic modules in PySA-Series. Instead of propagating global information to finer-scale grids and refining the coarse global features recursively, the reversed PySA-Series may overwrite the fine local information captured by

Table 3. Influence of reversing the order of basic modules in PySA-Series.

Method	Configuration (R, G)	mIoU	Δ GFLOPs
PySA-Series	(1, 2) (2, 2) (4, 2) (8, 2)	**78.8**	**0.24**
PySA-Series (reversed)	(8, 2) (4, 2) (2, 2) (1, 2)	78.6	**0.24**

previous basic modules with the coarse information captured by subsequent basic modules, which leads to the slightly worse performance shown in Table 3.

Efficiency of Our Pyramid Attention Mechanism. As we can see in Table 1, Table 2, and Table 3, the FLOPs of our series and parallel PySA modules are significantly reduced, which are more than two orders of magnitude less than the non-local self-attention module, while the mIoU scores of our PySA-Series and PySA-Parallel are even higher. These results reveal the effectiveness and efficiency of our PySA mechanism and the redundancy of traditional global pixel-wise self-attention mechanism.

4.3   Main Results

We compare PySA-Series and PySA-Parallel against some state-of-the-art methods in semantic segmentation task. The default settings of R and G are stated in Sect. 3.2 for PySA-Series and PySA-Parallel.

Table 4. Comparison with state-of-the-art methods on Cityscapes test set.

Method	Backbone	w. Validation Set	mIoU
Deeplab-v2 [2]	ResNet-101		70.4
RefineNet [15]	ResNet-101	✓	73.6
DUC [22]	ResNet-101	✓	77.6
PSPNet [31]	ResNet-101		78.4
BiSeNet [26]	ResNet-101	✓	78.9
PSANet [32]	ResNet-101	✓	80.1
DenseASPP [24]	DenseNet-161	✓	80.6
SVCNet [7]	ResNet-101	✓	81.0
DANet [9]	ResNet-101	✓	81.5
OCNet [29]	ResNet-101	✓	81.4
CCNet [12]	ResNet-101	✓	81.4
PySA-Series	ResNet-101	✓	80.5
PySA-Parallel	ResNet-101	✓	81.3

Table 5. Comparison with state-of-the-art methods on ADE20k validation set.

Method	Backbone	mIoU
RefineNet [15]	ResNet-101	40.20
RefineNet [15]	ResNet-152	40.70
PSPNet [31]	ResNet-101	43.29
PSPNet [31]	ResNet-152	43.51
PSANet [31]	ResNet-101	43.77
EncNet [30]	ResNet-101	44.65
GCU [14]	ResNet-101	44.81
OCNet [29]	ResNet-101	45.04
CCNet [12]	ResNet-101	45.22
PySA-Series	ResNet-101	45.04
PySA-Parallel	ResNet-101	**45.27**

Results on Cityscapes and ADE20K are summarised in Table 4 and Table 5 respectively. Following previous works, we adopt multi-scale testing and flip testing on Cityscapes dataset, and only multi-scale testing is applied on ADE20K. We can see that PySA-Parallel and PySA-Series outperform most of the other methods. Compared with previous self-attention based methods (*e.g.*, DANet, OCNet, and CCNet) whose computation cost grows geometrically with the input resolution increasing, the computation cost of PySA mechanism is not related to the resolution of input feature maps. Therefore, the required FLOPs of PySA mechanism will be much less when inputting a larger feature map.

5 Conclusion

In this work, we have presented a pyramid self-attention mechanism, which targeting at performing self-attention more efficiently and effectively. To reduce the redundant computation of previous global dense self-attention, we apply multi-scale region-level self-attention operations in a series or parallel manner. The ablation experiments demonstrate the efficiency and effectiveness of our PySA mechanism which can significantly reduce the computation cost and promote the performance in semantic segmentation task. Compared with previous self-attention based methods, our PySA-Series and PySA-Parallel networks achieve competitive performance on two popular large-scale benchmarks while being much more efficient. In the future, we would apply PySA for more dense prediction tasks and develop PySA-based vision Transformers.

Acknowledgement. This work was in part supported by NSFC (No. 61876212 and No. 61733007) and the Zhejiang Laboratory under Grant 2019NB0AB02.

References

1. Badrinarayanan, V., Kendall, A., Cipolla, R.: SegNet: a deep convolutional encoder-decoder architecture for image segmentation. IEEE Trans. Pattern Anal. Mach. Intell. **39**, 2481–2495 (2017)
2. Chen, L.C., Papandreou, G., Kokkinos, I., Murphy, K., Yuille, A.L.: DeepLab: semantic image segmentation with deep convolutional nets, Atrous convolution, and fully connected CRFs. IEEE Trans. Pattern Anal. Mach. Intell
3. Chen, L.C., Papandreou, G., Kokkinos, I., Murphy, K., Yuille, A.L.: Semantic image segmentation with deep convolutional nets and fully connected CRFs. arXiv preprint arXiv:1412.7062 (2014)
4. Chen, L.-C., Zhu, Y., Papandreou, G., Schroff, F., Adam, H.: Encoder-decoder with Atrous separable convolution for semantic image segmentation. In: Ferrari, V., Hebert, M., Sminchisescu, C., Weiss, Y. (eds.) ECCV 2018. LNCS, vol. 11211, pp. 833–851. Springer, Cham (2018). https://doi.org/10.1007/978-3-030-01234-2_49
5. Cordts, M., et al.: The cityscapes dataset for semantic urban scene understanding. In: Proceedings of the IEEE Conference on Computer Vision and Pattern Recognition
6. Devlin, J., Chang, M.W., Lee, K., Toutanova, K.: BERT: pre-training of deep bidirectional transformers for language understanding. arXiv preprint arXiv:1810.04805 (2018)
7. Ding, H., Jiang, X., Shuai, B., Liu, A.Q., Wang, G.: Semantic correlation promoted shape-variant context for segmentation. In: Proceedings of the IEEE/CVF Conference on Computer Vision and Pattern Recognition
8. Dosovitskiy, A., et al.: An image is worth 16x16 words: transformers for image recognition at scale. arXiv preprint arXiv:2010.11929 (2020)
9. Fu, J., et al.: Dual attention network for scene segmentation. In: Proceedings of the IEEE/CVF Conference on Computer Vision and Pattern Recognition (2019)
10. He, K., Zhang, X., Ren, S., Sun, J.: Deep residual learning for image recognition. In: Proceedings of the IEEE Conference on Computer Vision and Pattern Recognition
11. Hu, J., Shen, L., Sun, G.: Squeeze-and-excitation networks. In: Proceedings of the IEEE Conference on Computer Vision and Pattern Recognition (2018)
12. Huang, Z., Wang, X., Huang, L., Huang, C., Wei, Y., Liu, W.: CCNet: criss-cross attention for semantic segmentation. In: Proceedings of the IEEE/CVF International Conference on Computer Vision, pp. 603–612 (2019)
13. Lazebnik, S., Schmid, C., Ponce, J.: Beyond bags of features: spatial pyramid matching for recognizing natural scene categories. In: 2006 IEEE Computer Society Conference on Computer Vision and Pattern Recognition (CVPR 2006), vol. 2, pp. 2169–2178. IEEE (2006)
14. Li, Y., Gupta, A.: Beyond grids: learning graph representations for visual recognition. In: Proceedings of the 32nd International Conference on Neural Information Processing Systems (2018)
15. Lin, G., Milan, A., Shen, C., Reid, I.: RefineNet: multi-path refinement networks for high-resolution semantic segmentation. In: Proceedings of the IEEE Conference on Computer Vision and Pattern Recognition (2017)
16. Liu, W., Rabinovich, A., Berg, A.C.: ParseNet: looking wider to see better. arXiv preprint arXiv:1506.04579 (2015)
17. Liu, Z., Li, X., Luo, P., Loy, C.C., Tang, X.: Semantic image segmentation via deep parsing network. In: Proceedings of the IEEE International Conference on Computer Vision (2015)

18. Long, J., Shelhamer, E., Darrell, T.: Fully convolutional networks for semantic segmentation. In: Proceedings of the IEEE Conference on Computer Vision and Pattern Recognition
19. Noh, H., Hong, S., Han, B.: Learning deconvolution network for semantic segmentation. In: Proceedings of the IEEE International Conference on Computer Vision (2015)
20. Ronneberger, O., Fischer, P., Brox, T.: U-Net: convolutional networks for biomedical image segmentation. In: Navab, N., Hornegger, J., Wells, W.M., Frangi, A.F. (eds.) MICCAI 2015. LNCS, vol. 9351, pp. 234–241. Springer, Cham (2015). https://doi.org/10.1007/978-3-319-24574-4_28
21. Vaswani, A., et al.: Advances in Neural Information Processing Systems
22. Wang, P., et al.: Understanding convolution for semantic segmentation. In: 2018 IEEE Winter Conference on Applications of Computer Vision (WACV) (2018)
23. Wang, X., Girshick, R., Gupta, A., He, K.: Non-local neural networks. In: Proceedings of the IEEE Conference on Computer Vision and Pattern Recognition (2018)
24. Yang, M., Yu, K., Zhang, C., Li, Z., Yang, K.: DenseASPP for semantic segmentation in street scenes. In: Proceedings of the IEEE Conference on Computer Vision and Pattern Recognition (2018)
25. Yang, Z., Dai, Z., Yang, Y., Carbonell, J., Salakhutdinov, R., Le, Q.V.: XLNet: generalized autoregressive pretraining for language understanding. arXiv preprint arXiv:1906.08237 (2019)
26. Yu, C., Wang, J., Peng, C., Gao, C., Yu, G., Sang, N.: BiSeNet: bilateral segmentation network for real-time semantic segmentation. In: Ferrari, V., Hebert, M., Sminchisescu, C., Weiss, Y. (eds.) ECCV 2018. LNCS, vol. 11217, pp. 334–349. Springer, Cham (2018). https://doi.org/10.1007/978-3-030-01261-8_20
27. Yu, F., Koltun, V.: Multi-scale context aggregation by dilated convolutions. arXiv preprint arXiv:1511.07122 (2015)
28. Yuan, Y., Chen, X., Wang, J.: Object-contextual representations for semantic segmentation. arXiv preprint arXiv:1909.11065 (2019)
29. Yuan, Y., Huang, L., Guo, J., Zhang, C., Chen, X., Wang, J.: OCNet: object context network for scene parsing. arXiv preprint arXiv:1809.00916 (2018)
30. Zhang, H., et al.: Context encoding for semantic segmentation. In: Proceedings of the IEEE conference on Computer Vision and Pattern Recognition (2018)
31. Zhao, H., Shi, J., Qi, X., Wang, X., Jia, J.: Pyramid scene parsing network. In: Proceedings of the IEEE Conference on Computer Vision and Pattern Recognition (2017)
32. Zhao, H., et al.: PSANet: point-wise spatial attention network for scene parsing. In: Ferrari, V., Hebert, M., Sminchisescu, C., Weiss, Y. (eds.) ECCV 2018. LNCS, vol. 11213, pp. 270–286. Springer, Cham (2018). https://doi.org/10.1007/978-3-030-01240-3_17
33. Zheng, S., et al.: Conditional random fields as recurrent neural networks. In: Proceedings of the IEEE International Conference on Computer Vision
34. Zhou, B., Zhao, H., Puig, X., Fidler, S., Barriuso, A., Torralba, A.: Scene parsing through ADE20K dataset. In: Proceedings of the IEEE Conference on Computer Vision and Pattern Recognition (2017)

Re-Identify Deformable Targets for Visual Tracking

Runqing Zhang, Chunxiao Fan, and Yue Ming$^{(\boxtimes)}$

Beijing University of Posts and Telecommunications, Beijing 100876, China

Abstract. Visual tracking is an important topic in computer vision, where the methods extracting features from the target's appearance have made significant progress. However, deformation causes an entirely different appearance from the original target, which reduces the precision of target's position. In this paper, we propose a novel method, named re-identify deformable targets for visual tracking (ReDT), to localize the target with a reliable appearance. Specially, we utilize a detector to achieve the candidates in the keyframes, and these candidates have the possibility to contain the reliable appearance of the deformable target. After that, we propose a similarity estimation to evaluate the candidates and determine a new target appearance, which fuses appearance with spatial location and category information to improve both inter-class and within-class discrimination. Extensive experiments and results demonstrate the effectiveness and advancement of the proposed method on two long-term tracking datasets, LaSOT and VOT-2018 LTB35, and three routine challenging tracking datasets, GOT-10k, OTB2015, and UAV123.

Keywords: Visual tracking · Re-identify targets · Similarity estimation · Siamese network · Deformation

1 Introduction

Visual tracking is an important topic in computer vision with wide applications, such as robotics [18], autonomous driving [2], and the military [16]. However, the significant appearance changes of targets lead to the shift of the model, especially in long-term tracking. Most existing tracking algorithms utilized an update module to make the model adapt to the deformation. However, the samples for the continuous update are cropped by rectangle windows containing disturbing background information, which is cumulative to result in an inaccurate target's appearance model in a long sequence.

In order to solve the above issues, current tracking methods introduce re-detection and candidates evaluation to achieve a new target appearance and update the model. However, the re-detection with dense sampling and the candidates evaluation online leads to a heavy computation burden.

© Springer Nature Switzerland AG 2021
H. Ma et al. (Eds.): PRCV 2021, LNCS 13019, pp. 493–504, 2021.
https://doi.org/10.1007/978-3-030-88004-0_40

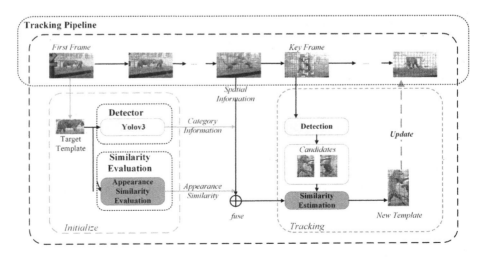

494   R. Zhang et al.

In this paper, we propose a novel tracking method, called re-identify deformable targets for visual tracking (ReDT), to update the appearance for deformation targets with high speed. ReDT includes the main tracking pipeline based on a Siamese Network, a detector, and a similarity evaluation module. An novel high-speed adaptive similarity evaluation module is applied to estimate the possibility of whether the candidate is the target, fusing category information and spatial information. The introduced detector avoids dense sampling based on Yolov3. The main contributions of this work can be concluded as follows:

1. The proposed similarity estimation module provides a reliable new appearance for targets, fusing the appearance information with spatial information. The spatial information improves the within-class discrimination, which improves the position precision between the deformation target and the similar objects.
2. The proposed tracking framework utilizes a high-speed detector based on Yolov3 to achieve category information. And we fuse the category information to improve the distinguishing ability for the inter-class objects, which improves the position precision of the targets with distractors.
3. The proposed method is trained with a small number of samples to estimate the appearance similarity by cross-correlation, which speeds up the re-detection process.

The experiment results show the effectiveness of the proposed method on five datasets. The rest of the paper is organized as follows: Related works on visual object tracking are surveyed in Sect. 2. The proposed algorithm is described in Sect. 3. Section 4 presents the results of the experimental evaluations. Our conclusions are given in Sect. 5.

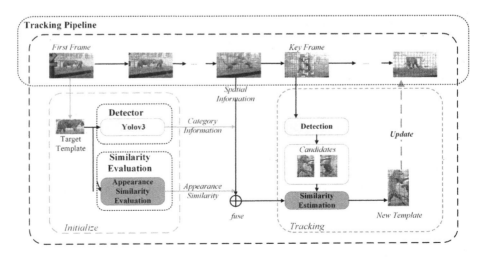

Fig. 1. The overview of our framework.

2   Related Work

This section discusses the tracking methods closely related to this work, including tracking methods with the update module and long-term tracking.

2.1   Tracking with the Update Module

Update module is used to address the target's appearance change in tracking methods. Existing tracking methods mainly including the Siamese networks, the deep learning classification/regression networks and correlation filters. The update modules they used are concluded as follows.

Tracking methods based on Siamese networks without the update module, such as SiamFC [1], are high-speed and efficient in short-term tracking tasks, but these methods may lose the targets when the target's appearance changes dramatically. DSiam [8] added layers for target appearance variation and background suppression transformations to improve the update module for the Siamese network. Most of the tracking methods based on Siamese networks lack the update module, such as SiamRPN++ [13], SiamFC++ [27], and Re-SiamNet [9]. Although the Siamese network had high efficiency, the end-to-end Siamese network lacking the update module leads to tracking failure when deformation occurs in the tracking tasks.

The online update is the typical strategy in tracking methods based on deep learning classification/regression networks. Nam et al. [22] proposed a multi-domain deep tracker by combing the pre-trained CNNs' layers and binary classification layers to fine-tune the classification layers for each sequence, which used the hard negative mining technique for online update. DLST [19] used the online update strategy for deep learning features and the incremental strategy for manual features. ATOM [20] utilized the IoU predictor to estimate a more accurate bounding box for the target, which only updates the classification network. Online update strategy makes the trackers keep learning the target's appearance. However, these methods applied dense sampling in each frame during the update, which increased the time cost.

Methods based on the correlation filters utilized the incremental update strategy for its high speed. The Kernel Correlation filter, KCF [10], formulated a circulant matrix to accelerate tracking speed, and the cross-correlation between the target's template and candidates was calculated in the frequency domain without dense sampling. Martin Danelljan et al. proposed Efficient Convolution Operation, ECO [5], to reduce the redundant feature responses caused by pre-trained CNN features, which improved both speed and target location precision. Dai et al. adopted adaptive spatial regularization to learn channel weights to track deformable targets [4]. Li et al. proposed online automatically and adaptively learn spatio-temporal regularization terms because predefined parameters failed to adapt to emerging issues [15]. Tracking methods based on correlation filters had high efficiency and precision in short-term tracking, but the circulant matrix generated unreal background samples, which lead to model drifting after cumulatively tracking a period of time.

2.2 Long-Term Tracking

Different from short-term tracking tasks, long-term tracking sequences include deformable targets, which have a huge appearance different from the targets given in the first frame. Liang et al. [17] exploited the prediction-detection-correlation tracking Framework, PDCT, which uses dense sampling and SVMs to re-train the model online. However, the dense samples brought high computational complexity, and the inaccurate candidates for model re-training lead to the model drift. Sui et al. used regularization terms for correlation filters to adapt the deformation targets [24]. However, the incremental update of correlation filters introduced cumulative background information into the template model, which reduced the accuracy of the appearance description. Yan et al. [28] proposed a 'Skimming-Perusal' long-term tracking framework, SPLT, applying a Siamese network to construct a tracker with secondary detection. However, the perusal model used cosine similarity to select the candidates as the predicted target results, which ignored the intensity of feature embedding and reduces the target's location precision. Qi et al. [23] used the Siamese network to calculate the similarity between the target template and candidates, but the dense samples predicted by correlation filter groups brought a long computing time. Tang et al. [25] exploited a re-detection module based on Kalman filters for dense samples. However, the Kalman filters focused on the spatial location information and ignored the appearance information of the targets, which reduced the precision of the re-detection module.

Different from the above method, we construct a detector to predict accurate candidates instead of dense sampling. Our similarity score estimation model utilizes spatial location information and the target's appearance to select the new target template, which is efficient for long-term tracking.

3 Methods

In this section, we first demonstrate the overview of our work. Second, we introduce the details of the proposed re-identify deformable target tracking method.

3.1 Overview

As shown in Fig. 1, the framework of our method mainly includes the tracking pipeline, a detector module and the similarity estimation module, and our work mainly pays attention to the similarity estimation module. The three modules are worked during the initializing process and tracking process separately.

We initialize the tracking model in the first frame. Firstly, we initialize the tracking pipeline with the given target appearance and location. Secondly, we crop the target's appearance image to train the similarity estimation module and achieve a template. Thirdly, we initialize the detector and detect the target's appearance in the first frame to achieve the target's category information.

During the tracking process, we re-initial the tracking pipeline based on the re-detection to achieve a new appearance for the deformable target. Firstly, we detect the keyframe to achieve candidates. Secondly, we calculate the similarity between the candidates and the template with the similarity estimation module. Thirdly, we use the candidate with the highest similarity score as the new target's appearance and re-initialize the tracking pipeline. The new appearance is reliable, and the tracking framework is robust for deformation targets.

The tracking pipeline is based on the high-speed tracker TransT(np) [3] because the self-attention network is used to focus the spatial relationship on the target's appearance. The training and parameter setting are the same as TransT [3]. The detector is based on Yolov3 [7] with the same setting. The proposed appearance similarity estimation is credited to the correlation filter [5], and only one frame is needed for training. The category information and spatial information are integrated into the score map produced by appearance similarity estimation, which improves the reliability of the new target's appearance.

3.2   Similarity Estimation Module

In this subsection, we first introduce the proposed appearance similarity estimation to generate the heat map and estimate the similarity between the candidates and the given target appearance. Then we integrate the category information and spatial information into the similarity estimation to achieve a more robust similarity score.

To train the proposed appearance similarity estimation module, we crop the target's appearance x_0 according to the given bounding box in the first frame and extract the feature $\psi(x_0)$, where $\psi(\cdot)$ is the function of the feature extraction network. Then the high dimension feature $\psi(x_0)$ is expanded into one dimension feature and transformed into $\Psi(x_0)$ after cyclic displacement. The corresponding label for the generated heat map is $\mathbf{Y_0}$. And the whole training process can be represented as:

$$min||\mathbf{Y_0} - \Psi(x_0) * \mathbf{W}|| \tag{1}$$

where the $\mathbf{W}$ is the learnable parameter of the appearance similarity model. According to the least square method:

$$\mathbf{W} = [\Psi(x_0)^H \cdot \Psi(x_0)]^{-1} \cdot \Psi(x_0)^H \cdot \mathbf{Y_0} \tag{2}$$

for the reason $[\Psi(x_0)^H \cdot \Psi(x_0)]$ is a circulant matrix, and the Eq. 2 can be formulated as:

$$\psi(x_0)^* \otimes \psi(x_0) \otimes \mathbf{W} = \psi(x_0)^* \otimes \mathbf{Y_0} \tag{3}$$

where $\otimes$ is convolution operator and $\psi(x_0)^*$ is the conjugate term of the repetend of $\Psi(x_0)$. Thus the parameter $\mathbf{W}$ can be calculated as:

$$\mathbf{W} = F^{-1} \frac{\psi(\hat{x}_0) \odot \hat{\mathbf{Y}}_0}{\psi(\hat{x}_0)^* \odot \psi(\hat{x}_0)} \tag{4}$$

where $\psi(\hat{x}_0)^*$ is conjugate term of the Fourier transform term $\psi(\hat{x}_0)$ and $\odot$ is the Hadamard product. F^{-1} is the inverse Fourier transform function. The inversion for the circulant matrix in Eq. 2 is transformed into the element-wise division, and the convolution operator in Eq. 3 is also transformed into the element-wise multiplication, which improve the solution efficiency.

When the keyframe is input, candidates are detected and estimated the similarity with the target template. For each candidates x_i, the feature $\psi(x_i)$ is extracted, and the cyclic displacement sample can be formulated as $\Psi(x_i)$. The heat map $\mathbf{Y_i}$ of each candidate x_i can be calculated as:

$$\mathbf{Y_i} = \Psi(x_i) * \mathbf{W} \tag{5}$$

In the heat map $\mathbf{Y_i}$, the location of peak value is the center location of the target in the candidate image, and the peak value P_i is the appearance similarity between the candidate and the target template.

If only the appearance similarity is estimated, there may be two issues. Firstly, the candidates with the same category may have a similar appearance, which leads to the unreliable maximum score. Secondly, there is a probability that the candidates achieve higher scores even if their categories are different from the target. To solve these two issues, first, we introduce the spatial location error as the spatial information to improve the robustness of the model. Second, we combined the category information. For issue 1, we introduce the spatial distance d_i, formulated as:

$$d_i = \frac{\rho_i}{(w_s + h_s)/2}. \tag{6}$$

where ρ_i is the Euclidean pixels distance between the candidate's center location in current frame x_i and the target's center location in the last frame, and w_s, h_s are the width and height of the input frame. For issue 2, we detect the candidate x_i and achieve its category cls_i, and compare it with the target's category cls_T detected in the first frame. Then, we can achieve the category parameter C_i:

$$C_i = \begin{cases} 0, & cls_i \neq cls_T, \\ 1, & cls_i = cls_T \end{cases} \tag{7}$$

We combine the Eq. 6 and Eq. 7 with the appearance similarity score P_i to calculate the final score:

$$S_i = (1 - C_i)e^{-d_i}P_i \tag{8}$$

From Eq. 8, we can find that the candidate whose location is closer to the target in the last frame achieves a higher similarity score. If the candidate's category is different from the target in the first frame, the similarity score is 0. We calculate all the candidates' similarities and select the candidate with the highest score as the new target template, and then the tracking pipeline is re-initialized.

4   Experiments

To evaluate the performance of the proposed visual object tracking algorithm, we conducted extensive experiments on 2 long-term tracking datasets, LaSOT [6] and VOT-2018 LTB35 [12], and 3 challenging public datasets: OTB-2015 [26], UAV123 [21], and GOT-10k [11]. We also present the results of ablation studies undertaken on OTB-2015 to confirm the model's efficiency.

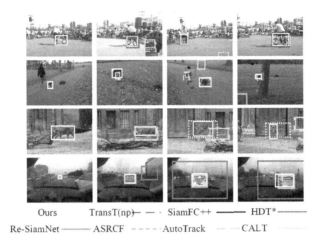

Ours   TransT(np)— — · SiamFC++ ———— HDT*———
Re-SiamNet———— ASRCF - - - - AutoTrack — — CALT

Fig. 2. The visualized results on LaSOT test dataset, including bicycle-2, dog-1, tiger-12, and car-2.

The proposed method is deployed on a single Titan X GPU with a speed of 36.83 FPS. We take the keyframe every 50 frames. The average number of candidates obtained from the detector is 2.6 on the LaSOT test dataset. We compare the proposed method with several long-term tracking methods and state-of-the-art tracking methods, including SiamFC++ [27], Re-SiamNet [9], HDT* [23], AutoTrack [15], SiamRPN++ [13], ASRCF [4], and ECO [5].

4.1 Long-Term Tracking Datasets

We compare the proposed method on 2 long-term tracking datasets, including the LaSOT test dataset and the VOT2018 LTB35 dataset. LaSOT test dataset contains 280 long-term sequences, and the VOT2018 LTB35 dataset contains 35 long-term sequences. Each of these sequences has more than 1000 frames, including different targets.

As shown in Fig. 3, the proposed method achieves the best performance of 0.608 precision and 0.623 success rate separately. In Fig. 4, the attributes include Deformation (DEF), Rotation (RO), Scale Variation (SV), Full Occlusion (FO), Fast Motion (FM), Out-of-View (OV), Low Resolution (LR), and Aspect Ration Change (ARC). The proposed method adopts a re-identification strategy to estimate the reliable target appearance, making the tracker performance more robust, especially in deformation issues. Our method benefits from

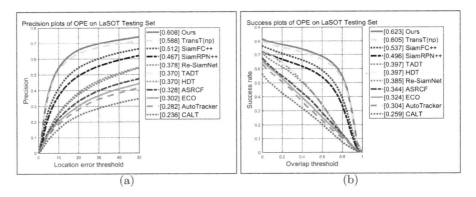

Fig. 3. The performance compared the proposed method with the state-of-the-art tracking methods on the LaSOT test dataset. (a) Precision; (b) Success rate.

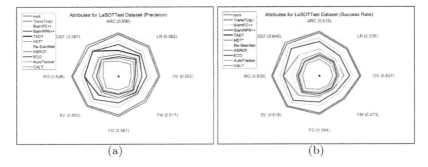

Fig. 4. The attributes performance compared the proposed method with the state-of-the-art tracking methods on the LaSOT test dataset. (a) Precision; (b) Success rate.

the detector and performs a better performance in deformation (DEF) with the 0.597 precision and 0.646 success rate. The visualized results are shown in Fig. 2.

The VOT2018 LTB35 dataset focuses on tracking in videos over 2 min long on average, where the object of interest can frequently disappear and reappear in the video. The evaluation results are shown in Table 1. The F-score is the long-term tracking measure and is used for ranking trackers. Pr and Re are the precision and tracking recalled separately. Our method obtains a competitive F-measure of 68.7 on this dataset and achieves a recall of 68.6 for the detector included in our re-identify model.

4.2 Normal Tracking Datasets

We also compare our method on the normal tracking datasets, including GOT-10k, OTB-2015, and UAV123. We report results on the Table 2. As shown in the table, we outperform other works in AO and SR on the GOT-10k dataset. The detector we utilized spends 0.71 s for every 50 frames, and the similarity

Table 1. Performance evaluation for algorithms on the VOT-2018 LTB35 dataset. The best result is marked in bold.

Method	F-score	Pr	Re
TransT(np) [3]	66.6	66.1	67.2
SiamFC++ [27]	62.1	62.0	62.3
SiamRPN++ [13]	61.9	61.0	62.9
TADT [14]	55.3	54.2	56.5
HDT* [23]	50.7	50.2	51.3
Re-SiamNet [9]	54.9	54.8	55.1
ASRCF [4]	54.0	53.3	54.7
ECO [5]	47.3	46.7	48.0
AutoTracker [15]	40.4	44.0	45.9
CALT [25]	41.0	40.4	41.6
Ours	**68.7**	**67.8**	**68.6**

estimation spends 0.68 s for every 50 frames. So that the proposed method is a little slower than the baseline method [3]. Tracking methods [4,5,15] run slowly because of the low-configure equipment. We also evaluate the proposed method on OTB-2015 and UAV123 datasets. As shown in Table 2, the proposed method obtains a better performance of 91.7 precision and 69.2 success rate in the OTB-2015 dataset. However, the videos in the UAV123 dataset are taken by UAV, which contains a mount of images with low resolution. In short-term tracking task, the proposed method improves the performance limitedly because some sequences are short than the keyframe number.

4.3 Ablation Study

To validate the effectiveness of the proposed method, we undertook ablation studies on the LaSOT test dataset. We combine other re-detection methods [23, 28] with the baseline for comparison. In Table 3, TransT(np) [3] is the baseline without re-detection. TransT-yolov3-cos is the method combining the baseline with the re-detection method based on Cosine distance [28]. TransT-yolov3-siamese is the method combining the baseline with the re-detection method based on a simple Siamese Network same as the method in HDT* [23].

From Table 3, we can find that the re-detection strategy is effective for long-term tracking tasks. Cosine distance is a useful to calculate the similarity, but Cosine distance attaches importance to the directional correlation of feature vectors and ignores the intensity value. Our method performs better than the TransT-yolov3-siamese because the Siamese Network can only estimate the appearance similarity, and we also take the spatial and category information into consideration. The proposed estimation method is faster than Siamese as shown in the table. We can find that the category information is essential to distinguish

Table 2. Performance comparison on OTB-2015, UAV123, and GOT-10k dataset, where AO means average overlap rate, and SR means success rate. 0.5 and 0.75 are the thresholds of success rate. Pre. means precision. The best result is marked in bold.

Method	OTB-2015		UAV123		GOT-10k			
	Pre.	SR	Pre.	SR	AO	$SR_{0.50}$	$SR_{0.75}$	Hz
TransT(np) [3]	89.4	65.1	**84.2**	**69.7**	0.614	0.707	0.537	39.2
SiamFC++ [27]	90.3	68.3	83.2	68.4	0.557	0.649	**0.428**	14.44
SiamRPN++ [13]	90.6	69.1	82.7	67.1	0.517	0.615	0.329	16.18
TADT [14]	78.4	65.0	70.3	53.2	0.374	0.404	0.144	34.63
HDT* [23]	91.3	69.8	–	–	0.426	0.563	0.299	25.31
Re-SiamNet [9]	85.4	61.2	72.9	55.1	0.433	0.581	0.317	18.14
ASRCF [4]	84.1	68.8	67.9	48.8	0.313	0.317	0.113	11.11
ECO [5]	83.3	65.3	77.1	58.9	0.316	0.309	0.111	2.62
AutoTracker [15]	79.5	63.7	71.3	51.6	0.286	0.277	0.109	15.37
CALT [25]	81.2	58.9	74.0	48.1	0.298	0.281	0.111	**54.13**
Ours	**91.7**	**69.2**	80.1	66.3	**0.616**	**0.708**	0.537	37.27

inter-class objects. Furthermore, spatial information is also important because the predicted target is usually closer to the target in the last frame than the within-class distractors. We test different keyframe in Table 3, the most suitable keyframe number is achieved and the frequency of detection will affect the speed of the algorithm.

Table 3. Ablation study on LaSOT test dataset.

Method	Precision	Success rate	Speed (FPS)
TransT(np) [3]	58.8	60.5	39.2
TransT-yolov3-cos	59.0	60.9	37.1
TransT-yolov3-siamese	59.6	61.4	33.7
Ours (lack spatial information)	60.2	61.8	37.1
Ours (lack category information)	59.9	61.5	36.9
Ours (keyframe = 25)	60.7	62.1	32.4
Ours (keyframe = 50)	**60.8**	**62.3**	**36.8**
Ours (keyframe = 70)	60.4	62.0	39.6

5   Conclusion

In this paper, we propose a novel re-identify deformable target for visual tracking, ReDT, that can effectively track deformable targets. A similarity estimation is also proposed fused with category and spatial information, reducing both inter-class and within-class interference. We utilize a detector to achieve sparse candidates instead of dense sampling and achieve a real-time speed. Experiments on 5 challenging datasets have demonstrated that the proposed algorithm can effectively improve the centering and precision for targets. In future research, we will further focus on optimizing the network structure for feature extraction, enabling a new, refined observation model to be used for complex scenes.

Acknowledgment. The work presented in this paper was partly supported by Natural Science Foundation of China (Grant No. 62076030), Beijing Natural Science Foundation of China (Grant No. L201023).

References

1. Bertinetto, L., Valmadre, J., Henriques, J.F., Vedaldi, A., Torr, P.H.S.: Fully-convolutional siamese networks for object tracking. In: Hua, G., Jégou, H. (eds.) ECCV 2016. LNCS, vol. 9914, pp. 850–865. Springer, Cham (2016). https://doi.org/10.1007/978-3-319-48881-3_56
2. Camara, F., et al.: Pedestrian models for autonomous driving part i: low-level models, from sensing to tracking. IEEE Trans. Intell. Transp. Syst. 1–15 (2020)
3. Chen, X., Yan, B., Zhu, J., Wang, D., Yang, X., Lu, H.: Transformer tracking. In: International Conference on Computer Vision and Pattern Recognition, pp. 1–11 (2021)
4. Dai, K., Wang, D., Lu, H., Sun, C., Li, J.: Visual tracking via adaptive spatially-regularized correlation filters. In: International Conference on Computer Vision and Pattern Recognition, pp. 4670–4679 (2019)
5. Danelljan, M., Bhat, G., Khan, F.S., Felsberg, M., et al.: ECO: efficient convolution operators for tracking. In: International Conference on Computer Vision and Pattern Recognition, pp. 3–14 (2017)
6. Fan, H., et al.: LaSOT: a high-quality benchmark for large-scale single object tracking. In: International Conference on Computer Vision and Pattern Recognition, pp. 5374–5383 (2019)
7. Farhadi, A., Redmon, J.: YOLOv3: an incremental improvement. In: International Conference on Computer Vision and Pattern Recognition, pp. 1–30 (2018)
8. Guo, Q., Wei, F., Zhou, C., Rui, H., Song, W.: Learning dynamic siamese network for visual object tracking. In: International Conference on Computer Vision, pp. 1–11 (2017)
9. Gupta, D.K., Arya, D., Gavves, E.: Rotation equivariant siamese networks for tracking. In: International Conference on Computer Vision and Pattern Recognition, pp. 1–10 (2021)
10. Henriques, J.F., Caseiro, R., Martins, P., Batista, J.: High-speed tracking with kernelized correlation filters. IEEE Trans. Pattern Anal. Mach. Intell. **37**(3), 583–596 (2014)
11. Huang, L., Zhao, X., Huang, K.: GOT-10k: a large high-diversity benchmark for generic object tracking in the wild. IEEE Trans. Pattern Anal. Mach. Intell. 1 (2019)
12. Kristan, M., et al.: The visual object tracking vot2018 challenge results. In: International Conference on Computer Vision Workshops, pp. 1–23 (2018)

13. Li, B., Wu, W., Wang, Q., Zhang, F., Xing, J., Yan, J.: SiamRPN++: evolution of siamese visual tracking with very deep networks. In: International Conference on Computer Vision and Pattern Recognition, pp. 4282–4291 (2019)
14. Li, X., Ma, C., Wu, B., He, Z., Yang, M.H.: Target-aware deep tracking. In: International Conference on Computer Vision and Pattern Recognition, pp. 1369–1378 (2019)
15. Li, Y., Fu, C., Ding, F., Huang, Z., Lu, G.: AutoTrack: towards high-performance visual tracking for UAV with automatic spatio-temporal regularization. In: International Conference on Computer Vision and Pattern Recognition, pp. 11923–11932 (2020)
16. Li, Y., Fu, C., Huang, Z., Zhang, Y., Pan, J.: Intermittent contextual learning for keyfilter-aware UAV object tracking using deep convolutional feature. IEEE Trans. Multimedia **23**, 810–822 (2020)
17. Liang, N., Wu, G., Kang, W., Wang, Z., Feng, D.D.: Real-time long-term tracking with prediction-detection-correction. IEEE Trans. Multimedia **20**(9), 2289–2302 (2018)
18. López-Nicolás, G., Aranda, M., Mezouar, Y.: Adaptive multirobot formation planning to enclose and track a target with motion and visibility constraints. IEEE Trans. Robot. **36**(1), 142–156 (2019)
19. Lu, X., Ma, C., Ni, B., Yang, X., Reid, I., Yang, M.H.: Deep regression tracking with shrinkage loss. In: European Conference on Computer Vision, pp. 353–369 (2018)
20. Danelljan, M., Bhat, G., Khan, F.S., Felsberg, M.: Accurate tracking by overlap maximization. In: International Conference on Computer Vision and Pattern Recognition, pp. 1–11 (2019)
21. Mueller, M., Smith, N., Ghanem, B.: A benchmark and simulator for UAV tracking. In: Leibe, B., Matas, J., Sebe, N., Welling, M. (eds.) ECCV 2016. LNCS, vol. 9905, pp. 445–461. Springer, Cham (2016). https://doi.org/10.1007/978-3-319-46448-0_27
22. Nam, H., Han, B.: Learning multi-domain convolutional neural networks for visual tracking. In: International Conference on Computer Vision and Pattern Recognition, pp. 4293–4302 (2016)
23. Qi, Y., et al.: Hedging deep features for visual tracking. IEEE Trans. Pattern Anal. Mach. Intell. **41**(5), 1116–1130 (2019)
24. Sui, Y., Wang, G., Zhang, L.: Correlation filter learning toward peak strength for visual tracking. IEEE Trans. Cybern. **48**(4), 1290–1303 (2017)
25. Tang, F., Ling, Q.: Contour-aware long-term tracking with reliable re-detection. IEEE Trans. Circuits Syst. Video Technol. **30**, 4739–4754 (2019)
26. Wu, Y., Lim, J., Yang, M.H.: Object tracking benchmark. IEEE Trans. Pattern Anal. Mach. Intell. **37**(9), 1 (2015)
27. Xu, Y., Wang, Z., Li, Z., Yuan, Y., Yu, G.: SiamFC++: towards robust and accurate visual tracking with target estimation guidelines. In: AAAI Conference on Artificial Intelligence, pp. 12549–12556 (2020)
28. Yan, B., Zhao, H., Wang, D., Lu, H., Yang, X.: 'Skimming-perusal' tracking: a framework for real-time and robust long-term tracking. In: International Conference on Computer Vision, pp. 2385–2393 (2019)

End-to-End Detection and Recognition of Arithmetic Expressions

Jiangpeng Wan[1]([✉]), Mengbiao Zhao[2,3], Fei Yin[2,3], Xu-Yao Zhang[2,3], and LinLin Huang[1]

[1] Beijing Jiaotong University, Beijing 100044, China
{19120021,huangll}@bjtu.edu.cn
[2] National Laboratory of Pattern Recognition, Institute of Automation of Chinese Academy of Sciences, Beijing 100190, China
zhaomengbiao2017@ia.ac.cn, {fyin,xyz}@nlpr.ia.ac.cn
[3] School of Artificial Intelligence, University of Chinese Academy of Sciences, Beijing 100049, China

Abstract. The detection and recognition of handwritten arithmetic expressions (AEs) play an important role in document retrieval [21] and analysis. They are very difficult because of the structural complexity and the variability of appearance. In this paper, we propose a novel framework to detect and recognize AEs in an End-to-End manner. Firstly, an AE detector based on EfficientNet-B1 [17] is designed to locate all AE instances efficiently. Upon AE location, the RoI Rotate module [11] is adopted to transform visual features for AE proposals. The transformed features are then fed into an attention mechanism based recognizer for AE recognition. The whole network for detection and recognition is trained End-to-End on document images annotated AE locations and transcripts. Since the datasets in this field are rare, we also construct a dataset named HAED, which contains 1069 images (855 for training, and 214 for testing). Extensive experiments on two datasets (HAED and TFD-ICDAR 2019) show that the proposed method has achieved competitive performance on both datasets.

Keywords: Arithmetic expression spotting · End-to-End training · Sequence-to-sequence

1 Introduction

Arithmetic Expression (AE) spotting aims to detect and recognize AEs in document images. It has attracted increasing attention in recent years, due to its wide application in document analysis and intelligent education systems. Although there exist many methods for scene text spotting, these methods cannot be directly applied to AE spotting because of the characteristics of AEs.

© Springer Nature Switzerland AG 2021
H. Ma et al. (Eds.): PRCV 2021, LNCS 13019, pp. 505–517, 2021.
https://doi.org/10.1007/978-3-030-88004-0_41

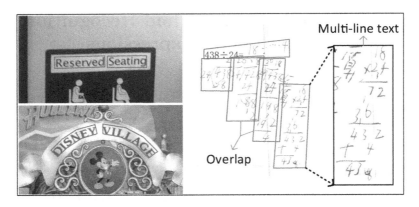

Fig. 1. Comparison of scene text and AEs. Scene text usually appear in straight line (Top left image) or curved line (Bottom left image). While AEs text always appear in multi-line (Right image) with complex layout.

Compared to scene text spotting, the difficulties of AE spotting mainly stem from these reasons: AEs have more complex structures, which are usually composed of printed and handwritten texts with arithmetical pattern; AEs may overlap each other (See the right of Fig. 1); Compared with printed texts, handwritten texts are usually scratchy, leading to zigzag boundaries and tangled rows. Therefore, AE spotting is more challenging than scene text spotting.

One most related method [4] performs AE spotting through two independent steps: a detector is firstly employed to detect all AE instances, then a recognizer is conducted on the detected regions. This method has two disadvantages: 1) multi-stage error accumulation, 2) heavy time consumption based on two modules.

In this paper, we propose an End-to-End trainable framework (See Fig. 2) for handwritten AE spotting, combining the detection network and the recognition network as a whole. By sharing the feature extractor, the shared feature maps from the input image are calculated only once, so that, the AE spotting framework can be effectively accelerated. Specifically, EfficientNet-B1 [17] is adopted as the backbone of the feature extractor to get the deep visual features of the input image. The extracted features are fed into the detection network to generate AE proposals. To connect the detection and recognition networks, a RoI feature extraction module is adopted to extract and rectify AE regions. After that, rectified AE features are fed into an attention based recognizer to generate the recognition results. To the best of our knowledge, this is the first End-to-End trainable framework for AE spotting. In addition, to promote the future researches in this direction, we collect documents of arithmetic exercises from a dozen of primary schools, and construct a new handwritten arithmetic expression dataset named HAED. The dataset contains 1069 images with 47308 AE instances. For evaluation and comparison, we use 855 images for training and 214 images for testing.

In summary, the main contributions of this work are as follows:

(1) A novel End-to-End AE spotting framework is proposed, which can detect and recognize the AE instances in an image simultaneously, and has achieved competitive performance on public benchmark (TFD-ICDAR 2019) and our own dataset (HAED).
(2) A handwritten arithmetic expression dataset (HAED) has been proposed. It contains 1069 exercise document images from more than a dozen primary schools, and the exercises were written by hundreds of students with various writing styles.

2   Related Works

This work focus on arithmetic expression (AE) spotting, including AE detection and AE recognition. However, to the best of our knowledge, very few works of AE spotting have bee published. Since AE is a special type of mathematical expressions, we also review the related works from the perspectives of mathematical expression detection and recognition.

2.1   Mathematical Expression Detection

Most of the mathematical expression detection approaches can be classified into two categories: traditional methods and deep learning based methods. Traditional methods [2,9] and [5] are based on handcraft features, whose pipelines have many intermediate steps. While, with the rapid development of technology, deep learning based methods have become the main stream. Ohyama et al. [15] used a U-net to detect mathematical expressions (MEs) and mathematical symbols in scientific document images. Mali et al. proposed the ScanSSD [14], which adopted an anchor based detector [10] to detect mathematical expressions and symbols.

2.2   Mathematical Expression Recognition

Before the era of deep learning, ME recognition was addressed mostly using symbol segmentation and structural analysis methods, but could not achieve high accuracies. Recent deep learning methods are mostly based on recurrent or attention networks. Le et al. [7] first use a convolution neural network to extract feature from the input image, then apply a bidirectional LSTM to encode the extracted features. Finally, an attention based LSTM decoder is used to generate LaTex strings. Zhang et al. proposed an End-to-End recognition model named WAP [20]. First, a convolution neural network was used to encode the input images, which was called the Watcher. Then, an attention based recurrent neural network (RNN) was used to parse the encoded features to LaTeX strings, which was called the Parser.

2.3   Arithmetic Expression Spotting

We only found one related work in this issue. Hu et al. [4] designed a system called arithmetical exercise checker (AEC), which can automatically correct primary arithmetic exercises. This system consists three independent branches: detection branch, recognition branch and evaluation branch. The disadvantages of the multi-stage system lie in heavy time cost and multi-stage error accumulation. In this paper, we propose an End-to-End trainable framework with an unified network, which have better efficiency and higher accuracy.

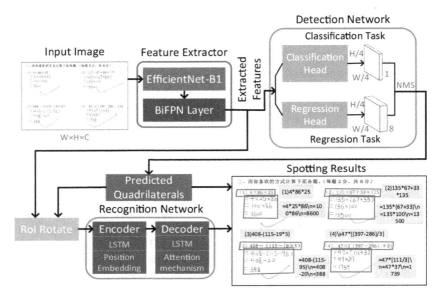

Fig. 2. Overview of the proposed framework, which consists of three major parts: feature extractor, detection network and recognition network.

3   Methodology

3.1   Framework Overview

An overview of the proposed framework is shown in Fig. 2, where it can be seen that the detection network and recognition network share the feature extractor. Inspired by [18], we use the EfficientNet-B1 [17] as the backbone network of the feature extractor, and we adopt a bi-directional feature pyramid network (BiFPN [18]) to fuse multi-level features. Then the detection network outputs dense per-pixel predictions of AE instances. With oriented AE proposals (bounded with quadrilaterals) produced by the detection network, the RoI feature extraction module converts quadrilaterals into fixed-height rectangles while maintaining the original region aspect ratios, and maps the visual features accordingly. Finally, the recognition network recognize AEs in all RoIs. In the following, we will describe these parts in details.

3.2   Feature Extractor

We adopt a modified EfficientDet-D1 [18] as our feature extractor. The original EfficientDet-D1 takes the level 3 to 7 features output by EfficientNet-B1 [17] as input, then outputs multi-level feature maps. Different from that, we use parts of the outputs (level 2 to 6) as extracted features, and apply a BiFPN [18] network to merge multi-level features. Finally, the feature extractor outputs a single feature map of $\frac{1}{4}$ size of the input image.

3.3   Detection Network

Inspired by the fast scene text detectors [3] and [22], we design a one stage detection network, which consists of two tasks: Pixel-wise Classification and Quadrilateral Regression. The detection network takes extracted features as input, and outputs dense per-pixel predictions in $\frac{1}{4}$ size of the input image. Finally, dense quadrilaterals are obtained by the post processor, where redundant quadrilaterals are removed by NMS.

Pixel-Wise Classification. This task is to extract AE regions from the input image and can be viewed as a down-sampled pixel-wise classification between AEs (positive category) and background (negative category). Because AEs may overlap each other, it is difficult for the detection network to locate each AE clearly. To alleviate this effect, only a shrunk version of AE region is considered as valid area. And the shrunk part is considered as "Ignore Region" and is not used for loss calculation.

In order to overcome the class-imbalance between AE and background pixels, we adopt the Focal Loss [8] with $\alpha = 0.25$ and $\gamma = 1.5$, where the α and γ are the hyper-parameters to control the balance of AE and background pixels. We denote the output of this task and the corresponding ground truth as $\hat{C}$ and C, respectively. Then the loss function of this task can be written as:

$$L_{cls} = \frac{1}{N} \sum_{y,x} \text{Focal Loss}(C_{y,x}, \hat{C}_{y,x}), \tag{1}$$

where N is the number of elements of C (ground-truthed pixels). $C_{y,x}$ is the value of C at pixel location (y, x).

Quadrilateral Regression. This task is to predict the offsets of positive samples (AE proposals) to the four vertices of the corresponding quadrilaterals. We denote the output of the quadrilateral regression task and the corresponding ground truth as $\hat{G}$ and G, respectively. The smooth L_1 loss [16] is adopted in this task, because it is more robust than L_1 loss and L_2 loss. Then the loss function of this task can be written as:

$$L_{reg} = \sum_{c,\,y,\,x} [C_{y,x} > 0] \cdot \text{smooth L}_1(G_{c,y,x} - \hat{G}_{c,y,x}), \qquad (2)$$

where c is the channel index, (x,y) are the coordinates of down-sampled feature map $\hat{G}$.

Thus the complete loss function of detection network can be defined as:

$$L_{det} = L_{cls} + \lambda_{reg}L_{reg}, \qquad (3)$$

where λ_{reg} is the hyper parameter to control the task balance of detection network.

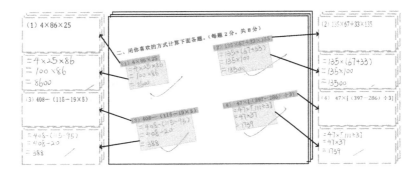

Fig. 3. Schematic diagram of RoI feature extraction.

3.4   RoI Feature Extraction

In order to reuse the visual features extracted by feature extractor (Sect. 3.2), we adopt affine transformation to extract RoI features from shared features based on the predicted AE proposals of the detection network. We denote the transformation matrices $\text{M} = \{M_1, M_2, ..., M_K\}$ if there are K AE instances. Then the K matrices are used to transform oriented quadrilateral AE features into axis aligned ones corresponding to fixed-height rectangles. In our experiments, the height is fixed to 8, and bilinear interpolation is used in this transformation. One schematic diagram of RoI feature extraction is shown in Fig. 3.

$$\begin{pmatrix} x_s \\ y_s \\ 1 \end{pmatrix} = M_i \begin{pmatrix} x_t \\ y_t \\ 1 \end{pmatrix} = \begin{bmatrix} \theta_i^{11} & \theta_i^{12} & \theta_i^{13} \\ \theta_i^{21} & \theta_i^{22} & \theta_i^{23} \\ 0 & 0 & 1 \end{bmatrix} \begin{pmatrix} x_t \\ y_t \\ 1 \end{pmatrix}, \qquad (4)$$

where (x_s, y_s) and (x_t, y_t) are the coordinates of a pixel on the shared feature map and the extracted RoI feature map respectively. And θs are the calculated parameters in affine transformation matrix M_i.

3.5 Recognition Network

The recognition network aims to predict the structured AE labels using the region features extracted by the feature extractor (Sect. 3.2) and transformed by the RoI feature extraction module (Sect. 3.4). Our recognition network consists of an encoder and a decoder. We denote the transformed feature map F and it's width W_F and height H_F.

First, the features F are fed into a 2-layer bi-directional LSTM network row by row. Then a simple but effective position embedding method introduced in [1] is used to capture the vertical information for the restoration of the structure information of AEs. In the encoding of each row, the row index is first fed into a trainable linear transformation matrix to generate the position weight, and then the position weight is added to the beginning of the row. And we denote the encoded features as $\bar{F}$.

Then the encoded features $\bar{F}$ are fed into decoder LSTMs. For the learning target $\{r_t\}$, the decoder predicts the probability of the next token given the history and the annotations. Thus the decoder can be defined as:

$$p(r_{t+1}|r_1, r_2, ..., r_t, \bar{F}) = \text{softmax}(W^{out}\tanh(W^c[\text{h}_t; \text{c}_t])), \qquad (5)$$

where W^{out}, W^c are linear transformation matrices to be learned, and h_t is the past decoded states. c_t is the context vector and its calculation procedure follows that in [12]. While for decoding, the global attention mechanism proposed in [12] is applied.

Finally, the loss function of recognition network can be formulated as:

$$L_{recog} = -\sum_{i=1}^{N} \log p(Y_i|R_i), \qquad (6)$$

where $p(Y_i|R_i)$ indicates the probability of the symbol sequence Y_i of i-th AE region R_i, and N indicates the number of AE regions.

For the End-to-End training of the proposed framework, we combine the detection loss with the recognition loss, then the complete loss function is:

$$L_{tot} = L_{cls} + \lambda_{reg}L_{reg} + \lambda_{recog}L_{recog}, \qquad (7)$$

where λ_{reg}, λ_{recog} are the hyper-parameters to control the balance among different tasks. In our experiments, λ_{reg} and λ_{recog} are set to 0.01 and 0.1, respectively.

4 HAED Dataset and Annotation

As there is no public dataset available in this field, we first propose a handwritten arithmetic expression dataset named HAED.

Dataset Introduction. The HAED dataset contains 1069 (855 for training and 214 for testing) document images with 47308 AE instances. All the images of various arithmetic papers were manually collected from more than a dozen primary schools, written by hundreds of students with different writing styles.

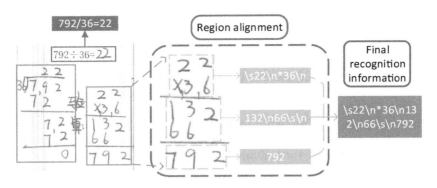

Fig. 4. The annotation details of HAED dataset.

Annotation Details. In terms of semantic information, all AEs can be divided into questions and answers, and each question can have multiple corresponding answers. Firstly, the bounding quadrilaterals are used to locate the questions and answers, which are then annotated transcript (sequence of symbols). The annotation of question is rather simple because the texts of question are always on a single line. While the texts of answers are always on multiple lines. We propose a LaTeX-like annotating rule to preserve the spatial information. More specifically, each answer region is divided into several parts through the calculation process, then we align the characters by vertical direction in each part to get the local recognition information, as shown in Fig. 4. In addition, several special symbols are used to represent the special mathematical symbols or structure information, e.g., "$\backslash s$", "$\backslash d$", and "$\backslash n$" represent blank, remainder, and line feed, respectively. Finally, we merge all parts' recognition information together to get the complete ground truth information.

5   Experiments

5.1   Datasets

To evaluate the performance of the proposed method, we conduct experiments on our own dataset (HAED) and a public benchmark (TFD-ICDAR 2019, for detection only).

TFD-ICDAR 2019 is a document image formula detection dataset. Its training set contains 36 PDF files (Total 569 pages), and the testing set contains 10 PDF files (Total 236 pages). This dataset provides both char-level and math-level annotations, and the attribution (character type) of the characters is also annotated.

5.2   Implementation Details

Training Stage. The whole experiments are implemented on PyTorch framework, and run on a workstation with 12-core CPU, 256G RAM and Titan RTX graphic cards. The whole network is optimized by stochastic gradient descent (SGD) with momentum 0.9 and weight decay 4e−5. The initial learning rate is set to 0.0001, and is linearly increased to 0.003 in the first 6 epochs. Then, the learning rate is divided by 2 every 50 epochs. The network is trained 300 epochs in 48 batches on 4 GPUs. The data augmentation is also important: (1) The input images are rotated randomly in the range of $[-10°, 10°]$; (2) The input images are resized randomly in the range of $[2^{-0.8}, 2^{0.8}]$.

Inference Stage. In the inference stage, we apply a multi-scale sliding window strategy, in which the window size, step and multi-scale set are set to 640×640, 480 and $\{2^{-3}, 2^{-2}, ..., 2^{1}\}$ respectively. The confidence threshold T_{cls} is set to 0.9. Results from all scales are combined and fed to NMS to get the final results. The evaluation criteria of detection follows that in ICDAR 2015 competition [6], which can evaluate the detection performance on quadrilateral targets, and the End-to-End recognition performance is measured by accuracy rate.

5.3   Experimental Results

Results on HAED. As shown in Table 1, the proposed method has achieved 97.1% of F-measure in AE detection, and 67.9% of accuracy rate in End-to-End AE spotting. To the best of our knowledge, there is no previous published methods on AE detection, so we compare our detection method with some scene text detection methods (EAST [22], and PSENet [19]). The detection performance of our method has outperformed them by a large margin using only single scale, which demonstrate the superiority of the proposed framework. More analysis of the proposed framework is studied in Sect. 5.4.

Table 1. Detection and End-to-End results on HAED dataset. MS represents multi-scale strategy. Both detection and recognition performance are measured in expression-level.

Methods	Detection			End-to-end accuracy	
	Precision	Recall	F-measure	With "\s"	Without "\s"
EAST [22]	87.6	85.7	86.6	–	–
PSENet [19]	90.3	88.4	89.3	–	–
Our detection	**99.1**	94.6	96.8	67.6	70.9
Our detection MS	99.0	**95.2**	**97.1**	**67.9**	**71.2**

Results on TFD-ICDAR 2019. We compare our method with other state-of-the-art methods in Table 2. With only single scale setting, our method achieves a F-measure of 80.2%. Using multi-scale setting, our method achieves a F-measure of 81.9%, which is a competitive performance (Note that the Samsung method is conducted with the help of character information, which is not suitable for comparison with our method).

Table 2. Comparison with other detection methods on TFD-ICDAR 2019 dataset. "*", "†" and "MS" indicate using character information, extra data and multi-scale strategy, respectively.

Methods	IoU ≥ 0.75			IoU ≥ 0.5		
	Precision	Recall	F-measure	Precision	Recall	F-measure
ScanSSD† [14]	78.1	69.0	73.3	84.8	74.9	79.6
RIT 1 [13]	63.2	58.2	60.6	74.4	68.5	71.3
RIT 2 [13]	75.3	62.5	68.3	83.1	67.0	75.4
Mitchiking [13]	19.1	13.9	16.1	36.9	27.0	31.2
Samsung* [13]	94.1	92.7	93.4	94.4	92.9	93.6
Our detection	80.3	72.2	76.0	84.2	76.6	80.2
Our detection MS	**81.5**	**73.7**	**77.4**	**85.6**	**78.5**	**81.9**

Table 3. Ablation studies of proposed method on HAED dataset. "att." indicates attention mechanism.

Backbone		End-to-end training	Detection			Recognition Accuracy	
VGG-16	EffNet-B1		Precision	Recall	F-measure	With att.	Without att.
✓	–	–	85.6	84.5	85.0	57.9	55.6
✓	–	✓	88.3	86.7	87.5	59.3	56.8
–	✓	–	94.4	93.8	94.1	65.2	63.4
–	✓	✓	**99.1**	**94.6**	**96.8**	**67.6**	**64.8**

5.4   Ablation Studies

For better understanding the strengths of the proposed framework, we provide the ablation studies from three aspects. All the ablation studies are conducted on the HAED dataset.

The Influence of Different Backbone Network. We compare the effects of different backbones (VGG-16 vs EfficientNet-B1) of the proposed method. As shown in Table 3, under the same setting, EfficientNet-B1 has outperformed VGG-16 by nearly 9.3%, which demonstrate the superiority of EfficientNet-B1 well. This is because the EfficientNet [17] proposes a compound scaling method which considers the width, depth and resolution of the convolution model, simultaneously. On this basis, the proposed EfficientNet-B1 has not only high performance but also high efficiency.

The Influence of End-to-End Training. As shown in Table 3, the performance of the model that trained End-to-End is better than that trained separately. End-to-End model performs better in detection because the recognition supervision helps the network to learn detailed character level features. Without End-to-End training, the detection network may miss some AE regions or misclassify AE-like background.

The Influence of Attention Mechanism. As shown in Table 3, the model trained with attention mechanism achieved the best performance of 67.6%, while the model trained without the attention mechanism only achieved 64.8% (under the same setting), which confirmed the importance of attention mechanism.

6   Conclusion

In this work, we present an novel End-to-End trainable framework for arithmetic expression spotting. A RoI feature extraction module is adopted to unify detection network and recognition network. By sharing the feature extractor, the computational cost of the whole system is largely reduced, and end-to-end training yields superior performance of AE detection and recognition. In addition, we proposed the first handwritten arithmetic expression dataset named HAED, which could promote future researches in this direction. Our experiments on standard benchmark (TFD-ICDAR 2019) and HAED show that our method achieves competitive performance compared with the previous methods.

Acknowledgements. This work has been supported by the National Key Research and Development Program under Grant No. 2020AAA0109702.

References

1. Deng, Y., Kanervisto, A., Ling, J., Rush, A.M.: Image-to-markup generation with coarse-to-fine attention. In: International Conference on Machine Learning, pp. 980–989. PMLR (2017)
2. Drake, D.M., Baird, H.S.: Distinguishing mathematics notation from English text using computational geometry. In: Eighth International Conference on Document Analysis and Recognition (ICDAR 2005), pp. 1270–1274. IEEE (2005)

3. He, W., Zhang, X.Y., Yin, F., Liu, C.L.: Deep direct regression for multi-oriented scene text detection. In: Proceedings of the IEEE International Conference on Computer Vision, pp. 745–753 (2017)
4. Hu, Y., Zheng, Y., Liu, H., Jiang, D., Ren, B.: Accurate structured-text spotting for arithmetical exercise correction. In: Proceedings of the AAAI Conference on Artificial Intelligence, vol. 34, no. 1, pp. 686–693 (2020)
5. Kacem, A., Belaïd, A., Ahmed, M.B.: Automatic extraction of printed mathematical formulas using fuzzy logic and propagation of context. Int. J. Doc. Anal. Recogn. 4(2), 97–108 (2001)
6. Karatzas, D., Gomez-Bigorda, L., Nicolaou, A., Ghosh, S., Valveny, E.: ICDAR 2015 competition on robust reading. In: International Conference on Document Analysis and Recognition (2015)
7. Le, A.D., Nakagawa, M.: Training an end-to-end system for handwritten mathematical expression recognition by generated patterns. In: 2017 14th IAPR International Conference on Document Analysis and Recognition (ICDAR), vol. 1, pp. 1056–1061. IEEE (2017)
8. Lin, T.Y., Goyal, P., Girshick, R., He, K., Dollár, P.: Focal loss for dense object detection. In: Proceedings of the IEEE International Conference on Computer Vision, pp. 2980–2988 (2017)
9. Lin, X., Gao, L., Tang, Z., Baker, J., Sorge, V.: Mathematical formula identification and performance evaluation in pdf documents. Int. J. Doc. Anal. Recogn. (IJDAR) 17(3), 239–255 (2014)
10. Liu, W., et al.: SSD: single shot MultiBox detector. In: Leibe, B., Matas, J., Sebe, N., Welling, M. (eds.) ECCV 2016. LNCS, vol. 9905, pp. 21–37. Springer, Cham (2016). https://doi.org/10.1007/978-3-319-46448-0_2
11. Liu, X., Liang, D., Yan, S., Chen, D., Qiao, Y., Yan, J.: FOTS: fast oriented text spotting with a unified network. In: Proceedings of the IEEE Conference on Computer Vision and Pattern Recognition, pp. 5676–5685 (2018)
12. Luong, M.T., Pham, H., Manning, C.D.: Effective approaches to attention-based neural machine translation. arXiv preprint arXiv:1508.04025 (2015)
13. Mahdavi, M., Zanibbi, R., Mouchere, H., Viard-Gaudin, C., Garain, U.: ICDAR 2019 CROHME+ TFD: competition on recognition of handwritten mathematical expressions and typeset formula detection. In: 2019 International Conference on Document Analysis and Recognition (ICDAR), pp. 1533–1538. IEEE (2019)
14. Mali, P., Kukkadapu, P., Mahdavi, M., Zanibbi, R.: ScanSSD: scanning single shot detector for mathematical formulas in pdf document images. arXiv preprint arXiv:2003.08005 (2020)
15. Ohyama, W., Suzuki, M., Uchida, S.: Detecting mathematical expressions in scientific document images using a U-Net trained on a diverse dataset. IEEE Access 7, 144030–144042 (2019)
16. Ren, S., He, K., Girshick, R., Sun, J.: Faster R-CNN: towards real-time object detection with region proposal networks. arXiv preprint arXiv:1506.01497 (2015)
17. Tan, M., Le, Q.: EfficientNet: rethinking model scaling for convolutional neural networks. In: International Conference on Machine Learning, pp. 6105–6114. PMLR (2019)
18. Tan, M., Pang, R., Le, Q.V.: EfficientDet: scalable and efficient object detection. In: Proceedings of the IEEE/CVF Conference on Computer Vision and Pattern Recognition, pp. 10781–10790 (2020)
19. Wang, W., et al.: Shape robust text detection with progressive scale expansion network. In: Proceedings of the IEEE/CVF Conference on Computer Vision and Pattern Recognition, pp. 9336–9345 (2019)

20. Zhang, J., et al.: Watch, attend and parse: an end-to-end neural network based approach to handwritten mathematical expression recognition. Pattern Recogn. **71**, 196–206 (2017)
21. Zhang, L., He, Z., Yang, Y., Wang, L., Gao, X.B.: Tasks integrated networks: joint detection and retrieval for image search. IEEE Trans. Pattern Anal. Mach. Intell. 1 (2020). https://doi.org/10.1109/TPAMI.2020.3009758
22. Zhou, X., et al.: EAST: an efficient and accurate scene text detector. In: Proceedings of the IEEE Conference on Computer Vision and Pattern Recognition, pp. 5551–5560 (2017)

FD-Net: A Fully Dilated Convolutional Network for Historical Document Image Binarization

Wei Xiong[1,2(✉)], Ling Yue[1], Lei Zhou[1], Liying Wei[1], and Min Li[1,2]

[1] School of Electrical and Electronic Engineering, Hubei University of Technology,
Wuhan 430068, Hubei, China
xw@mail.hbut.edu.cn
[2] Department of Computer Science and Engineering, University of South Carolina,
Columbia, SC 29201, USA

Abstract. Binarization is a key step in document analysis and archiving. The state-of-the-art models for document image binarization are variants of the encoder-and-decoder architecture, such as *fully convolutional network* (FCN) and U-Net. Despite their success, they still suffer from two challenges: (1) max-pooling or strided convolution reduces the spatial resolution of the intermediate feature maps, which may lead to information loss, and (2) interpolation or transposed convolution attempts to restore the feature maps to the desired spatial resolution, which may also result in pixelation. To overcome these two limitations, we propose a *fully dilated convolutional network*, termed FD-Net, using atrous convolutions instead of downsampling or upsampling operations. We have conducted extensive experiments on the recent DIBCO (*document image binarization competition*) and H-DIBCO (*handwritten document image binarization competition*) benchmark datasets. The experimental results show that our proposed FD-Net outperforms other state-of-the-art techniques by a large margin. The source code and pretrained models are publicly available at https://github.com/beargolden/FD-Net.

Keywords: Historical document image binarization · Document image segmentation · *Fully dilated convolutional network* (FD-Net) · Dilated convolution · Atrous convolution

1 Introduction

Document image binarization (also referred to as segmentation or thresholding) aims to extract text pixels from complex document background, which plays an important role in *document analysis and recognition* (DAR) systems. It converts a color or grayscale image into a binary one, essentially reducing the information contained within the image, and thus greatly reducing the disk storage capacity as well as network transmission bandwidth. It is widely considered to be one of

© Springer Nature Switzerland AG 2021
H. Ma et al. (Eds.): PRCV 2021, LNCS 13019, pp. 518–529, 2021.
https://doi.org/10.1007/978-3-030-88004-0_42

the most important pre-processing steps, and the performance of binarization will directly affect the accuracy of subsequent tasks, such as page layout analysis, machine-printed or handwritten character recognition. It also helps to resolve the conflict between document conservation and cultural heritage.

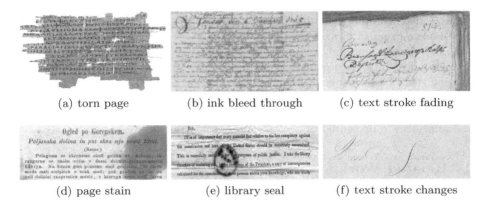

(a) torn page   (b) ink bleed through   (c) text stroke fading

(d) page stain   (e) library seal   (f) text stroke changes

Fig. 1. Historical document image samples from recent DIBCO and H-DIBCO benchmark datasets

The thresholding of high-quality images is simple, but the binarization of historical document images is quite challenging. The reason is that the latter suffers from severe degradation, such as torn pages, ink bleed through, text stroke fading, page stains, and artifacts, as shown in Fig. 1. In addition, variations in the color, width, brightness, and connectivity of text strokes in degraded handwritten manuscripts further increase the difficulty of binarization.

The *state-of-the-art* (SOTA) models for document image binarization are variants of the encoder-and-decoder architecture, such as *fully convolutional network* (FCN) [1] and U-Net [2]. These segmentation models have 3 key components in common: an encoder, a decoder, and skip connections. In the encoder, consecutive of convolutions and downsampling (e.g., max-pooling or strided convolution) are performed. This helps extract higher-level features, but reduces the spatial resolution of intermediate feature maps, which may lead to information loss. In the decoder, repeated combination of upsampling (e.g., bilinear interpolation) and convolutions are conducted to restore feature maps to the desired spatial resolution, which may also result in pixelation or texture smoothing. Therefore, after each upsampling operation, feature maps with the same level are merged by skip connections, which transfer localization information from the encoder to the decoder. In addition, sampling operations like max-pooling and bilinear interpolation are not learnable.

To overcome the aforementioned problems, we present FD-Net, a *fully dilated convolutional network* for degraded historical document image binarization. The proposed segmentation model removes all the downsampling and upsampling operations, and employs dilated convolutions (also known as atrous convolutions) instead. Therefore, the proposed segmentation model contains only convolutional and dilated convolutional layers, which are fully trainable. In this way, the spatial resolutions of all the intermediate layers are identical, but without significantly increasing the number of model parameters.

Our contributions are two folds. First, we propose a new paradigm that replaces downsampling or upsampling operations with dilated convolutions. It can achieve promising pixel-wise labeling results on various degraded historical document images. Second, we investigate hybrid dilation rate settings to alleviate the grid effect in dilated convolution.

The rest of this paper is organized as follows. Section 2 briefly reviews the related work on document image binarization. Section 3 describes our proposed fully dilated convolutional neural network in detail. The experimental results and analysis are presented in Sect. 4, and Sect. 5 concludes the paper.

2   Related Work

Existing document image binarization methods can be classified as global thresholding, local adaptive thresholding, and hybrid approaches [3].

The global thresholding method uses a single threshold to classify the pixels of a document image into two classes, namely text and background. The Otsu's [4] method is one of the best known global thresholding techniques. It uses the grayscale histogram of an image to select an optimal threshold that makes the variance within each class as small as possible and the variance between the two classes as large as possible. The Otsu's method is fast but ineffective and has poor noise immunity when dealing with low-quality images.

The local adaptive thresholding method can handle more complex cases, and automatically computes local thresholds based on the grayscale distribution within a neighborhood window around a pixel. The Niblack's [5] method uses a smoothing window mechanism, where the local threshold is determined by the mean and standard deviation of the grayscale values within the window centered at each pixel. This method is good at segmenting low-contrast text, but since only local information is considered, it is more likely to treat background noise as foreground text as well. The Sauvola's [6] and Wolf's [7] methods overcome the drawbacks of Niblack's counterpart. They are based on the assumption that the gray value of text pixels is close to 0 and the gray value of background pixels is close to 255. It makes the threshold smaller for background points with higher gray values and the same standard deviation, thus filtering out some distracting textures and noises in the background, but the binarization is still not effective in the case of low contrast between the foreground and background.

Hybrid methods for the binarization of historical document images have also been developed. Su et al. [8] present a document image binarization method using *local maximum and minimum* (LMM). The document text is segmented by constructing a contrast image and then detecting high-contrast pixels that typically lie around text stroke boundaries and using local thresholds that are estimated from the detected high-contrast pixels within a local neighborhood window. Jia et al. [9] propose a document image binarization method based on *structural symmetric pixels* (SSPs), which are located at the edges of text strokes and can be extracted from those with large gradient values and opposite gradient directions. Finally, a multiple local threshold voting-based framework is used to further determine whether each pixel belongs to the foreground or background. The contrast or edge-based segmentation methods do not work well for binarization of degraded images with complex document background, e.g., low contrast, gradients, and smudges.

Howe [10] presents an energy-based segmentation method, which treats each image pixel as a node in a connected graph, and then applies the max-flow/min-cut algorithm to partition the connected graph into two regions to determine the text and background pixels. Mesquita et al. [11], Kligler et al. [12], and Xiong et al. [13,14] propose different document enhancement techniques, followed by Howe's binarization method, to provide guidance for text and background segmentation, respectively. In addition, Chen et al. [15] and Xiong et al. [16] propose the use of *support vector machines* (SVMs) for statistical learning-based segmentation. Bhowmik et al. [17] introduce a *document image binarization inspired by game theory* (GiB). However, the main drawback of these methods is that only handcrafted features are employed to obtain segmentation results. Therefore, it is difficult to design representative features for different applications, and handcrafted features work well for one type of image, but may fail on another.

Deep learning-based binarization of degraded document images is a hot topic and trend of current research. Tensmeyer and Martinez [18] present a multi-scale FCN architecture with pseudo F-measure loss. Zhou et al. [19] also explore a multi-scale deep contextual convolutional neural network with densely connected *conditional random fields* (CRFs) for semantic segmentation. Vo et al. [20] propose a supervised binarization method for historical document images based on hierarchical *deep supervised networks* (DSNs). Calvo-Zaragoza and Gallego [21] present a *selectional auto-encoder* (SAE) approach for document image binarization. Bezmaternykh et al. [22] present a historical document image binarization method based on U-Net [2], a convolutional neural network originally designed for biomedical image segmentation. Zhao et al. [23] consider binarization as an image-to-image generation task and propose a method for historical document image binarization using *conditional generative adversarial networks* (cGANs). Peng et al. [24] propose a deep learning framework to infer the probabilities of text regions by a multi-resolution attention-based model, and then fed into a *convolutional conditional random field* (ConvCRF) to obtain the final binary images. Xiong et al. [25] present an improved semantic segmentation model, called DP-LinkNet, which adopts *hybrid dilated convolution* (HDC) and *spatial pyramid pooling* (SPP) modules between the encoder and the decoder, for more accurate binarization of degraded historical document images.

3   Proposed Network Architecture: FD-Net

The proposed fully dilated convolutional network model, referred to as FD-Net, is shown in Fig. 2. As can be seen from the figure, it consists of 3 main components, namely an encoder, a decoder, and skip connections. What distinguishes

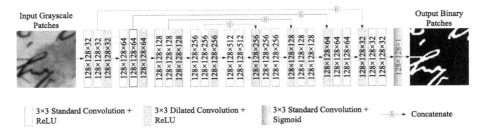

Fig. 2. The proposed FD-Net architecture

our proposed FD-Net from other SOTA models for document image binarization is that the proposed model does not contain downsampling or upsampling layers. The pooling operation or strided convolution reduces the spatial resolution of intermediate feature maps, which may lead to internal data structure missing or spatial hierarchical information loss; while the interpolation operation or strided deconvolution attempts to restore the feature maps to the desired spatial resolution, which may also result in pixelation or checkerboard artifacts. In addition, the pooling and interpolation operations are deterministic (a.k.a. not learnable or trainable).

To overcome the above problems, an intuitive approach is to simply remove those downsampling and upsampling layers from the model, but this will also decrease the receptive field size and thus severely reduce the amount of context. For instance, a stack of three 3×3 convolutional layers is equivalent to the regularization of a 7×7 convolutional layer. That's why pooling operations exist for increasing the receptive field size, and upsampling for pixel-wise prediction. Fortunately, dilated convolution can compensate for this deficiency, and it has been proven to be effective in semantic segmentation [26]. For this reason, we replace all the downsampling and upsampling layers with dilated convolutions. Therefore, the spatial resolution of all the intermediate convolutional layers in the proposed model is identical.

The encoder subnetwork comprises 4 consecutive convolutional blocks. Each encoding block includes two 3×3 standard convolutional layers followed by a 3×3 dilated convolutional layer (the dilation rate settings will be discussed in Subsect. 3.1). The number of channels or feature maps after each encoding block is doubled, so that the network can effectively learn higher-level abstract feature representations. The central block contains two 3×3 standard convolutional layers, and connects the encoder and decoder. The decoder subnetwork, similar to the encoder counterpart, also consists of 4 nearly symmetrical convolutional blocks. Each decoding block includes a 3×3 dilated convolutional layer, a merge layer that concatenates the feature maps of the decoder with those of the corresponding encoder by skip connections, and two 3×3 standard convolutional layers. The number of feature maps before each decoding block is halved. The ReLU (*rectified linear unit*) activation function is used in all the aforementioned convolutional layers. At the end, a 3×3 standard convolutional layer with Sigmoid activation function is adopted to generate the resulting binary image patches.

3.1 Hybrid Dilation Rate Settings

We introduce a simple hybrid dilation rate solution by setting different dilation rates to avoid the gridding effect [27,28]. In addition, choosing an appropriate dilation rate can effectively increase the receptive field size and also improve the segmentation accuracy. The purpose of our hybrid dilation rate setting is to make the final receptive field size of all the successive convolutions (including dilated ones) completely cover a specific local neighborhood. The maximum distance M between two non-zero kernel weights is defined as:

$$M_i \Leftarrow max[M_{i+1} - 2r_i, M_{i+1} - 2(M_{i+1} - r_i), r_i] \tag{1}$$

with $M_n = r_n$.

Instead of using the same dilation rates or those with a common factor relationship among all the convolutional layers, we set the dilation rates of the 3 layers in each encoding block to $[1, 1, r_i]$, where the values of r_i used in the 4 encoding blocks are set to $[2, 3, 5, 7]$, respectively. The dilation rates of the decoder subnetwork are set in the reverse order of the corresponding encoder subnetwork. Since spatial resolutions of all feature maps are the same, skip connections are essentially merging the features with different receptive field sizes.

3.2 Implementation Details

Given a color antique document image, it is first converted to its grayscale counterpart, then cropped to a patch size of 128×128 and fed into our proposed FD-Net model. The output binary patches are seamlessly stitched together to generate the resulting binary image.

We combine the Dice loss with the standard *binary cross-entropy* (BCE) loss. Combining these two metrics allows for some diversity in the loss function, while benefiting from the stability of the BCE. The overall loss function is defined as:

$$L = 1 - \underbrace{\frac{2\sum_{n=1}^{N} y_n \hat{y}_n}{\sum_{n=1}^{N} y_n^2 + \sum_{n=1}^{N} \hat{y}_n^2}}_{L_{Dice}} - \underbrace{\frac{1}{N} \sum_{n=1}^{N} [y_n \log \hat{y}_n + (1 - y_n) \log(1 - \hat{y}_n)]}_{L_{BCE}} \tag{2}$$

where N is the number of image pixels, y_n and $\hat{y}_n$ are the *ground truth* (GT) and predicted segmentation, respectively.

The Adam optimization algorithm is adopted in our deep learning model. The initial learning rate defaults to 0.001, and the exponential decay rates for the first and second moment estimates are set to 0.9 and 0.999, respectively. We monitor the cost function values of the validation data, and if no improvement is seen for 10 epochs, the learning rate is reduced to half. We also use an early stop strategy once the learning stagnates for 20 consecutive epochs.

We collect 50 degraded historical document images from the *recognition and enrichment of archival documents* (READ) project[1] as training data. The Bickley Diary dataset is used for the ablation study, while the DIBCO and H-DIBCO 2009–2019 benchmark datasets are used as test data.

Table 1. Ablation study of FD-Net on the Bickley Diary dataset with varying dilation rate settings (image patch size: 128×128, and batch size: 32)

Network model	Dilation rates	# of 1st layer channels	Validation loss	Validation accuracy	# of model parameters
U–Net	–	32	0.0577	0.9903	8,630,177
U–Net	–	64	0.0541	0.9917	34,512,705
FD-Net	2,2,2,2	32	0.0600	0.9899	9,414,017
FD-Net	2,3,5,7	32	**0.0514**	**0.9931**	9,414,017
FD-Net	2,4,8,16	32	0.0524	0.9914	9,414,017

During the training and testing phases, the traditional color-to-gray method is performed, with no other pre-processing or post-processing. Our implementation does not apply any data augmentation techniques either.

4 Experiments

4.1 Ablation Study

In this study, we use the Bickley Diary dataset to evaluate the impact of hybrid dilation rates on the performance of our proposed FD-Net. It consists of 92 badly degraded handwritten travel diary documents, 7 of which have GT images. We crop these 7 historical document images into 1764 patches with corresponding GT ones, 20% of which to be used as validation data.

The BCE-Dice loss and accuracy metrics are used to measure the performance of our deep learning model. The loss is a sum of the errors made for each sample in the training or validation set. The loss value implies how poorly or well a model performs after each iteration of optimization, so higher loss is the worse for any model. The accuracy is usually determined after the model parameters and is calculated in the form of a percentage. It is a measure of how accurate your model's prediction is compared to the true data.

The experimental results of the ablation study are shown in Table 1. As can be seen from the table, the segmentation performance of our proposed FD-Net is basically similar to or even better than that of the vanilla U-Net architecture with the same number of layers, but the number of parameters of our FD-Net model is almost the same or much less than that of the U-Net. Among all the compared models, the FD-Net with our proposed hybrid dilation rates performs the best. The results of our ablation experiments suggest that the correct setting of the dilated convolutions can not only help maintain the receptive field size, but also further improve the segmentation accuracy.

[1] https://read.transkribus.eu/.

4.2   More Segmentation Experiments

In this experiment, we use 10 document image binarization competition datasets to evaluate the segmentation performance of our proposed FD-Net. The DIBCO 2009 [29], 2011 [30], 2013 [31], 2017 [32], 2019 [33] and H-DIBCO 2010 [34], 2012 [35], 2014 [36], 2016 [37], 2018 [38] benchmark datasets consist of 90 handwritten, 36 machine-printed, 10 Iliad papyri document images and their corresponding GT images. The 10 datasets contain representative historical document degradation, such as fragmented pages, ink bleed through, background texture, page stains, text stroke fading, and artifacts.

Table 2. Performance evaluation results of our proposed method against SOTA techniques on the 10 DIBCO and H-DIBCO test datasets

Method	FM↑ (%)	pFM (%)	PSNR (dB)	NRM (%)	DRD	MPM (‰)
Gallego's SAE [21]	79.221	81.123	16.089	9.094	9.752	11.299
Bhowmik's GiB [17]	83.159	87.716	16.722	8.954	8.818	7.221
Jia's SSP [9]	85.046	87.245	17.911	5.923	9.744	9.503
Peng's woConvCRF [24]	86.089	87.397	18.989	6.429	4.825	4.176
Zhao's cGAN [23]	87.447	88.873	18.811	5.024	5.564	5.536
Vo's DSN [20]	88.037	90.812	18.943	6.278	4.473	3.213
Bezmaternykh's UNet [22]	89.290	90.534	21.319	5.577	3.286	1.651
Proposed FD-Net	**95.254**	**96.648**	**22.836**	**3.224**	**1.219**	**0.201**

We adopt evaluation metrics used in DIBCO and H-DIBCO competitions to evaluate the performance of our proposed method. The evaluation metrics are FM (*F-measure*), pFM (*pseudo F-measure*), PSNR (*peak signal-to-noise ratio*), NRM (*negative rate metric*), DRD (*distance reciprocal distortion metric*), and MPM (*misclassification penalty metric*). The first 2 metrics (FM and pFM) reach the best value at 1 and the worst at 0. The PSNR measures how close a binarized image to the GT image, and therefore, the higher the value, the better. In contrast to the former 3 metrics, the binarization quality is better for lower NRM, DRD, and MPM metrics. Due to space limitations, we omit definitions of those evaluation metrics, but readers can refer to [30,33,34] for more details.

The proposed FD-Net is compared with Jia's SSP [9], Bhowmik's GiB [17], Vo's DSN [20], Gallego's SAE [21], Bezmaternykh's U-Net [22], Zhao's cGAN [23], and Peng's attention-based[24] techniques on all the 10 DIBCO and H-DIBCO datasets, and the evaluation results are listed in Table 2. It can be seen from the table that all the evaluation measures of our proposed method achieve the best results on all the 10 test datasets. Compared with Bezmaternykh's U-Net [22], the FM, pFM, PSNR, NRM, DRD, and MPM of our proposed method are 5.963%, 6.115%, 1.516dB, 2.353%, 2.067 and 1.451‰better than those of the second best technique, respectively. This also implies that our proposed FD-Net architecture outperforms U-Net, and produces more accurate segmentation.

Figure 3 further displays the resulting binary images generated by different techniques. From the figure we can see that Gallego's SAE [21] tends to produce

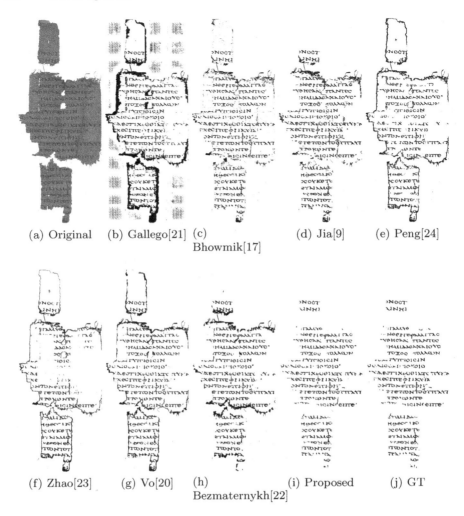

(a) Original   (b) Gallego[21]   (c) Bhowmik[17]   (d) Jia[9]   (e) Peng[24]

(f) Zhao[23]   (g) Vo[20]   (h) Bezmaternykh[22]   (i) Proposed   (j) GT

Fig. 3. Binarization results of all evaluation techniques for CATEGORY2_20 in DIBCO 2019 dataset

ghost text pixels in the true background region. Bhowmik's GiB [17], Peng's attention-based segmentation (woConvCRF) [24], Zhao's cGAN [23], and Vo's DSN [20] have difficulty in removing the edges of fragmented pages. Compared to Jia's SSP [9] and Bezmaternykh's U-Net [22], our proposed FD-Net can produce better visual quality by preserving most text strokes and eliminating possible noise.

5   Conclusion

In this paper, we present a fully dilated convolutional neural network, termed FD-Net, for more accurate binarization of degraded historical document images. The superior performance is attributed to its dilated convolutional architecture and skip connection, which is designed to address two major challenges faced by current segmentation models: (1) internal data structure missing or spatial hierarchical information loss, and (2) max-pooling and interpolation operations are not learnable or trainable. We conducted extensive experiments to evaluate the performance of our proposed FD-Net on the recent DIBCO and H-DIBCO benchmark datasets. Results show that the proposed method outperforms other SOTA techniques by a large margin.

Acknowledgement. This work was supported in part by the National Natural Science Foundation of China (61571182, 61601177), the Natural Science Foundation of Hubei Province, China (2019CFB530), and China Scholarship Council (201808420418).

References

1. Shelhamer, E., Long, J., Darrell, T.: Fully convolutional networks for semantic segmentation. IEEE Trans. Pattern Anal. Mach. Intell. **39**(4), 640–651 (2017). https://doi.org/10.1109/tpami.2016.2572683
2. Ronneberger, O., Fischer, P., Brox, T.: U-Net: convolutional networks for biomedical image segmentation. In: Navab, N., Hornegger, J., Wells, W.M., Frangi, A.F. (eds.) MICCAI 2015. LNCS, vol. 9351, pp. 234–241. Springer, Cham (2015). https://doi.org/10.1007/978-3-319-24574-4_28
3. Eskenazi, S., Gomez-Krämer, P., Ogier, J.M.: A comprehensive survey of mostly textual document segmentation algorithms since 2008. Pattern Recogn. **64**, 1–14 (2017). https://doi.org/10.1016/j.patcog.2016.10.023
4. Otsu, N.: A threshold selection method from gray-level histograms. IEEE Trans. Syst. Man Cybern. **9**(1), 62–66 (1979). https://doi.org/10.1109/tsmc.1979.4310076
5. Niblack, W.: An Introduction to Digital Image Processing. Prentice-Hall International Inc., Englewood Cliffs (1986)
6. Sauvola, J., Pietikäinen, M.: Adaptive document image binarization. Pattern Recogn. **33**(2), 225–236 (2000). https://doi.org/10.1016/s0031-3203(99)00055-2
7. Wolf, C., Jolion, J.M.: Extraction and recognition of artificial text in multimedia documents. Pattern Anal. Appl. **6**(4), 309–326 (2003). https://doi.org/10.1007/s10044-003-0197-7
8. Su, B., Lu, S., Tan, C.L.: Binarization of historical document images using the local maximum and minimum. In: 9th IAPR International Workshop on Document Analysis Systems (DAS 2010), pp. 159–165. https://doi.org/10.1145/1815330.1815351
9. Jia, F., Shi, C., He, K., Wang, C., Xiao, B.: Degraded document image binarization using structural symmetry of strokes. Pattern Recogn. **74**, 225–240 (2018). https://doi.org/10.1016/j.patcog.2017.09.032
10. Howe, N.R.: Document binarization with automatic parameter tuning. Int. J. Doc. Anal. Recogn. **16**(3), 247–258 (2013). https://doi.org/10.1007/s10032-012-0192-x
11. Mesquita, R.G., Silva, R.M.A., Mello, C.A.B., Miranda, P.B.C.: Parameter tuning for document image binarization using a racing algorithm. Expert Syst. Appl. **42**(5), 2593–2603 (2015). https://doi.org/10.1016/j.eswa.2014.10.039

12. Kligler, N., Katz, S., Tal, A.: Document enhancement using visibility detection. In: 31st IEEE/CVF Conference on Computer Vision and Pattern Recognition (CVPR 2018), pp. 2374–2382. https://doi.org/10.1109/cvpr.2018.00252
13. Xiong, W., Jia, X., Xu, J., Xiong, Z., Liu, M., Wang, J.: Historical document image binarization using background estimation and energy minimization. In: 24th International Conference on Pattern Recognition (ICPR 2018), pp. 3716–3721. https://doi.org/10.1109/icpr.2018.8546099
14. Xiong, W., Zhou, L., Yue, L., Li, L., Wang, S.: An enhanced binarization framework for degraded historical document images. EURASIP J. Image Video Process. **2021**(1), 1–24 (2021). https://doi.org/10.1186/s13640-021-00556-4
15. Chen, X., Lin, L., Gao, Y.: Parallel nonparametric binarization for degraded document images. Neurocomputing **189**, 43–52 (2016). https://doi.org/10.1016/j.neucom.2015.11.040
16. Xiong, W., Xu, J., Xiong, Z., Wang, J., Liu, M.: Degraded historical document image binarization using local features and support vector machine (SVM). Optik **164**, 218–223 (2018). https://doi.org/10.1016/j.ijleo.2018.02.072
17. Bhowmik, S., Sarkar, R., Das, B., Doermann, D.: GiB: a game theory inspired binarization technique for degraded document images. IEEE Trans. Image Process. **28**(3), 1443–1455 (2019). https://doi.org/10.1109/tip.2018.2878959
18. Tensmeyer, C., Martinez, T.: Document image binarization with fully convolutional neural networks. In: 14th IAPR International Conference on Document Analysis and Recognition (ICDAR 2017), pp. 99–104. https://doi.org/10.1109/icdar.2017.25
19. Zhou, Q., et al.: Multi-scale deep context convolutional neural networks for semantic segmentation. World Wide Web **22**(2), 555–570 (2018). https://doi.org/10.1007/s11280-018-0556-3
20. Vo, Q.N., Kim, S.H., Yang, H.J., Lee, G.: Binarization of degraded document images based on hierarchical deep supervised network. Pattern Recogn. **74**, 568–586 (2018). https://doi.org/10.1016/j.patcog.2017.08.025
21. Calvo-Zaragoza, J., Gallego, A.J.: A selectional auto-encoder approach for document image binarization. Pattern Recogn. **86**, 37–47 (2019). https://doi.org/10.1016/j.patcog.2018.08.011
22. Bezmaternykh, P.V., Ilin, D.A., Nikolaev, D.P.: U-Net-bin: hacking the document image binarization contest. Comput. Optics **43**(5), 825–832 (2019). https://doi.org/10.18287/2412-6179-2019-43-5-825-832
23. Zhao, J., Shi, C., Jia, F., Wang, Y., Xiao, B.: Document image binarization with cascaded generators of conditional generative adversarial networks. Pattern Recogn. **96** (2019). https://doi.org/10.1016/j.patcog.2019.106968
24. Peng, X., Wang, C., Cao, H.: Document binarization via multi-resolutional attention model with DRD loss. In: 15th IAPR International Conference on Document Analysis and Recognition (ICDAR 2019), pp. 45–50. https://doi.org/10.1109/icdar.2019.00017
25. Xiong, W., Jia, X., Yang, D., Ai, M., Li, L., Wang, S.: DP-LinkNet: a convolutional network for historical document image binarization. KSII Trans. Internet Inf. Syst. **15**(5), 1778–1797 (2021). https://doi.org/10.3837/tiis.2021.05.011
26. Chen, L.C., Papandreou, G., Kokkinos, I., Murphy, K., Yuille, A.L.: DeepLab: semantic image segmentation with deep convolutional nets, Atrous convolution, and fully connected CRFs. IEEE Trans. Pattern Anal. Mach. Intell. **40**(4), 834–848 (2018). https://doi.org/10.1109/tpami.2017.2699184

27. Wang, P., Chen, P., Yuan, Y., Liu, D., Huang, Z., Hou, X., Cottrell, G.: Understanding convolution for semantic segmentation. In: 2018 IEEE Winter Conference on Applications of Computer Vision (WACV), pp. 1451–1460. https://doi.org/10.1109/wacv.2018.00163

28. Wang, Z., Ji, S.: Smoothed dilated convolutions for improved dense prediction. Data Min. Knowl. Disc. **35**(4), 1470–1496 (2021). https://doi.org/10.1007/s10618-021-00765-5

29. Gatos, B., Ntirogiannis, K., Pratikakis, I.: ICDAR 2009 document image binarization contest (DIBCO 2009). In: 10th International Conference on Document Analysis and Recognition (ICDAR 2009), pp. 1375–1382. https://doi.org/10.1109/icdar.2009.246

30. Pratikakis, I., Gatos, B., Ntirogiannis, K.: ICDAR 2011 document image binarization contest (DIBCO 2011). In: 11th International Conference on Document Analysis and Recognition (ICDAR 2011), pp. 1506–1510. https://doi.org/10.1109/icdar.2011.299

31. Pratikakis, I., Gatos, B., Ntirogiannis, K.: ICDAR 2013 document image binarization contest (DIBCO 2013). In: 12th International Conference on Document Analysis and Recognition (ICDAR 2013), pp. 1471–1476. https://doi.org/10.1109/icdar.2013.219

32. Pratikakis, I., Zagoris, K., Barlas, G., Gatos, B.: ICDAR 2017 competition on document image binarization (DIBCO 2017). In: 14th International Conference on Document Analysis and Recognition (ICDAR 2017), pp. 1395–1403. https://doi.org/10.1109/icdar.2017.228

33. Pratikakis, I., Zagoris, K., Karagiannis, X., Tsochatzidis, L., Mondal, T., Marthot-Santaniello, I.: ICDAR 2019 competition on document image binarization (DIBCO 2019). In: 15th International Conference on Document Analysis and Recognition (ICDAR 2019). https://doi.org/10.1109/icdar.2019.00249

34. Pratikakis, I., Gatos, B., Ntirogiannis, K.: H-DIBCO 2010 - handwritten document image binarization competition. In: 12th International Conference on Frontiers in Handwriting Recognition (ICFHR 2010), pp. 727–732. https://doi.org/10.1109/icfhr.2010.118

35. Pratikakis, I., Gatos, B., Ntirogiannis, K.: ICFHR 2012 competition on handwritten document image binarization (H-DIBCO 2012). In: 13th International Conference on Frontiers in Handwriting Recognition (ICFHR 2012), pp. 817–822. https://doi.org/10.1109/icfhr.2012.216

36. Ntirogiannis, K., Gatos, B., Pratikakis, I.: ICFHR 2014 competition on handwritten document image binarization (H-DIBCO 2014). In: 14th International Conference on Frontiers in Handwriting Recognition (ICFHR 2014), pp. 809–813. https://doi.org/10.1109/icfhr.2014.141

37. Pratikakis, I., Zagoris, K., Barlas, G., Gatos, B.: ICFHR 2016 handwritten document image binarization contest (H-DIBCO 2016). In: 15th International Conference on Frontiers in Handwriting Recognition (ICFHR 2016), pp. 619–623. https://doi.org/10.1109/icfhr.2016.110

38. Pratikakis, I., Zagoris, K., Kaddas, P., Gatos, B.: ICFHR 2018 competition on handwritten document image binarization (H-DIBCO 2018). In: 16th International Conference on Frontiers in Handwriting Recognition (ICFHR 2018), pp. 489–493. https://doi.org/10.1109/icfhr-2018.2018.00091

Appearance-Motion Fusion Network for Video Anomaly Detection

Shuangshuang Li, Shuo Xu, and Jun Tang[✉]

School of Electronics and Information Engineering, Anhui University,
Anhui 230601, Hefei, China

Abstract. Detection of abnormal events in surveillance video is an important and challenging task, which has received much research interest over the past few years. However, existing methods often only considered appearance information or simply integrated appearance and motion information without considering their underlying relationship. In this paper, we propose an unsupervised anomaly detection approach based on deep auto-encoder, which can effectively exploit the complementarity of both appearance and motion information. Two encoders are used to extract appearance features and motion features from RGB and RGB difference frames, respectively, and then a feature fusion module is employed to fuse appearance and motion features to produce discriminative feature representations of regular events. Finally, the fused features are sent to their corresponding decoders to predict future RGB and RGB differential frames for determining anomaly events according to reconstruction errors. Experiments and ablation studies on some public datasets demonstrate the effectiveness of our approach.

Keywords: Video anomaly detection · Feature fusion · Deep auto encoder

1 Introduction

Abnormal events detection in video sequences has been extensively studied by many researchers due to its important role in surveillance video. It is one of the most challenging problems in the field of computer vision. On one hand, abnormal events rarely occur in daily life and their types are diverse, which bring great difficulties for data collection. On the other hand, the definition of abnormal events may be quite different in diverse scenarios (for example, it is normal to hold a knife in the kitchen, but it is abnormal to hold a knife in the park), so the annotation of abnormal data is extremely complicated. Therefore, video anomaly detection (VAD) is usually considered as an unsupervised learning problem, which aims to learn a model that can only describe what the normal is. During test, events that cannot be reasonably described by the learned model are considered abnormal events[12].

Supported by the Natural Science Foundation of China under grant 61772032.

H. Ma et al. (Eds.): PRCV 2021, LNCS 13019, pp. 530–541, 2021.
https://doi.org/10.1007/978-3-030-88004-0_43

With the prevalence of deep learning, deep autoencoder (AE) has been one of the most frequently-used frameworks for reconstruction-based VAD[3,15,18]. Generally speaking, an AE consists of an encoder and a decoder. The encoder is used to encode input data into compact feature representations, and the decoder is employed to reconstruct input data from the learned features. Reconstruction-based models are basically built upon the assumption that abnormal data cannot be well well-reconstructed as only normal data are available during training.

Under this setting, how to model normal data is a fundamental yet important problem for reconstruction-based VAD. Notably, motion information is as important as appearance information for the task of VAD as most of abnormal events are related to movement. For example, an anomaly event happens when a person throws a bag. If only appearance information is considered, both of people and bag are regular objects without any change of unusual appearance, and consequently it is unlikely to identify this event as abnormal. In other words, it is difficult to effectively probe anomaly events for an anomaly detector if it only pays attention to appearance information while ignoring motion information. However, how to use appearance information in conjunction with motion information to enhance performance is a long-standing challenge for designing an anomaly detector. Some previous works directly combine these two kinds of information together, but they are unable to discover the underlying relationship between two modalities of information, even possibly leading to performance degradation. Therefore, it is crucial to find a solution to mining the complementarity between appearance and motion information so as to jointly improve the accuracy of anomaly detection. To address this issue, we propose an appearance-motion fusion framework based on the deep AE structure for video anomaly detection, which has two branches focusing on learning appearance and motion features, respectively. And the learned two kinds of features are fed into a fusion module to integrate them adaptively to play the complementarity between the two features.

In summary, our contributions are three-fold:

(1) We propose a deep AE based framework for unsupervised VAD that can effectively use both appearance and motion information.
(2) A novel feature fusion module is designed to enable the proposed framework to effectively utilize the complementarity between appearance and motion features.
(3) Extensive experiments demonstrate that our approach achieves competitive performance on several benchmark datasets in comparison with some stat-of-the-art unsupervised VAD methods.

2   Related Work

Recently, a high number of methods have been proposed to tackle the problem of VAD. In this section, we briefly review some typical reconstruction-based VAD approaches that are closely related to our work. Conv-AE[4] is the pioneer work to use deep AE to reconstruct input video sequences. Conv3D-AE[14] uses 3D convolutional neural network to extract appearance features of input video sequences and correlation information between adjacent frames, and input video sequences are reconstructed via deconvolutional neural network. In order to alleviate the identity mapping in the reconstruction task, [10,11] have proposed a proxy task to train deep AE. These approaches treat video frames as a time sequence, and the goal is to learn a generative model that uses the input past frames to predict the future ones. In [11], the input video sequence is first passed onto a convolution neural network, and then the output is fed into a convolutional LSTM to learn feature representations. After that, a deconvolutional layer uses the learned features to reconstruct the output with the same resolution as the input. In addition, in consideration of the complex and diverse characteristics of abnormal events, some methods attempted to use memory modules to improve the robustness of the learned models. In [3], the decoder does not directly use the features extracted by the encoder for reconstruction, but takes the features as the query items, and finds out the features closest to the query items from the memory module for reconstruction. [12] further optimizes the update strategy of memory. Some methods aim to improve the compactness of feature space through clustering. In [1], an additional clustering layer is used to constrain the features extracted by the encoder.

The methods most relevant to our proposed approach are [2,16]. These two methods use both appearance and motion information to deal with the problem of anomaly detection. However, in the process of feature fusion, only a simple concatenation operation was used, and less efforts were made to exploit the complementarity between two kinds of information in depth. Compared with the above methods, our proposed model can effectively utilize the complementarity between appearance information and motion information, and thus it is beneficial to improve the accuracy of anomaly detection.

3   Method

3.1   Overview

The overall structure of the algorithm is shown in Fig. 1. The input data is the original video sequence. The RGB difference sequence is obtained by making a difference between successive frames. The video and RGB difference sequences are sent to the corresponding encoder to produce appearance and motion features, respectively, and then the two kinds of features are passed onto the feature fusion module to generate the final features. Finally, the fusion features are fed into the decoders to predict the future appearance and motion, respectively.

3.2   Encoder and Decoder

We choose U-Net [13], which is commonly used in the reconstruction task, to extract the feature representations of the input video sequence and reconstruct future frames. Following [12], we remove the final batch normalization and ReLU

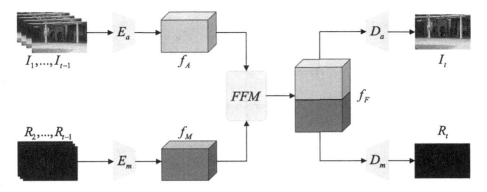

Fig. 1. Overview of our framework for predicting a future video frame. Our model mainly consists of three parts: two encoders, a feature fusion module, and two decoders. The two encoders extract appearance features and motion features from the input video sequence and the RGB difference sequence, respectively. Feature fusion module further mines the underlying relationship information between appearance and motion features and integrates them together. The decoders then use the fused features to predict the future appearance and motion, respectively.

layers of the encoder and replace them with an L2 normalized layer. This is because ReLU truncates negative values and consequently weakens the expression of features. And L2 normalization can make features have a common scale. We use two autoencoders with the same structure to process appearance (video sequence) and motion (RGB difference sequence), respectively. Let I_t represent the video frame at time t, and R_t denote the difference between the video frame at time t and time $t-1$. The video sequence $\{I_{t-l}, I_{t-l-1}, ..., I_{t-1}\}$ and the RGB difference sequence $\{R_{t-l-1}, R_{t-l-2}, ..., R_{t-1}\}$ are separately passed onto the E_a (appearance encoder) and E_m (motion encoder) to produce the corresponding f_A (appearance feature) and f_M (motion feature), where l represents the length of the input video sequence. Afterwards, f_A and f_M are sent to the feature fusion module, which will be elaborated in the following section, and the fused features are used to reconstruct video frame I_t and R_t by D_a(appearance decoder) and D_m(motion decoder), respectively.

3.3   Feature Fusion Module

In the previous work of [2,16], the commonly used feature fusion method is to directly concatenation the appearance feature f_a and the motion feature f_m. Although this way is simple and easy to operate, it cannot make full use of the complementarity of two kinds of features. Inspired by the idea of using the Squeeze-and-Excitation (SE) module to process multi-modality information[6], [5] designs a feature fusion block named Squeeze-and-Excitation feature fusion module (SEFM). The structure of SEFM is shown in Fig. 2. SEFM uses the SE module to explore the correlation between appearance and motion features and

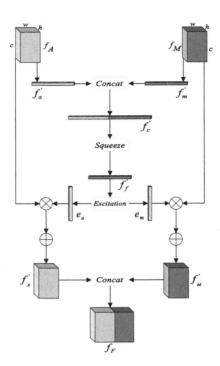

Fig. 2. Architecture of SEFM, where $\oplus$ denotes element-wise addition, and $\otimes$ denotes Hadamard product.

then fuse two kinds of features with the aid of the obtained correlation. The fusion process is described as follows:

(1) The appearance feature map f_A and the motion feature map f_M have the same dimension $R^{h \times w \times c}$ due to their same encoding structures. The global appearance feature $f_a' \in R^{1 \times 1 \times c}$ and global motion feature $f_m' \in R^{1 \times 1 \times c}$ are obtained by pooled down-sampling of the two feature maps respectively. Then the two features are concatenated together to produce spliced features $f_c' \in R^{1 \times 1 \times 2c}$.

(2) The spliced feature is mapped to a low-dimensional space to generate the final fused feature. This operation can be denoted by:

$$f_f' = W f_c' + b, \tag{1}$$

where W is the weight, and b is the deviation; $f_f' \in R^{1 \times 1 \times c_e}$, and c_e denotes the number of feature channels after mapping. This process is called Squeeze, and c_e is set to $2c/4$.

(3) The fusion features f_f' are mapped to the global appearance feature space and the global motion feature space to obtain the excitation values of the appearance feature map e_a and the motion feature map e_m, respectively.

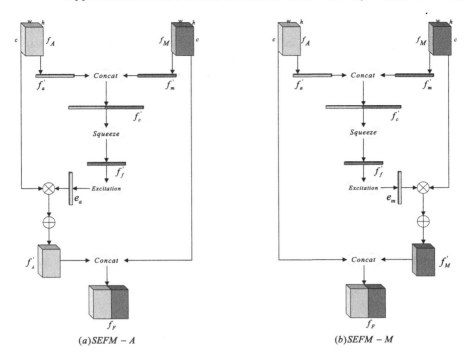

(a)$SEFM - A$ (b)$SEFM - M$

Fig. 3. Architecture of SEFM, $\oplus$ denotes element-wise addition, $\otimes$ denotes Hadamard product.

This process named Excitation is represented as:

$$e_a = \sigma(W_a f_f' + b_a),\tag{2}$$

$$e_m = \sigma(W_m f_f' + b_m),\tag{3}$$

where $\sigma(\cdot)$ is the sigmoid function and $e_a, e_m \in R^{1\times1\times c}$. The parameters of two branches are not shared, and they are trained independently.

(4) Taking the excitation value e_a and e_m as the weights on the channel of the appearance feature mapping and the motion feature mapping respectively, the weighted feature mappings f_A' and f_M' can be calculated. This process is somewhat similar to apply attention mechanism to the channel dimension. Afterwards, f_A' and f_M' are concatenated together to produce the final feature.

Altogether, we can see that SEFM extracts the correlation between appearance and motion features with the aid of the squeeze-excitation process, and finally realizes adaptive fusion. From another perspective, SEFM imposes transformations to both appearance and motion information to produce the fused features, and the transformations are obtained from the interaction between appearance features and motion features. However, we are unable to realize such transformations can play a positive role in boosting the discrimination of original features.

Base on these considerations, we attempt to design an alternative fusion approach that fuses one modality into the other given modality while keeping the given modality unchanged. Accordingly, we modify the structure of the original SEFM and design two asymmetric fusion modules, i.e., SEFM-A and SEFM-M. As shown in Fig. 3, SEFM-A only integrates motion features into appearance features, while keeping motion features unchanged. Similarly, SEFM-M only fuses appearance features into motion features, and appearance features remain unchanged. In the following experiments, we will demonstrate that such a fusion strategy can yield better results.

3.4   Loss Function

For the appearance branch, we use the predictive loss to force the video frame predicted from the decoder to be similar to its ground truth. Specifically, we minimize the $L2$ distance between the decoder output $\hat{I}_t$ and the ground truth I_t in the intensity space:

$$L_a = |\hat{I}_t - I_t|_2 \tag{4}$$

For the motion branch, in Eqs. (5) and (6), We define the smooth $L1$ loss between the RGB difference frame $\hat{R}_t$ produced by the decoder and the ground-truth R_t as the predicted loss, because the smooth $L1$ loss is more suitable for the highly sparse characteristic of RGB difference frames.

$$L_m = S_{L1}(\hat{R}_t - R_t) \tag{5}$$

$$S_{L1} = \begin{cases} 0.5x^2 & |x| < 1 \\ |x| - 0.5 \ otherwise \end{cases} \tag{6}$$

The overall loss is a weighted combination of L_a and L_m:

$$L = L_a + \lambda L_m, \tag{7}$$

where λ is the balance parameter.

3.5   Abnormality Score

In order to quantify the normal (abnormal) degree of test video frames, following [8], we computer the PSNR between the predicted video frames and the ground-truth frames:

$$P(\hat{I}_t, I_t) = 10 \log_{10} \frac{max(\hat{I}_t)}{||\hat{I}_t - I_t||_2^2/N}, \tag{8}$$

where N represents the number of pixels in a video frame. As mentioned earlier, if the predicted frame I_t is abnormal, the prediction result will have a relatively large error. Therefore, the calculated PSNR value is lower and vice versa. Following [3,8,12], we use the normalized function to process the values calculated by Eq. (8). The final abnormal score is defined as:

$$S_t = 1 - g(P(\hat{I}_t, I_t)), \quad (9)$$

where $g(\cdot)$ represents the min-max normalization.

$$g(P) = \frac{P - min(P)}{max(P) - min(P)} \quad (10)$$

4 Experiments

4.1 Datasets

To evaluate the effectiveness of our approach, we conduct comprehensive experiments on three benchmark datasets and compare its performance with some typical and state-of-the-art methods.

(1) UCSD Ped2 [7]. The dataset contains 16 training videos and 12 test videos, with a total of 2,550 training video frames and 2,010 test video frames. And it includes 12 abnormal events such as riding a bicycle and driving a vehicle.

(2) CUHK Avenue [9]. The dataset consists of 16 training videos and 21 test videos, including 15,328 training video frames and 15324 test video frames. And it includes 47 abnormal events such as riding a bicycle and driving a vehicle.

(3) ShanghaiTech [10]. The dataset contains 330 training videos and 107 test videos collected from 13 scenarios. By preprocessing the videos, we can obtain 274,515 training video frames and 40,898 test video frames. It is the largest data set in the existing anomaly detection benchmark.

4.2 Implementation Detail

During training, We set the size of each video frame as 256×256 and normalize the pixel value to the range of $[-1, 1]$. We set the learning rate to $2e - 4$ and trained 60, 60 and 10 epochs on UCSD Ped2, CUHK Avenue and ShanghaiTech, respectively by using the Adam optimizer with the batch size of 4. We implemented our method with the PyTorch framework, and run all the experiments on a workstation with i7-7700K 4.2 GHz CPU and GTX 2080Ti GPU.

4.3 Results

In this section, we present the results of comparative experiments. For quantitative analysis, we calculate the frame-level Area Under Curve(AUC) to measure model performance. Our results are reported in Table 1. We adopt SEFM-A as the fusion module in this experiment, which will be discussed in the next subsection. These evaluation results show that our method surpass the typical methods by a useful margin, such as AE-Conv3D, AMDN and StackRNN. Meanwhile, we can observave that our method is highly competitive with the state-of-the-art methods, especially on the challenging ShanghaiTech dataset.

Table 1. The frame-level AUC performance of different comparison methods.

	Ped2	Avenue	ShanghaiTech
ConvLSTM-AE (2017 ICME) [11]	0.881	0.770	–
AE-Conv3D (2017 ACMMM) [19]	0.912	0.771	–
AMDN (TwoStream, 2017 CVIU) [2]	0.908	–	–
TSC(2017 ICCV) [10]	0.910	0.806	0.679
StackRNN(2017 ICCV) [10]	0.922	0.817	0.680
MemAE(2019 CVPR) [3]	0.941	0.833	0.712
DDGAN(2020 PRL) [17]	0.949	0.856	0.737
MNAD(2020 CVPR) [12]	0.970	0.885	0.705
Ours	**0.963**	**0.856**	**0.731**

4.4 Ablation Study

In this section, we first conduct experiments to investigate the effects of several different feature fusion modules, and the experimental results are reported in Table 2. We also perform experiments to explore the influence of the hyper-parameter λ, and the results are summarized in Fig. 4.

Table 2. The frame-level AUC on Ped2, Avenue dataset with different feature fusion methods.

	Ped2	Avenue
Baseline	0.941	0.845
Concat	0.951	0.850
SEFM	0.957	0.849
SEFM-M	0.955	0.846
SEFM-A	**0.963**	**0.856**

Different Feature Fusion Modules. The first row of Table 2 shows the AUC of a single branching network only using appearance features, and the second row shows the AUC of a two-branch network that directly concatenate appearance and motion features. Through comparison, It can be found that after features concat, the AUC is better than the appearance branch network, which proves that the appearance features and motion features are complementary, and the feature obtained after fusion of the two are more accurate. The last three lines show the AUC obtained from the presented three feature fusion methods in this paper. From the results, it is easy to find that the addition of SEFM and SEMF-M only gets small gain or even negative gain, while the addition of SEFM-A feature fusion module significantly improves AUC. The reason for this phenomenon is that the motion features extracted from the RGB difference map are not that distinct compared with the appearance features extracted from the image, and the integration of appearance features into motion features will lead to inaccurate motion description due to the introduction of noises. According to the comparison of AUC, the proposed feature fusion method can further enhance the complementarity of appearance features and motion features. The results also validate the effectiveness of the feature fusion module designed in our work.

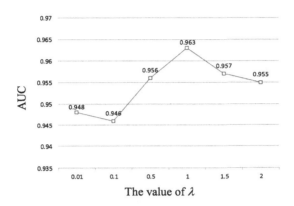

Fig. 4. The frame-level AUC on Ped2 dataset with different hyper-parameter λ.

The value of λ. Figure 4 shows the frame-level AUC on the Ped2 dataset with different values of λ. It can be seen that we obtain the best results when $\lambda = 1$. When λ is greater or less than 1, the performance degrades to some extent. It also demonstrates that motion information is as important as appearance information in our framework.

5 Conclusions

In this paper, we presented an unsupervised video anomaly detection method based on deep AE, which extracts appearance and motion features respectively by using a two-stream network, and employs a feature fusion module to fuse appearance and motion features to better model the normal. Experimental results have verified that our proposed method can effectively integrate two kinds of features. We have also demonstrated that our method achieves competitive performance in comparison with some typical and state-of-the-art methods by experiments.

References

1. Chang, Y., Tu, Z., Xie, W., Yuan, J.: Clustering driven deep autoencoder for video anomaly detection. In: Vedaldi, A., Bischof, H., Brox, T., Frahm, J.-M. (eds.) ECCV 2020. LNCS, vol. 12360, pp. 329–345. Springer, Cham (2020). https://doi.org/10.1007/978-3-030-58555-6_20
2. Dan, X.A., Yan, Y.D., Erb, C., Ns, A : Detecting anomalous events in videos by learning deep representations of appearance and motion. In: Computer Vision and Image Understanding, pp. 117–127. ScienceDirect (2017)
3. Gong, D., et al.: Memorizing normality to detect anomaly: memory-augmented deep autoencoder for unsupervised anomaly detection. In: 2019 IEEE International Conference on Computer Vision (ICCV) (2019)
4. Hasan, M., Choi, J., Neumann, J., Roy-Chowdhury, A.K., Davis, L.S.: Learning temporal regularity in video sequences. In: 2016 IEEE Conference on Computer Vision and Pattern Recognition (CVPR) (2016)
5. Jie, H., Li, S., Gang, S.: Squeeze-and-excitation networks. In: 2018 IEEE Conference on Computer Vision and Pattern Recognition (CVPR) (2018)
6. Joze, H.R.V., Shaban, A., Iuzzolino, M.L., Koishida, K.: MMTM: multimodal transfer module for CNN fusion. In: 2020 IEEE Conference on Computer Vision and Pattern Recognition (CVPR) (2020)
7. Li, W., Mahadevan, V., Vasconcelos, N.: Anomaly detection and localization in crowded scenes. IEEE Trans. Pattern Anal. Mach. Intell. **36**, 18–32 (2013)
8. Liu, W., Luo, W., Lian, D., Gao, S.: Future frame prediction for anomaly detection - a new baseline. In: 2017 IEEE Conference on Computer Vision and Pattern Recognition (CVPR) (2017)
9. Lu, C., Shi, J., Jia, J.: Abnormal event detection at 150 FPS in MATLAB. In: 2014 IEEE International Conference on Computer Vision (ICCV) (2014)
10. Luo, W., Wen, L., Gao, S.: A revisit of sparse coding based anomaly detection in stacked RNN framework. In: 2017 IEEE International Conference on Computer Vision (ICCV) (2017)
11. Luo, W., Wen, L., Gao, S. Remembering history with convolutional LSTM for anomaly detection. In: 2017 IEEE International Conference on Multimedia and Expo (ICME) (2017)
12. Park, H., Noh, J., Ham, B.: Learning memory-guided normality for anomaly detection. In: 2020 IEEE Conference on Computer Vision and Pattern Recognition (CVPR) (2020)
13. Ronneberger, O., Fischer, P., Brox, T.: U-Net: convolutional networks for biomedical image segmentation. In: Navab, N., Hornegger, J., Wells, W.M., Frangi, A.F. (eds.) MICCAI 2015. LNCS, vol. 9351, pp. 234–241. Springer, Cham (2015). https://doi.org/10.1007/978-3-319-24574-4_28
14. Sabokrou, M., Fathy, M., Hoseini, M.: Video anomaly detection and localisation based on the sparsity and reconstruction error of auto-encoder. Electron. Lett. **52**, 1122–1124 (2016)
15. Song, Q.: Deep autoencoding gaussian mixture model for unsupervised anomaly detection. In: 2018 International Conference on Learning Representations (ICLR) (2018)
16. Yan, S., Smith, J.S., Lu, W., Zhang, B.: Abnormal event detection from videos using a two-stream recurrent variational autoencoder. IEEE Trans. Cogn. Dev. Syst. **12**, 30–42 (2020)

17. Yao, T., Lin, Z., Szab, C., Chen, G., Gla, B., Jian, Y.: Integrating prediction and reconstruction for anomaly detection. Pattern Recogn. Lett. **129**, 123–130 (2020)
18. Paffenroth, R.C., Chong, Z.: Anomaly detection with robust deep autoencoders. In: The 23rd ACM SIGKDD International Conference (2017)
19. Zhao, Y. Deng, B. Shen, C., Liu, Y., Lu, H., Hua, X.: Spatio-temporal autoencoder for video anomaly detection. In: 2017 ACM International Conference on Multimedia (ACM MM) (2017)

Can DNN Detectors Compete Against Human Vision in Object Detection Task?

Qiaozhe Li[1(✉)], Jiahui Zhang[1], Xin Zhao[1], and Kaiqi Huang[1,2]

[1] Center for Research on Intelligent System and Engineering, CASIA, Beijing, China
`liqiaozhe2015@ia.ac.cn`, {`xzhao,kqhuang`}`@nlpr.ia.ac.cn`
[2] CAS Center for Excellence in Brain Science and Intelligence Technology,
Shanghai, China

Abstract. Object detection is the most fundamental problem in computer vision, which receives unprecedented attention in past decades. Although state-of-the-art detectors achieve promising results on public evaluation datasets, they may still suffer performance degradation problems dealing with some specific challenging factors. In contrast, evidences suggest that human vision system exhibits greater robustness and reliability compared with the deep neural networks. In this paper, we study and compare deep neural networks and human visual system on generic object detection task. The purpose of this paper is to evaluate the capability of DNN detectors benchmarked against human, with a proposed object detection Turing test. Experiments show that there still exists a tremendous performance gap between object detectors and human vision system. DNN detectors encounter more difficulties to effectively detect uncommon, camouflaged or sheltered objects compared with human. It's hoped that findings of this paper could facilitate studies on object detection. We also envision that human visual capability is expected to become a valuable reference to evaluate DNN detectors in the future.

Keywords: Visual turing test · Object detection · Comparison between DNNs and Human

1 Introduction

Object detection, which is one of the most fundamental and challenging problems in computer vision, has been widely studied [29,47,52]. Owing to the development of deep learning techniques, together with the increase of GPU's computing power, remarkable breakthroughs have been made in the field of object detection. Object detection has now been widely used in many real-world applications, such as autonomous driving, human computer interaction, video surveillance, etc.

On one hand, DNN detectors have already reported very competitive results on a series of public evaluation datasets, e.g., MS COCO [28]. It has become increasingly difficult to further improve the performance of detection algorithms compared with state-of-the-art models. On the other hand, there still exist insurmountable difficulties caused by multiple challenging factors that can not be well

© Springer Nature Switzerland AG 2021
H. Ma et al. (Eds.): PRCV 2021, LNCS 13019, pp. 542–553, 2021.
https://doi.org/10.1007/978-3-030-88004-0_44

handled by DNN detectors. For example, mainstream object detectors may fail to recognize objects when snow appears in the image, even though the objects are clearly visible to a human eye [34]. Furthermore, it's not surprising that DNN detectors may also exhibit vulnerability to subtle image corruptions [4], which seems to be an universal problem for most DNN models [20].

As is argued [2], the visual recognition of objects by humans in everyday life is typically rapid and effortless, as well as largely independent of viewpoint and object orientation. Besides, from the perspective of biological visual mechanism, Huang and Tan [23] encapsulated that human visual system greatly surpasses visual computing models based on statistical learning. Due to the prominence of DNNs, Studies to compare DNNs and human visual system have gained increasing attention. Experiments reveal that the robustness of human visual system outperforms DNNs [3,7,15,16,49]. However, most these experiments focus on image classification problem, few is conducted to explore the performance gap between DNNs and human on object detection task. Therefore, it's desirable to study and compare DNN detectors and human performance on object detection task.

Visual Turing test (VTT) is an important approach to evaluate deep learning performance benchmarked against human. Researches [22,48,51] also raised that Visual Turing test is considered a potential way to enhance human-like machine intelligence by constructing a "human-in-the-loop" framework. In this paper, an object detection Turing test is proposed to verify the robustness and generalization ability of DNN object detectors compared against human. A novel VTT object detection dataset is constructed containing 11 challenging factors, including complex appearance, complex background, camouflaged object, infrared image, low resolution, noise, sheltered, low-light, small object, motion blur, and uncommon object. 26 human participants are involved into the object detection test. The human experiments are designed to mimic the training and testing stages of DNN detectors. At the training stage, participants are required to freely view the training samples. At the testing stage, the participants are asked to click all the objects of one particular category following the system instructions image-by-image. The DNN detectors are trained using various datasets and then perform object prediction on the proposed VTT detection dataset. The performance of DNN detectors and human are compared. Experiments verify that there still exists a tremendous performance gap between object detectors and human vision system. It's might still be a long march before DNN detectors can achieve comparable performance with human, specifically when detecting uncommon, camouflaged or sheltered objects.

2 Related Work

2.1 Object Detection

Past decades witnessed that object detection methods evolve from traditional methods into current DCNN methods. Most early methods [5,11,12,43] designed sophisticated feature representation policies for object detection, due to the limited computing resources. The emergence of deep learning [26], which performs as a powerful feature representation method, has significantly promoted the revolution of object detection algorithms. In deep learning era, the milestone approaches of object detection can be usually organized into two categories: two-stage detection framework and one-stage detection framework. In two-stage detection framework, category-independent region proposals are firstly generated from images, and then category-specific classifiers are used to determine category labels of the proposals. Some important methods include RCNN [19], SPPNet [21], Fast RCNN [18], Faster RCNN [39], etc. In contrast, one-stage detectors directly predict class probabilities and bounding box offsets from full images with a single feed-forward CNN. Related work includes YOLO [36] and its variants [37,38], SSD [30], RetinaNet [27], FCOS[42]. These researches are of vital importance to the research of object detection, as well as other vision tasks.

Along with the progress of detection models, a variety of datasets and benchmarks have also been published in the past 20 years. These datasets not only provide public platforms to measure and compare object detection algorithms, but also push forwards the research into the fields of increasing complex and challenging problems. For example, PASCAL VOC [8,9], ImageNet [40], MS COCO [28] and Open Images [24] are famous datasets for generic object detection. In these datasets, a large portion of the images is obtained from web data with diversified categories. Besides the generic object detection problem, some researches focus on some specific challenges in object detection, such as small object detection [50], low-light object detection[31], and camouflaged object detection [10]. Studies on these practical challenging factors are advantageous to object detection applications.

2.2   Deep Neural Networks vs. Human

By far, comparing the performance on well recognized fixed datasets between state-of-the-art methods is still the most important way to verify the superiority of the DNN detectors. According to the competing experiments, DNN detectors seem to achieve promising results [29,47,52]. However, few work has been conducted to compare the detection capability between DNN detectors and human.

On the other side, Visual Turing test (VTT) [17], which is designed to evaluate deep learning machines in terms of human-like intelligence and robustness [22,48], has gradually gain attention. In [17,35], a Visual Turing test framework is proposed to evaluate whether computers can effectively understand natural images like human. Given a test image, The VTT system generates a stochastic sequence of story-line based binary questions, and administered to computer vision systems. Inspiringly, this work has greatly promoted the following-up computer vision researches, such as visual description and visual question answering (VQA) [14,32,33]. However, this VTT design emphasizes on evaluating comprehensive image understanding ability of machines, especially spatial and causal reasoning [35]. Therefore, it's hard to conclude that if machines can detect object

Fig. 1. The proposed VTT object detection dataset. 11 challenging factors are considered, including complex appearance, complex background, camouflaged object, infrared image, low resolution, noise, sheltered, low-light, small object, motion blur, and uncommon object.

instances as well as human by this test. Besides the recognition tasks, Visual Turing test has also been adopted in image generation tasks, including image style transfer [45], image colorization [46] and image rendering [41]. In this kind of VTT, human participants are required to judge if one image is natural or computer-created [25,41,45,46]. How well the generated samples can fool human perception becomes an increasingly important metric for performance evaluation of these tasks.

The Achilles Heel of deep neural networks is their vulnerability to adversarial examples [20]. This motivates researchers to compare the difference between DNNs and the human visual system [13]. In [3,7,15,16,49], experiments are conducted to compare the robustness and generalization ability of current DNNs and human via image manipulations. However, most these studies focus on image classification task. Further research is required on more vision tasks, e.g., object detection.

3   Methods

The purpose of this research is to compare current DNN detectors with human performance on object detection task. We argue that human performance gains potential to become a valuable reference to evaluate future DNN detectors.

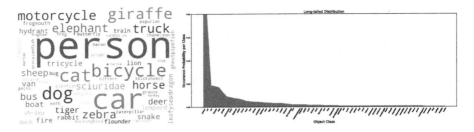

Fig. 2. The proposed VTT dataset contains 63 object categories. The probability of object occurrence is long-tailed distributed.

3.1 Dataset

To measure the performance of human and detectors, a novel VTT object detection dataset is constructed containing 11 challenging factors, including complex appearance, complex background, camouflaged object, infrared image, low resolution, noise, sheltered, low-light, small object, motion blur, and uncommon object. Illustrations of the proposed dataset can be viewed in Fig. 1. Images are originated from the validation and test sets of 6 representative object detection datasets, including MS COCO dataset [28], Visdrone dataset [50], Exclusively Dark dataset [31], ImageNet VID [6], COD-10k dataset [10], and FLIR Thermal dataset [1], according to the image characteristics of each dataset. For low resolution images, the selected images are firstly down-scaled to 160×120 pixels and then re-scaled to their original sizes by bilinear sampling. For noise images, Pepper noise and Gaussian noise are randomly selected to achieve image corruptions.

The entire dataset consists of 488 testing images, with each challenging factor regarding to approximate 45 images. It's worth noting that a small portion of images may be inevitably tangled with multiple challenging factors. For convenience of statistics, the images are categorized non-repetitively by their most salient attributes. Object class labels of the proposed dataset are unified, resulting to 63 classes. The probability of object occurrence is long-tailed distributed, which is shown in Fig. 2.

3.2 Human Experiments

26 human participants are involved into the object detection test. The human experiments are designed to mimic the training and testing stages of DNN detectors. In the training stage, participants are required to freely view the training samples. Totally 112 training images are provided, with rectangular bounding boxes cropped and object categories annotated onto the images. The training set for human covers all 11 challenging factors, and each object category has at least one corresponding image. The training stage allows the participants to get familiar with all object categories. This procedure is analogous to the training stage used for deep neural networks (Fig. 3).

After manual confirmation of finishing the training stage, the system enters the testing stage. At the testing stage, for each image, the participants are asked to click all the objects of one particular category following the system instructions. Once participants confirm that task has been completed for current image, the system will automatically show the next image. This process continues until the testing stage finished. During testing, not all the prompted objects appear in corresponding images. The purpose is to verify if human can be obfuscated by similar object categories, for example, "frogfish" and "ghost pipefish". The ratio of the true and false prompts is set to 7:3. Each subject is randomly assigned with 100 images from the proposed dataset in the testing stage.

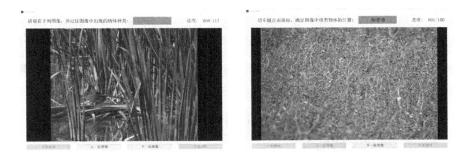

Fig. 3. Interface of the training (left) and testing (right) stages of the designed system for Visual Turing test. At the training stage, participants are required to freely view the training samples with detection annotations. At the testing stage, the participants are asked to click all the objects of one particular category following the system instructions.

3.3   DNN Detectors

Inspired by the great breakthroughs of deep convolutional neural networks in image recognition task, DNN detectors are widely studied and have become the mainstream in object detection. In this paper, Faster RCNN family is adopted as the baseline model, as it's the most popular benchmark to evaluate detector results. The training sets of MS COCO, Visdrone, Exclusively Dark, ImageNet VID, COD-10k, and FLIR Thermal are integrated to train a monolithic detector, after merging the same object categories between these datasets. The detectors are implemented using the famous Detectron2 framework [44], with default network structure and parameter settings. After fine-tuning, standard prediction procedure is conducted on the proposed dataset using the trained detectors.

3.4   Evaluation Metrics

Since human provide definite predictions, the standard Average Precision (AP) [28] metrics might not be suitable to compare human and DNN detectors in this paper. As an alternative, precision and recall are employed for performance evaluation. **For human participants**, if one of his (her) clicks locates inside the groundtruth bounding boxes of the correct object class instructed by VTT system, then it's counted as a true positive (TP) result. If one of the prompted objects is missed by subjects, then it's counted as a false negative (FN) result. If one click locates outside any bounding box of the correct object class or one correct object is clicked repeatedly, then it's counted as a false positive (FP) result. **For object detectors**, metrics are calculated by only considering those samples that have already tested human. For example, given an image, only if VTT system has tested human to find out objects of a particular category, then these objects in this image are included as statistical samples. Otherwise, they are abandoned. TP samples are calculated considering objects that are correctly located (with IOU > 0.5) and classified by detectors. FN samples refer to those objects missed by detectors. FP samples are objects that are well localized but incorrectly classified. To achieve fair comparison between DNN detectors and human, the evaluation metrics of this paper might slightly differ from classical ones.

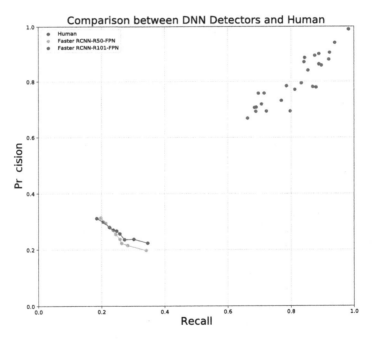

Fig. 4. Performance comparison between Faster RCNN detectors and human.

4   Results

Figure 4 reports the performance comparison between Faster RCNN detectors and human. Faster RCNN-R50-FPN and Faster RCNN-R101-FPN are employed as they are most widely referred. Precision and recall are utilized as evaluation metrics, since standard AP may not be suitable to evaluate human. For DNN detectors, the confidence score is adjusted from 0.1 to 0.9, with an interval of 0.1. As is illustrated, due to the trait of the DNN detectors, reaching higher precision rate means the sacrifice of recall rate, and vice versa. Interestingly, the precision and recall of human participants are positively correlated, the average ratio of which is around 1.0. The confidence scores are 0.5 for RCNN-R50-FPN ($\frac{precision}{recall} = \frac{0.253}{0.246}$) and 0.4 for RCNN-R101-FPN ($\frac{precision}{recall} = \frac{0.257}{0.257}$), when the ratios of precision and recall are closest to human. In general, compared with human visual system, the DNN detectors still lag far behind in performance.

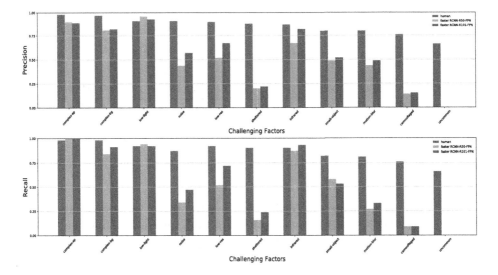

Fig. 5. Influence of 11 challenging factors to DNN detectors and human.

The proposed detection VTT dataset contains a variety of challenging factors. Thus, it's desirable to analyze the influence of each challenging factor to DNN detectors and human (Fig. 5). The confidence scores of Faster RCNN-R50-FPN and Faster RCNN-R101-FPN are respectively set to 0.5 and 0.4, to make the ratios of precision and recall closer to human. The results show that with a stronger baseline for feature extraction, Faster RCNN-R101-FPN slightly outperforms Faster RCNN-R50-FPN. DNN detectors have achieved comparable performance when dealing with complex appearance and complex background factors, and even exceed human by a small margin to detect objects in low-light conditions. However, experiments also suggest that it's extremely difficult for DNN detectors to predict uncommon, camouflaged, and sheltered objects. In contrast, although human participants are affected to a certain extent when detecting uncommon and camouflaged objects, they still exceed DNN detectors by a marked margin. It's implied that human vision system exhibits greater robustness and reliability compared with the deep neural networks in object detection.

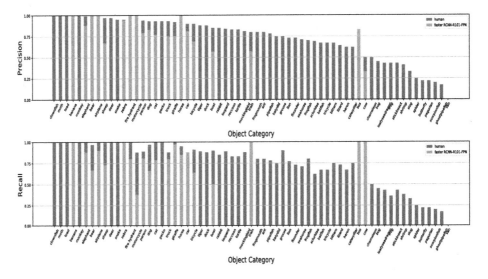

Fig. 6. The performance of DNN detectors and human on different object categories.

Furthermore, influence of object categories to object detection is also compared between DNN detectors and human (Fig. 6). Experiments show that Faster RCNN-R101-FPN can achieve promising results when detecting objects from some specific categories, such as banana, horse and airplane. Note that sufficient training samples of these object categories can be obtained from the training set of DNN detectors. It cannot be ignored that Faster RCNN also fails on a considerable portion of object categories, which may have fewer samples compared with the former. As a comparison, the performance of human participants also varies on different object categories. Yet the variance of human performance on different categories is much smaller than object detectors.

5   Conclusion

In this paper, an object detection Turing test is conducted to study and compare the performance of DNN detectors and human visual system. 26 human participants are involved into the object detection test. As a comparison, the well-known Faster RCNN family is adopted to represent the DNN detectors.

Experiments show that there may still exist a huge gap between the performance of state-of-the-art DNN detectors and human, especially when detecting uncommon, camouflaged or sheltered objects. It's hoped that human visual capability could provide a novel valuable benchmark to evaluate DNN detectors, and our findings could facilitate the study to improve the robustness of DNN detectors in the future.

References

1. https://www.flir.com/oem/adas/adas-dataset-form/
2. Biederman, I.: Recognition-by-components: a theory of human image understanding. Psychol. Rev. **94**(2), 115 (1987)
3. Borji, A., Itti, L.: Human vs. computer in scene and object recognition. In: Proceedings of the IEEE Conference on Computer Vision and Pattern Recognition, pp. 113–120 (2014)
4. Chow, K.H., Liu, L., Gursoy, M.E., Truex, S., Wei, W., Wu, Y.: Understanding object detection through an adversarial lens. In: European Symposium on Research in Computer Security, pp. 460–481 (2020)
5. Dalal, N., Triggs, B.: Histograms of oriented gradients for human detection. In: 2005 IEEE Computer Society Conference on Computer Vision and Pattern Recognition, vol. 1, pp. 886–893 (2005)
6. Deng, J., Dong, W., Socher, R., Li, L.J., Li, K., Fei-Fei, L.: Imagenet: a large-scale hierarchical image database. In: 2009 IEEE Conference on Computer Vision and Pattern Recognition, pp. 248–255 (2009)
7. Dodge, S., Karam, L.: Can the early human visual system compete with deep neural networks? In: Proceedings of the IEEE International Conference on Computer Vision Workshops, pp. 2798–2804 (2017)
8. Everingham, M., Eslami, S.A., Van Gool, L., Williams, C.K., Winn, J., Zisserman, A.: The pascal visual object classes challenge: a retrospective. Int. J. Comput. Vis. **111**(1), 98–136 (2015)
9. Everingham, M., Van Gool, L., Williams, C.K., Winn, J., Zisserman, A.: The pascal visual object classes (voc) challenge. Int. J. Comput. Vis. **88**(2), 303–338 (2010)
10. Fan, D.P., Ji, G.P., Sun, G., Cheng, M.M., Shen, J., Shao, L.: Camouflaged object detection. In: Proceedings of the IEEE/CVF Conference on Computer Vision and Pattern Recognition, pp. 2777–2787 (2020)
11. Felzenszwalb, P., McAllester, D., Ramanan, D.: A discriminatively trained, multi-scale, deformable part model. In: 2008 IEEE Conference on Computer Vision and Pattern Recognition, pp. 1–8 (2008)
12. Felzenszwalb, P.F., Girshick, R.B., McAllester, D., Ramanan, D.: Object detection with discriminatively trained part-based models. IEEE Trans. Pattern Anal. Mach. Intell. **32**(9), 1627–1645 (2009)
13. Firestone, C.: Performance vs. competence in human-machine comparisons. Proc. Natl. Acad. Sci. **117**(43), 26562–26571 (2020)
14. Gao, H., Mao, J., Zhou, J., Huang, Z., Wang, L., Xu, W.: Are you talking to a machine? dataset and methods for multilingual image question. Adv. Neural Inf. Process. Syst. **28**, 2296–2304 (2015)
15. Geirhos, R., Janssen, D.H., Schütt, H.H., Rauber, J., Bethge, M., Wichmann, F.A.: Comparing deep neural networks against humans: object recognition when the signal gets weaker. arXiv preprint arXiv:1706.06969 (2017)

16. Geirhos, R., Temme, C.R.M., Rauber, J., Schütt, H.H., Bethge, M., Wich-mann, F.A.: Generalisation in humans and deep neural networks. arXiv preprint arXiv:1808.08750 (2018)
17. Geman, D., Geman, S., Hallonquist, N., Younes, L.: Visual turing test for computer vision systems. Proc. Natl. Acad. Sci. **112**(12), 3618–3623 (2015)
18. Girshick, R.: Fast r-cnn. In: Proceedings of the IEEE International Conference on Computer Vision, pp. 1440–1448 (2015)
19. Girshick, R., Donahue, J., Darrell, T., Malik, J.: Rich feature hierarchies for accurate object detection and semantic segmentation. In: Proceedings of the IEEE Conference on Computer Vision and Pattern Recognition, pp. 580–587 (2014)
20. Goodfellow, I.J., Shlens, J., Szegedy, C.: Explaining and harnessing adversarial examples. arXiv preprint arXiv:1412.6572 (2014)
21. He, K., Zhang, X., Ren, S., Sun, J.: Spatial pyramid pooling in deep convolutional networks for visual recognition. IEEE Trans. Pattern Anal. Mach. Intell. **37**(9), 1904–1916 (2015)
22. Hu, B.G., Dong, W.M.: A design of human-like robust ai machines in object identification. arXiv preprint arXiv:2101.02327 (2021)
23. Huang, K.Q., Tan, T.N.: Review on computational model for vision. Pattern Recogn. Artif. Intell. **26**(10), 951–958 (2013)
24. Kuznetsova, A., et al.: The open images dataset v4. Int. J. Comput. Vis. **128**, 1–26 (2020)
25. Lake, B.M., Salakhutdinov, R., Tenenbaum, J.B.: Human-level concept learning through probabilistic program induction. Science **350**(6266), 1332–1338 (2015)
26. LeCun, Y., Bengio, Y., Hinton, G.: Deep learning. Nature **521**(7553), 436–444 (2015)
27. Lin, T.Y., Goyal, P., Girshick, R., He, K., Dollár, P.: Focal loss for dense object detection. In: Proceedings of the IEEE International Conference on Computer Vision, pp. 2980–2988 (2017)
28. Lin, T.Y., et al.: Microsoft COCO: common objects in context. In: Fleet, D., Pajdla, T., Schiele, B., Tuytelaars, T. (eds.) ECCV 2014. LNCS, vol. 8693, pp. 740–755. Springer, Cham (2014). https://doi.org/10.1007/978-3-319-10602-1_48
29. Liu, L., et al.: Deep learning for generic object detection: a survey. Int. J. Comput. Vis. **128**(2), 261–318 (2020)
30. Liu, W., et al.: SSD: single shot multibox detector. In: Leibe, B., Matas, J., Sebe, N., Welling, M. (eds.) ECCV 2016. LNCS, vol. 9905, pp. 21–37. Springer, Cham (2016). https://doi.org/10.1007/978-3-319-46448-0_2
31. Loh, Y.P., Chan, C.S.: Getting to know low-light images with the exclusively dark dataset. Comput. Vis. Image Underst **178**, 30–42 (2019)
32. Malinowski, M., Fritz, M.: A multi-world approach to question answering about real-world scenes based on uncertain input. In: Proceedings of the 27th International Conference on Neural Information Processing Systems, pp. 1682–1690 (2014)
33. Malinowski, M., Rohrbach, M., Fritz, M.: Ask your neurons: a neural-based approach to answering questions about images. In: Proceedings of the IEEE International Conference on Computer Vision, pp. 1–9 (2015)
34. Michaelis, C., et al.: Benchmarking robustness in object detection: autonomous driving when winter is coming. arXiv preprint arXiv:1907.07484 (2019)
35. Qi, H., Wu, T., Lee, M.W., Zhu, S.C.: A restricted visual turing test for deep scene and event understanding. arXiv preprint arXiv:1512.01715 (2015)
36. Redmon, J., Divvala, S., Girshick, R., Farhadi, A.: You only look once: unified, real-time object detection. In: Proceedings of the IEEE Conference on Computer Vision and Pattern Recognition, pp. 779–788 (2016)

37. Redmon, J., Farhadi, A.: Yolo9000: better, faster, stronger. In: Proceedings of the IEEE Conference on Computer Vision and Pattern Recognition, pp. 7263–7271 (2017)
38. Redmon, J., Farhadi, A.: Yolov3: an incremental improvement. arXiv preprint arXiv:1804.02767 (2018)
39. Ren, S., He, K., Girshick, R., Sun, J.: Faster r-cnn: towards real-time object detection with region proposal networks. IEEE Trans. Pattern Anal. Mach. Intell. **39**(6), 1137–1149 (2017)
40. Russakovsky, O., et al.: Imagenet large scale visual recognition challenge. Int. J. Comput. Vis. **115**(3), 211–252 (2015)
41. Shan, Q., Adams, R., Curless, B., Furukawa, Y., Seitz, S.M.: The visual turing test for scene reconstruction. In: International Conference on 3D Vision, pp. 25–32 (2013)
42. Tian, Z., Shen, C., Chen, H., He, T.: Fcos: fully convolutional one-stage object detection. In: Proceedings of the IEEE/CVF International Conference on Computer Vision, pp. 9627–9636 (2019)
43. Viola, P., Jones, M.: Rapid object detection using a boosted cascade of simple features. In: Proceedings of the 2001 IEEE Computer Society Conference on Computer Vision and Pattern Recognition, vol. 1, pp. I (2001)
44. Wu, Y., Kirillov, A., Massa, F., Lo, W.Y., Girshick, R.: Detectron2 (2019). https://github.com/facebookresearch/detectron2
45. Xue, A.: End-to-end Chinese landscape painting creation using generative adversarial networks. In: Proceedings of the IEEE/CVF Winter Conference on Applications of Computer Vision, pp. 3863–3871 (2021)
46. Zhang, R., Isola, P., Efros, A.A.: Colorful image colorization. In: Leibe, B., Matas, J., Sebe, N., Welling, M. (eds.) ECCV 2016. LNCS, vol. 9907, pp. 649–666. Springer, Cham (2016). https://doi.org/10.1007/978-3-319-46487-9_40
47. Zhao, Z.Q., Zheng, P., Xu, S.T., Wu, X.: Object detection with deep learning: a review. IEEE Trans. Neural Netw. Learn. Syst. **30**(11), 3212–3232 (2019)
48. Zheng, N.N., et al.: Hybrid-augmented intelligence: collaboration and cognition. Front. Inf. Technol. Electron. Eng. **18**(2), 153–179 (2017)
49. Zhou, Z., Firestone, C.: Humans can decipher adversarial images. Nature Commun. **10**(1), 1–9 (2019)
50. Zhu, P., et al.: Visdrone-det2018: the vision meets drone object detection in image challenge results. In: Proceedings of the European Conference on Computer Vision (ECCV) Workshops (2018)
51. Zhu, W., Wang, X., Gao, W.: Multimedia intelligence: when multimedia meets artificial intelligence. IEEE Trans. Multimedia **22**(7), 1823–1835 (2020)
52. Zou, Z., Shi, Z., Guo, Y., Ye, J.: Object detection in 20 years: a survey. arXiv preprint arXiv:1905.05055 (2019)

Group Re-Identification Based on Single Feature Attention Learning Network (SFALN)

Xuehai Liu[1], Lisha Yu[1,2], and Jianhuang Lai[1,2,3(✉)]

[1] The School of Computer Science and Engineering, Sun Yat-sen University,
Guangzhou 510006, China
{liuxh76,yulsh6,stsljh}@mail2.sysu.edu.cn
[2] The Guangdong Key Laboratory of Information Security Technology,
Guangzhou 510006, China
[3] The Key Laboratory of Machine Intelligence and Advanced Computing,
Ministry of Education China, Beijing, China

Abstract. People often go together in groups, and group re-identification (G-ReID) is an important but less researched topic. The goal of G-ReID is to find a group of people under different surveillance camera perspectives. It not only faces the same challenge with traditional ReID, but also involves the changes in group layout and membership. To solve these problems, we propose a Single Feature Attention Learning Network (SFALN). The proposed network makes use of the abundant ReID datasets by transfer learning, and extracts effective feature information of the groups through attention mechanism. Experimental results on the public dataset demonstrate the state-of-the-art effectiveness of our approach.

Keywords: Group Re-identification · Transfer learning · Attention mechanism

1 Introduction

In recent years, Person Re-Identification (ReID) has become a hot topic in the field of image retrieval for its extensive application in security and monitoring problems [10,11,22]. The goal is to integrate an individual from different surveillance camera views. While a large number of studies have focused on re-identification of individuals, few studies have focused on multiple individuals within a group at the same time. In real application scenarios, people often appear together with other people [2,25]. The goal of group re-Identification is to match groups in disjointed camera images. The challenge of group re-identification is different from person re-identification. First, the research object

This project is supported by the NSFC(62076258).

H. Ma et al. (Eds.): PRCV 2021, LNCS 13019, pp. 554–563, 2021.
https://doi.org/10.1007/978-3-030-88004-0_45

becomes a group of people. In this case, the relative positions of the group members will change, and new members may be added to the group or members may leave and regroup. Therefore, the existing person re-identification methods cannot effectively deal with the problems of group re-identification. In addition, person re-identification has been studied for many years and there are a large number of datasets [24]. However, the group re-identification problem makes it difficult to obtain datasets, and it is difficult to train effective neural networks using the existing datasets. In order to resolve the problems of group re-identification, we propose a Single Feature Attention Learning Network(SFALN). The proposed network makes use of the abundant ReID datasets, and extracts effective the feature of groups through attention mechanism. Our contribution are as follows:

(1) We proposed a SFALN attention neural network that can accurately re-identify a group of person.
(2) The proposed network contains attention mechanism, which can effectively extracts the feature of groups.
(3) Compared with the state-of-the-art methods, our proposed method achieved the best result in the public datasets.

2   Related Work

2.1   Person Re-Identification

Person re-Identification has attracted great attention in computer vision in the recent years..The goal of person re-identification (ReID) is to determine whether there is a specific person in an image or video sequence by computer vision technology. There are mainly two steps: obtaining feature representation and perform matching with distance metrics [1,3,6,18,21]. Although person re-identification has become a hot topic in recent years, the methods designed for one person is difficult to apply to the group. Therefore, group re-identification is a less studied but crucially important complement of person re-identification.

2.2   Group Re-Identification

Group re-identification(G-ReID) is an extensive form of person ReID. Compared with ReID, G-ReID aims to match groups of people instead of single person across disjoint cameras. In this task, obtaining effective feature representation and performing matching with distance metrics are still necessary steps and could be much harder than ReID. The reason is that new challenges are proposed in G-ReID: the lack of datasets, severe member occlusion, changes of membership and layout changes. Part of existing methods mainly focus on regarding the group as a whole and extracting global or patch features. For example, Zheng et al. [25] decide a group into several patches and learn semi-global features. Cai et al. [2] introduce covariance descriptors to represent a group, and obtain both global and covariance descriptors . Most of them are based on hand-crafted features.

To address the challenge of group layout and membership changes, Xiao et al. [15] propose to consider a group as a multi-grained object, and associate group members while performing group matching. In order to make use of the abundant person ReID datasets, Huang et al. [13] propose to use CycleGAN network to transfer the style of ReID datasets to group ReID datasets. They divide a group into two grains: single individual and couple subgroups, and perform feature fusion to make prediction. This method achieves a relatively high accuracy, but its network is relatively simple and training couple subgroups could be extremely time-consuming, considering the cartesian product of pictures is utilized in generating couple subgroups dataset.

Different from above works, this paper mainly considers the group as a set of individuals, in order to enhance the robustness against group layout and membership changes. Moreover, we introduce a single feature attention learning network(SFALN). Experiments in Chap. 4 demonstrate that our method achieves a better result on Road dataset than the traditional methods, and cost less training time than above works.

2.3 Style Transfer

In recent years, style transfer has been proved successful in image-to-image translation area. In 2016, Gatys et al. [9] separated the image content from the image style and recombined it so that the style of one image can be transformed into another image. TaiGman et al. [20] proposed a style transmission network, which can transfer the image to another style while retaining the original identity. Deng et al. [5] put forward the use of style transfer network to realize the transfer of label domain with re-identification by persons, which effectively solves the problem of style difference. Through these style transfer techniques, we can extend the G-ReID dataset with the rich ReID datasets, so that our deep learning model can train effective models on this dataset, and further improve the performance of our method.

2.4 Attention Model

The success of visual attention is largely due to the reasonable assumption that human vision does not process the whole image at once, but instead focuses on selective parts of the entire visual space. In 2017, An end-to-end residual-learning approach based on Attention was proposed by FeI, W. et al. [7], in which a new branch was added in the forward process to serve as the Attention of the extraction model. Moreover, attention model is used in ReID task and is proved to make effect: for example, [14] proposed a Harmonious Attention CNN (HA-CNN) for joint learning of soft pixel attention and hard area attention to optimize the re-identification of misaligned pedestrians in an image. [19] proposed a dual attention matching network based on an inter-class and an intra-class attention module to capture the context information of video sequences for person ReID. [4] proposed an attentive but diverse net, focusing on channel aggregation and position awareness.

In this paper, we innovatively introduce attention mechanism in group re-identification task and focus on the position attention model.

3   Proposed Method

In the G-ReID task, we have a Query image p containing N people. Our goal is to find the corresponding group image p in the gallery set $G = \{g_t\}$, where t represents the tth group image in gallery G, let g_t^j represent the jth person in the image g_t. Our proposed framework consists of two main parts. First, the ReID-style training set is transferred to the G-ReID style using the style transfer method. Then, we send the images to the single feature attention learning network (SFALN) to extract features. During testing, these features will be used to calculate similarity of group images. The overall architecture of our proposed single feature attention learning network is shown in Fig. 1.

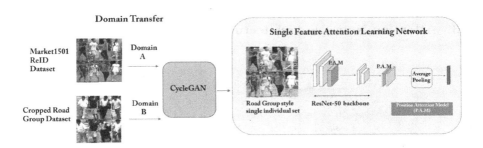

Fig. 1. Our proposed model of SFALN

3.1   Style Transfer

In this chapter, we need to solve the dataset problem of group re-identification (G-ReID): the existing limited group re-identification datasets are insufficient to train an effective deep learning neural network. However, the abundant researchs on individual person re-identification (ReID) make enough ReID datasets available, and we should make full use of these useful information. Notice that a group re-identification task can be considered as a combination of individual re-identification, inspired by [13], we propose to use ReID dataset to help train feature extraction network that could be used on G-ReID dataset. However, there is a domain gap(i.e. light condition, angle of camera) between ReID dataset and G-ReID dataset, which will result in the limited performance of the feature extraction. Therefore, in this chapter, the method of style transfer is adopted, which is named CycleGAN, to effectively solve the problem of data style differences, and thus obtain a large number of datasets that can be used for subsequent training of group feature extraction.

3.2   Single Feature Attention Learning Network

Our method proposes that a group can be described using its individual characteristics of each person that makes up the group, regardless of the number of members. Therefore, we treat a group as a multiple object and train an attention learning network, SFALN, to focus on the useful information that encodes the group and extract individual features.

For individual feature extraction network SFALN, we use the CNN neural network based on Resnet-50 to build a model to effectively re-identify each person in the group, and therefore achieve the ambition of re-identifying the whole group. Here, we add an attention module which will be later discussed to make the network focus more on discriminative features of the persons in the group. As methods that consider global or semi-global factors would be easily impacted by group layout or membership changes[2,25], our method pay more attention to the encoded information of each group member, and can therefore effectively ensure the robustness of the trained network.

3.3   Attention Network Architecture Overview

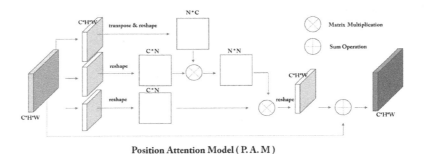

Position Attention Model (P. A. M)

Fig. 2. Structure of Position Attention Model (P.A.M.)

Our proposed single feature attention learning network is depicted in Fig. 1. Because of the well performance of ResNet-50 in ReID task, we use it as the default backbone of our feature learning network. We add a P.A.M. (position attention model) on the outputs of resnet_conv2 block. Next, after the res_conv4 block, the output is regularized and a P.A.M block is added. Finally, we add an average pooling layer and obtain the feature of each individual in the group. The structure of P.A.M. is depicted in Fig. 2, in which the input feature $\mathbf{A} \in \mathbb{R}^{C \times H \times W}$ is first fed into ResNet-50 convolution layers with ReLU activation to produce feature maps as $\mathbb{R}^{C \times H \times W}$. Then we use matrix multiplication to compute the pixel affinity matrix $\mathbf{S} \in \mathbb{R}^{N \times N}$, where $N = H \times W$. Finally, the attentive feature is generated with sum operation of original feature and produced feature map.

3.4   Online Prediction

Finally, we use the trained feature extraction network to make online prediction. Let $\left\{ \mathcal{F}^{\mathrm{sin}}\left(p^{i}\right)\right\}_{i=1}^{N_{s}}$ represents the single feature of all persons $\left\{ p^{i} \mid i = 1, 2...N_{s}\right\}$ in query image q, where Ns represents the number of persons in the query image q, and D($\cdot$) function represents the feature distance function, as we use Euler Distance in this paper. Then the feature distance between query image q and gallery image g_{t} can be defined as follow.

$$d_{t}^{\mathrm{sin}} = \frac{1}{N_{s}} \sum_{i=1}^{N_{s}} \min \left\{ D\left(\mathcal{F}^{\mathrm{sin}}\left(p^{i}\right), \mathcal{F}^{\mathrm{sin}}\left(g_{t}^{j}\right)\right) \mid j = 1, 2, \ldots, N_{t}\right\} \tag{1}$$

Thus, the best matched picture in gallery can be obtained as:

$$d_{p} = \min \left\{ d_{t}^{\mathrm{sin}} \mid t = 1, 2, ...N_{t}\right\} \tag{2}$$

4   Experiment

4.1   Datasets

Our approach was evaluated on the public available group re-identification (G-ReID) dataset **Road Group**. ReID dataset **Market-1501** is also involved in our experiment of transfer learning.

Market-1501 [24] is a widely used ReID dataset due to its large amount of training instances: 15,936 images for 751 individuals. For this resson, it is chosen to be used as the source domain ReID dataset. In our work, we transfer the style of Market-1501 dataset to those of Road Group, in order to generate a large amount of effective dataset for our deep learning network.

Road Group dataset is widely used in the G-ReID task and proposes high challenges: it contains severe occlusion, layout changes, and group membership changes, etc. Road Group dataset contains 177 group pictures, including the group images from the perspectives of six cameras numbered A-F. Following the protocol [16], we divide the dataset into train, query, and gallery sets in a 2:1:1 ratio and validate our model on query and gallery. In addition, we used CMC(Cumulative Matching Characteristic) accuracy and MAP (Mean Average Precision) to measure our network performance.

4.2   Implementation Details

Before training the feature extraction network, we transfer the style of the existing ReID dataset(Market-1501) to the target group re-identification dataset (Road Group dataset). We use CycleGAN to achieve this goal, and therefore obtain Market-1501 dataset in Road Group-style. In the process of training CycleGAN, we resize the size of all input images to 256×128 and use Adam optimizer. We set the batch_size as 10, and set the learning rate of generator and discriminator as 0.0002 and 0.0001 in the experimental environment. Tensorflow2.0 was chosen as the framework and 4 GTX1080Ti GPUs were used in the experimental environment to train 50 epochs for CycleGAN.

As for the construction of feature extraction network SFALN, we use Resnet-50 as our network framework. The learning rate of training is set to 0.01 and the dropout rate is set to 0.5. We used SGD(stochastic gradient descent) as the optimizer, and carried out training on the server of our laboratory. The experimental environment used 4 GTX1080Ti GPUs, and 27 epochs are trained for SFALN to achieve convergence.

4.3   Ablation Experiment

Table 1. Performance Evaluation of Transfer Learning and Attention Mechanism on Road Group dataset. The suffix "with T.L." refers that transfer learning is used, and SFALN(with attention) performs attention mechanism comparing to SFLN

Method	Road Group			
	CMC-1	CMC-5	CMC-10	CMC-20
SFLN(baseline) [13]	78.4	88.7	94.5	98.8
SFLN(with T.L.)	82.7	93.8	96.3	98.8
SFALN(with attention)	83.6	94.5	98.8	98.8
SFALN(with attention & T.L.)	**86.4**	**100**	**100**	**100**

We conduct several experiments to prove the effectiveness of proposed attention mechanism and transfer learning. Table 1 shows the performance evaluation of our ablative experiment on Road group dataset. From the table, we can find that transfer learning can make a stable improvement on the network with or without attention mechanism. Moreover, network with attention mechanism(i.e. SFALN(with attention) and SFALN(with attention & T.L.)) outperforms those without(i.e. SFLN and SFLN(with T.L.)). The above observation help us draw the conclusion that attention mechanism can achieve a 5.2% CMC-1 accuracy improvement on our feature learning network, and transfer learning enhances the SFLN network by 4.3% CMC-1 accuracy on Road Group. Finally, both of the mechanisms together makes a great improvement of 8.0% CMC-1 accuracy on our baseline.

4.4   Comparisons with the State-of-the-art Methods

he methods for comparison include crro-bro [25], Covariance [2], PREF [17], BSC+CM [8], MGR [16], DotGNN [12], GCGNN [26], MACG [23], DotSCN [13]. Comparison of performance is shown in Table 2.

Table 2. Comparison of Performance with state-of-the-art Methods on Road Group

Method	Road croup			
	CMC-1	CMC-5	CMC-10	CMC-20
CRRRO-BRO[25]	17.8	34.6	48.1	62.2
Covariance [2]	38.0	61.0	73.1	82.5
PREF [17]	43.0	68.7	77.9	85.2
BSC+CM [8]	58.6	80.6	87.4	92.1
MGR [16]	72.3	90.6	94.1	97.5
DotGNN [12]	74.1	90.1	92.6	98.8
GCGNN [26]	53.6	77.0	91.4	94.8
MACG [23]	84.5	95.0	96.9	98.1
DotSCN [13]	84.0	95.1	96.3	98.8
SFALN (Ours)	**86.4**	**100**	**100**	**100**

It can be seen that our method achieves better results than the above optimal state-of-the-art method, and achieves a performance improvement of 1.9% CMC-1 accuracy on Road Group. The success of method is inseparable from the strong ability of the feature extraction of deep learning network. At the same time, the powerful performance of transfer learning(CycleGAN) also makes a great influence, which provides sufficient and effective training set for our deep learning method. Moreover, attention mechanism is proved to be effective in group re-identification task in this paper, as focusing on characteristics of each individual member could help learn a more comprehensive representation of the whole group.

5   Conclusion

In this paper, we mainly discuss an important but poorly studied task: group re-identification (G-ReID). G-ReID task is an extended form of person re-identification task (ReID), whose re-identification target change from individuals to groups, which brings new challenges such as group member changes, member layout changes, frequent intra-group occlusion, and so on. In addition, the lack of datasets makes it difficult to train effective neural networks. In this paper, we still use style transfer learning (CycleGAN) method to expand the group re-identification datasets, thus providing sufficient data for training deep learning network. At the same time, we propose the deep learning feature extraction network: Single Feature Attention Learning Network (SFALN) to extract the features of each person, effectively easing the impact of challenges such as changes in group members and layout. Compared with the previous work, we introduce an attention mechanism, and do not use the time-consuming couple feature extraction network. Our experimental results strongly demonstrate the effectiveness of the proposed SFALN framework.

References

1. Ahmed, E., Jones, M., Marks, T.K.: An improved deep learning architecture for person re-identification. In: 2015 IEEE Conference on Computer Vision and Pattern Recognition (CVPR) (2015)
2. Cai, Y., Takala, V., Pietikainen, M.: Matching groups of people by covariance descriptor. In: International Conference on Pattern Recognition (2010)
3. Chen, S.Z., Guo, C.C., Lai, J.H.: Deep ranking for person re-identification via joint representation learning. IEEE Trans. Image Process. **25**(5), 2353–2367 (2016)
4. Chen, T., et al.: Abd-net: attentive but diverse person re-identification. arXiv (2019)
5. Deng, W., Zheng, L., Ye, Q., Kang, G., Yang, Y., Jiao, J.: Image-image domain adaptation with preserved self-similarity and domain-dissimilarity for person re-identification. In: 2018 IEEE/CVF Conference on Computer Vision and Pattern Recognition (2018)
6. Dong, Y., Zhen, L., Liao, S., Li, S.Z.: Deep metric learning for person re-identification. In: International Conference on Pattern Recognition (2014)
7. Fei, W., Jiang, M., Chen, Q., Yang, S., Tang, X.: Residual attention network for image classification. In: 2017 IEEE Conference on Computer Vision and Pattern Recognition (CVPR) (2017)
8. Feng, Z., Qi, C., Yu, N.: Consistent matching based on boosted salience channels for group re-identification. In: IEEE International Conference on Image Processing (2016)
9. Gatys, L.A., Ecker, A.S., Bethge, M.: Image style transfer using convolutional neural networks. In: 2016 IEEE Conference on Computer Vision and Pattern Recognition (CVPR) (2016)
10. Hou, R., Ma, B., Chang, H., Gu, X., Chen, X.: Iaunet: global context-aware feature learning for person reidentification. IEEE Trans. Neural Netw. Learn. Syst., 99 (2020)
11. Hou, R., Ma, B., Chang, H., Gu, X., Chen, X.: Feature completion for occluded person re-identification. IEEE Trans. Pattern Anal. Mach. Intell. **99**, 1 (2021)
12. Huang, Z., Wang, Z., Hu, W., Lin, C.W., Satoh, S.: Dot-gnn: domain-transferred graph neural network for group re-identification. In: the 27th ACM International Conference (2019)
13. Huang, Z., Wang, Z., Satoh, S., Lin, C.W.: Group re-identification via transferred single and couple representation learning (2019)
14. Li, W., Zhu, X., Gong, S.: Harmonious attention network for person re-identification. In: 2018 IEEE/CVF Conference on Computer Vision and Pattern Recognition (2018)
15. Lin, W., Li, Y., Xiao, H., See, J., Mei, T.: Group reidentification with multigrained matching and integration. IEEE Trans. Cybern. **99**, 1–15 (2019)
16. Lin, W., et al.: Group reidentification with multigrained matching and integration. IEEE Trans. Cybern. **51**(3), 1478–1492 (2021). https://doi.org/10.1109/TCYB.2019.2917713
17. Lisanti, G., Martinel, N., Bimbo, A.D., Foresti, G.L.: Group re-identification via unsupervised transfer of sparse features encoding. In: IEEE International Conference on Computer Vision (2017)
18. Paisitkriangkrai, S., Shen, C., Hengel, A.: Learning to rank in person re-identification with metric ensembles. In: 2015 IEEE Conference on Computer Vision and Pattern Recognition (CVPR) (2015)

19. Si, J., et al.: Dual attention matching network for context-aware feature sequence based person re-identification. In: CVPR (2018)
20. Taigman, Y., Polyak, A., Wolf, L.: Unsupervised cross-domain image generation (2016)
21. Wei, L., Rui, Z., Tong, X., Wang, X.G.: Deepreid: deep filter pairing neural network for person re-identification. In: Computer Vision & Pattern Recognition (2014)
22. Xu, F., Ma, B., Chang, H., Shan, S.: Isosceles constraints for person re-identification. IEEE Trans. Image Process. (2020)
23. Yan, Y., Qin, J., Ni, B., Chen, J., Shao, L.: Learning multi-attention context graph for group-based re-identification. IEEE Trans. Pattern Anal. Mach. Intell. (2020)
24. Zheng, L., Shen, L., Lu, T., Wang, S., Qi, T.: Scalable person re-identification: a benchmark. In: 2015 IEEE International Conference on Computer Vision (ICCV) (2015)
25. Zheng, W.S., Gong, S., Xiang, T.: Associating groups of people. In: Active Range Imaging Dataset for Indoor Surveillance (2009)
26. Zhu, J., Yang, H., Lin, W., Liu, N., Zhang, W.: Group re-identification with group context graph neural networks. IEEE Trans. Multimedia **99**, 1 (2020)

Contrastive Cycle Consistency Learning for Unsupervised Visual Tracking

Jiajun Zhu[1], Chao Ma[2(✉)], Shuai Jia[2], and Shugong Xu[1]

[1] Shanghai Institute for Advanced Communication and Data Science,
Shanghai University, Shanghai 200444, China
{jiajun_zhu,shugong}@shu.edu.cn
[2] MoE Key Lab of Artificial Intelligence, AI Institute,
Shanghai Jiao Tong University, Shanghai 200240, China
{chaoma,jiashuai}@sjtu.edu.cn

Abstract. Unsupervised visual tracking has received increasing attention recently. Existing unsupervised visual tracking methods mainly exploit the cycle consistency of sequential images to learn an unsupervised representation for target objects. Due to the small appearance changes between consecutive images, existing unsupervised deep trackers compute the cycle consistency loss over a temporal span to reduce data correlation. However, this causes the learned unsupervised representation not robust to abrupt motion changes as the rich motion dynamics between consecutive frames are not exploited. To address this problem, we propose to contrastively learn cycle consistency over consecutive frames with data augmentation. Specifically, we first use a skipping frame scheme to perform step-by-step cycle tracking for learning unsupervised representation. We then perform unsupervised tracking by computing the contrastive cycle consistency over the augmented consecutive frames, which simulates the challenging scenarios of large appearance changes in visual tracking. This helps us make full use of the valuable temporal motion information for learning robust unsupervised representation. Extensive experiments on large-scale benchmark datasets demonstrate that our proposed tracker significantly advances the state-of-the-art unsupervised visual tracking algorithms by large margins.

Keywords: Unsupervised visual tracking · Contrastive learning · Cycle consistency.

1 Introduction

Visual tracking is one of the most fundamental computer vision problems. It is challenging as target objects often undergo large appearance changes caused by scale variation, occlusion, and non-rigid deformation, etc. Benefiting from the powerful representation of deep convolutional neural networks (CNNs), visual

---

J. Zhu—Student.

© Springer Nature Switzerland AG 2021
H. Ma et al. (Eds.): PRCV 2021, LNCS 13019, pp. 564–576, 2021.
https://doi.org/10.1007/978-3-030-88004-0_46

tracking has made much progress in the last few years. State-of-the-art visual tracking methods [5,11,23] often require a large number of labeled data for training CNNs. As manually labeled data are expensive and time-consuming, unsupervised visual tracking methods [16,20,21] have attracted increasing attention recently.

Existing unsupervised visual tracking methods build on a cycle consistency tracking framework [12]. They do not make use of the temporal smoothness between consecutive frames because small appearance changes between consecutive frames yield intractable computation of the cycle consistency loss. The pioneering unsupervised tracker UDT [21] randomly samples a number of frames from video sequences for training unsupervised representation, where the valuable temporal motion information is discarded. Such a set of randomly sampled frames likely includes repeated frames, resulting in smaller appearance variations between consecutive images. Furthermore, existing unsupervised deep trackers directly crop sampled regions of target objects from input images as training data, considerably less attention has been paid to capturing rich appearance changes via data augmentation. This limits tracking performance due to the learned unsupervised representation not robust to abrupt appearance changes.

To address the above issues, we propose an unsupervised tracking framework with contrastive cycle consistency learning. To achieve this, we propose a skipping frame strategy to decrease the correlation of training data and maintain the temporal consistency simultaneously. Different from the random sampling strategy in UDT [21], we construct each training set with fixed inter skipping video frames between different sets to capture the temporal motion of target objects. For a single training set, we choose a proper intra skipping interval to exploit richer temporal information and streamline training data. In order to further learn from the motion information for unsupervised learning, we propose a step-by-step cycle tracking strategy including forward and backward tracking. To capture the variations of target objects for learning robust visual representations, we consider that data augmentation has been one of the most effective techniques to increase appearance changes over the temporal span. Different from using common data augmentations directly in the tracking process [14], we integrate an augmentation branch into the template branch for learning from the optimized data augmentation operator. Inspired by the recent contrastive learning algorithm [3], we reformulate the entire tracking framework by a contrastive cycle consistency learning process. We systematically study the impact of data augmentation operators for unsupervised tracking and sample the optimal operator to perform our contrastive cycle consistency learning. Our final tracking framework consists of a cycle consistency loss, a data augmentation loss, and a contrastive loss.

We summarize the main contributions of this work as follows:

- We propose a skipping frame strategy to reduce data correlation and maintain the temporal motion of target objects.
- We propose a step-by-step cycle tracking strategy to capture the temporal motion dynamics in an unsupervised manner.

- We incorporate contrastive learning with the optimized data augmentation scheme into our unsupervised tracking framework to handle rich appearance variations.
- We conduct extensive experiments on benchmark datasets. Compared with the baseline (i.e., UDT [21]), our tracker obtains competitive results with precision and success rate increasing by 6.0% and 3.8% respectively on OTB-2015.

2   Related Work

Contrastive Learning. Contrastive learning, as one of the effective methods for unsupervised representation learning, has shown great promising results. Tian et al. [18] propose to perform contrastive learning under multiple perspectives to obtain a multi-view representation. Bachman et al. [1] propose to learn representation by maximizing mutual information between features, which are extracted from multiple views of a shared context. Chen et al. [3] simplify the recently contrastive self-supervised learning algorithms. In this work, we construct the contrastive learning by applying the data augmentation to obtain a rich appearance information for our unsupervised learning.

Unsupervised Visual Tracking. There have been great interests in applying unsupervised representation learning to visual tracking recently. Existing unsupervised visual tracking methods resort to pretext tasks and unsupervised representations to improve tracking performance. The typical pretext task is learning pixel-level or object-level correspondences with unsupervision. For example, Vondrick et al. [20] propose to learn pixel-level correspondences by preserving the energy of color feature representations during transformations. Some methods [21,24] propose unsupervised visual tracking methods on the Siamese framework by performing forward-backward matching, belonging to learning object-level correspondences. Li et al. [16] propose to learn pixel-level and object-level correspondences jointly among video frames in an unsupervised manner.

3   Proposed Method

We build our method based on the unsupervised visual tracking framework introduced by UDT [21]. In the following, we first introduce the skipping frame strategy and the step-by-step cycle tracking strategy. Following that, we present the contrastive learning tracking framework, and study the effectiveness of individual data augmentation operators to select the most effective data augmentation operator for contrastive learning.

3.1   Skipping Frame Strategy

Due to the temporal smoothness between neighboring frames, the cropped training data from neighboring frames maintain fewer appearance variations, leading to the computation of cycle consistency loss intractable. UDT randomly samples frames to increase appearance changes over the temporal span at the price of the valuable temporal motion information. In this work, we aim to investigate how to best exploit video sequences to advance unsupervised trackers. To this end, we propose a skipping frame strategy to reduce data correlation and make use of the temporal motion information as well.

For a single training iteration, we select a set of video frames as inputs to perform forward and backward tracking. Note that there is a trade-off between increasing appearance variations and exploiting temporal motion information. That is, random samples like UDT capture large appearance variations but discard consecutive trajectories, while using dense consecutive frames as inputs are highly correlated with little appearance variations. A proper skipping interval can help to keep a good balance between reducing the correlation of training data and maintaining the temporal motion information among video sequences. Based on this observation, we propose to use a set of skipping video frames $\{I_{t+nq}\}$ as inputs with the intra skipping interval q. We empirically set the hyper-parameter $n = 0, 1, \ldots, 4$, and the intra skipping interval q in a single set to be 2. In addition, we observe that how to sample the initial frame among a set of skipping video frames is critical. If randomly sampling the frames in the entire training data, it may cause repeated or missing frames. In order to make full use of training data, we select different sets of frames $\{I_{t+p+nq}\}$ in the same video sequence using an inter skipping interval p. Empirically, we set the inter skipping interval p to be 5, which avoids using the same frames repeatedly.

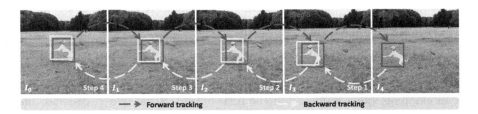

Fig. 1. Step-by-step cycle tracking strategy with a set of skipping video frames. The red boxes and yellow boxes are the forward and backward tracking results respectively.

3.2   Step-by-Step Cycle Tracking Strategy

To exploit rich motion dynamics, we propose a step-by-step cycle tracking scheme. As shown in Fig. 1, unlike UDT that only includes step-by-step process in the forward tracking and not in the backward tracking, the step-by-step cycle tracking scheme includes forward and backward tracking steps on a set of skipping video frames. For a concise notation, we reorder the set of skipping frames as $\{I_k\}$, $k = 0, 1, \ldots, N - 1$, where N is the length of the set.

When performing forward tracking, firstly we take I_k, $k = 0$, as the template patch and I_{k+1} as the search patch, and extract their features T_k and S_{k+1} by a Siamese CNN network φ, which is composed of two convolutional layers with ReLUs and a local response normalization (LRN) layer:

$$T_k = \varphi_\theta(I_k), \tag{1}$$

$$S_{k+1} = \varphi_\theta(mask(I_{k+1}))., \tag{2}$$

where $mask(\cdot)$ is a boundary response truncation module in order to alleviate the boundary effect following [10], and θ is the parameters of proposed tracker. Then, the discriminative correlation filter (DCF) W_k is learned by the template feature T_k with the initial Gaussian label Y_k:

$$W_k = \mathcal{F}^{-1}\left(\frac{\mathcal{F}(T_k) \odot \mathcal{F}^*(Y_k)}{\mathcal{F}^*(T_k) \odot \mathcal{F}(T_k) + \lambda}\right), \tag{3}$$

where $\mathcal{F}(\cdot)$ is the Discrete Fourier Transform (DFT) and $\mathcal{F}^{-1}(\cdot)$ is the inverse DFT. $\odot$ is an element-wise product and $*$ denotes the complex-conjugate operation. We then apply the learned filter W_k and the search feature S_{k+1} to generate the corresponding search response map $R(I_k, I_{k+1})$:

$$R(I_k, I_{k+1}) = W_k * S_{k+1} = \mathcal{F}^{-1}(\mathcal{F}^*(W_k) \odot \mathcal{F}(S_{k+1})), \tag{4}$$

In the following of forward tracking, we take I_{k+1} as the template patch and I_{k+2} as the search patch to extract features T_{k+1} and S_{k+2} respectively. Different from the first step, the search response map $R(I_k, I_{k+1})$ from the previous step is assumed as a pseudo Gaussian label, which resembles the Gaussian label Y_k in the first step. Then, we use the template feature T_{k+1} and the pseudo label $R(I_k, I_{k+1})$ to update the filter W_{k+1}. Similarly, the search response map $R(I_{k+1}, I_{k+2})$ can be computed by the search feature S_{k+2} and the filter W_{k+1}. Following the above steps, the forward tracking process tracks the entire set of skipping video frames, and the final output is the search response map $R(I_{N-2}, I_{N-1})$ from the search patch I_{N-1}.

For backward tracking, different from UDT that directly tracks back from the last frame to the first frame in only one step, we track backward frame by frame and step by step. Specifically, we take I_k, $k = N - 1$, as the template patch, I_{k-1} as the search patch, and $R(I_{k-1}, I_k)$ as the pseudo Gaussian label. Then we use the same tracking framework as forward tracking to obtain the search response map $R(I_k, I_{k-1})$:

$$T_k = \varphi_\theta(I_k), \tag{5}$$

$$S_{k-1} = \varphi_\theta(mask(I_{k-1})), \tag{6}$$

$$W_k = \mathcal{F}^{-1}\left(\frac{\mathcal{F}(T_k) \odot \mathcal{F}^*(R(I_{k-1}, I_k))}{\mathcal{F}^*(T_k) \odot \mathcal{F}(T_k) + \lambda}\right), \tag{7}$$

$$R(I_k, I_{k-1}) = W_k * S_{k-1} = \mathcal{F}^{-1}(\mathcal{F}^*(W_k) \odot \mathcal{F}(S_{k-1})), \tag{8}$$

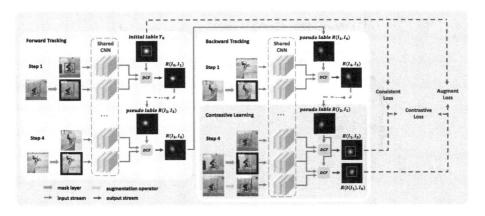

Fig. 2. Pipeline of the contrastive learning tracking framework with step-by-step cycle tracking strategy. In forward tracking, we perform four steps tracking from I_0 to I_4. In backward tracking, we perform four steps tracking from I_4 to I_0 step by step.

Similarly, we perform backward tracking continuously until $k = 1$. Finally, we get the search response map $R(I_k, I_{k-1})$ of the search patch I_0, which is the corresponding output of the initial Gaussian map. Note that $R(I_{k+1}, I_k)$ and $R(I_k, I_{k+1})$ are two completely different response maps in backward and forward tracking.

3.3 Contrastive Learning Tracking Framework

Inspired by SimCLR [3], we develop a novel contrastive learning tracking framework. The difference is that we do not learn representation consistency between differently augmented views of the same data input, but learn consistency of response maps between differently augmented template inputs with the same search input in the last step. However, we do not use data augmentation in each step, which may lead to unreasonable pseudo Gaussian labels and deteriorate the temporal consistency among frames. Meanwhile, it causes slow convergence during training. We instead add an extra data augmentation branch with the same template input to construct a contrastive learning tracking framework. The pipeline of contrastive cycle consistency tracking framework is shown in Fig. 2.

Specifically, we have three different branches in the last step of backward tracking, where the input of the two branches is the template patch I_k, and the other input is the search patch I_{k-1}. We implement a data augmentation operator on one template branch and transform the patch I_k to the correlated patch $t(I_k)$. In other words, we add another data augmentation branch parallel to the original template patch branch, yielding a contrastive learning tracking framework. When we take I_k as the template patch and I_{k-1} as the search patch, the search response map $R(I_k, I_{k-1})$ can be obtained. Similarly, we can obtain the augmented search response map $R(t(I_k), I_{k-1})$ via the augmented template

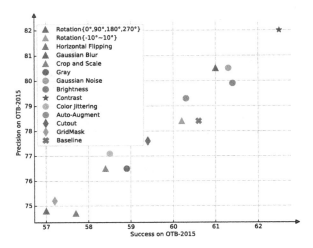

Fig. 3. Tracking results on the OTB-2015 dataset [25] of models with individual data augmentation operators.

branch. Finally, we formulate a novel loss function combined with the consistent loss, the augmentation loss and the contrastive loss. The consistent loss can be defined as follows:

$$L_{cs} = \| R(I_k, I_{k-1}) - Y_{k-1} \|, \tag{9}$$

The augment loss can be defined as follows:

$$L_{aug} = \| R(t(I_k), I_{k-1}) - Y_{k-1} \|, \tag{10}$$

The contrastive loss can be defined as follows:

$$L_{ct} = \| R(t(I_k), I_{k-1}) - R(I_k, I_{k-1}) \|, \tag{11}$$

We formulate the final loss function as follows:

$$L = \lambda_1 L_{cs} + \lambda_2 L_{aug} + \lambda_3 L_{ct}, \tag{12}$$

where λ_1, λ_2 and λ_3 balance the weights among these three losses.

3.4 Individual Data Augmentation Operators

In order to systematically study the impact of data augmentation for our unsupervised tracking framework, we consider several representative data augmentation operators, and conduct comparative experiments. Figure 3 shows the controlled tracking results on the OTB-2015 dataset [25] under different data augmentation operators. We observe that not all data augmentation operators are beneficial to our unsupervised visual tracking algorithms, such as large-scale rotation, color jittering, and GridMask. We infer that excessively transformed data augmentation operators can destroy the motion information for cycle consistent learning. And the tracking performance can obtain a great improvement with Gaussian blur, Gaussian noise, and brightness. Especially, the *contrast* operator performs favorably against all data augmentation operators. We infer that contrasting the images forces the model to learn features with high-level semantics rather than only learning simple color distribution.

4   Experimental Results

4.1   Dataset and Implementation Details

Dataset. For training, we adopt ILSVRC2015 [17]. With the skipping frame strategy, we only sample 0.22 million frames compared to 1.12 million frames in UDT [21]. The size of the template and search patches is 125×125 pixels. For evaluation, we validate the proposed tracker on three single object tracking benchmarks, including the OTB-2015 [25], VOT-2018 [13], and LaSOT [8] datasets.

Network Architecture. Throughout experiments, we follow UDT [21] to use a shallow Siamese network for the training and inference settings. The network takes a pair of 125×125 RGB patches as inputs. The template branch has three hidden layers including two convolutional layers combined with ReLUs and a local response normalization (LRN) layer. The filter sizes of these convolutional layers are 3×3×3×32 and 3×3×32×32, respectively. The search branch uses a boundary response truncation module to alleviate the boundary effect as preprocessing in the template branch.

Optimizations. Our tracker is trained with stochastic gradient descent (SGD) with a momentum of 0.9 and a weight decay of 0.005. We use SGD on a single GPU with a total of 64 pairs per minibatch, which costs 12 h to converge. There are 50 epochs in total and the learning rate decreases from 10^{-2} to 10^{-5} gradually. All experiments are implemented using PyTorch on a PC with Intel(R) Xeon(R) CPU E5-2687W v4 @ 3.00GHz and NVIDIA Tesla P100-PCIE-16GB.

4.2   Comparisons with the State-of-the-Art

OTB-2015. The OTB-2015 dataset consists of 100 sequences. The performance is evaluated in two metrics: precision and success plots by one-pass evaluation (OPE). We compare our tracker with the representative real-time trackers, as shown in Table 1. Compared with UDT, we obtain a precision score of 82.0% and success rate of 62.5%, with a 6.0% increase and a 3.8% increase respectively. Meanwhile, the performance of our unsupervised tracker is far superior to some supervised trackers based on correlation filters (e.g., CFNet, and SiamFC). However, our unsupervised tracker has not yet outperformed the advanced supervised trackers (e.g., SiamRPN++ [14], and SiamRPN [15]) since we do not use

Table 1. Tracking results on the OTB-2015 dataset [25].

Trackers	Ours	UDT [21]	JSLTC [16]	SiamRPN ++[14]	SiamRPN [15]	SiamFC [2]	DCFNet [22]	CFNet [19]
Supervised	×	×	×	√	√	√	√	√
Success(%)	**62.5**	59.4	59.2	69.6	63.6	58.2	58.0	56.8
Precision(%)	**82.0**	76.0	-	91.4	85.0	77.1	75.0	77.0

Table 2. Tracking results on the VOT-2018 dataset [13].

Trackers	Accuacy (↑)	Robustness (↓)	Failures (↓)	EAO (↑)
DSST [6]	0.395	1.452	310	0.079
SRDCF [7]	0.490	0.974	208	0.119
KCF [9]	0.447	0.773	165	0.134
UDT [21]	0.497	0.787	168	0.149
SiamFC [2]	0.503	0.585	125	0.187
ATOM [5]	0.590	0.200	-	0.401
TREG [4]	0.612	0.098	-	0.496
Ours	**0.499**	**0.716**	**153**	**0.161**

accurate data annotations during training. Besides, our tracker also achieves a real-time speed of ̃68 FPS.

VOT-2018. The VOT-2018 dataset consists of 60 sequences. Different from the OTB-2015 dataset, the performance is evaluated using accuracy, robustness, failures, and Expected Average Overlap (EAO). EAO measures the overall performance of trackers, including accuracy and robustness. The results on the VOT-2018 dataset are reported in Table 2. Compared with UDT, our tracker achieves the similar accuracy, but with large performance gains in robustness (+7.1%) and EAO (+1.2%), which illustrates that our tracker not only accurately locates the target but also maintains great robustness.

LaSOT. The LaSOT dataset consists of 1,400 sequences totally with an average sequence length of more than 2,500 frames. We evaluate our tracker on 280 challenging videos using precision, normalized precision, and success. In this experiment, we compare our tracker with the baseline unsupervised tracker (e.g., UDT) and the other 35 supervised trackers. Figure 4 shows the precision, norm precision, and success plots of the compared trackers. For presentation clarity, we only show the top 20 trackers. Compared with UDT, our tracker achieves large improvements of 1.7%, 2.1%, and 1.3% on OPE respectively. Our model involving more temporal information obtains better performance for long-term tracking.

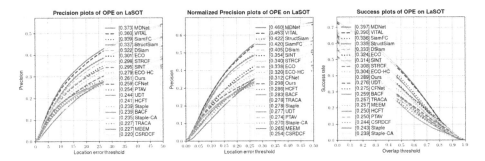

Fig. 4. Precision, norm precision and success plots on the LaSOT dataset [8] using OPE for recent representative trackers.

Table 3. Tracking results on the OTB-2015 [25] with different tracking strategies.

Tracking Strategy	Precision (%)	Success (%)
Ours	**82.0**	**62.5**
Ours w/o step-by-step cycle	78.7	60.4
UDT w/ step-by-step cycle	77.7	60.0
UDT	76.0	59.4

4.3   Ablation Studies and Analysis

Step-by-step cycle tracking strategy. To validate the effectiveness of our step-by-step cycle tracking strategy, we implement a limited method by disabling this scheme. We also add this scheme to UDT. The compared results are shown in Table 3. Our method obtains large increases in precision and success respectively compared with the method without the step-by-step cycle tracking scheme. With the application of the step-by-step cycle tracking, the vanilla UDT tracker achieves large performance gains as well.

Contrastive Learning Tracking Framework. To investigate the effectiveness of the contrastive learning tracking framework, we experiment our method with or without the contrastive learning module, as shown in Table 4. In addition, we apply our contrastive learning module to UDT, which achieves large improvements in terms of precision and success rates. Similarly, we apply the contrastive learning to the vanilla UDT tracker and get some improvement as well.

Loss Function. To further study the effectiveness of each component of our loss function, we control our tracker trained with the consistent loss, the augment loss, and the contrastive loss, which are denoted as L_{cs}, L_{aug} and L_{ct}, respectively. Table 5 shows the detailed results of the performance contributions of each loss. We observe that using a single loss function improves the baseline tracker slightly. It is worth noting that integrating different loss functions obtain

Table 4. Tracking results on the OTB-2015 [25] with different tracking framework.

Tracking Framework	Precision (%)	Success (%)
Ours	**82.0**	**62.5**
Ours w/o contrastive learning	78.4	60.6
UDT w/ contrastive learning	78.7	60.0
UDT	76.0	59.4

Table 5. Tracking results on the OTB-2015 dataset [25] with different loss functions.

L_{cs}	L_{aug}	L_{ct}	Precision (%)	Success (%)
$\checkmark$	$\checkmark$	$\checkmark$	82.0	**62.5**
$\checkmark$	$\checkmark$	–	81.5	62.4
–	$\checkmark$	–	79.6	61.1
$\checkmark$	-	–	78.4	60.0

favorable performance. Especially, when we consider all of the loss functions together, the tracker achieves the highest success rate of 62.5% and the highest precision score of 82.0%, improving the baseline by large margins.

5   Conclusion

In this paper, we propose a contrastive learning tracking framework with a skipping frame strategy and a step-by-step cycle tracking strategy. The skipping frame strategy not only reduces the correlation of training data but also captures the valuable motion dynamics. The step-by-step cycle tracking strategy makes full use of the rich temporal features among frames via step-by-step forward and backward tracking. In the proposed contrastive learning framework for visual tracking, we use the optimized data augmentation to learn robust representation against large appearance variations in an unsupervised manner. The extensive experiments on five benchmarks demonstrate that our tracker performs favorably against state-of-the-art trackers and sets new state-of-the-art results for unsupervised visual object tracking.

References

1. Bachman, P., Hjelm, R.D., Buchwalter, W.: Learning representations by maximizing mutual information across views. In: Advances in Neural Information Processing Systems, pp. 15535–15545 (2019)
2. Bertinetto, L., Valmadre, J., Henriques, J.F., Vedaldi, A., Torr, P.H.: Fully-convolutional siamese networks for object tracking. In: European Conference on Computer Vision, pp. 850–865 (2016)

3. Chen, T., Kornblith, S., Norouzi, M., Hinton, G.: A simple framework for contrastive learning of visual representations. arXiv preprint arXiv:2002.05709 (2020)
4. Cui, Y., Jiang, C., Wang, L., Wu, G.: Target transformed regression for accurate tracking. arXiv preprint arXiv:2104.00403 (2021)
5. Danelljan, M., Bhat, G., Khan, F.S., Felsberg, M.: Atom: Accurate tracking by overlap maximization. In: The IEEE Conference on Computer Vision and Pattern Recognition (CVPR) (June 2019)
6. Danelljan, M., Häger, G., Khan, F., Felsberg, M.: Accurate scale estimation for robust visual tracking. In: British Machine Vision Conference (2014)
7. Danelljan, M., Hager, G., Shahbaz Khan, F., Felsberg, M.: Learning spatially regularized correlation filters for visual tracking. In: The IEEE International Conference on Computer Vision (ICCV) (December 2015)
8. Fan, H., et al.: Lasot: A high-quality benchmark for large-scale single object tracking. In: The IEEE Conference on Computer Vision and Pattern Recognition (CVPR) (June 2019)
9. Henriques, J.F., Caseiro, R., Martins, P., Batista, J.: High-speed tracking with kernelized correlation filters. IEEE Trans. Pattern Anal. Mach. Intell. **37**(3), 583–596 (2014)
10. Hu, Q., Zhou, L., Wang, X., Mao, Y., Zhang, J., Ye, Q.: Spstracker: sub-peak suppression of response map for robust object tracking. In: Thirty-Fourth AAAI Conference on Artificial Intelligence (2020)
11. Huang, Z., Fu, C., Li, Y., Lin, F., Lu, P.: Learning aberrance repressed correlation filters for real-time uav tracking. In: The IEEE International Conference on Computer Vision (ICCV) (October 2019)
12. Kalal, Z., Mikolajczyk, K., Matas, J.: Forward-backward error: automatic detection of tracking failures. In: 2010 20th International Conference on Pattern Recognition, pp. 2756–2759 (2010)
13. Kristan, M., Leonardis, A., Matas, J., Felsberg, M., Pflugfelder, R., Cehovin Zajc, L.: The sixth visual object tracking vot2018 challenge results. In: The European Conference on Computer Vision (ECCV) Workshops (September 2018)
14. Li, B., Wu, W., Wang, Q., Zhang, F., Xing, J., Yan, J.: Siamrpn++: evolution of siamese visual tracking with very deep networks. In: The IEEE Conference on Computer Vision and Pattern Recognition (CVPR) (June 2019)
15. Li, B., Yan, J., Wu, W., Zhu, Z., Hu, X.: High performance visual tracking with siamese region proposal network. In: The IEEE Conference on Computer Vision and Pattern Recognition (CVPR) (June 2018)
16. Li, X., Liu, S., De Mello, S., Wang, X., Kautz, J., Yang, M.H.: Joint-task self-supervised learning for temporal correspondence. In: Advances in Neural Information Processing Systems, pp. 317–327 (2019)
17. Russakovsky, O., et al.: Imagenet large scale visual recognition challenge. Int. J. Comput. Vision **115**(3), 211–252 (2015)
18. Tian, Y., Krishnan, D., Isola, P.: Contrastive multiview coding. arXiv preprint arXiv:1906.05849v1 (2019)
19. Valmadre, J., Bertinetto, L., Henriques, J., Vedaldi, A., Torr, P.H.S.: End-to-end representation learning for correlation filter based tracking. In: The IEEE Conference on Computer Vision and Pattern Recognition (CVPR) (July 2017)
20. Vondrick, C., Shrivastava, A., Fathi, A., Guadarrama, S., Murphy, K.: Tracking emerges by colorizing videos. In: The European Conference on Computer Vision (ECCV) (September 2018)

21. Wang, N., Song, Y., Ma, C., Zhou, W., Liu, W., Li, H.: Unsupervised deep tracking. In: The IEEE Conference on Computer Vision and Pattern Recognition (CVPR) (June 2019)
22. Wang, Q., Gao, J., Xing, J., Zhang, M., Hu, W.: Dcfnet: Discriminant correlation filters network for visual tracking. arXiv preprint arXiv:1704.04057 (2017)
23. Wang, Q., Zhang, L., Bertinetto, L., Hu, W., Torr, P.H.: Fast online object tracking and segmentation: a unifying approach. In: The IEEE Conference on Computer Vision and Pattern Recognition (CVPR) (June 2019)
24. Wang, X., Jabri, A., Efros, A.A.: Learning correspondence from the cycle-consistency of time. In: The IEEE Conference on Computer Vision and Pattern Recognition (CVPR) (June 2019)
25. Wu, Y., Lim, J., Yang, M.H.: Object tracking benchmark. IEEE Trans. Pattern Anal. Mach. Intell. **37**(9), 1834–1848 (2015)

Group-Aware Disentangle Learning for Head Pose Estimation

Mo Zhao[1] and Hao Liu[1,2(✉)]

[1] School of Information Engineering, Ningxia University, Yinchuan 750021, China
liuhao@nxu.edu.cn
[2] Collaborative Innovation Center for Ningxia Big Data and Artificial Intelligence
Co-founded by Ningxia Municipality and Ministry of Education,
Yinchuan 750021, China

Abstract. In this paper, we propose a group-aware disentangle learning (GADL) approach for robust head pose estimation. Conventional methods usually utilize landmarks or depth information to recover the head poses, likely leading to bias prediction due to that the performance of head pose estimation highly relies on landmark detection accuracy. Instead of using external information, our GADL estimates the head poses directly from facial images in a group-supervised manner. Specifically, we first map a group of input images that share a common orientation onto a latent space, then swap the shared orientation subspaces across images to reorganize the representation. Next, we use the decoder to synthetic the original images via the reorganized representation. Thus, we explicitly disentangle the pose-relevant features shared by a group of facial images. Extensive experimental results on the challenging widely-evaluated datasets indicate the superiority of our GADL compared with the state of the arts.

Keywords: Head pose estimation · Group-supervised learning · Feature disentangle

1 Introduction

Head pose estimation aims to predict the continuous orientations (pitch, yaw and roll) of the human head based on a given facial image, which plays an important role in many applications of visual attribute analysis tasks, such as face recognition [6,20], expression recognition [25], gaze estimation [15], etc. In addition, the head pose is regarded as a key meta-information in human-computer interaction since the changes of head pose convey rich information, such as the attention of a person [16]. Recently, although efforts have been devoted to head pose estimation, the performance still remains limited due to the ambiguity of head pose labels and the non-linear relationship of face data caused by expression, occlusion, gender, race, etc.

Student paper.

© Springer Nature Switzerland AG 2021
H. Ma et al. (Eds.): PRCV 2021, LNCS 13019, pp. 577–588, 2021.
https://doi.org/10.1007/978-3-030-88004-0_47

Conventional methods estimate head poses by facial landmark detection, which typically recovers the head poses by establishing correspondence between landmarks and a 3D head model [2]. However, as a by-product of facial landmark detection, the performance of head pose estimation highly relies on the accuracy of landmark detection and requires more computation than necessary. To address this issue, deep learning [19,20] has been applied to learn discriminative features directly from image pixels to exploit the complex and nonlinear relationship between the facial image and head pose labels, which achieves significant improvements for the head pose estimation. Unfortunately, these extracted features undergo much underlying irrelevant information other than pose, e.g. illumination, appearance, gender, background, race, etc. Besides, assigning the face image with three single angles is inaccurate in practice since the head orientation is intrinsically continuous. Hence, the major challenge is to be able to disentangle the information most related to head poses while suppressing the other irrelevant ones and to model the ambiguity among head pose labels for better fitting the real-world situation.

In this paper, we propose a group-aware disentangle learning method for head pose estimation to address the aforementioned issues. Motivated by the fact that head pose should be robust to the expression, gender, race, and so on, our model aims to disentangle the pose-relevant information from intermingled ones. To achieve this, we carefully design a fine-grained feature disentangle network to extract three different orientation features in a group-supervised manner. Technically, we first map a group of samples which share a common orientation value (e.g. pitch) onto a latent representation space, then we swap the shared orientation subspace of different samples to obtain the reorganized latent representation. Finally, the reorganized latent representation of different samples is sent to the decoder to reconstruct the original samples. Thus, we not only disentangle pose-relevant information but extract the fine-grained orientation-related features in this group-supervised manner. Besides, to model smoothness and correlation among head pose labels, we encode a range of head pose labels to a mean-variance label distribution for each orientation for better fitting the real-world situation. The network parameters of the designed feature disentangle architecture are optimized by back-propagation in an end-to-end manner. Figure 1 details the flowchart of our proposed GADL. To further evaluate the effectiveness of our approach, we conduct extensive experiments on widely-used benchmarks. Experimental results show that our method achieves superior performance compared with the state of the arts.

2   Related Work

In this section, we briefly review existing head pose estimation methods to position our contribution.

Over the past few years, head pose estimation has become a widely studied task in computer vision and a lot of methods have been proposed to improve the accuracy. Existing head pose estimation methods could be roughly divided into two types. One type aims to recover the head pose utilizing geometric facial information. Some recover the head pose through detecting the facial landmarks. They first detect facial landmarks from the target face, then solve the 2D to 3D correspondence problem via a mean human head model [2]. However, as a by-product of facial landmark detection, the performance of head pose estimation heavily relies on the accuracy of landmark detection, the external head model and requires more computation than necessary. The other type estimates the head pose by learning the complex and nonlinear relationship from images to pose directly. Anh [1] was the first one that uses a convolutional neural network to regress the head pose information directly. Patacchiola and Cangelosi [18] evaluated different CNN architectures and adaptive gradient methods for head pose estimation.

Since additional face attributes (e.g., gender, expression, race, and so on) other than head pose are often available, some studies [6,13,19,20] leverage multi-task learning to jointly model multiple face attribute estimation. For example, Chang [6] predicted facial keypoints and head pose jointly using a ResNet architecture. Ranjan [20] proposed a CNN architecture that simultaneously performs face detection, landmarks localization, pose estimation, gender recognition, smile detection, age estimation and face identification, thus the influence of other face attributes to head estimation is modeled implicitly. Nevertheless, multi-analysis approaches only coarsely evaluated pose estimation performance. Besides, these approaches do not distinguish between the main task and auxiliary tasks, which can be sub-optimal for the main task. To address this issue, more deep-learning based methods were devoted to learn discriminative features directly from image pixels to exploit the complex and nonlinear relationship between facial image and head pose labels, which achieves significant improvements for the head pose estimation performance. Nataniel Ruiz [21] proposed to predict intrinsic Euler angles (yaw, pitch, and roll) directly from image intensities through joint binned pose classification and regression. To further extract features more relevant to head pose, FSA-Net [24] proposed a feature aggregation method, which utilizes the spatial information through spatially group pixel-level features of the feature map.

While encouraging performance has been obtained depending on the feature extraction ability of deep neural networks, these methods hardly explicitly exploit the information related to different orientations of the head also cannot eliminate the effects of other underlying factors. We cope with this issue by a group-aware disentangle learning framework, which obtains the fine-grained pose-relevant features through disentangling the complex influence.

3   Proposed Method

Our basic idea of this paper is to disentangle the fine-grained pose-relevant information from the complex information, in parallel to encode the label correlation for more robust head pose estimation. Since the three orientations of a head are equally important and our model infers the final angles in the same way, thus, we will give a detailed description about the process of disentangling features and inferring the final angle related to pitch for example in the following. In this section, we introduce the problem formulation and the optimization objectives in detail.

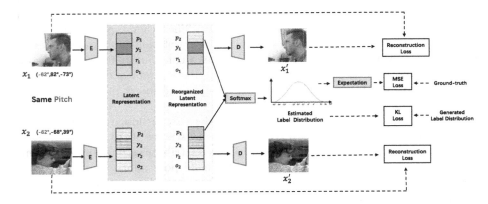

Fig. 1. Flowchart of our GADL. Our architecture starts with a group of input faces which have a common orientation value (Here, we take the orientation of pitch for example, and the other two orientations have the same process with pitch.). We first map the group of samples onto a latent representation space to obtain the corresponding features, then we swap the shared orientation subspace of different samples to get the reorganized representation. Finally, the reorganized latent representation of different samples is sent to the decoder to reconstruct the original samples captured by the reconstruction loss. In this way, we disentangle the fine-grained features related to the orientation of pitch from the mingle ones. In the meantime, we encode the correlation of labels and get the corresponding label distribution which measured by the KL divergence and MSE loss.

3.1   Problem Formulation

Let $X = \{(x_i, y_i)\}_{i=1}^{N}$ be the training set which contains N samples in total, where $x_i \in R^d$ specifies the i-th face sample with size $H \times W$, and y_i is a three-dimensional vectors composed by the corresponding angles of yaw, pitch and roll, respectively. We let $N(\cdot)$ specify the core function of our approach, which aims to output the final pose predictions by exploiting the disentangled fine-grained features. Hence, our objective is formulated as minimizing the following optimization:

$$ J(X) = \frac{1}{N} \sum_{i=1}^{N} \| N(x_i) - y_i \|_1 . \tag{1} $$

Obviously, the core step in the formulation (1) mainly lies on learning the parameters of $N(\cdot)$, typically disentangling the fine-grained pose-relevant information from the complex ones in a group-supervised manner.

3.2   Reconstruction Loss

For an input image x_i, our model first selects images which share an angle with x_i according to their labels y. The selected three facial images x_j, x_k, and x_l should meet the following conditions: $y_j = (p_i, y_j, r_j)$, $y_k = (p_k, y_i, r_k)$ and $y_l = (p_l, y_l, r_i)$. The above conditions mean that we aim to select a group of facial images which share one common pose information with the input x_i. Utilizing these images, we disentangle the pose information in a group-supervised manner.

As shown in Fig. 1, our GADL consists of an encoder network E and a decoder network D. Before training, we fix partitioning of the latent space $h = E(X)$. Let vector $h = [f_{pitch}, f_{yaw}, f_{roll}]$ be the concatenation of four vectors. Our architecture starts with the group of selected inputs. The encoder E maps the group of inputs onto a latent representation space thus we get the latent representation of the group samples: $h_1 = E(x_1) = \left[f_{pitch}^1, f_{yaw}^1, f_{roll}^1, f_{other}^1\right]$ and $h_2 = E(x_2) = \left[f_{pitch}^2, f_{yaw}^2, f_{roll}^2, f_{other}^2\right]$. Then, in order to exploit pose-relevant information, we reorganized the latent representation via swapping the common shared representation subspace and retaining the others in the latent space, and we get the reorganized representation as follow: $h_{new}^1 = \left[f_{pitch}^2, f_{yaw}^1, f_{roll}^1, f_{other}^1\right]$ and $h_{new}^2 = \left[f_{pitch}^1, f_{yaw}^2, f_{roll}^2, f_{other}^2\right]$. Finally, the decoder D synthesizes the original images with the reorganized representation: $x_1' = D\left(h_{new}^1\right)$ and $x_2' = D\left(h_{new}^2\right)$. It means that the swapped subspace storage the information related to the orientation of pitch, thus we obtain the fine-grained pose-relevant information. The reconstruction loss is computed as follow:

$$L_{rec} = \left\| x_1 - x_1' \right\|_1 . \qquad (2)$$

3.3   KL Divergence

To encode the correlation and ambiguity between labels, we adopt the label distribution to obtain the final prediction. Specifically, we first generate a Gaussian distribution based on the ground-truth. In the inference stage, we predict the label distribution using the Softmax function. To optimize the prediction, we use the KL-divergence to measure the ground-truth distribution and the predicted distribution.

Label Distribution Generation. During the training phase, given a facial image associated with the labeled pitch p, we first generate a Gaussian distribution with the mean of p and standard deviation of σ as follow:

$$g(l) = \frac{1}{\sqrt{2\pi}\sigma} exp\left(-\frac{(l-p)^2}{2\sigma^2}\right), \qquad (3)$$

where $l = (l_1, l_2, \cdots, l_m)$ means the binned pitch angles in the range of $\pm99°$. Here σ is the finest granularity of the pose angles and we set it as 2. The final generated distribution $P_l = [p(l_1), p(l_2), \cdots, p(l_m)]$ is generated by normalizing $g(l)$ as follow to make sure $\sum_{i=1}^{m} p(l_i) = 1$:

$$p(l_i) = \frac{g(l_i)}{\sum_{j=1}^{m} g(l_j)}, i = 1, 2, \cdots, m. \tag{4}$$

Learning from Label Distribution. In order to learn from the generated distribution P_l, we map the learned representation f_{pitch}^1 into the predicted pitches vector as v, where v has the same size with the P_l. Then, after the softmax activation on vector v, we obtain the final predicted distribution as $P_{pre} = [p_{pre}(l_1), p_{pre}(l_2), \cdots, p_{pre}(l_m)]$. Finally, the learning loss is calculated as the Kullback-Leibler(KL) divergence between P_{pre} and P_l:

$$D_{KL}(P_l \parallel P_{pre}) = \sum_{i=1}^{m} p(l_i) log \frac{p(l_i)}{p_{pre}(l_i)}. \tag{5}$$

The final prediction angle $\hat{p}$ is obtained by calculating the expectation of P_{pre}, we further use the MSE loss to measure the final prediction and the ground-truth. Thus, the final loss is computed as follows:

$$L_{dis} = D_{KL}(P_l \parallel P_{pre}) + \lambda MSE(\hat{p}, p). \tag{6}$$

3.4   Training Objective

Motivated by the above elements, the training objective of our method for exploiting the pose-relevant information and predicting the final angles is formulated as follows:

$$L_{total} = L_{rec} + L_{dis}. \tag{7}$$

By minimizing the L_{total}, we disentangle the pose-relevant information from the complex ones and get the fine-grained features related to each orientation.

4   Experiments

To evaluate the effectiveness of the proposed method, we conducted experiments on three widely-used datasets including 300W-LP [29],AFLW2000 [29] and BIWI [7], where challenging cases of large poses, occlusions are exposed. Furthermore, we also conducted an ablation experiment on the AFLW2000 [29] dataset to evaluate the effectiveness of our method.

Fig. 2. Examples from three datasets. The first row is from 300W-LP dataset which is rendered from the AFW dataset across large poses. The second row is from AFLW2000 dataset which is gathered in-the-wild with various backgrounds and different light conditions. The third row is from BIWI dataset which is collected in a laboratory setting. From the samples, we see that there exists a domain gap between these datasets.

4.1 Evaluation Datasets and Metric

Evaluation Datasets. *300W-LP* [29]*:* The 300W-LP dataset is derived from the 300W dataset [22] which contains several subsets including AFW [28], LFPW [3], HELEN [27] and IBUG [22]. These face samples were captured in unconstrained environments. Zhu *et al.* [29] used face profiling and a face model to fit on each image to generate 61, 225 samples across large poses and further expanded to 122, 450 samples with flipping. The synthesized dataset is named the 300W across Large Poses, dubbed 300W-LP.

AFLW2000 [29]*:* AFLW2000 contains the first 2000 images of the in-the-wild AFLW dataset [12] with accurate fine-grained pose annotations. This is a challenging dataset since the face samples in this dataset usually undergo challenging cases due to different illumination, occlusion conditions, and various backgrounds.

BIWI [7]*:* BIWI collects roughly 15, 000 frames derived from 24 videos of 20 subjects. This dataset was gathered in a controlled laboratory environment by recording RGB-D video of different subjects across different facial poses using Kinect v2 device. 20 people (some were recorded twice- 6 women and 14 men) were recorded while turning their heads, sitting in front of the sensor, at roughly one meter of distance. In addition to RGB frames, the dataset also provides the depth data for each frame. In our experiments, we only use the color frames instead of the depth information to prove the effectiveness of our model.

For fair comparisons, we strictly followed the common protocols employed in [24]. For the first protocol, we trained our model on the synthetic

300W-LP dataset while test on two real-world datasets, the AFLW2000 and
BIWI datasets, respectively. For the other protocol, we randomly choose 70% of
videos (16 videos) in the BIWI dataset for training and the left (8 videos) for
testing.

Evaluation Metric. In the experiments, we leveraged the mean absolute error
(MAE) which computes the discrepancy between the estimated head pose angles
and the ground-truths. We reported the MAEs of each orientation and the average of them as the results. Obviously, the lower the MAE value, the better
performance it achieves.

4.2   Implementation Details

For all input images, we randomly cropped them to 224 × 224 including the
whole head, and then the images are normalized by ImageNet mean and standard
deviation following [21]. We employed the ResNet-50 [9] as our backbone encoder
network. We trained the networks using Adam optimization with the initial
learning rate 0.0001. The training process lasts for 50 epochs and the batch size
is 64. For parallel acceleration, we trained our model with PyTorch [17] on 2 T
V100 GPUS.

Table 1. Comparisons of MAEs of our approach compared with different state-of-the-
art methods on BIWI dataset under the first protocol.

Method	Yaw	Pitch	Roll	MAE	Year
Dlib [11]	16.8	13.8	6.19	12.2	2014
3DDFA [29]	36.2	12.3	8.78	19.1	2016
KEPLER [13]	8.80	17.3	16.2	13.9	2017
FAN [4]	8.53	7.48	7.63	7.89	2017
Hopenet [21]	4.81	6.61	3.27	4.90	2018
FSA-Net [24]	4.27	4.96	2.76	4.00	2019
FDN [26]	4.52	4.70	2.56	3.93	2020
MNN [23]	3.34	4.69	3.48	3.83	2020
GADL	**3.24**	4.78	**3.15**	**3.72**	–

4.3   Comparisons with State of the Arts

We compared our approach with folds of state-of-the-art methods. We carefully conducted the experiments of the state-of-the-art methods [21,24] by their
released source codes. For other methods, we strictly reported their performance
by cropping the results from the original papers.

Table 1 and Table 2 show the MAEs on BIWI and AFLW2000 datasets under
the first protocol, respectively. From the results, we see that our model achieves

Table 2. Comparisons of MAEs of our approach compared with different state-of-the-art methods on AFLW2000 dataset under the first protocol.

Method	Yaw	Pitch	Roll	MAE	Year
Dlib [11]	23.1	13.6	10.5	15.8	2014
3DDFA [29]	5.40	8.53	8.25	7.39	2016
FAN [4]	6.36	12.3	8.71	9.12	2017
Hopenet [21]	6.47	6.56	5.44	6.16	2018
FSA-Caps-Fusion [24]	4.50	6.08	4.64	5.07	2019
QuatNet [10]	3.97	5.61	3.92	4.50	2019
FDN [26]	3.78	5.61	3.88	4.42	2020
GADL	3.79	**5.24**	4.13	**4.37**	–

Table 3. Comparisons of MAEs of our approach compared with different state-of-the-art methods on BIWI dataset under the second protocol. † was used to indicate that these methods use depth information.

Method	Yaw	Pitch	Roll	MAE	Year
Martin † [14]	3.6	2.5	2.6	2.9	2014
DeepHeadPose † [15]	5.32	4.76	–	–	2015
DeepHeadPose [15]	5.67	5.18	–	–	2015
VGG16 [8]	3.91	4.03	3.03	3.66	2017
Hopenet [21]	3.29	3.39	3.00	3.23	2018
FSA-Caps-Fusion [24]	2.89	4.29	3.60	3.60	2019
FDN [26]	3.00	3.98	2.88	3.29	2020
GADL	**2.36**	**3.25**	**2.72**	**2.77**	–

competitive performance compared with the state-of-the-art methods and even achieves better performance than some feature extracted learning methods. Moreover, we made three-fold conclusions: (1) The traditional methods, such as FAN [4], Dlib [11] and 3DDFA [29], which utilize the facial landmarks or a 3D face model to help to recover the head pose, introduce too much external information. Thus, the accuracy of head pose estimation highly relies on the landmark detection performance and the external 3D head model. While our GADL directly predicts the head pose label from the RGB images without using external information which alleviates the influence of introducing external information and is more convenient in real-world situations. (2) Unfortunately, the landmark-free method such as Hopenet [21], directly exploits the nonlinear relation between features and head pose labels without searching information mostly related to the head poses, resulting in poor performance. (3) Differently, the other landmark-free method, FSA-Net [24], employs the way of feature aggregation by grouping the pixel-level features encoded with spatial information. While, although they utilize the way of feature aggregation, they can not explicitly disentangle the pose relevant information from the complex one. Benefiting from the data-dependent feature disentangle learning manner, our GADL captures more fine-grained information related to each orientation and achieves better performance. Figure 3 compares our model with FSA-Net by showing a few example results and we achieve more robust results.

Fig. 3. Head pose estimation on the AFLW2000 dataset under the first protocol. From top to bottom are the ground-truth, the results of FSA-Net and our GADL. The blue line indicates the front of the face; the green line points the downward direction and the red one for the side. (Color figure online)

Table 3 tabulates the results on BIWI dataset under the second protocol. Compared with the first protocol where the training set 300W-LP and testing set AFLW2000 and BIWI are homologous, there exists no domain gap between training and test set. Examples of three datasets are shown in Fig. 2. From the results, we find that our GADL achieves superior performance compared with the state-of-the-art methods, which also proves the effectiveness of our approach.

4.4   Ablation Study

We conducted the ablation study on AFLW2000 dataset under the first protocol to observe the influence of pose irrelevant information. We re-partition the latent representation space which only composed of three parts indicate the three different orientations respectively and retain the other elements of our GADL. Table 4 reports the experimental results. From Table 2 and Table 4, we figure out that our GADL achieves better results. The reason lies that GADL has four parts of subspace which indicate that besides the three orientations there also the subspace belongs to pose-irrelevant information such as appearance, expression, and so on. The comparison demonstrates that suppressing the pose-irrelevant information and utilizing the fine-grained pose-relevant information could help to improve the classification accuracy.

Table 4. Ablation study results on the AFLW2000 dataset. The results average the MAEs of different orientations.

Method	Yaw	Pitch	Roll	MAE
GADL	4.19	5.30	4.73	4.74

5   Conclusion

In this paper, we have proposed a group-aware disentangle learning method (GADL) for head pose estimation. The proposed GADL has explicitly disentangled the pose-relevant features from the complex facial information in a group-supervised manner. Experiments on widely-used datasets have shown the effectiveness of the proposed approach. In the future, we will focus on the domain adaption [5] to transfer more information between the in-the-wild and under-controlled datasets.

Acknowledgement. This work was supported in part by the National Science Foundation of China under Grant 61806104 and 62076142, in part by the West Light Talent Program of the Chinese Academy of Sciences under Grant XAB2018AW05, and in part by the Youth Science and Technology Talents Enrollment Projects of Ningxia under Grant TJGC2018028.

References

1. Ahn, B., Park, J., Kweon, I.: Real-time head orientation from a monocular camera using deep neural network. In: ACCV, vol. 9005, pp. 82–96 (2014)
2. Baltrusaitis, T., Robinson, P., Morency, L.: 3d constrained local model for rigid and non-rigid facial tracking. In: CVPR, pp. 2610–2617 (2012)
3. Belhumeur, P.N., Jacobs, D.W., Kriegman, D.J., Kumar, N.: Localizing parts of faces using a consensus of exemplars. TPAMI **35**(12), 2930–2940 (2013)
4. Bulat, A., Tzimiropoulos, G.: How far are we from solving the 2d & 3d face alignment problem? (and a dataset of 230, 000 3d facial landmarks). In: ICCV, pp. 1021–1030 (2017)
5. Busto, P.P., Iqbal, A., Gall, J.: Open set domain adaptation for image and action recognition. TPAMI **42**(2), 413–429 (2020)
6. Chang, F., et al.: Faceposenet: making a case for landmark-free face alignment. In: ICCV Workshops, pp. 1599–1608 (2017)
7. Fanelli, G., Dantone, M., Gall, J., Fossati, A., Gool, L.V.: Random forests for real time 3d face analysis. IJCV **101**(3), 437–458 (2013)

8. Gu, J., Yang, X., Mello, S.D., Kautz, J.: Dynamic facial analysis: from bayesian filtering to recurrent neural network. In: CVPR, pp. 1531–1540 (2017)
9. He, K., Zhang, X., Ren, S., Sun, J.: Deep residual learning for image recognition. In: CVPR, pp. 770–778 (2016)
10. Hsu, H., Wu, T., Wan, S., Wong, W.H., Lee, C.: Quatnet: quaternion-based head pose estimation with multiregression loss. TMM **21**(4), 1035–1046 (2019)
11. Kazemi, V., Sullivan, J.: One millisecond face alignment with an ensemble of regression trees. In: CVPR, pp. 1867–1874 (2014)
12. Köstinger, M., Wohlhart, P., Roth, P.M., Bischof, H.: Annotated facial landmarks in the wild: a large-scale, real-world database for facial landmark localization. In: ICCV Workshops, pp. 2144–2151 (2011)
13. Kumar, A., Alavi, A., Chellappa, R.: KEPLER: keypoint and pose estimation of unconstrained faces by learning efficient H-CNN regressors. In: FG, pp. 258–265 (2017)
14. Martínez, M., van de Camp, F., Stiefelhagen, R.: Real time head model creation and head pose estimation on consumer depth cameras. In: 3DV, pp. 641–648 (2014)
15. Mukherjee, S.S., Robertson, N.M.: Deep head pose: gaze-direction estimation in multimodal video. TMM **17**(11), 2094–2107 (2015)
16. Murphy-Chutorian, E., Trivedi, M.M.: Head pose estimation in computer vision: a survey. TPAMI **31**(4), 607–626 (2009)
17. Paszke, A., et al.: Automatic differentiation in pytorch (2017)
18. Patacchiola, M., Cangelosi, A.: Head pose estimation in the wild using convolutional neural networks and adaptive gradient methods. PR **71**, 132–143 (2017)
19. Ranjan, R., Patel, V.M., Chellappa, R.: Hyperface: a deep multi-task learning framework for face detection, landmark localization, pose estimation, and gender recognition. TPAMI **41**(1), 121–135 (2019)
20. Ranjan, R., Sankaranarayanan, S., Castillo, C.D., Chellappa, R.: An all-in-one convolutional neural network for face analysis. In: FG, pp. 17–24 (2017)
21. Ruiz, N., Chong, E., Rehg, J.M.: Fine-grained head pose estimation without keypoints. In: CVPR Workshops, pp. 2074–2083 (2018)
22. Sagonas, C., Tzimiropoulos, G., Zafeiriou, S., Pantic, M.: 300 faces in-the-wild challenge: the first facial landmark localization challenge. In: ICCV Workshops, pp. 397–403 (2013)
23. Valle, R., Buenaposada, J.M., Baumela, L.: Multi-task head pose estimation in-the-wild. TPAMI **43**, 2874–2881 (2020)
24. Yang, T., Chen, Y., Lin, Y., Chuang, Y.: Fsa-net: learning fine-grained structure aggregation for head pose estimation from a single image. In: CVPR, pp. 1087–1096 (2019)
25. Zhang, F., Zhang, T., Mao, Q., Xu, C.: Joint pose and expression modeling for facial expression recognition. In: CVPR, pp. 3359–3368 (2018)
26. Zhang, H., Wang, M., Liu, Y., Yuan, Y.: FDN: feature decoupling network for head pose estimation. In: AAAI, pp. 12789–12796 (2020)
27. Zhou, E., Fan, H., Cao, Z., Jiang, Y., Yin, Q.: Extensive facial landmark localization with coarse-to-fine convolutional network cascade. In: ICCV Workshops, pp. 386–391 (2013)
28. Zhu, X., Ramanan, D.: Face detection, pose estimation, and landmark localization in the wild. In: CVPR, pp. 2879–2886 (2012)
29. Zhu, X., Lei, Z., Liu, X., Shi, H., Li, S.Z.: Face alignment across large poses: a 3d solution. In: CVPR, pp. 146–155 (2016)

Facilitating 3D Object Tracking in Point Clouds with Image Semantics and Geometry

Lingpeng Wang, Le Hui, and Jin Xie[(✉)]

PCA Lab, Key Lab of Intelligent Perception and Systems for High-Dimensional Information of Ministry of Education, Nanjing University of Science and Technology, Nanjing, China
{cslpwang,le.hui,csjxie}@njust.edu.cn

Abstract. Recent works have shown remarkable success in 3D object tracking in point clouds. However, these methods may fail when tracking distant objects or objects interfered by similar geometries in point clouds. We aim to use high-resolution images with rich textures to help point cloud based tracking to deal with the above-mentioned failures. In this paper, we propose an end-to-end framework, which effectively uses both image semantic features and geometric features to facilitate tracking in point clouds. Specifically, we design a fusion module to establish the correspondence between image and point cloud features in a point-to-point manner and use attention-weighted image features to enhance point features. In addition, we utilize geometric transformation to convert 2D image geometric features inferred by deep layer aggregation network (DLA) to 3D as extra tracking clues for 3D voting. Quantitative and qualitative comparisons on the KITTI tracking dataset demonstrate the advantages of our model.

Keywords: 3D object tracking · Multimodal fusion · 3D voting · 3D deep learning

1 Introduction

3D object tracking is essential for autonomous vehicles to understand the surrounding environment and make reliable responses. Its goal is to accurately track the target in the 3D scene frame to frame. With the progress of deep learning on point clouds, several works [3,4,16] that only use point clouds for 3D object tracking have gradually appeared, and tracking performance has been significantly improved compared with works [8,12] based on RGB-D. However, point cloud data defects (sparse and disordered) also limit task performance. For example, for objects with sparse points on the surface, we cannot extract valuable 3D geometric features for tracking. [4] and [3] attempt to alleviate point sparseness by shape completion, which may bring two flaws: 1) shape prior reduces

© Springer Nature Switzerland AG 2021
H. Ma et al. (Eds.): PRCV 2021, LNCS 13019, pp. 589–601, 2021.
https://doi.org/10.1007/978-3-030-88004-0_48

generalization ability; 2) proposal-wise matching limits the inference speed. Subsequently, P2B [16] regresses search area seeds augmented by target-specific features to potential object center for proposal generation. However, the sparse point cloud scene is still a huge challenge for it.

All in all, relying only on point clouds cannot meet the needs of 3D object tracking in the cases of complex scenes. By analyzing the defects of point cloud data, a natural idea is to supplement it with images. The image contains rich texture information, which can provide additional semantic and geometric information for the point cloud.

However, how to effectively fuse the information of image and point clouds for 3D object tracking is a key problem. On the one hand, there is a huge gap between the image and the point cloud modality, on the other hand, images are susceptible to environmental influences such as illumination and occlusion. Zhang et al. [22] propose a robust fusion module to self-learn the weights of different modal features, and perform weighted fusion on proposal-wise multimodal features. In [22], there is no clear master-slave relationship between the modalities and the proposal-wise feature fusion is relatively rough. In addition, [13] use the trained detector to obtain image geometric feature and attach it to the point feature as a fused feature in a proposal-to-point manner. However, it does not explicitly consider the interference information that may exist in the image and the network is not end-to-end.

In this paper, we propose an end-to-end 3D object tracking framework, which efficiently fuses image semantics and LiDAR point features to define the tracked target in a more discriminative representation and use image geometric features to generate pseudo 3D offset to guide 3D voting [14]. Specifically, we input the template and search area image into the image encoder while feeding the template and search area point cloud into the point cloud encoder. Then we develop the double judgment fusion (DJ-Fusion) module to fuse corresponding different modal features in a point-to-point manner in the template and search area branches respectively. After that, we embed template target information and tracking clues in the obtained search area seeds to generate 3D votes and use the seed-wise pseudo 3D offset yielded from the search area image to guide this process. Further, in VoteNet, we cluster local votes to generate target proposals and proposal-wise scores. Finally, we choose the proposal with the highest score as the tracking result box.

To summarize, the main contributions of this paper are as follows:

1. To the best of our knowledge, this is the first work that efficiently uses image semantics and geometric features to facilitate 3D single object tracking in point clouds.
2. We propose a DJ-Fusion module to adaptively incorporate valuable image semantic features into LiDAR point features in a point-to-point manner.
3. We introduce the pseudo 3D offset module, which uses 2D geometric information to guide 3D voting via geometric transformation and can be trained end-to-end.

2   Related Work

2.1   3D Object Tracking

Several early 3D object tracking studies were based on RGB-D data. They try to reconstruct 3D objects through image and depth for object tracking. Klein and Murray [10] proposed a camera pose tracking method, which divides tracking and mapping into two tasks. Specifically, it tracks the key points between image frames while using past frames to build a 3D map of point features. Kart *et al.* [9] proposed a framework that uses 3D reconstruction to constrain a set of view-specific discriminative correlation filters.

With the development of 3D deep learning on point clouds,, several 3D object tracking works [3,4,16] based on point clouds have appeared. Giancola *et al.* [4] first proposed a point cloud based 3D object tracking network (we call it SC3D in simplified form) with shape constraints and proposal-wise template matching. But the proposals generated by Kalman filtering [5] are not accurate. Subsequently, Qi *et al.* [16] proposed an end-to-end point-to-box network called P2B, which locates the potential center of the target in the 3D search area embedded with target's 3D geometric information, thus avoiding time-consuming 3D exhaustive search. Feng *et al.* [3] then proposed a two-stage 3D object re-track framework, using Kalman filtering and 3D proposal generation network to jointly obtain high-quality proposals for 3D template matching. None of the above methods can properly handle the tracking failure caused by sparsity of point clouds.

2.2   Camera-LiDAR Fusion

Multi-modal fusion has always been a difficult task, especially for modalities with large differences such as image and point cloud. In the 3D single object tracking task, there is no work on image and point cloud fusion. We refer to some works in similar fields such as 3D object detection and multi-object tracking. These works are roughly divided into three categories: 2D-driven, 3D-driven and joint-driven. 2D-driven methods [1,11] finetune the detection results by fusing corresponding ROI region on the feature maps of the LiDAR multi-view and the image. 3D-driven methods [7,17] mainly operate on LiDAR data to establish the point-wise correspondence between point cloud and image features. In addition, joint-driven methods [13,22] can utilize any modal to complete the task and train a multi-modal network in a gradient blending manner [18] to balance the information of different modalities.

2.3   Tracking by Detection Scheme

Tracking by detection methods [6,20,24] have achieved satisfactory results in the 2D object tracking task. These methods embed the target clues of the template in the search area and obtain the high-quality proposal of objects via the detector.

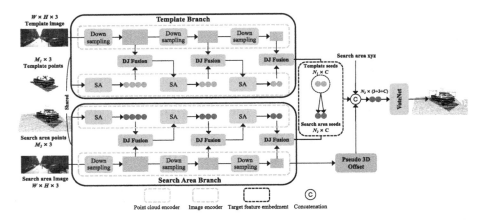

Fig. 1. Overview of the image facilitate point clouds network. The network inputs the paired multi-modal template and search area of the tracked target, and outputs the tracking result.

[6,20] apply the regional proposal network on each pixel. In addition, an anchor-free detector is used in [24]. Recent 3D object tracking methods [2,16] have learned from 2D tracking. P2B [16] solves the problem of 3D target-specific feature generation through symmetric functions. It further uses Hough voting [14] to obtain a seed-wise 3D vote and clusters local context to generate high-quality proposals. In contrast, we mainly focus on how to use images to promote 3D tracking, so we apply the global correlation module mentioned in P2B and only use VoteNet [14] as our detector.

3   Method

We design a novel framework that uses images to facilitate 3D object tracking based on point clouds. We are committed to integrating the corresponding image semantic features into the template and search area seeds through the DJ-Fusion module to define the tracked target in a more discriminative representation. Then we use geometric cues from the 2D image to guide search area seeds embedded with target tracking cues to perform Hough voting [14] and cluster votes to generate proposals and score. As shown in Fig. 1, our network is mainly composed of three parts: two-branch feature extraction, pseudo 3D offset generation and target clue embedment and Hough voting. We will detail each part separately in following sections.

3.1   Two-Branch Feature Extraction

Two-branch feature extraction is based on the Siamese network architecture which has been successful in the 2D tracking task. We respectively input the template image and point cloud into corresponding encoders, and use multiple

DJ-Fusion modules to enhance point features with corresponding image semantic features at different scales. After that, we get template seeds. For the search area branch, we apply the same operation to get search area seeds.

Multimodal Network. Multi-modal backbone network consists of a point cloud encoder and an image encoder. The point cloud encoder contains three Set Abstraction (SA) layers [15] for extracting 3D geometric features. The image encoder contains three downsampling modules. Each downsampling module consists of three convolutional blocks containing residual structure. Each branch receives a pair of image and point cloud and uses corresponding point-wise image semantic features extracted by downsampling module to enhance point features extracted by SA layer at multiple scales.

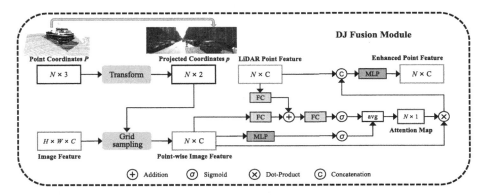

Fig. 2. DJ Fusion module internal structure. Red points in image represent the corresponding position of the point cloud projected on the image. (Color figure online)

Double Judgment Fusion. Image features can not only provide additional regional semantic understanding, but also help distinguish geometrically similar targets. However, images are easily affected by the environment, such as illumination and occlusion. Simply fusing point clouds and images will introduce interference information and hinder tracking. Here, we propose a DJ-Fusion module to make better use of image semantic information.

As shown in Fig. 2, the DJ-Fusion module has two parts: point-wise image feature extraction and double judgment fusion. We first project the 3D points onto the image through the transformation module. The transformation module contains a 3×4 size camera intrinsic matrix T and an interval standardization. Specifically, given a 3D point P located in the camera coordinate system, we can get the corresponding point p projected onto the image by left multiplying T. After that, we normalize the projected coordinates to [-1.0,1.0] and feed projected point coordinates and image feature maps into the grid sampling module to obtain point-wise image features. Since the projected coordinates are not integers, bilinear interpolation is selected for grid sampling to get continuous image features. Specifically, for a projected point $\boldsymbol{p}(u, v)$, we have four adjacent pixels $\boldsymbol{p}_1(u_1, v_1)$, $\boldsymbol{p}_2(u_2, v_1)$, $\boldsymbol{p}_3(u_1, v_2)$, $\boldsymbol{p}_4(u_2, v_2)$ with features $\boldsymbol{f}_1^I$, $\boldsymbol{f}_2^I$, $\boldsymbol{f}_3^I$, and $\boldsymbol{f}_4^I \in \mathbb{R}^C$. The image feature $\boldsymbol{f}^I$ corresponding to $\boldsymbol{p}$ can be expressed as:

$$
\begin{aligned}
\boldsymbol{f}^I \approx \ &\boldsymbol{f}_1^I \left(u_2 - u\right)\left(v_2 - v\right) \oplus \boldsymbol{f}_2^I \left(u - u_1\right)\left(v_2 - v\right) \\
&\oplus \boldsymbol{f}_3^I \left(u_2 - u\right)\left(v - v_1\right) \oplus \boldsymbol{f}_4^I \left(u - u_1\right)\left(v - v_1\right),
\end{aligned}
\tag{1}
$$

where $\oplus$ means element-wise addition.

Multi-modal feature fusion needs to consider multiple factors. We aim to utilize valuable image features to mitigate tracking failures caused by sparse point clouds or similar geometries. Therefore, we propose a double judgment fusion to adaptively judge image quality and importance relative to point features. We first input the point-wise image feature $\boldsymbol{f}^I$ into the multi layer perceptron (MLP) to adaptively perceive image feature quality, and then use the sigmoid function to normalize the quality to $[0, 1]$. In addition, we feed the $\boldsymbol{f}^I$ and $\boldsymbol{f}^P$ (LiDAR point feature) into fully connected layer and add them together to establish the feature correlation. Subsequently, we further extract more complex correlations and normalize correlations. We apply average pooling to balance the quality of $\boldsymbol{f}^I$ and its importance relative to $\boldsymbol{f}^P$. Finally, we multiply the obtained attention map with the point-wise image feature $\boldsymbol{f}^I$ and concatenate it with the point feature $\boldsymbol{f}^P$ to yield the enhanced point feature.

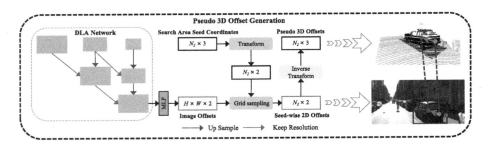

Fig. 3. Pseudo 3D Offsets Generation. The input of the DLA network in the figure is the three-scale feature map of the image encoder.

3.2   Pseudo 3D Offset Generation

It is not easy to search the target center directly in 3D space, especially for search area objects with few points. In contrast, searching in high-resolution images is much simpler. Pseudo 3D offset generation can provide additional geometric clues from the search area image for 3D voting. First, we input multi-scale image feature maps from the search area branch into deep layer aggregation

(DLA) network [23] to get an image offset map. After that, seed-wise 2D offsets are obtained via projection transform and grid sampling, and then converted into a 3D offset via inverse transform. The detailed process is shown in the Fig. 3.

2D Image Offsets. In order to adapt to different scale targets and alleviate the alignment problem of multi-scale feature map aggregation, we adopt a variant of the DLA network [23]. Based on the original DLA [21], it replace all convolution layer in the up-sampling module with deformable convolution layers. DLA network takes the multi-scale feature map from the search area image to generate the $H \times W \times 2$ size offset map. The values on each pixel in map represent the offsets between it and 2D target center.

2D-3D Offsets Transform. We project search area seed coordinates to the image physical coordinate system through the transformation module. After that, the grid sampling module takes projected coordinates to yield the seed-wise 2D offset by bilinear interpolation on the offset map. This process is similar to point-wise image feature generation in Sect. 3.1. Finally, we use a inverse transformation based on the camera intrinsic and seed depth to convert the 2D offset to 3D.

Specifically, we define the 3D target center in the camera coordinate system as $C = (x_c, y_c, z_c)$ and its projected point on the image as $c = (u_c, v_c)$. In addition, a seed on the 3D target surface and its projected point are respectively denoted as $S = (x_s, y_s, z_s)$ and $s = (u_s, v_s)$. Therefore, the 2D offset can be expressed as a vector:

$$\vec{sc} = (\Delta u, \Delta v) = (u_c - u_s, v_c - v_s), \tag{2}$$

and the ground-truth(GT) 3D offset is denoted as:

$$\vec{SC} = (x_c - x_s, y_c - y_s, z_c - z_s). \tag{3}$$

For a pinhole camera with focal length f, according to the perspective projection relationship, we have $(u, v) = (f\frac{x}{z}, f\frac{y}{z})$. Equation 2 is transformed into:

$$\vec{sc} = \left(f\left(\frac{x_c}{z_c} - \frac{x_s}{z_s} \right), f\left(\frac{y_c}{z_c} - \frac{y_s}{z_s} \right) \right). \tag{4}$$

Since z_c is unknown, we follow [13] to assume that z_c is similar to z_s, which is reasonable for most targets that are not too close to the camera optical center. After that, we can get pseudo-3D offset

$$\widehat{SC} = f^{-1}(z_s\Delta u, z_s\Delta v, 0) \approx \vec{SC} \tag{5}$$

with error (along the X-axis)

$$\Delta x - \widehat{\Delta x} = x_c(1 - \frac{z_s}{z_c}). \tag{6}$$

Real projection involves focal length, principal point shift and lens distortion, which is a 3×4 size intrinsic matrix $\boldsymbol{T}$. Therefore, we replace the focal length f with $\boldsymbol{T}$. Finally, given intrinsic matrix $\boldsymbol{T}$, seed-wise depth and seed-wise 2D offset, we generate a pseudo-3D offset

$$\widehat{SC} = \boldsymbol{T}^{-1} \left(z_s \Delta u, z_s \Delta v, 0 \right), \tag{7}$$

where $\boldsymbol{T}^{-1}$ represents the $\boldsymbol{T}$'s generalized inverse matrix with size 4×3.

3.3   Target Clue Embedment and Hough Voting

Following [16], we embed global target clues and tracking clues in the search area by establishing the correlation between each search area seed and all template seeds. After that, all search area seeds with space coordinates and pseudo-3D offsets are fed into VoteNet [14] to perform 3D voting and output target proposals with scores.

Target Feature Embedment. After feature extraction, we get N_1 template seeds and N_2 search area seeds with sufficient 2D semantic and 3D geometric features. Then the global target feature and similarity (tracking clue) are embedded in each search area seed. In detail, for a seed $\boldsymbol{S}_s$ in search area, we calculate cosine similarity between it and each template seed $\boldsymbol{S}_t$ in feature space to obtain a $N_1 \times 1$ size tensor. Then, the corresponding template seed features and coordinates are introduced into the similarity relationship to get a $N_1 \times (1 + 3 + C)$ size tensor. We apply MLP on this tensor and then perform max-pooling on N_1 channel to obtain a C-dimensional feature.

Deep Hough Voting. VoteNet [14] obtains point-wise vote via 3D voting and further samples and clusters in votes. Finally, it uses vote clusters to generate proposals and scores. Specifically, our VoteNet only retains 3D voting and target proposal and classification process. We first input seed coordinates with feature $\boldsymbol{f}^P \in \mathbb{R}^{3+C}$ embedded with pseudo-3D offsets and target clues into vote layer, which is composed of MLP with weight sharing, to predict the coordinate offset $\widetilde{SC}_i \in \mathbb{R}^3$ between each seed $\boldsymbol{S}_i$ and GT target center and feature residual $\Delta \boldsymbol{f}_i^P \in \mathbb{R}^C$ for $\boldsymbol{f}_i^P[3 : 3 + C]$. Further, each vote V_i is represented as $[\boldsymbol{S}_i + \widetilde{SC}_i; \boldsymbol{f}_i^P[3 : 3 + C] + \Delta \boldsymbol{f}_i^P]$. After that, we input the votes into an SA layer [15] with a sampling number K and a ball query radius R, which further clusters the local votes to generate K clusters t_j containing regional context. Finally, we get a set of target proposal boxes $\{\boldsymbol{b}_j \in \mathbb{R}^4\}_{j=1}^{K}$ with score $cla_j \in \mathbb{R}^1$. $\boldsymbol{b}_j$ has four parameters: 3D offset between cluster t_j and GT target center C and the rotation angle around the Y axis of the camera coordinate system. The proposal with the highest classification score is used as the final tracking result.

3.4 Loss Functions

2D Offset Loss. In order to get accurate pseudo 3D offset, the 2D offset loss is defined as:

$$L_{2D} = \frac{1}{N_{ts}} \sum_i \left\| \hat{sc}_i - \overrightarrow{sc}_i \right\| \cdot \mathbb{I}\left[s_i \text{ on target} \right], \tag{8}$$

where $\hat{sc}_i$ represents the predicted 2D offset from s_i to c, $\mathbb{I}\left[s_i \text{ on target} \right]$ indicates that we only supervise the projected point s_i of the seed on GT target surface and N_{ts} denotes the number of seeds used for training.

Voting Loss. For the 3D voting process, we use a regression loss to supervise the 3D offset from the seed to the GT target center, which can implicitly affect the feature residual.

$$L_{vote} = \frac{1}{N_{ts}} \sum_i \left\| \widetilde{SC}_i - \overrightarrow{SC}_i \right\| \cdot \mathbb{I}\left[S_i \text{ on target} \right]. \tag{9}$$

Here, $\mathbb{I}\left[S_i \text{ on target} \right]$ indicates that we only supervise seeds locating on the GT target surface.

Proposal and Classification Loss. We follow P2B [16] to constrain proposal and score generation. We consider proposals with an offset of less than $0.3\,\text{m}$ from the GT target center as positive and more than $0.6\,\text{m}$ as negative, and apply binary cross entropy loss L_{cla} to cla_j. In addition, we only use smooth-L1 loss L_{pro} for the parameters of the positives' proposal b_j.

Joint Loss. We aggregate the above losses to train the network, defined as:

$$L = \lambda_{2D}L_{2D} + \lambda_{vote}L_{vote} + \lambda_{cla}L_{cla} + \lambda_{pro}L_{pro}, \tag{10}$$

where λ_{2D}, λ_{vote}, λ_{cla} and λ_{pro} are balance factors.

4 Experiments

In this section, we first introduce the data set and evaluation metrics, and then present the implementation details of the framework. After that, we compare with the existing 3D single object tracking work and obtain state-of-the-art results. Finally, we conduct the ablation study to prove the effectiveness of our network.

4.1 Datasets and Evaluation Metric

Following the previous work, we train and test on the KITTI training set (the only data set for 3D single object tracking), which contains 21 outdoor scenes and 8 types of targets. We use scenes 0–16 in the dataset for training, 17–18 for validation, and 19–20 for testing. We use Success and Precision proposed in One Pass Evaluation (OPE) [19] to measure the tracking results. Success represents the IOU of the prediction box and the GT box, and Precision represents the closeness of two boxes' centers.

4.2   Implementation Details

During training, we merged the points in previous GT box with random offset and first GT box, and then randomly sampled $M_1 = 512$ points as template points. We apply a similar random offset to current GT box and expand it by 2 m as the target search area. We randomly sampled $M_2 = 1024$ points in search area as the search area points. We use the previous image as image template and the current image as search area image, both of which are resized to 512×160.

During testing, we have the first GT box, first use it as a template, and then get the search area and use the trained network to infer the target result box frame by frame along the video sequence. Note that starting from the third frame, the template points are the aggregation of the points in the first GT box and the previous result box. The difference from the training period is: 1) There was no random offset to the box. 2) The search area was replaced by the previous result box expanded by two meters.

For the point cloud encoder, half of the points are sampled for each SA. Finally, it outputs $N_1 = 64$ template seeds and $N_2 = 128$ search area seeds with $C = 128$ dimensions feature. For the image encoder, the feature map is reduced by half each time through a down-sampling module. The output feature map size is 64×20. The input of the DLA network is three feature maps of size $256 \times 80 \times 64$, $128 \times 40 \times 128$, $64 \times 20 \times 128$, and the shape of the 2D offset map is $256 \times 80 \times 128$.

We use Adam optimizer and set the learning rate to 0.001, and the learning rate decays by four-fifths every 12 epochs. Moreover, we set the batch size as 36 and balance factors λ_{2D}, λ_{vote}, λ_{cla} and λ_{pro} is 0.1, 1.0, 1.5, and 0.2.

4.3   Quantitative and Qualitative Comparisons

Following the previous works, we use Success and Precision to conduct quantitative analysis on car tracking. All methods in Table 1 rely on the previous tracking result boxes for tracking the current frame, instead of using the current GT box. The results of quantitative comparison are shown in Table 1. Our method achieves state-of-the-art results on both evaluation metrics. Despite the use of additional image information, our method can still perform real-time tracking.

Table 1. Quantitative comparison results of different methods. To be fair, all experiments are performed on single TITAN RTX GPU.

Methods	Modality	Success (%)	Precision (%)	FPS
SC3D [4]	LiDAR	41.3	57.9	2
P2B [16]	LiDAR	56.2	72.8	17
Re-Track [3]	LiDAR	58.4	73.4	–
3D-SiamRPN [2]	LiDAR	58.2	76.2	–
Ours	RGB+LiDAR	**60.8**	**77.0**	11

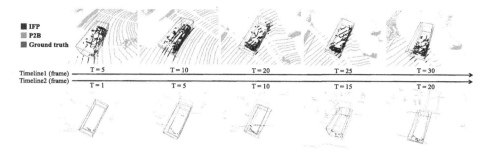

Fig. 4. Qualitative comparison in the visualized tracking process. The first line shows tracking in dense scenes, while the second line shows tracking in sparse scenes.

In addition, we also visualize the tracking results of different methods in sparse scenes and dense scenes in Fig. 4 for qualitative comparison. From the results, we can see that our network can track targets more accurately than P2B for dense scenes. More importantly, P2B will gradually lose the target along with the video sequence in the sparse scene, but our network can successfully track the target using image information.

4.4 Ablation Study

In order to demonstrate the effectiveness of DJ-fusion module, we compare tracking effects of three different ways, as shown in Table 2. Simple concatenate means that we directly concatenate the point-wise LiDAR point feature and point-wise image feature and directly obtain the enhanced point feature through MLP. Without fusion denotes that we directly use LiDAR point feature instead of point-wise feature to. We can find that compared with DJ-fusion, simply concatenate fusion performance drops by about 2%, and without fusion drops by about 3%, which shows that DJ-fusion can better use image semantic features to enhance point features.

Table 2. Different ways for image and point cloud feature fusion.

Ways for fusion	Success (%)	Precision (%)
DJ-fusion	**60.8**	**77.0**
Simply concatenate	58.8	75.4
Without fusion	57.8	74.2

In Sect. 3.2, we obtain seed-wise pseudo 3D offsets and concatenate them with corresponding seed features to guide 3D voting. Here we test network without this concatenation or the whole branch of pseudo 3D offsets generation. From

Table 3. Effectiveness of pseudo 3D offset.

Ways for using pseudo 3D offset	Success (%)	Precision (%)
Our default setting	**60.8**	**77.0**
Without concatenation	59.2	75.1
Without the whole pseudo 3D offset	58.6	74.4

the results in Table 3, we can see that removing concatenation reduces the performance by about 1.8% and removing the whole branch decreased performance by about 2.5%. This demonstrates that pseudo 3D offsets from image geometry information can further promote subsequent depth Hough voting to obtain accurate proposals and scores.

5   Conclusion

In this paper, we proposed an end-to-end network that makes full use of image information to facilitate point cloud tracking. Specifically, we developed a DJ-Fusion module and a pseudo 3D offset generation module, which use the texture and geometric information of images respectively. Experiments demonstrate that our framework can achieve state-of-the-art results in KITTI tracking dataset and show robust tracking performance in both dense and sparse point cloud tracking scenarios.

References

1. Chen, X., Ma, H., Wan, J., Li, B., Xia, T.: Multi-view 3D object detection network for autonomous driving. In: CVPR, pp. 6526–6534 (2017)
2. Fang, Z., Zhou, S., Cui, Y., Scherer, S.: 3D-siamrpn: an end-to-end learning method for real-time 3D single object tracking using raw point cloud. IEEE Sens. J. **21**, 4995–5011 (2021)
3. Feng, T., Jiao, L., Zhu, H., Sun, L.: A novel object re-track framework for 3D point clouds. In: ACM MM, pp. 3118–3126 (2020)
4. Giancola, S., Zarzar, J., Ghanem, B.: Leveraging shape completion for 3D siamese tracking. In: CVPR, pp. 1359–1368 (2019)
5. Gordon, N., Ristic, B., Arulampalam, S.: Beyond the Kalman Filter: Particle Filters for Tracking Applications, pp. 1–4. Artech House, London (2004)
6. Han, W., Dong, X., Khan, F., Shao, L., Shen, J.: Learning to fuse asymmetric feature maps in siamese trackers. In: CVPR, pp. 16570–16580 (2021)
7. Huang, T., Liu, Z., Chen, X., Bai, X.: Epnet: enhancing point features with image semantics for 3D object detection. In: ECCV, pp. 35–52 (2020)
8. Kart, U., Kämäräinen, J., Matas, J.: How to make an rgbd tracker? In: ECCV Workshops (2018)
9. Kart, U., Lukezic, A., Kristan, M., Kämäräinen, J., Matas, J.: Object tracking by reconstruction with view-specific discriminative correlation filters. In: CVPR, pp. 1339–1348 (2019)

10. Klein, G., Murray, D.: Parallel tracking and mapping for small ar workspaces. In: ISMAR, pp. 225–234 (2007)
11. Ku, J.T., Mozifian, M., Lee, J., Harakeh, A., Waslander, S.L.: Joint 3D proposal generation and object detection from view aggregation. In: IROS, pp. 1–8 (2018)
12. Liu, Y., Jing, X., Nie, J., Gao, H., Liu, J., Jiang, G.P.: Context-aware three-dimensional mean-shift with occlusion handling for robust object tracking in rgb-d videos. IEEE Trans. Multimedia **21**, 664–677 (2019)
13. Qi, C., Chen, X., Litany, O., Guibas, L.: Imvotenet: boosting 3D object detection in point clouds with image votes. In: CVPR, pp. 4403–4412 (2020)
14. Qi, C., Litany, O., He, K., Guibas, L.: Deep hough voting for 3D object detection in point clouds. In: ICCV, pp. 9276–9285 (2019)
15. Qi, C., Yi, L., Su, H., Guibas, L.: Pointnet++: deep hierarchical feature learning on point sets in a metric space. In: NIPS (2017)
16. Qi, H., Feng, C., Cao, Z., Zhao, F., Xiao, Y.: P2b: point-to-box network for 3D object tracking in point clouds. In: CVPR, pp. 6328–6337 (2020)
17. Simon, M., et al.: Complexer-yolo: real-time 3D object detection and tracking on semantic point clouds. In: CVPR Workshops, pp. 1190–1199 (2019)
18. Wang, W., Tran, D., Feiszli, M.: What makes training multi-modal classification networks hard? In: CVPR, pp. 12692–12702 (2020)
19. Wu, Y., Lim, J., Yang, M.H.: Online object tracking: a benchmark. In: CVPR, pp. 2411–2418 (2013)
20. Yan, B., Zhang, X., Wang, D., Lu, H., Yang, X.: Alpha-refine: boosting tracking performance by precise bounding box estimation. In: CVPR, pp. 5289–5298 (2021)
21. Yu, F., Wang, D., Darrell, T.: Deep layer aggregation. In: CVPR, pp. 2403–2412 (2018)
22. Zhang, W., Zhou, H., Sun, S., Wang, Z., Shi, J., Loy, C.C.: Robust multi-modality multi-object tracking. In: ICCV, pp. 2365–2374 (2019)
23. Zhang, Y., Wang, C., Wang, X., Zeng, W., Liu, W.: A simple baseline for multi-object tracking, p. 6 (2020). arXiv preprint arXiv:2004.01888
24. Zhang, Z., Peng, H.: Ocean: object-aware anchor-free tracking. In: ECCV, pp. 771–787 (2020)

Multi-criteria Confidence Evaluation for Robust Visual Tracking

Siqi Shi[1($\boxtimes$)], Nanting Li[2], Yanjun Ma[2], and Liping Zheng[1]

[1] School of Automation and Information Engineering,
Xi'an University of Technology, Xi'an 710048, China
sqshi@xaut.edu.cn
[2] School of Electrical Engineering, Xi'an University of Technology, Xi'an 710048, China

Abstract. To solve the challenge of visual tracking in complex environment, a multi-criteria confidence evaluation strategy is proposed in this paper. Three kinds of criteria are introduced to comprehensively evaluate and analyze the confidence of those tracking results obtained by adaptive spatially-regularized correlation filters (ASRCF). The evaluation result is further utilized to establish the sample management and template updating mechanism, which aims to obtain the best template in the tracking process. The obtained template is used to update both scale filters and position filters in ASRCF. Experimental results on OTB100 and TC128 verify that the proposed method is more robustness compared with other similar algorithms.

Keywords: Object tracking · Correlation filtering · Confidence evaluation

1 Introduction

Visual object tracking is the research highlights in computer vision and has widespread applications in the fields of intelligent video surveillance, human-computer interaction, intelligent transportation systems etc. [1, 2]. Those tracking algorithms based on correlation filter (CF), which capture the target by seeking the maximal correlation between the tracker and the candidates in the scene, have received great attention for its superiority in feasibility and real-time [3, 4]. However, the performance of those tracker has declined obviously and may lead to the failed tracking because of the inappropriate tracker and the tracking template in the complex environment, such as occlusion, scale changing, fast motion and so on. Thus, it is urgent demands for the CF-based tracker to improve the robustness and accuracy in real applications.

The performance of the tracker is closely related with the CF model. Some solutions are proposed to improve the updating efficiency [5–7]. SRDCF introduces spatial regularization term to penalize filter coefficients with respect to their special location

---

Electronic supplementary material The online version of this chapter (https://doi.org/10.1007/978-3-030-88004-0_49) contains supplementary material, which is available to authorized users.

H. Ma et al. (Eds.): PRCV 2021, LNCS 13019, pp. 602–613, 2021.
https://doi.org/10.1007/978-3-030-88004-0_49

[5]. Adaptive spatially-regularized correlation filters (ASRCF) introduced adaptive spatial regularization to obtain reliable filter coefficients by learning the change of object appearance [6]. Unfortunately, those unreliable samples, which are inevitably contaminated in the complex environment, may bring about the training error, model drifting and target losing.

To eliminate the influence of unreliable samples, some decontamination methods were proposed. The sample quality was improved via directly dropping those with low value of peak to sidelobe ratio (PSR) [8]. SRDCFdecon overcame the model drifting via jointly optimizing parameters of CF and weights of the whole sample set [9]. An adaptive decontamination method in our previous work [10] classified each sample into various subset according to its pollution level, and dynamically updated the specific samples and their corresponding filter, which efficiently increased the decontamination ability in the severely polluted case. In addition, Gaussian mixture model (GMM) in ECO [11], which is used to model the distribution of sample space and decreased the risk of overfitting, are important to improve the reliability of samples. In all, the reliable sample is crucial to obtain the optimal tracker.

The confidence of tracking result is beneficial to judge the state of tracker and the reliability of samples set. Some methods for confidence evaluation were proposed [12–16], such as average peak-to-correlation energy [12], Hamming distance [13]. PSR [14], etc. Those methods above, which only adopted single evaluation criterion, cannot accurately and robustly evaluate the tracking result in complex environment. For example, the PSR of response map reflected the change of object appearance to some extent, but failed to distinguish the reason of such change (More details can be found in Sect. 3.1). Multiply criteria is necessary and required to evaluate the confidence of tracking results in complex environment.

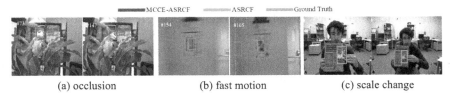

(a) occlusion   (b) fast motion   (c) scale change

Fig. 1. Tracking example of MCCE-ASRCF and ASRCF.

Motivated by the above consideration, a multi-criteria confidence evaluation strategy with ASRCF (MCCE-ASRCF) is proposed for robust visual tracking. In Fig. 1, the proposed MCCE-ASRCF algorithm is more robust and accurate than ASRCF in various challenge scenes. In sum, contributions of this paper are mainly as follow:

- Proposed a multi-criteria confidence evaluation strategy. Three kinds of criteria, similarity, smoothness and coincidence, are incorporated properly to evaluate the tracking state more detailed and comprehensive.
- Proposed a collaborative process to improve the performance of tracker. Sample management and template updating are combined to reduce the tracker drifting and object losing.

- Proposed a robust visual tracking algorithm in complex environment. The test results on OTB100 and TC128 can testify the proposed algorithm over other similar algorithms.

2 Related Work

2.1 ASRCF

Tracker based on correlation filter has superior computational efficiency. MOSSE [8] was the first CF-based tracker. KCF [17] used correlation filter in a kernel space with multichannel histogram of oriented gradient (HOG) [18] features to achieve better performance. SAMF used multi-scale searching strategy to solve the scale adaptation problem [19]. However, those algorithms failed to accurately update filters in complex environment.

Recently, ASRCF [6] improved the locating precision at the cost of relative low computation, which utilized two types of CFs to estimate the scale and position of target within searching area. Its objective function is described as follow:

$$
E(\boldsymbol{H}, w) = \frac{1}{2} \left\| y - \sum_{k=1}^{K} x_k * (\boldsymbol{P}^T h_k) \right\|_2^2 + \frac{\lambda_1}{2} \sum_{k=1}^{K} \| w \odot h_k \|_2^2 + \frac{\lambda_2}{2} \| w - w^r \|_2^2 \quad (1)
$$

where $X = [x_1, x_2, \ldots, x_K]$ and $\boldsymbol{H} = [h_1, h_2, \ldots, h_K]$ respectively denote the templates and the correlation filters. K denotes the channel number of filters. Parameter w is regularization weight, P is a diagonal binary matrix to make the correlation operator directly apply on the true foreground and background samples. The first term is ridge regression term, which aims to fit the ideal label with Gaussian distribution y. The second term introduces an adaptive spatial regularization term into the filter and optimize the spatial weight. The third term introduces reference weight w^r to prevent the model degenerating. λ_1 and λ_2 are the regularization parameters. The alternating direction method of multipliers (ADMM) [20] is used to obtain a local optimal solution of the objective function. All calculations are performed in the frequency domain.

In the updating process of two CFs, ASRCF took consideration of tracking results but cannot accurately evaluate the unfavorable effect caused by the unreliable samples, which eventually resulted in the degradation of tracking performance.

2.2 Confidence Evaluation for Tracking Result

Confidence evaluation is important to analyze tracking states and obtain better tracking performance. Here, several evaluation criteria are briefly introduced.

The criterion of similarity usually measures the variation of object appearance. The similarity is generally measured by [8]. The PSR is defined as follow:

$$
PSR = \frac{|g_{\max} - g_{mean}|^2}{\frac{1}{N_g} \left(\sum_{x,y}^{N} (g_{x,y} - g_{mean})^2 \right)} \quad (2)
$$

where $g_{\max}$, g_{mean} and $g_{(x,y)}$ respectively indicate the maximum, average and coordinate value (x, y) of response map, N_g is the number of matrix elements.

The criterion of coincidence, which is similar to intersection over union (IoU), usually measures the changing of object size and movement. The definition of coincidence is as follow:

$$o_{i,j} = \frac{area(b_i \cap b_j)}{area(b_i \cup b_j)} \tag{3}$$

where b_i and b_j indicate rectangular boxes of the same object in different frame.

The criterion of smoothness, which indicates tracking reliability to some extent, usually measures the continuity of tracking trajectory [21]. The definition of smoothness is as follow:

$$s_{(i,j)} = e^{-kd_{(i,j)}^2} \tag{4}$$

where $k = 2(W + H)^{-2}$, W and H are respectively the width and height of object in initial frame, $d_{i,j}$ is the Euclidean distance of object position in frame i and j.

In complex environment, PSR is sensitive to the changing of object appearance (caused by background clutters and occlusion) and usually evaluates the quality and reliability of samples. Coincidence and smoothness are sensitive the quick changing of object location (caused by fast motion or disappearance out of sight) and are prone to evaluate the reliability of tracking results.

3   Method

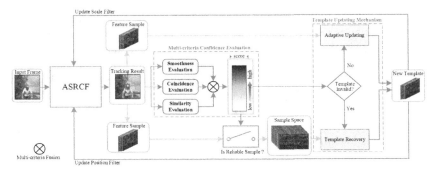

Fig. 2. The framework of MCCE-ASRCF. Tracking result obtained by ASRCF is analyzed by the module of confidence evaluation, which integrates three kinds of evaluation criteria. The evaluate score reflects the reliability of tracking and sample, and determine whether the template is adaptive updating with feature sample or recovered from sample space. In addition, the feature sample evaluated as reliable is stored in sample space. Finally, the new template is used to update the position and scale filters in ASRCF.

To improve the tracking performance in complex environment, the adverse effects of unreliable samples and tracking results must be eliminated as much as possible. In Fig. 2, the framework of our method MCCE-ASRCF combines the multi-criteria confidence evaluation strategy and template update mechanism with ASRCF.

3.1 Multi-criteria Confidence Evaluation

To obtain the optimal tracker in complex environment, it is crucial to accurately estimate the state of tracker and analyze the reliability of tracking result and samples. Here, a multi-criteria integration scheme for confidence evaluation is proposed to overcome the shortcomings of single criterion.

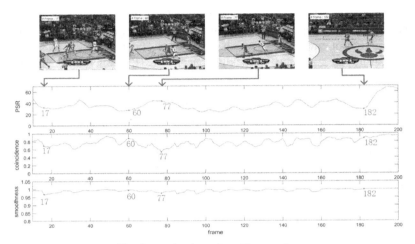

Fig. 3. Evaluation of tracking results.

Evaluation of tracking results for various criteria are shown in Fig. 3. At frame #17, the tracker drifting is resulted by the occluded object and the clutter background. The greatly decreased evaluation scores of smoothness, similarity and coincidence indicate the low reliability of tracking result. At frame #60 and #77, the gradually decreased evaluation scores of smoothness and coincidence indicate that the object had fast motion at frame #77, and the high evaluation score of similarity indicates that the tracking result is reliable and accurate at frame #77. At frame #188, the object had pose change largely, the score of similarity is decreased, and the score of smoothness and coincidence still high, it indicates tracking result is reliable at frame #188. Thus, multi-criteria are more accurate to evaluate the tracking process comprehensively.

According to the above analysis, the confidence of tracking result can be evaluated by the confidence score. A linear fusion scheme of the confidence evaluation shown as follows.

$$C_t = \xi_s C_t^s + \xi_o C_t^o + \xi_p C_t^p. \tag{5}$$

In Eq. (5), C_t is the evaluation score of confidence. C_t^s, C_t^o and C_t^p are respectively the evaluation score of smoothness, coincidence and similarity, which can be obtained by various evaluation functions in Sect. 3.2. Parameters ξ_s, ξ_o and ξ_p are respectively the weights of smoothness, coincidence and similarity, and meeting $\xi_s+\xi_o+\xi_p = 1$.

Three weights above describe the contribution of various criteria to the confidence evaluation. Score of similarity can directly reflect the reliability of tracking result. When the score of similarity is low, the scores of smoothness and coincidence can jointly reflect the reliability of tracking result. The score of smoothness usually fluctuates slightly, except that the position of object changes greatly. Thus, the value of three weights needs met the rule as $\xi_s < \xi_o < \xi_p \leq 0.5$.

3.2 Evaluation Functions

Considering that the difference in evaluation scores of similarity, coincidence and smoothness, three evaluation functions are presented to measure their contributions to tracking confidence more reasonably. Furthermore, inter-frame information is used to ensure the stability of evaluation result. Temporal context is used to assign weights of previous frames with respect to the current frame [22]. Temporal weight sequence λ is a monotone increasing geometric sequence and shown in Eq. (6).

$$\lambda = [\lambda_{t-m}, \lambda_{t-m+1}, \cdots, \lambda_t], \tag{6}$$

where m indicates the number of previous frames, The previous frame is closer to the current frame t, the greater the temporal weight value.

Smoothness Evaluation. Smoothness score fluctuates greatly in the case of serious tracker drifting. To obtain the accurate evaluation result and avoid the overfitting to the newest sample, the smoothness score at the current frame $C_t^s(s)$ is calculated by using nonlinear Gaussian function to integrate those scores at various frames.

$$C_t^s(s) = exp\left(-\left(1 - \lambda s^{\mathrm{T}} \|\lambda\|_1^{-1}\right)^2\right), \tag{7}$$

where $s = [s_{t-m}, s_{t-m+1} \cdots, s_t]$ is sequence of smoothness score with m frames.

Coincidence Evaluation. Coincidence score varies little in the case of high tracking accuracy. However, the lower the coincidence score is, the more unreliable tracking result. so, the coincidence score $C_t^o(o)$ is calculated by Eq. (8).

$$C_t^o(o) = \begin{cases} \eta_o^{-3} e^{-(1-\eta_o)^2} (\lambda o^{\mathrm{T}} \|\lambda\|_1^{-1})^3, 0 \leq \lambda o^{\mathrm{T}} \|\lambda\|_1^{-1} < \eta_o \\ e^{-(1-\lambda o^{\mathrm{T}} \|\lambda\|_1^{-1})^2}, \eta_o \leq \lambda o^{\mathrm{T}} \|\lambda\|_1^{-1} \leq 1 \end{cases}, \tag{8}$$

where $o = [o_{t-m}, o_{t-m+1} \cdots, o_t]$ is sequence of coincidence score with m frames, η_0 is coincidence threshold.

Similarity Evaluation. Similarity score can be measured with PSR. To avoid the drastic fluctuation of PSR caused by small changing of object appearance, the normalization of PSR is shown in Eq. (9).

$$p_{norm\_t} = min(p_t, p_0)/p_0, \tag{9}$$

where p_0 is used to limit the value range of PSR. The similarity score $C_t^p(\boldsymbol{p})$ is calculated by Eq. (10).

$$
C_t^p(\boldsymbol{p}) = \begin{cases} e^{-(1-\lambda \boldsymbol{p}^{\mathrm{T}}\|\lambda\|_1^{-1})^2}, \ \lambda \boldsymbol{p}^{\mathrm{T}}\|\lambda\|_1^{-1} \geq \eta_1 \\ \dfrac{e^{-(1-\eta_1)^2}}{\eta_1^3}\left(\lambda \boldsymbol{p}^{\mathrm{T}}\|\lambda\|_1^{-1}\right)^3, \ \lambda \boldsymbol{p}^{\mathrm{T}}\|\lambda\|_1^{-1} < \eta_1 \end{cases}, \tag{10}
$$

where η_1 is similarity threshold, $\boldsymbol{p} = [p_{norm\_t-m}, p_{norm\_t-m+1} \cdots, p_{norm\_t}]$, which is the sequence of PSR score with normalization in m frames.

3.3 Mechanisms of Management and Updating

Sample Management. Reliable sample set is beneficial to template recovery when object tracking fails. Compared with the trained GMM sample space in ECO [11], our GMM sample space is continuously and dynamically updated in the tracking process, in which our trackers are more adaptable to new sample and more robust in complex environment. The sample management mechanism is described as follows: At first, storage reliable samples ($C_t \geq \eta_3$, the threshold η_3 is predefined) into GMM sample space until the sample space memory is full. Then, calculating the distance between the new sample and those stored samples, and merging into its nearest stored sample. When $C_t < \eta_3$, GMM sample space will not be updated.

Template Updating. To ensure the effectiveness of trackers, a dynamic template updating mechanism is designed. The validity of template is evaluated by Eq. (11):

$$
\begin{cases} \overline{C}_t = \lambda C^{\mathrm{T}}\|\lambda\|_1^{-1} \\ \delta_t^c = \dfrac{1}{m}\sqrt{\sum_{n=1}^{m}(C_{t-m} - \overline{C}_t)^2} \end{cases}, \tag{11}
$$

where $\boldsymbol{C} = [C_{t-m}, C_{t-m+1} \cdots, C_t]$ is confidence scores with m frames, δ_t^c is overall standard deviation, δ_0 and η_2 are the threshold of template failure, When $\delta_t^c > \delta_0$ and $C_t < \eta_2$, the template is invalid, otherwise it is valid.

Template Recovery. For the invalid template, the sample template is recovered by those samples of GMM sample space. Template recovery formula is as follows:

$$
X_{model}^{new} = \sum_{n=1}^{N} \pi_n \mathcal{N}(x_n|\mu_n, \textstyle\sum_n), \tag{12}
$$

where X_{model}^{new} is a new sample template, N is the number of samples stored in GMM, $\mathcal{N}(x_n|\mu_n, \sum_n)$ is the Gaussian function corresponding to GMM sample space, π_n is the weight of each Gaussian function.

Adaptive Updating. For the valid template, the tracker adaptively adjusts the learning rate of sample template according to evaluation results. A bi-linear interpolation operation is used to update the sample template as follows:

$$X_{model}^{new} = (1 - l)X_{model}^{old} + lX^*, \tag{13}$$

where X_{model}^{old} is the last frame sample template, X_{model}^* is the current frame sample, l denotes adaptive learning rate. Updating speed of template should be adjusted automatically according to the reliability of the samples, adaptive learning rate l is shown in Eq. (14):

$$l = k_1 C_t + b_1, \tag{14}$$

where $k_1 = (\zeta_2 - \zeta_1)/(1 - \eta_2)$, $b_1 = (\zeta_1 - \zeta_2\eta_2)/(1 - \eta_2)$, ζ_1 and ζ_2 are the upper and lower limits of learning rate respectively.

Finally, the updated template X_{model}^{new} is employed to updating the scale filters and the position filters in ASRCF.

4 Experiment

4.1 Dataset and Setting

Here, the datasets of OTB100 [23] and TC128 [24] are used for experiment. These datasets represent the tracking difficulties in real scene, including motion blur, size variation, occlusion, fast motion, deformation, background clutters and so on. The experimental simulation operates on MATLAB2018a, and run on a PC with an Inter i5-8300H CPU, 16GB RAM and a single NVIDIA GTX 1060 GPU.

In the experiment, features of HOG and color name (CN) [25] are adopted for object representation. Position filters are trained with HOG and CN, and scale filters are trained with HOG. The hyperparameters of MCCE-ASRCF are set as follows: coincidence threshold $\eta_0 = 0.4$, similarity threshold $\eta_1 = 0.1$, number of consecutive sampling frames $m = 5$, PSR limiting value $p_0 = 100$, template failure parameters $\delta_0 = 0.2, \eta_2 = 0.55$, number of stored samples $N = 50$, threshold for reliable sample $\eta_3 = 0.7$, the upper and lower limits of learning rate $\zeta_1 = 0.0185$, $\zeta_2 = 0.0180$. The weights of three criteria are selected as $\xi_s = 0.2, \xi_o = 0.3, \xi_p = 0.5$.

4.2 Effectiveness

Confidence Evaluation Validation. Different strategies are used to further demonstrate the effectiveness of multi-criterion fusion in Table 1, Single criterion has lower intersection over union (IOU) than multi criterion, with each criteria evaluation function added to MCCE-ASRCF, the IOU is smoothly improved. The proposed fusion strategy (VII) is the highest IOU score at 82.2%.

Template Updating Validation. The ablation study on template updating mechanism is presented in Table 2. Template adaptive updating and template recovery both contribute to the substantial improvement over template fixed learning rate, the combination of Template adaptive updating and template recovery can achieve the best accuracy.

Table 1. Ablation study of confidence evaluation on OTB100. Values in bracket represent e weights of corresponding to various evaluation criteria.

Number	Integration strategy	IOU (%)
I	Smoothness	80.67
II	Coincidence	80.01
III	Similarity	75.17
IV	Smoothness (0.5) + coincidence (0.5)	80.30
V	Smoothness (0.5) + similarity (0.5)	81.25
VI	Coincidence (0.5) + similarity (0.5)	79.53
VII	Smoothness (0.2) + coincidence (0.3) + similarity (0.5)	**82.20**

Table 2. Ablation study of template updating mechanism on OTB100. The learning rate of template updating is fixed and set as 0.0186 by ASRCF.

Number	Update strategy	IOU (%)
I	Fixed learning rate	80.55
II	Fixed learning rate + template recovery	81.13
III	Adaptive updating	82.00
IV	Adaptive updating + template recovery	**82.20**

4.3 Comparison

Quantitative Comparison. In order to evaluate the tracking performance, the proposed MCCE-ASRCF is compared with state-of-the-art trackers which include CSK [3], SRDCF [5], ASRCF [6], BACF [7], SRDCFDecon [9], ECOHC [11], KCF [17], SAMF [20], MCCT [21], DSST [26], Staple [27], AutoTrack [28]. Figure 4 shows precision and success plots of different tracker in term of OPE rule. Overall, our proposed algorithm is the top-1 in these datasets.

In Table 3, we compared the tracking speed of some trackers in the same experiment environment. Compared with ASRCF, our method slightly decreases in frame per-second (FPS) but obtain the better tracking accuracy.

Qualitative Comparison. In Fig. 5, the performance of various algorithms in complex environment are compared and displayed. In Basketball sequence, the object is influenced by occlusion, background interference and fast motion. MCCF-ASRCF, which adaptively decreases the template learning rate, can locate the object more accurate than other algorithms. In Freeman3 and Bike_ce2 sequence, where object scale is changed, MCCF-ASRCF can stably track the object, whereas ECOHC and MCCTHC are failed. In KiteSurf sequence, the violent shake of object easily leads to the object lost in tracking. MCCF-ASRCF can recapture the object due to template recover mechanism. Overall, MCCF-ASRCF has better tracking performance in complex environment.

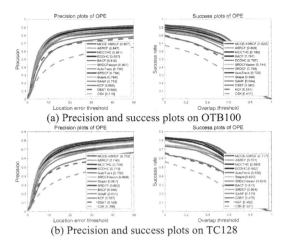

(a) Precision and success plots on OTB100

(b) Precision and success plots on TC128

Fig. 4. Quantitative comparison of various algorithms

Table 3. Comparison with mean-FPS of each algorithm

	ECOHC	BACF	ASRCF	Ours	MCCTHC	SRDCF	SRDCFDecon
OTB100	29.5	26.0	17.6	16.2	9.2	5.0	1.8
TC128	43.1	41.6	19.9	17.9	10.0	9.6	3.5

Fig. 5. Qualitative comparison of various algorithms. Sequences from top to bottom: Basketball, Freeman3, Bike_ce2 and KiteSurf

5 Conclusions

This paper proposes a new strategy of confidence evaluation for robust visual tracking in complex environment. The reliability of tracking result and template are comprehensively analyzed by multi-criteria evaluation functions, and the template updating mechanism is introduced to eliminate the influence of unreliable sample and ensure the accurate updating of tracking template. Experimental results testified the effectiveness of the proposed MCCF-ASRCF algorithm. In addition, the proposed confidence evaluation strategy has good transportability and can be integrated with other trackers.

Acknowledgement. This work was partially supported by National natural science fund (61643318) and Key research and development program of Ningxia Hui Autonomous Region (2018YBZD0923).

References

1. Ke, T., Li, Y.: Research of object detection and tracking algorithm on the video surveillance in electric power system. Electr. Power Sci. Eng. **30**(1), 42–46 (2014)
2. Lu, M., Xu, Y.: A survey of object tracking algorithms. Acta Autom. Sin. **45**(7), 1244–1260 (2019)
3. Henriques, J., Caseiro, R., Martins, P., et al.: Exploiting the circulant structure of tracking-by-detection with kernels. Lect. Notes Comput. Sci. **7575**(1), 702–715 (2012)
4. Yan, Y., Guo, X., Tang, J., et al.: Learning spatio-temporal correlation filter for visual tracking. Neurocomputing **436**, 273–282 (2021)
5. Danelljan, M., Häger, G., Khan F.S., et al.: Learning spatially regularized correlation filters for visual tracking. In: IEEE International Conference on Computer Vision, pp. 4310–4318 (2015)
6. Dai, K., Wang, D., Lu, H., et al.: Visual tracking via adaptive spatially-regularized correlation filters. In: IEEE Conference on Computer Vision and Pattern Recognition, pp. 4670–4679 (2019)
7. Galoogahi, H.K., Fagg, A., Lucey, S.: Learning background-aware correlation filters for visual tracking. In: IEEE International Conference on Computer Vision, pp. 1144–1152 (2017)
8. Bolme, D.S., Beveridge, J.R., Draper, B.A., et al.: Visual object tracking using adaptive correlation filters. In: IEEE Conference on Computer Vision and Pattern Recognition, pp. 2544–2550 (2010)
9. Danelljan, M., Häger, G., Khan F.S., et al.: Adaptive decontamination of the training set: a unified formulation for discriminative visual tracking. In IEEE Conference on Computer Vision and Pattern Recognition, pp. 1430–1438 (2016)
10. Shi, S., Ma, Y., Li, N., et al.: Adaptive decontamination algorithm based on PSR sample classification. Comput. Eng. Appl., 1–9 (2021). http://kns.cnki.net/kcms/detail/11.2127.TP. 20210414.1441.016.html
11. Danelljan, M., Bhat, G., Khan, F.S., et al.: ECO: efficient convolution operators for tracking. In: IEEE Conference on Computer Vision and Pattern Recognition, pp. 6931–6939 (2017)
12. Wang, M., Liu, Y., Huang, Z.: Large margin object tracking with circulant feature maps. In: IEEE Conference on Computer Vision and Pattern Recognition, pp. 4800–4808 (2017)
13. Gao, J., Ling, H., Hu, W., et al.: Transfer learning based visual tracking with Gaussian process regression. In: European Conference on Computer Vision, pp. 188–203 (2014)

14. Wei, B., Wang, Y., He, X.: Confidence map based KCF object tracking algorithm. In: IEEE Conference on Industrial Electronics and Applications, pp. 2187–2192 (2019)
15. Cai, D., Yu, L., Gao, Y.: High-confidence discrimination via maximum response for object tracking. In: International Conference on Software Engineering and Service Science, pp. 1–4 (2018)
16. Song, Z., Sun, J., Duan, B.: Collaborative correlation filter tracking with online re-detection. In: IEEE Information Technology, Networking, Electronic and Automation Control Conference, pp. 1303–1313 (2019)
17. Henriques, J., Caseiro, R., Martins, P., et al.: High-speed tracking with kernelized correlation filters. IEEE Trans. Pattern Anal. Mach. Intell. **37**(3), 583–596 (2014)
18. Boyd, S., Parikh, N., Chu, E., et al.: Distributed optimization and statistical learning via the alternating direction method of multipliers. Now Found. Trends **3**(1), 1–122 (2011)
19. Dalal, N., Triggs, B.: Histograms of oriented gradients for human detection. In: IEEE Conference on Computer Vision and Pattern Recognition, pp. 886–893 (2005)
20. Li, Y., Zhu, J.: A scale adaptive kernel correlation filter tracker with feature integration. In: European Conference on Computer Vision, pp. 254–265 (2014)
21. Wang, N., Zhou, W., Tian, Q., et al.: Multi-cue correlation filters for robust visual tracking. In: IEEE Conference on Computer Vision and Pattern Recognition, pp. 4844–4853 (2018)
22. Li, J., Wang, J., Liu, W.: Moving target detection and tracking algorithm based on context information. IEEE Access **7**, 70966–70974 (2019)
23. Wu, Y., Lim, J., Yang, M.: Object tracking benchmark. IEEE Trans. Pattern Anal. Mach. Intell. **37**(9), 1834–1848 (2015)
24. Liang, P., Blasch, E., Ling, H.: Encoding color information for visual tracking: algorithms and benchmark. IEEE Trans. Image Process. **24**(12), 5630–5644 (2015)
25. Weijer, J., Schmid, C., Verbeek, J., et al.: Learning color names for real-world applications. IEEE Trans. Image Process. **18**(7), 1512–1523 (2009)
26. Danelljan, M., Häger, G., Khan F.S.: Discriminative scale space tracking. IEEE Trans Pattern Anal. Mach. Intell. **39**(8), 1561–1575 (2017)
27. Bertinetto, L., Valmadre, J., Golodetz, S., et al.: Staple: complementary learners for real-time tracking. In: IEEE Conference on Computer Vision and Pattern Recognition, pp. 1401–1409 (2016)
28. Li, Y., Fu, C., Ding, F., et al.: AutoTrack: towards high-performance visual tracking for UAV with automatic spatio-temporal regularization. In: IEEE Conference on Computer Vision and Pattern Recognition, pp. 11920–11929 (2020)

Author Index

Printed in the United States
by Baker & Taylor Publisher Services